EXERCISE PHYSIOLOGY

American Physiological Society

People and Ideas Series

EXERCISE PHYSIOLOGY

People and Ideas

Edited by

Charles M. Tipton

Published for the American Physiological Society by

OXFORD
UNIVERSITY PRESS

2003

OXFORD
UNIVERSITY PRESS

Oxford   New York
Auckland   Bangkok   Buenos Aires   Cape Town   Chennai
Dar es Salaam   Delhi   Hong Kong   Istanbul   Karachi   Kolkata
Kuala Lumpur   Madrid   Melbourne   Mexico City   Mumbai
Nairobi   São Paulo   Singapore   Taipei   Tokyo   Toronto

Copyright © 2003 by the American Physiological Society

Published for the American Physiological Society by Oxford University Press, Inc.
198 Madison Avenue, New York, New York, 10016
http://www.oup-usa.org

Oxford is a registered trademark of Oxford University Press

Library of Congress Cataloging-in-Publication Data
Exercise physiology / edited by Charles M. Tipton.
p. ; cm. — (People and ideas series)
Includes bibliographical references and index.
ISBN 0-19-512527-4 (cloth)
1. Exercise—Physiological aspects.
I. Tipton, Charles M., 1927–   II. Series.
QP301 .E975 2003   612'.044—dc21   2002029293

9 8 7 6 5 4 3 2 1
Printed in the United States of America
on acid-free paper

Dedicated to the memory of
Carl V. Gisolfi, Ph.D. (1942–2000),
physiologist, author,
colleague, and friend

PREFACE

This book joins the People and Ideas series of the American Physiological Society (APS), and we thank the Society for the opportunity to present various perspectives on the history of exercise physiology. Although APS acknowledgment of the existence of exercise physiology as a subdiscipline of physiology is a relatively recent development (1977), it evolved more than a century ago from the classrooms and laboratories of physical education departments whose leaders were usually physicians (2,3).

The People and Ideas series currently consists of five volumes covering the history of the circulation of the blood (edited by Fishman and Richards), renal physiology (edited by Gottschalk, Berliner, and Giebisch), endocrine physiology (edited by McCann), membrane physiology (edited by Tosteson), and respiratory physiology (edited by West). Since exercise elicits integrated physiological responses from virtually all systems of the body, the sixth book of the series was devoted to the history of exercise physiology with an emphasis on how contemporary ideas evolved within the respective systems.

Before the responses of major body systems are discussed in the following chapter, Berryman provides a historian's perspective on ideas about exercise and its physiological responses from the time of Hippocrates (460–370 B.C.) to that of Edward Smith (1818–1874). Since the central command associated with the initiation of exercise elicits motor, circulatory, respiratory, and autonomic responses (1), the early chapters follow this sequence. However, the focus on systems created problems of selectivity for the authors. Thus, it is not surprising that many were apologetic to their contemporaries for not citing their important findings. Recall that the recent *Handbook of Physiology* volume on exercise contained references to more than 8200 studies in 1183 pages concerned with the physiological responses from five different systems (1). The integrative nature of exercise does not elicit equal involvement from all systems. Some are more important than others. This consideration plus a constraint on space led to various emphases within the text: namely, a separate chapter on maximal oxygen consumption, a focus within the autonomic nervous system

vii

chapter on cardiorespiratory responses, an emphasis in the endocrine chapter on metabolic reactions, and a reference to muscle function in the chapter on aerobic metabolism.

Scientific societies have few historians within their membership, and the American Physiological Society is no exception. Authors invited to present a historical perspective in this book were selected for their demonstrated expertise in the system being discussed and for their experience as physiologists. Every chapter was reviewed by two or more physiologists with equal expertise and experience. The purpose was to retain the individual perspective of the author while verifying the historical accuracy of the material. Time will tell if we were successful.

Besides the History Subcommittee of the Publications Committee of APS, thanks are rendered to Jeffrey House and his colleagues at Oxford University Press for their efforts in helping make an idea become a reality.

Tucson, Arizona   C.M.T.

REFERENCES

1. Rowell, L.R. and J.T. Shepherd, Editors, *Handbook of Physiology.* Section 12, exercise: regulation and integration of multiple systems. New York: Oxford University Press, 1996.
2. Tipton, C.M., Exercise physiology, part II: a contemporary historical perspective. In: *A History of Exercise and Sport Sciences*, edited by R.W. Swanson and J. Massengale. Champaign, IL: Human Kinetics Publishers, 1997, pp. 396–428, 431–438.
3. Tipton, C.M. Contemporary exercise physiology: Fifty years after the closure of the Harvard Fatigue Laboratory. In: *Exercise and Sport Sciences Reviews*, edited by John O. Holloszy. Baltimore: Williams and Wilkins, A Waverly Company, 1998, Vol. 26, pp. 315–340.

CONTENTS

CONTRIBUTORS

R. James Barnard, Ph.D.
Department of Physiological Science
University of California, Los Angeles
Los Angeles, California

Jack W. Berryman, Ph.D.
Department of Medical History and
    Ethics
University of Washington
School of Medicine
Seattle, Washington

George A. Brooks, Ph.D.
Exercise Physiology Laboratory
Department of Integrative Biology
University of California, Berkeley
Berkeley, California

Elsworth R. Buskirk, Ph.D.
Noll Physiological Research Center
The Pennsylvania State University
University Park, Pennsylvania

Jerome A. Dempsey, Ph.D.
Department of Preventive Medicine
University of Wisconsin Medical
    School
Madison, Wisconsin

Carl V. Gisolfi, Ph.D. (deceased)
Department of Exercise Science
University of Iowa
Iowa City, Iowa

L. Bruce Gladden, Ph.D.
Department of Health and Human
    Performance
Auburn University
Auburn, Alabama

John O. Holloszy, M.D.
Division of Geriatrics and Gerontology
Washington University
School of Medicine
St. Louis, Missouri

Alan J. McComas, M.B., F.R.C.P.(C.)
Professor Emeritus
Department of Medicine
Division of Neurology
McMaster University
Hamilton, Ontario
Canada

Jere H. Mitchell, M.D.
Harry S. Moss Heart Center
University of Texas Southwestern
    Medical Center
Dallas, Texas

Jacques R. Poortmans, Ph.D.
Professor Emeritus
Department of Physiological Chemistry
Institut Supérieur d'Education Physique
    et de Kinésithérapie
Université Libre de Bruxelles
Brussels, Belgium

Michael C. Riddell, Ph.D.
Department of Kinesiology and Health
    Science
York University
Faculty of Pure and Applied Science
Toronto, Ontario
Canada

Loring B. Rowell, Ph.D.
Department of Physiology and
    Biophysics
University of Washington
School of Medicine
Seattle, Washington

Neil B. Ruderman, M.D., D. Phil.
Department of Medicine
Boston University School of Medicine
Boston, Massachusetts

Bengt Saltin, M.D., Ph.D.
The Copenhagen Muscle Research
    Centre
Rigshospitalet and University of
    Copenhagen
Copenhagen, Denmark

Charles M. Tipton, Ph.D.
Department of Physiology
University of Arizona
Tucson, Arizona

Evangelia Tsiani, Ph.D.
Department of Community Health
    Sciences
Brock University
Faculty of Applied Health Science
Brock University
St. Catharines, Ontario
Canada

Mladen Vranic, M.D., D.Sc., F.R.S.C.,
    F.R.C.P.(C.)
Department of Physiology
University of Toronto
Toronto, Ontario
Canada

Brian J. Whipp, Ph.D., D.Sc.
Division of Respiratory and Critical
    Care Physiology and Medicine
Harbor-UCLA Medical Center
Torrance, California

Edward J. Zambraski, Ph.D.
Department of Cell Biology and
    Neuroscience
Rutgers University
Piscataway, New Jersey

EXERCISE PHYSIOLOGY

chapter 1

ANCIENT AND EARLY INFLUENCES

Jack W. Berryman

IDENTIFYING and analyzing the ancient and early influences on what would become the discipline known today as exercise physiology are formidable tasks. The range of disciplines involved and the time period of potential influences are both expansive, and the fact that no identifiable field of exercise physiology existed before the mid-nineteenth century is particularly troublesome. Accordingly, questions concerning what to read, where to look, and what years to examine loom large. Whereas the other authors in this volume could focus on a particular organ system, I did not have that luxury.

This chapter begins with the contributions of Hippocrates (460–370 B.C.), the most influential physician from Greek antiquity, and then moves chronologically through relevant geography. Specific sections are devoted to the ancient period, the medieval period, the sixteenth century, the seventeenth century, the eighteenth century, and to part of the nineteenth century. Most of the general thought and study concerning physiology during these years took place within the larger context of medicine and medical schools. For example, it was not until the sixteenth century that the French physician Jean Fernel (1497–1558) became the first to apply the term "physiology" to the science of the functions of the body. Concerning the physiological aspects of exercise, most of the relevant information published appeared in literature dealing with the value of exercise in promoting health and preventing disease. Some of it also appeared in the early literature relating to athletic performance. Generally, then, those engaged in the physiological study of exercise were physicians by training.

This continued to be the case in the seventeenth, eighteenth, and nineteenth centuries in France, Germany, Italy, England, and the United States. Those who did most of the early research related to the effects of exercise on the body were physicians. Some also held appointments at the most prestigious medical schools.

The chapter concludes in the mid- to late nineteenth century when the term "physiology of exercise" began to appear. The American physician William Byford (1817–1890) first used the term in 1855 in an article published in *The American Journal of Medical Sciences*. He was followed in the 1880s by another American physician, Edward Hartwell (1850–1922), who published two articles on the physiology of exercise in the *Boston Medical and Surgical Journal*. Also, Fernand Lagrange (1846–1909), a French physician, authored a book entitled the *Physiology of Bodily Exercise* in 1888.

Because of the unique nature of much of the early material, several passages in the actual words of the writer are reproduced below. In these writings illustrations were not very plentiful, so only paintings of individual scientists and title pages from the more prominent books are available to embellish the material within this chapter.

ANCIENT PERIOD

To grasp the underlying conceptions of physiological processes upon which the ancient Greek and Roman physicians based their medical theory, and consequently their practice, is ultimately a frustrating endeavor. The most prominent and influential physician from Greek antiquity was Hippocrates (460–370 B.C.). Unfortunately, the collection of medical writings attributed to his influence, the Corpus Hippocraticum, is filled with physiological explanations that are mostly inconsistent and, from a modern perspective, completely nonsensical. Translator W. H. S. Jones pointed out that "We must not expect of them [Hippocratic works] too much consistency, too much conformity with experience, too much scientific method. We must realise that they are in part works of imagination, often figurative, allusive and metaphorical. They portray truth, or what the writers consider to be truth, in an allegorical disguise" (43; p. xxiv). He also noted that "the arts were distinguished from the sciences only when Greek thought was past its zenith" and that the Greeks were known for their "luxuriant" imagination (43; p. xxiii). However, it is fascinating to consider that although physicians from this era relied heavily upon a luxuriant imagination for their view of the systems of the body, the therapies they prescribed and the advice they provided for the preservation and restoration of health were based almost solely on experience and observation—clear, empirical knowledge. So, although their explanations of exact physiological processes were often quite fantastical, their careful observations and their reliance on experience made their practice better than their theory.

Hippocrates is considered "the founder of Western scientific medicine," (65; p. 4). Though the term "physiology" was coined only in the sixteenth century, the early Hippocratic physicians were intent on understanding the functioning of the human body and its specific organic processes. But they pursued knowledge of the functioning of the human body in order to improve their ability to preserve and re-store the health of their patients. And they examined exercise in relation to training for athletic competition in light of what were deemed excesses or deviations from normalcy.

The ancient Greek and Roman physicians' application of close observation distinguished them as the first practitioners of the scientific method, embryonic though it may have been. According to Rothschuh, because of their "purely empirical procedures and practices . . . medicine was the first discipline to come into its own," and consequently served as a starting point for other scientific disciplines, including physiology (65; pp. 2–5).

It appears that Hippocrates authored two separate works on healthful living—*Regimen in Health*, with nine very short chapters, and *Regimen*, composed of four long sections or books. The first seven chapters of *Regimen in Health* offered advice on the preservation of health and were directed to the layman. Advice was given on what to drink and eat at certain times of the year. Rapid walking was suggested in winter and slow in summer, emetics and clysters were recommended for the bowels, and a chapter was devoted to athletic training.

Regimen's four books expressed many of the same physiological principles and considerations as *Regimen in Health*. In Book I, Hippocrates emphasized that when establishing a regimen, one must exert careful attention "to proportion exercise to bulk of food, to the constitution of the patient, to the age of the individual, to the season of the year, to the changes in the winds, to the situation of the region in which the patient resides, and to the constitution of the year." Hippocrates clearly had some notion of what we call metabolic processes in relation to exercise:

> eating alone will not keep a man well; he must also take exercise. For food and exercise, while possessing opposite qualities, yet work together to produce health. For it is the nature of exercise to use up material, but of food and drink to make good deficiencies. And it is necessary, as it appears, to discern the power of the various exercises, both natural exercises and artificial, to know which of them tends to increase flesh and which to lessen it. (43; p. 229)

In their article "Hippocratic and Galenic Concepts of Metabolism," C. E. A. Winslow and R. R. Bellinger concluded from this statement that Hippocrates "postulated [the] relation between the need for fuel foods and muscular activity, bringing us almost to the threshold of quantitative metabolism" (84; p. 130).

In Book II, Hippocrates also explained the excretory effects of exercise:

> Early morning walks too reduce [the body], and render the parts about the head light, bright and of good hearing, while they relax the bowels. They reduce because the body as it moves grows hot, and moisture is thinned and purged, partly by the breath, partly when the nose is blown and the throat cleared, partly being consumed by the heat of the soul for the nourishment thereof. (43; pp. 351–353)

According to Winslow and Bellinger, this statement suggests a connection between respiration and exercise (84; p. 132). The statement that "running in a circle dissolves flesh least, but reduces and contracts the belly most, because, as it causes the most rapid respiration, it is the quickest to draw moisture to itself," more explicitly supports this connection (43; p. 355).

Hippocrates emphasized the reduction of "flesh" through exercise in Book II. He observed that athletes with "physiques of a less firm flesh and inclined to be hairy are more capable of forcible feeding and of fatigue," and that "their good condition is of longer duration" (43; pp. 55–57). With regard to hydration and excessive exercise, Hippocrates advised:

> When people are attacked by thirst, diminish food and fatigue, and let them drink their wine well diluted and as cold as possible. Those who feel pains in the abdomen after exercise or after other fatigue are benefited by resting without food; they ought also to drink that of which the smallest quantity will cause the maximum of urine to be passed, in order that the veins across the abdomen may not be strained by repletion. (43; p. 57)

In *Regimen*, Hippocrates also discussed "trained bodies" and "untrained people," especially in relation to "fatigue pains," or the potential for muscle soreness:

> The fatigue pains that arise in the body are as follow. Men out of training suffer these pains after the slightest exercise, as no part of their body has been inured to any exercise; but trained bodies feel fatigue pains after unusual exercises, some even after usual exercises if they be excessive. These are the various kinds of fatigue pains; their properties are as follow. Untrained people, whose flesh is moist, after exercise undergo a considerable melting, as the body grows warm. Now whatever of this melted substance passes out as sweat, or is purged away with the breath, causes pain only to the part of the body that has been emptied contrary to custom; but such part of it as remains behind causes pain not only to the part of the body emptied contrary to custom, but also to the part that has received the moisture, as it is not congenial to the body but hostile to it. It tends to gather, not at the fleshless, but at the fleshy parts of the body, in such a way as to cause them pain until it has passed out. (43; pp. 359–361).

His explanation for those in training suffering "fatigue pains from unaccustomed exercises" was that

> any unexercised part of the body must of necessity have its flesh moist, just as persons out of training are moist generally throughout. So the flesh must of necessity melt, secrete itself and collect itself, as in the former case. ... Accustomed exercises should be practised, so that the collected humour may grow warm, become thin, and purge itself away, while the body generally may become neither moist nor yet unexercised. (43; pp. 361–363)

For accustomed exercises causing pain, Hippocrates said that "moderate toil is not followed by pain; but when immoderate it dries the flesh overmuch, and this flesh, being emptied of its moisture, grows hot, painful and shivery, and falls into a longish fever" (43; p. 363).

Clearly, the most important figure in medicine and physiology since Hippocrates was Claudius Galenus—or Galen (ca. 129–200 A.D.). His theories and writings dominated medicine and were unchallenged at least through the later Middle Ages, and were still widely accepted during the Renaissance. He was born in Pergamum—a city in the Roman empire on the western coast of Asia Minor, now

Turkey—and went on to acquire an extensive education in philosophy and medicine in the cities of Smyrna, Corinth, and Alexandria. During this time, he did extensive study in anatomy based largely on animal structures (71). Around 158–159 A.D., Galen returned to Pergamum to serve as the physician for the gladiators, performing his duties in the Asclepeion. Here he not only learned anatomy, physiology, and surgery, but also experimented with various training regimens to produce overall health, strength, and endurance for the combatants (67).

Galen was a prominent author as well as a physician, and wrote about four hundred treatises on mostly medical and philosophical subjects. Some of his works were commentaries on the Hippocratic writings—which he respected and used—but the majority were unique and innovative. Galen's extensive systematic medical theory also had its basis in the Hippocratic wisdom, but he moved far beyond it because he based his beliefs on anatomical and physiological facts acquired through observation and dissection. While his anatomy had some shortcomings, he far exceeded his predecessors and went on to establish anatomy as the basis for medicine as well as physiology. In fact, Galen's physiological concepts dealt with the physiology related to specific organs as well as to a general theory of vitality.

Central to what became known as "Galenic medical theory" or in some cases "humoral theory" was Galen's identification of the "naturals" (referring to physiology), the "non-naturals" (things not innate—hygiene), and the "contra-naturals" (against nature—pathology). The non-naturals were six in number and included those things which humans were exposed to in daily life: (1) air and environment, (2) food and drink, (3) motion and rest, (4) sleep and wake, (5) excretions and retentions, and, (6) passions of the mind (emotions). The non-naturals needed to be utilized in moderation as to quantity, quality, time, and order; for if they were taken in excess or put into imbalance, Galen believed disease would result. In particular, the role of motion (exercise) and rest in his theory was related to their respective effects on the qualities and humors. Galen believed that excessive rest would increase cold and moisture, whereas excessive exercise would at first heat the body and then produce cold and dryness. Moderate exercise would maintain warmth. Since causes for disease were thought to be due to heat, cold, dryness, or moisture, exercise and the other five non-naturals played important roles in therapy. Therefore, regulating the non-naturals along with drug prescription and surgery were the key components of the physician's activities. However, the non-naturals played a more significant role in hygiene. At the time, hygiene referred to the science of health and its preservation. Consequently, those liable to illness or those wishing to maintain their healthfulness were given a regimen by their physician whereby the non-naturals were regulated to provide a proper balance (4).

Most of Galen's material on exercise was in the first three books of *On Hygiene*. Book 1, "The Art of Preserving Health," was composed of 15 chapters. Chapter 8 was entitled, "The Use and Value of Exercise" and dealt mainly with the need for motion at all ages. Whether by sailing, riding on horseback, driving, or via cradles, swings, and arms, everyone, even infants, Galen said, needed exercise (32). Other chapters pertained to bathing and massage, fresh air, beverages, and evacuations. "Exercise and

Massage" was the title of Book 2 and it comprised 12 chapters. Chapter 2, "Purposes, Time, and Methods of Exercise and Massage," included some very important material on the role exercise played in Galen's conception of hygiene. For example, in reference to the type and definition of exercise, Galen said:

> To me it does not seem that all movement is exercise, but only when it is vigorous. But since vigor is relative, the same movement might be exercise for one and not for another. The criterion of vigorousness is change of respiration; those movements which do not alter the respiration are not called exercise. But if anyone is compelled by any movement to breathe more or less faster, that movement becomes exercise for him. This therefore is what is commonly called exercise or gymnastics. (32; pp. 53–54)

He saw the uses and values of exercise as follows:

> The uses of exercise, I think, are twofold, one for the evacuation of the excrements, the other for the production of good condition of the firm parts of the body. For since vigorous motion is exercise, it must needs be that only these three things result from it in the exercising body—hardness of the organs from mutual attrition, increase of the intrinsic warmth, and accelerated movement of respiration. These are followed by all the other individual benefits which accrue to the body from exercise; from hardness of the organs, both insensitivity and strength for function; from warmth, both strong attraction for things to be eliminated, readier metabolism, and better nutrition and diffusion of all substances, whereby it results that solids are softened, liquids diluted, and ducts dilated. And from the vigorous movement of respiration the ducts must be purged and the excrements evacuated. (32; p. 54)

Space does not permit elaboration, but Galen's ideas about the proper time for exercise, factors to consider before exercise, the varieties of exercise, the different qualities of exercises, and the places for exercise are very perceptive.

In *The Pulse for Beginners*, Galen discussed the impact of exercise on the pulse in a section on "the result of non-natural causes":

> Exercise to begin with—and so long as it is practised in moderation—renders the pulse vigorous, large, quick, and frequent. Large amounts of exercise, which exceed the capacity of the individual, make it small, faint, quick, and extremely frequent. In cases of great excess, whereby the subject is scarcely still able to move, and only at great intervals if at all, and there is a considerable loss of power, the pulse becomes very small, faint, slow, and sparse. If a state of dissolution of the faculty is reached, then the pulse will be that specific to this state; the nature of this pulse will be discussed in due course. (30; p. 332)

In *Good Condition*, Galen identified "good condition" as "a kind of excellence of health, and is thus found in those bodies which have the best constitutions." Athletes, as Galen saw in the writings of Hippocrates, were viewed as being purveyors of "the extremes of good condition" and were therefore seen as being "dangerous." Galen himself argued that "athletes' health, for example, is so far from being highly desirable that it was rightly criticized" and explained that athlete's strive for "the acquisition not only of good mixture, but also of physical mass—which cannot take place without an ill-balanced type of filling. And thus the state is rendered both dangerous and, from the point of view of public service, valueless" (29; pp. 296–297). Good condition involved "a good state of humours" and a "good proportion in the blood and in all the physical mass of the solid bodies." Athletes, Galen said:

> have poor proportion in these same bodies, especially in those of the fleshy type, and therefore are necessarily attended by danger when this state reaches its peak. When such individuals eat according to necessity, and their stomachs digest vigorously, and distribution follows readily from the digestive process, so that blood production and addition, as well as outgrowth and nutrition, are the result, there is a risk that the condition is overfilled, and that there remains no natural place for the addition. (29; p. 298)

Galen exhibited his thorough understanding of the various degrees of healthfulness or "condition" in his treatise *To Thrasyboulos: Is Healthiness a Part of Medicine or of Gymnastics?* Presented in a continuum format, the scale begins with "impairment of function which occurs in illness," goes to "health in state" (weakness and poor performance), moves on to "health in condition" as the midpoint (not weak but "still not achieving strength"), continues to "good condition" ("position of a kind of excellence in functions"), and concludes at the top of perfection of functions, which Galen called the "peak of good condition." Then, in a statement similar to today's identification of a super health category, Galen said: "The term 'health' applies to a certain state; the term 'good condition' to excellence within that state, and to its stability" (31; pp. 57–58, 62). In reference to strength, Galen noted that "strength has the same relationship to function as good condition has to health." His rationale for a physician to prescribe exercise was: "he will give gymnastic training when wishes to strengthen the faculties within us and to clean out the fine pores, but will command rest when he sees that the subject is either tired by the exertions or dissipated more than appropriate" (31; p. 68). Galen was clear to explain that "there are two types of material which produce and maintain the good condition: regimen and exercise."

The last author from the ancient period who wrote seriously about the physiological aspects of exercise was Flavios Philostratos (ca. 170–244 A.D.), "the Athenian." He traveled between Athens and Rome on several occasions, but it was in Athens, the literary capitol of the world, that Philostratos wrote his treatise *Concerning Gymnastics*, sometime between 219 and 230 A.D. Professional athletics flourished in Athens, and Philostratos's text was devoted to this subject rather than to exercise for ordinary persons or for the aged and diseased. It was an essay by an able and well-known sophist who was critical of the unscientific and unenlightened practices used in athletic training. Philostratos argued that training could become a true science if trainers had a good knowledge of each athlete's individual capacities, if they adapted exercise, food, and rest to individual states, and if they used sunbaths, dust, jumping weights, and other gymnasium facilities properly (86).

Philostratos divided athletic contests into heavy, light, and mixed. The pankration (a form of extreme fighting with no rules), wrestling, and boxing were in the heavy category; different kinds of running were classified as light; and the mixed was the pentathlon. (Wrestling and boxing were heavy and throwing the javelin, jumping, and running were light.) The running events were the stade (a short foot race the length of a stadium; the word is derived from the word "stadium"), distance race, race in armor, and the double stade. Philostratos then discussed training regimens for each of the particular athletic events. For example, he noted that "the distance runner practices running some eight to ten laps" and other athletes "exercised themselves by carrying heavy burdens; others by competing in speed with horses and hares, bending or straightening thick iron plates, or by having themselves yoked with powerful oxen, and finally, by subduing bulls or even lions." He also discussed the use of "tetrads," a 4-day cycle of training; alluded to the detrimental effects of "sexual indulgence" on athletes in training; and explained the value of lifting the jumping weights, or "halteres," for all athletes:

> The halter is an invention of the pentathletes, but was invented for the jump, from which it receives its name; for the rules of the game consider the jump a difficult type of contest and inspire the jumper with music of the flute and give him wings with the jumping weight; ... The oblong jumping weights, however, exercise the arms and shoulders, the roundish ones also the fingers. Light as well as heavy athletes ought to employ them along with all exercises. (86; pp. 25–26)

This appears to be one of the earliest recommendations of weight lifting and resistance training to develop strength beyond normal use.

MEDIEVAL PERIOD

The Hippocratic and Galenic traditions in medicine remained dominant throughout the Middle Ages, first in the Islamic world and later in the Christian West, largely through Arabic and Latin translations of the classic Greek works. These mainstays of formal medicine were joined by a few new medical texts mostly published during the 1200s and 1300s. Here, as earlier, exercise discussions usually appeared in sections devoted to the non-naturals, a concept developed by Galen. Information about physiology was subsumed within the highly accepted Galenic system of physiology based on the elements, qualities, humors, members, faculties, operations, and spirits. And, as Siraisi has argued, "medical writers identified themselves as the heirs and exponents of a primarily Galenic philosophical tradition" (72). She also noted that physiological theory "provided a general conceptual underpinning for explanations of illness and prescriptions for treatment," and both were "couched in terms derived from the theory of *complexio*, or temperament, that is, of the role played by the balance of the elementary qualities of hot, wet, cold, and dry in the body" (72).

From the tenth to the fifteenth centuries, it was the medical writers and universities that furthered the idea that studying the human body is in fact an honorable and scientific enterprise. While physiological knowledge was only part of medical theory and consequently only a segment of the university medical curriculum, it nevertheless attracted its share of attention. Human dissection also began as early as the twelfth century at Salerno and spread to other medical schools—but not as an independent research tool. Instead, the practice provided visual assistance for learning the physiology and anatomy found in the major texts.

University-related medical schools and their faculties provided the leadership in medical writing and teaching, and accordingly, dealt with the issues of exercise and physiology. Medical schools were formed during this era at Salerno (ca. 10th century), Paris (1110), Bologna (1158), Oxford (1167), Montpellier (1181), Cambridge (1209), Padua (1222), and Tübingen (1481), among others. The most important medical texts of the medieval era were the *Pantegni* of Ali Habbas and the *Isagoge* of Johannitius (both translated into Latin around the late 1070s), the *Canon* by Avicenna (translated into Latin around 1180), and the *Compendium Medicine* by Gilbertus de Aquila (ca. 1240) (13, 33). University-trained physicians in Paris, Montpellier, and Bologna studied *Isagoge* during the 1250s, *Pantegni* was popular in French and British universities in the first half of the 1200s, and the *Canon* exhibited several centuries of popularity in most medical schools from the late 1100s into the 1400s. All were based on the doctrine of Galenism, and in fact each provided the reader with specific ideas relating to the maintenance of health and how to recover it when lost, hallmarks of the formal study of medicine.

The key medical and physiological treatise of the period was the magnificent compilation *Canon Medicinae* written by the Persian physician Ibn Sina—also known as Avicenna (ca. 980–1037). A prominent historian of physiology, Rothschuh, described the *Canon* as a "veritable code of medical principles which had attained the certainty of laws," and "a thorough systemization of all the medical knowledge available since antiquity, an attractive compendium known for its clarity, logical formulation and substantial content" (65). Siraisi called it "an encyclopedia of Greco-Arabic medicine" and noted that Avicenna wrote it because "neither among the Greeks nor among the Arabs was there any single, complete, continuous book that taught the art of medicine" (72).

An aspect of Galenist doctrine with direct bearings on exercise that was formalized in the *Canon* was the concept of principal members—the heart, brain, and liver each provided the controlling "principle" of a separate group of organs and functions. Each of these systems had their own set of virtues, operations, and faculties. Virtues referred to general powers of action or sensation, operations were functions of particular organs, and faculties were specific abilities limited only to certain parts of the body. This scheme of separate systems was both anatomical and physiological, but the portions of the *Canon* on virtues focused more on physiological function. Avicenna described vital, animal, and natural virtues and associated each with a particular physiological system or systems relating to particular organs. Each set of virtues was governed by its own principal member. Vital virtues were associated with "spiritus" (the Greek pneuma) and their manifestations were the heartbeat, pulse, and respiration. Their organs were those of the thoracic cavity and arteries (they disseminated a mixture of blood and "spiritus" throughout the body), and the heart was the principal member of the system. Note the connection to the term "vitality," often used to refer to energy or vigor (36). It should also be noted that the vital virtues were the seat of the body's innate heat, a concept that received much attention in general physiological discussions as well as in attempts to understand the effects of exercise on the human body. Mendelsohn has argued that although Avicenna continued an important tradition by linking exercise to increased body heat, this led to confusion "as later interpreters tried to use this as proof that innate heat was generated by a mechanical process" (56).

The discussions of exercise in the *Canon* are both detailed and numerous. In the discussion of animal virtues, the *Canon* presents "the influence of exercise and repose" as one of the "corporeal causes unavoidable because physiological." The effect of exercise on the human body varies "according to its degree (strong, weak), amount (little, much), and according to the amount of rest taken, [and] the movement of the humours associated." In particular, Avicenna wrote:

> All degrees of exercise (strong, weak, little, or alternating with rest) agree in increasing the innate heat. It makes little difference whether the exercise be vigorous or weak and associated with much rest or not, for it makes the body very hot, and even if exercise should entail a loss of the innate heat, it does so only to a small amount. The dissipation of heat is only gradual, whereas the amount of heat produced is greater than the loss. If there be much of both exercise and repose, the effect is to cool the body, because the natural heat is now greatly dispersed, and consequently the body becomes dry. (36; p. 210)

In connection with rest or inactivity, Avicenna suggested that "repose always has a cooling effect because the invigorating life-giving heat passes away and the innate heat is confined." He also noted that "it has a choking and moistening effect, because of the lack of dispersal of waste matters." Later in the *Canon*, exercise was discussed as a heat producer (a "calefacient"), "but not in excess," and only if "not too vigorous or beyond the right measure and duration." If in excess, exercise would "disperse the innate heat unduly." Conversely, excessive repose "aggregates and strangles the innate heat, thereby having an infrigidant effect" (36).

Avicenna discussed "the pulse during (rigorous) athletic exercise" as part of his "doctrine of the pulse" in the section of the *Canon* on "the evidences of diseases." Here, exercise included "athletic sports of all kinds: running, endurance tests, sprinting, gymnastics of all kinds, military exercises, laborious manual or physical work, work in the fields, necessary exercise and walking exercise taken for health's sake and recreation." For rigorous exercise or athletics, he explained that

> At the outset, as long as the exertion is moderate, the pulse is large and strong. This is because the innate heat increases, and is strong. The pulse is also swift and brisk. This is because the resistance becomes greatly increased by the exertion. As exertion continues and increases, even if it be intense for only a short time, the pulse weakens, and, with the dispersal of the innate heat, becomes small. The pulse remains swift and brisk for two reasons: (1) the degree of resistance (i.e. blood-pressure) is further increased; (2) the vital power progressively fails until it is insufficient. After this, the swiftness steadily and progressively lessens; and the briskness increases correspondingly to the lessening of vital power. Still further prolongation of the exertion weakens the pulse until it becomes formicant and very brisk. (36; p. 317)

Finally, if exercise were continued to an "extremely excessive extent," Avicenna said it would lead to "a state akin to death, acting like all resolvents—that is, it renders the pulse vernicular, very brisk, slow, weak, and small" (36).

Avicenna also argued that the best time to start exercising was "when the body is free from impurities in the internal organs and blood-vessels, so that there is no risk of unhealthy chyme being dispersed through the body by exercise." A person about to take exercise was advised to "first get rid of the effete matters of the body by way of the intestines and bladder." Exercise should be continued as long "as the skin goes on becoming florid" [red or ruddy] and the "movement is moderated." It should cease if the members "show any puffiness" or should the "insensible perspiration lessen and the visible sweating stop." Finally, for the adult whose limbs were "weakly and undersized," Avicenna recommended that they may be strengthened and caused to grow by massage and "a suitable form of exercise, steadily persisted in." In this regard, he prescribed running for underdeveloped legs, making it a longer run each day, and suggested arm exercises, deep breathing, and uttering loud sounds for those experiencing respiratory weakness. The need to stretch, it was explained, came "when effete substances have accumulated in the muscles" (36).

SIXTEENTH CENTURY—THE EARLY RENAISSANCE

Throughout the 1500s, university-trained physicians and their medical schools continued to practice and teach medicine based upon the classical authority of Hippocrates and Galen—a tradition in which the human body was described and studied in humoral and qualitative terms. The followers of Galen in particular had great influence at the major universities, and numerous compilations of Hippocrates, Galen, and Avicenna were popular among learned scholars. In fact, at least 481 texts of Galen alone were edited during the Renaissance. For physiology, then, major conceptions of the body's functions did not develop beyond the Galenic doctrines already discussed. Yet new advances were taking place in anatomy, and through their detailed work anatomists provided the basis for much of the physiological progress that occurred in the following century. Toward the end of the fifteenth century and into the sixteenth, new writing on human anatomy was appearing and more and more dissections were being performed by medical faculty at the major universities (65).

The most stunning development in anatomy occurred in 1543 when the Flemish physician and anatomist Andreas Vesalius (1514–1564) published his treatise *De Humani Corporis Fabrica*. This book, based on careful personal dissections, described various body structures and provided brilliant illustrations by Flemish artist Jan Stephan (1499–1546). Several of the drawings depicted a series of large muscular figures that exhibited the muscles in a state of contraction and thus represented an early form of "living anatomy." As a professor at the University of Padua in Italy, Vesalius derived his views from his own studies rather than what he could infer from Galen's anatomical works. He also realized that much of Galen's work had been based on animal dissections, although Vesalius did include a chapter called "Some Remarks on the Vivisection of Animals." The work of Vesalius has been recognized as an "overthrow of the ancients," especially Galen. It established anatomy as the basis for medicine and the standards for anatomical methods. The historian of physiology Sir Michael Foster believed that *Fabrica* was "the beginning not only of modern anatomy but of modern physiology." Others have regarded Vesalius as the most commanding figure in European medicine after Galen and before Harvey (27, 65).

The anatomists' accomplishments were joined by additional scientific developments representing the work of other physicians during the 16th century who helped undermine and dilute the influence of Galenic doctrine. The Swiss physician Bombastus von Hohenheim, known as Paracelsus (1493–1541), preceded Vesalius with his *Paragranum* (1530), *Grosse Wundarznei* (1536), and *Defensiones* and *Labyrinthus Medicorum* (1538). His contributions to physiology were related to his view of vital actions as chemical phenomena. In taking this view, he broke with the classical explanations of physiological function based on humors and elements and provided a rationale for chemical interpretations of bodily functions like metabolism and digestion. Paracelsus was followed by the French physician Jean Fernel (1497–1558), who authored *De Naturali Parte Medicinae* in 1542 while he was a professor of medicine at the College de Cornouailles. A reedited edition was published as a textbook titled *Medicina* in 1554. In this book, Fernel coined the term physiology to describe the healthy body that belongs to "the things natural," as compared to "non-natural" or "contra-natural." This was the first time that "physiology" was used in the modern sense to refer to the science of the functions of the body. Fernel's colleague in Paris, Jacobo Sylvius, quickly adopted the term the following year, but it was not until 1597 that the term in this new sense appeared in English in Guillemeau's *Surgery*: "Physiologia handleth and treateth of the structure and situation of Man's bodye." Fernel's book went through 31 posthumous reissues and remained unaltered until the late 1600s. Eventually Harvey's discovery of circulation and Boerhaave's lack of faith in any physiological treatise written before Harvey led to the demise of his ideas (69). Fernel's biographer, Sir Charles Sherrington, claimed that the book "set forth the theory of the healthy body and mind, as introductory to medicine." Yet Fernel's "physiology" was still humoral medicine. He devoted chapters to the elements, temperaments, spirits, innate heat, faculties, and humors. Still, his book had lasting influence on the language of medicine. The physiology section was followed by two other sections on "pathology" and "therapeutics," also new terms for medicine (35, 65, 69).

Exercise played a significant role in the sixteenth century revival of Greek hygiene, or the regimen of the old "non-naturals." In fact, Galen's influence became so strong in the 1500s that one historian referred to the period as "the golden age of Galenism." He observed that "Galen's works were more widely available, and in more complete and accurate form, than previously, and thanks to the dedication of his followers they were more highly admired, more thoroughly studied, and probably better understood than at anytime before or since" (12). A good case in point was Thomas Linacre's Latin translation of Galen's *Hygiene* in 1517. As the founder and first president of England's College of Physicians the following year, Linacre helped introduce exercise into mainstream medicine via the Galenic tradition of the non-naturals. On the footsteps of this revival was the growing popularity of self-help and longevity literature written in the vernacular (as opposed to Latin), where proper behavior was advocated for the improvement of one's health and life expectancy (5).

SEVENTEENTH CENTURY

It was during the seventeenth century that we began to see the coming of the scientific revolution and, accompanying it, the downfall of Galenism. Developments in science and medicine during the 1600s not only changed medicine but also presented new opportunities in the unique subjects studied and the methodologies utilized. Accordingly, scientific interest shifted to individual problems and the application of the experimental method as well as to the collection of information on the observation of events and unusual objects. For medicine, this translated into the use of scientific instruments, more effective methods of treatment, the introduction of new medicaments, and an improvement in the method of studying disease. For the study of physiological topics, most often embedded in medicine and done by physicians, an emphasis was placed on investigating cardiac activity, vascular blood flow, muscular contraction, the source of animal heat, and the role of the lungs and liver. Additionally, quantitative methods, as well as the inductive method and experimental analysis, were sometimes used by those studying physiological problems. It is interesting too, that those doing most of the physiological research were "famous doctors," but not famous, as physicians. Indeed, they were pioneers in biology, physics, mathematics, and chemistry. And, although they had received a formal and traditional medical school education, their scientific contributions tended to be associated more with institutions outside of conservative universities such as the Academia del Cimento (Italy, 1657), the Royal Society (London, 1662), and the Academie des Sciences (Paris, 1666) (65, 85).

Physiology was significantly influenced by the application of chemical principles to the study of the body's functions. As the alchemy of the medieval years gave way to modern chemistry, individuals such as Paracelsus and Robert Boyle (1627–1691), contributed greatly to the understanding of vital actions. The new chemical approach to physiological questions, known as iatrochemistry, eventually led to new discoveries and refined concepts. Its undisputed founder was Jean B. van Helmont, the Belgian physician who was the first to recognize the importance of "ferments" and gases. In looking at digestion in particular, van Helmont was able to increase doubt regarding the classical physiology of Galen. He was joined by the Dutch physician, Franciscus de le Boë (1614–1672), better known as Sylvius, who explained all vital actions in chemical terms as a series of fermentations (27, 65).

A second major development that affected physiological study during this century was William Harvey's (1578–1667) discovery of the circulation of the blood. He studied under Fabricius at the University of Padua and received his medical degree in 1602. In his classic treatise *Exercitatio Anatomica De Motu Cordis et Sanguinis in Animalibus*, published in 1628, Harvey expressed doubt regarding the Galenic conceptions of blood movement and realized that Galen and Vesalius, among others, were wrong. Accordingly, in Chapter 14, he wrote that

> It has been shown by reason and experiment that blood by the beat of the ventricles flows through the lungs and heart and is pumped to the whole body. ... It must therefore be concluded that the blood in the animal body moves around in a circle continuously, and that the action or function of the heart is to accomplish this by pumping. This is the only reason for the motion and beat of the heart. (42; p. 104)

Foster called Harvey's discovery "the death-blow to the doctrine of the 'spirits'" and Rothschuh claimed it was "of epoch-making significance for the subsequent development of physiological thought." Rothschuh also explained that Harvey's innovation "forced a total re-orientation about almost all other organic functions, especially cardiac activity, the vascular flow, the role played by the lungs and the liver, as well as the source of animal heat." He also called it "the starting point for modern physiology." Harvey's contribution is also unique since it was based almost entirely on observations of anatomy and vivisection without any consideration of chemical or physical viewpoints (27, 65).

Harvey's *De Motu Cordis* included several references to exercise and its effect on the heart and circulation. In Chapter 10, Harvey concluded that "the blood is transmitted sometimes in a larger amount, other times in a smaller, and that the blood circulates sometimes rapidly, sometimes slowly, according to temperament, age, external or internal causes, normal or abnormal factors, sleep, rest, food, exercise, mental condition, and such like." In Chapter 12, in a discussion of ligation, he explained that the "best subject is one who is lean, with large veins, warm after exercise when more blood is going to the extremities and the pulse is stronger, for then all will be more apparent." He further explained the importance of movement and muscle contraction to the circulation of the blood in Chapter 15 and noted that the blood "tends to move from the tiny veins to the intermediate branches and then to the larger veins because of the movements of the extremities and the compression of muscles." Later, in the final chapter, when Harvey discussed the "braces" or "fleshy and fibrous bands" in the heart walls and septum, he observed that "in man there are more in the left than the right ventricle, and more in the ventricles than the auricles. ... In large, muscular, peasant-type individuals there are many, in more slender frames, and in women, few." On the same topic, he also noticed that "in some men of heavier and huskier build, the right auricle is so robust and so well braced inside by bands and various connecting fibers that it approximates in strength the ventricle of other subjects." Finally, Harvey concluded that "the stronger, more muscular, and more substantial the build of men, the thicker, heavier, more powerful and fibrous the heart, and the auricles and arteries are proportionally increased in thickness, strength, and all other respects" (42).

A third and final development took place during the seventeenth century that had a significant impact on the growth of physiological thought. This was the application of mechanical principles to solve physiological questions, referred to as iatrophysics or iatromathematics. The early proponent of this approach was the French philosopher René Descartes (1596–1650), who attempted to explain all bodily functions by purely mechanical laws in *De Homine Liber* (1662). He argued that the human body was a machine, mechanical like a clock, which was inhabited by a soul that commanded voluntary motions. Yet, Descartes also understood the roles of experiment and hypothesis in establishing scientific knowledge (18, 27, 65). Another major contributor to this school of thought was the Italian physician Santorio Santorio (1561–1636), who measured the skin's "invisible perspiration" in an attempt to determine its relationship to health and disease. He also invented an instrument to count the pulse, used a thermometer to measure the body's temperature, and pursued metabolic questions by comparing body weight to the differing weights of food that was eaten and the excrements—namely, feces and urine. (See additional information in Chapter 10.) All of his research, including that pertaining to "insensible perspiration," was published in *Ars de Statica Medicina* (1614). By using a large balance with a chair on one of the scales, Santorio was able to record variations in body weight under a variety of conditions—when awake or sleeping, before and after meals, during exercise and rest, and under a variety of emotional situations. His results were published in a series of aphorisms on exercise and rest arranged according to the six non-naturals (65, 66, 85).

However, the true founder of the iatrophysics or iatromechanics school was the famous Italian physicist and student of Galileo, Giovanni Borelli (1608–1679). In his *De Motu Animalium* (1680–1681), which he wrote in two volumes during the 1670s, Borelli applied mechanical principles to several physiological problems. He investigated the flight of birds, the swimming of fish, and the locomotion and motion of the human body. Also discussed were the mechanics of muscular contraction, respiration, and circulation. While Borelli's experiments, measurements, and calculations had errors, his mechanical interpretations of organic functions were very important to the progress of physiology (27, 65). The famous Swiss mathematician Johann Bernoulli (1667–1748) was another contributor to the "physico-mechanical" school of physiological thought. He published *On Effervescence and Fermentation* in 1690 and *On the Movement of Muscles* in 1694 as dissertations to earn his degree in medicine. Bernoulli paid tribute to Borelli in the Foreword to *On the Movement of Muscles* and noted that he had adopted "his hypothesis." Here, Bernoulli applied integral calculus to describe muscles as small machines and looked at the pressure of liquids to calculate the raising force of muscles as well as their vigor and tiredness (3).

Further developments in physiological thought during the seventeenth century were achieved largely through the application of both chemical and physical principles by a number of physicians residing in England. Addressing physiological questions by observation and experimentation was of course exemplified by William Harvey, as discussed earlier. Others, such as Francis Glisson (1597–1677), professor of medicine at Cambridge University, made contributions to the study of muscle contraction. In his *De Ventriculo et Intestinis* (1677), Glisson used the term "irritability" to describe a biological property not dependent on consciousness or the nervous system. He argued that the living substance had its own independent dynamic principle and suggested muscle fibers were dependent on irritability. Whereas some others, most notably Thomas Willis (1621–1666), believed muscular contraction was a sort of explosion resulting in the inflation of the muscle, Glisson suggested just the opposite. To prove his point, he put the arm of a well-developed muscular man in a totally submerged glass container. Upon muscular contraction, Glisson discovered that the water level actually decreased, therefore showing that muscles did not inflate upon contraction (65). He described the experiment in *De Ventriculo*, which he wrote as early as 1662:

> From which it is clear that muscles are not inflated or swollen at the time that they are contracting, but on the contrary are lessened, shrunk, and subsided. ... From this therefore we may infer that the fibres are shortened by an intrinsic vital movement and have no need of any abundant afflux of spirits, either animal or vital, by which they are inflated, and being so shortened carry out the movements ordered by the brain. (27; pp. 290–291)

It has also been suggested by some historians that Glisson was simply describing the experiment conducted by the physician Jonathan Goddard (1616–1675) and explained to the Royal Society of London on April 1, 1669 (6, 28). Goddard actually performed the experiment before the members of the Royal Society on December 16, 1669 (6).

Questions of muscular contraction interested several other physicians during the seventeenth century. Thomas Willis (1621–1666), educated at Oxford University, became interested in the iatrochemistry of van Helmont and consequently attempted to understand the human organism both as a machine and as the subject of fermentations. His view of muscular contraction was published in *De Motu Musculari* (1670) and involved an "explosion" which dilated and shortened the muscle fibers. But, as noted earlier, the experiments described by Glisson negated his hypothesis (65). Niels Stensen (1638–1686) the Danish physician, authored two works relating to the muscles. In 1664, he published *De Musculis et Glandulis Observationum Specimen* and in 1667 expanded it and added illustrations. The *Elementorum Myologicae Specimen seu Musculorum Descripto Geometrica* included information about the direction of the muscle fibers in the heart and was unique in the presentation of muscular contraction based on a mechanical and geometrical treatment. Stensen said the muscle fibers were arranged in a parallelogrammatic manner and that during contraction a purely geometrical change took place in the fibers, causing them to decrease in total length while enlarging their width (49, 65). William Croone (1633–1684), a physician and one of the founders of the Royal Society of London, published *De Ratione Motus Musculorum* in 1664 and no doubt influenced Willis. Croone believed that a chemical reaction caused the muscle to swell and contract and identified the "spirit" as "a rectified and enriched juice," thereby suggesting it was an observable and material substance (57, 83). In the final section of his book, Croone briefly discussed exercise and the muscles and presented his explanation for muscle soreness following exercise:

> No sooner does the muscle begin to swell at its boundary than are the fibres also contracted, and so everything occurs at one and the same time in the twinkling of an eye.

However, that effervescence, which I have mentioned, ceases almost immediately and the very active spirits are dissipated through the membranes of the muscle nearly in an instant and, unless a new impulse arrives at once, the muscle is immediately pulled back and made flaccid by the inequality of the circulation of the blood through the fibres of the antagonist, and the blood flows out in greater quantity through the vein LMN. Hence violent exercises also remove the spirits and make them break out through sweats. This is also the reason why the limbs are very painful and stiff after violent movements of the body (especially in those who are not accustomed to them). (83; p. 163)

The process of respiration was another bodily function examined and studied by a group of physicians, physicists, and scientists in London and at Oxford University during the second half of the 17th century. Robert Boyle (1627–1691), a founder of the Royal Society, rejected many of the ideas of Paracelsus as well as the iatrochemical concepts regarding acid and alkaline humors. As a chemist and physicist, he argued that "the study of physiology is not only delightful, as it teaches us to know nature, but also as it teaches us in many cases to Master and Command her" (65). He experimented with an air pump and used it to study the effects of a vacuum on combustion and animal life. And, in his *New Experiments, Physicomechanical Touching the Spring and Weight of Air and Its Effects* (1660), Boyle showed that animals died of asphyxiation once a flame in the testing chamber had gone out (27). Boyle was assisted by Robert Hooke (1635–1703) in some of his experiments on respiration. (For additional information on both Boyle and Hooke, see Chapter 4.) Hooke served as curator of experiments for the Royal Society from 1664 until his death, and with Boyle, constructed his air pump. Hooke's experiments were published in 1667 in *Micrographia*. Later in his career, Hooke discussed exercise in a manuscript composed in the mid-1670s titled, *Philosophical Algebra*. He noted that he had evidence to show that "animals that breathe great amounts of air were very hot," and that "exercise that caused the blood to circulate more, to make the animal breathe more, also increased the heat and copiousness of the expired vapors" (28, 65).

Boyle and Hooke were joined in their quest to understand respiration by two Oxford University physicians, Richard Lower (1631–1693) and John Mayow (1641–1679). Boyle introduced Lower to the Royal Society in 1667 and he also did some cardiopulmonary experiments with Hooke. His other experiments were reported in his most significant work, *Tractatus de Corde* (1669) (7, 27, 28, 65). Lower's *De Corde* included a substantial amount of information relating to health and exercise. For example, in his explanation of the outward sound of the heart's beat, Lower used the example of a race horse: "There is also the fact that one can hear from afar the individual heart-beats in horses returning from a long ride. The blood is at this time being sent through the vessels with such force and violence that you can count the individual beats from far off, and can proclaim them as surely as if you had your finger on the artery" (54; pp. 84–85). Lower further explained how the movement of the heart was changed by a "variation in the inflow of spirits" by describing "violent exercise":

The movement of the heart is accelerated in violent exercise in proportion as the blood is driven and poured into its ventricles in greater abundance as a result of the movement of the muscles. The heart must pass on the blood as fast as it receives it, and so it distributes it in larger amount to the brain as well as to the other organs. To discharge a mutual obligation, the spirits are likewise sent out in larger amount to hasten the movement of the heart. (54; p. 122)

Later, after analyzing the heart and circulation, Lower provided his interpretation of the physiological value of physical activity:

> From these facts it is obvious how useful exercises and movements of the body are as an aid to health; for the more often the blood is shaken up within the heart and thrown against the walls of the vessels, and is moved and activated in the body by contraction of the muscles, and finally driven through the pores of the body; the more must it be thinned and freed from those stagnations, to which the nutrient portion of the blood is otherwise over-subject. (54; p. 130)

John Mayow (1641–1679), a physician and collaborator with Hooke and Lower, was admitted to the Royal Society in 1678. His first publication, *Tractatus Duo*, was published in 1668. In 1674, Mayow expanded his *Tractatus Duo* into *Tractatus Quinque* by the addition of essays on nitre and nitro-aereal spirit, respiration of the fetus, and muscular motion. His work was unique in that he was able to build upon the previous contributions of Harvey, Boyle, Willis, Malpighi, Hooke, Lower, and Stensen (8, 27, 28, 65). Like Lower before him, Mayow included extensive information on exercise and the efficacy of exercise for good health in his *Tractatus Quinque*. Every section except that dealing with the respiration of the fetus included references to exercise. For example, in his essay on the nitro-aerial spirit, Mayow explained the rise in temperature from exercise:

> To this I add that the very intense heat which animals experience when urged to violent motion, arises partly because in violent movements there is very great need of increased respiration, and thus the nitro-aërial particles introduced into the blood in greater abundance will produce greater effervescence and heat than usual; for the friction of the limbs in the most violent movements is not so great as to be able to excite so fervid a heat. Nay, if any one breathes, even when at rest, but a little more intensely, he will soon feel himself in an unusual glow of warmth. However the heat excited in animals by violent exercise is in part also due to the effervescence of nitro-aërial particles and sulphureous particles, originating in the motor parts, as will be pointed out elsewhere. (55; p. 105)

He also included a lengthy passage noting the importance of nitro-aerial salt for muscle contraction during violent exercise.

Mayow's essay on muscular motion and animal spirits included some interesting information on muscle fibers, strength, muscular contraction, and exercise. In a discussion of strength, he concluded "that the shortness of the fibrils and their almost infinite number contribute to the strength of the muscles and to the more effective performance of their pull. Certainly the fibrils, whether we consider their number, their size, or their position, would seem much more fit for bringing about muscular contraction than the fleshy fibres" (55; pp. 237–238). Later, he discussed the prerequisites for sustained exercise and the source of heat during violent exercise, which included his view on the importance of nitro-aerial particles:

> From what has been said, we may seek the reason why so intense a heat is excited in the motor parts by violent exercise. That heating is commonly attributed to the motion of the body itself, but indeed in animal motion there is no such friction of the parts (from which alone heat arises) as could account for so intense a fervour. We must, therefore, believe that the heat of strongly contracting muscles comes from nitro-aërial particles, at that time much agitated in the muscles; as I have endeavoured elsewhere to show that every kind of heat arises from their motion. ... Further, from the foresaid hypothesis a reason can be deduced why the sweat given out in violent movements is of a saline character, and very penetrating. (55; pp. 248–249)

In explaining the "need of more intense respiration for some time after violent exercise," Mayow said the reason "seems to be that the blood returned from the brain to the heart is to a great extent deprived of nitro-aërial particles" (55; p. 257). He also provided an elaborate description of how the muscles aid the body in jumping.

EIGHTEENTH CENTURY—THE ENLIGHTENMENT

The spirit of the Enlightenment commenced at the beginning of the eighteenth century first in Europe, specifically in France. As the century developed, the new science and philosophy of the era spread into England and Germany. Science displaced the earlier dependence on theology and faith and philosophers were joined by scientists in a celebration of empirical methodology. Facts were collected, they were interpreted critically and rationally, and conclusions were tested via concrete experiments structured in the real world of phenomena. It was during the 1700s that scientists achieved a high societal standing, the first time they had surpassed poets and writers in public admiration.

The day-to-day usefulness of the various scientific discoveries appeared in most fields of endeavor, including medicine, physiology, and exercise. Specifically, the exact natural sciences of physics and chemistry furthered the evolution of physiology. These two fields provided new methods and ideas for the solution of physiological problems, and yet, in the first half of the eighteenth century there were few significant findings. Instead, several physicians devoted considerable time and attention to ordering and systematizing the current knowledge relating to the physiology of the body and to incorporating the most applicable and relevant information into the theory and practice of medicine. In the field of exercise, physicians and scientists published major treatises summarizing the knowledge to date regarding exercise, health, and longevity and also wrote books applying exercise principles to physical activities like dance and running. At least one major treatise was written that recomended exercise as an important therapy or cure for specific ailments. Still others conducted experiments where questions relating to the affects of exercise on the human body were asked (65).

The two most notable eighteenth century systematists of physiological knowledge were Friedrich Hoffmann (1660–1742) and George Stahl (1660–1734), both physicians. Hoffmann was the son of a well-known physician in Halle, Germany. He received his medical degree from the University of Jena, where George Stahl was a fellow student, and Hoffmann went on to become the first professor of medicine at the new University of Halle in 1693. He was a fellow of the Royal Society of London, the Berlin Academy of Sciences, and the Imperial Russian Academy of Sciences in St. Petersburg. In his major works, *Fundamenta Medicinae* (1695) and *Medicina Rationalis Systematica* (1718), Hoffmann tried to incorporate the information learned from the new physics, chemistry, anatomy, and microscopy (45, 51, 63).

Although Hoffmann did not use the term "institutes" (Boerhaave first did in 1708), his *Fundamenta Medicinae* was an early example where physiology and pathology especially, were presented as the fundamental principles or "institutes" of medical science. In his section on "physiologia," Hoffmann presented a survey of the principles of physics and chemistry upon which the bodily economy rested. He also discussed the principles of bodily structure and function in both health and disease. Emphasis was placed on respiration, circulation, digestion, excretion, sensation, neuromuscular activity, secretion, and reproduction. Exercise was addressed in the part of his book devoted to "medical hygiene" (45).

George Stahl studied medicine at Jena in Germany, receiving his degree in 1684. Ten years later, because of Hoffmann, Stahl was appointed the second professor of medicine at the University of Halle. Here, he developed an "animistic" view of vitality much different from the mechanical view popular in physiological thought at the time. This idea was expressed in *Theoria Medica Vera* (1708) and further elaborated in his *Negotium seu Skiamachia* (1720). Although both of these chemical "theories" were eventually proven false, animism helped expose some of the weaknesses of the mechanistic view. In particular, Stahl believed the "anima" exerted control over the body through the motion of the heart and the circulation of the blood. And, like other physicians of his era, the goal of medicine for him was to maintain health, keep the body free from ailments, and prevent disease. Accordingly, in both his "animistic" thought and in his philosophy of medicine, there was a need for exercise (52, 65).

Sir John Floyer (1649–1734), a famous English physician, concluded an era in medicine begun by Harvey in which measurement in biological investigation formed a new starting point for the explanation of vital phenomenon. Until this time, patients had been examined rather than measured. In Floyer's era, medicine increasingly depended on the natural sciences and experimental research. Floyer published the first volume of *The Physician's Pulse-watch: Or, an Essay to Explain the Old Art of Feeling the Pulse and to Improve It by the Help of a Pulse-watch* in 1707. The second volume appeared 3 years later. Basically, Floyer invented a watch with a second hand which allowed him to accurately count the pulse, something Harvey, Lower, and others before him could not do. He categorized the pulse by numbers and humors and by qualities and inequalities and followed that with information on the influence of disease and various external factors on the pulse. For example, a pulse of

75–80 beats per minute was referred to as "hot in the first degree," was associated with the choleric disposition, and was caused by such things as "hot seasons, hot air, exercise, passions." He also noted that "After half an hour's moderate walking in a minute I have counted 112 pulses." In fact, Lower estimated that there were 2000 pulse beats in an hour and Harvey believed there were 1000 pulses in a half hour and that the number could increase to 4000 (82).

Floyer's most detailed explanation of the pulse's response to exercise was in his chapter dealing with alterations of the pulse by external causes (the non-naturals). Here, he noted that bodily heat stimulated the heart "to make a vehement contraction and quick circulation, which forces the Blood to return oftner, and stimulates the Heart to a more frequent contraction." Similarly, as the heat of the blood increased "a little above the Natural, the Pulse becomes greater; but if it increases more, the Pulse becomes not only greater, but quicker; but if the Heat increases to the highest degree, the Pulse becomes very great and quick, and frequent" (26). Floyer's studies of exercise yielded the following findings:

> The Pulse by moderate Exercise labours more, and becomes more vehement; and because the Heat and Rarification of Humours increases, it becomes great and quick, and at last very frequent. If Exercise be much, and it exceeds, the Spirits are exhausted, and the Pulse becomes languid and small, and very frequent, by reason of the Heat. If the Exercise be Immoderate with great Weariness, the Spirits and Heat are very much evaporated, and the Body cooled; and then the Pulse is very languid and slow, small and rare. After half an Hours moderate Walking, in a Minute I have counted 112 Pulses; so that the Pulses was accelerated 20 or 30 Strokes in a Minute, and it fell again not long after the Exercise ceas'd; and before Dinner I counted but 70 Beats in a Minute; but after Dinner they were 90, before the Exercise in the Morning the Pulse was 76. I rid 16 Miles in a cool Day, and the Pulse beat 90 in a Minute; therefore Walking is a greater Exercise than Riding, because it makes the Pulse beat faster. The shortness of Breath upon Exercise stops the Pulse, and makes it irregular, smaller and weaker for some time. A fat Man by Walking had 90 Pulses in a Minute, a thin Girl had a 110 by the same Walk, tho' the Morning Pulse was 75, and the Pulse sunk to the same Number before Dinner; after which the Girl of 12 Years had 83 Pulses as soon as she rose from the Table. (26; pp. 86–87)

Another London physician and lecturer on anatomy at Oxford and Cambridge, James Keill (1673–1719), authored *An Account of Animal Secretion, the Quantity of Blood in the Humane Body, and Muscular Motion* in 1708. Of particular importance for exercise physiology was Keill's chapter, "Of Muscular Motion." Here, he discussed blood supply to the muscles, muscle fiber size and structure, and how they contracted. In reference to muscle strength and their ability to lift heavy weights, Keill said "the Weight a Muscle can raise, will be always as the Number of its Fibres, that is, as its Thickness supposing the Distention of the Vesicles equal." He also noted "the absolute Strength of one Muscle is to the absolute Strength of another, as their Bulks" (50).

In what historians of physiology have labeled "the most important step after Harvey and Malpighi in elucidating the physiology of the circulation," Stephen

Hales (1677–1761), an English clergyman without formal scientific or medical training, conducted a series of experiments with animals on the hydraulics of the vascular system during the early 1700s. He provided a scheme for the study of both the
systemic and pulmonary circulations by relying on direct measurements, observations, simple computations, and logical deductions. In fact, his research was an extension of the "statical" investigations on humans done previously by Sanctorius
and James Keill. Hales was also the first to record accurate measurements of blood
pressure under a variety of physiological conditions, including deep breathing, blood
loss, and exercise. He also studied the influence of venous return on the heart's output under the influence of deep breathing and muscular strain and was a precursor
of Lavoisier in measuring the volume of air during the exchange of air and blood in
the lungs. Most of his findings on animal circulation were published under the title
*Haemastaticks; or an Account of Some Hydraulic and Hydrostatical Experiments
Made on the Blood and Blood-Vessels of Animals* (1733), which was then combined
with an earlier work on *Vegetable Staticks* (1727) as *Statical Essays* (1738–1740)
(17, 37).

In *Haemastaticks*, Hales discussed exercise in an experiment with a dog (Experiment XII) in which he examined the dilatation of the lungs through an incision
between the ribs into the thorax (39; p. 82). In another discussion related to the same
experiment after remarking on the harmful effects of "intemperance and excesses in
eating and drinking" on the passage of blood through the lungs, Hale provided the
following summary of the physiological value of exercise:

> And here it may not be improper to observe, as it naturally occurs from the foregoing
> Considerations, the great Benefit of Exercise even to the temperate Liver; for it not only
> by the meer Effect of Motion agitates the Blood in all Parts, but also gives it a brisk Cir-
> culation, not only by its increasing the Number of the Systoles of the Heart, but also
> by giving it a freer Course thro' the more dilated and agitated Lungs: Which Dilatation
> exercise also makes more free and easy, by promoting Digestion and the Descent and
> Evacuation of the Contents of the Bowels, whereby not only the Midriff can more
> freely act and dilate the Thorax and Lungs, but the Blood also can have a freer Passage
> thro' the Coats of the Stomach and Guts. Thus in whatever View we consider the Ani-
> mal Oeconomy, many cogent Arguments for Temperance and Exercise do always occur.
> (39; p. 88)

Additional information on Hales can be found in Chapter 3.

Irish physician Bryan Robinson (1680–1746) followed Sir John Floyer's research in the study of the proportional relation of the pulse to respiration. In fact, he
proposed a mathematical formula for determining pulse rate (82). Robinson was educated at Trinity College in Dublin and was president of the College of Physicians in
Ireland. He was also a member of the Royal College of Surgeons and had a very successful practice in Dublin (68). Descriptions of his various experiments and their results were published in *A Treatise of the Animal Oeconomy* in 1732. In the Preface
to his second edition, published 2 years later, Robinson explained that "I have added
a Section concerning the Effects of various Fluids, of Age, of different kinds of
Weather, and of Exercise, on animal Fibres" (64). In particular, with regard to exer-

cise, he discussed muscle size and strength and also detailed experiments he had done to determine how quantities of perspiration and urine changed according to seasons of the year, time of day, age, exercise intensity, and body size. For the muscles, Robinson believed that they "grow larger and stronger by moderate Exercise: For the expansive Force of the Ether must be encreased, before it can move the Muscles; and a frequent Increase of this Force in Muscles much moved, must of Necessity increase both their Magnitudes and Strengths." He then concluded that "labouring Persons have larger and stronger Muscles, than Persons who lead a sedentary and inactive Life" (64; p. 101).

Robinson's findings relating to exercise in his perspiration and urine experiments were summarized as follows:

> At the Beginning of the Exercise of Walking I have observed, that Urine has been increased as well as Perspiration; but on continuing the Exercise, Urine in a very little Time has decreased, and grown less than it was before the Exercise, from the large Discharge which was made by the Skin. If we suppose the Quantity of Urine not to be lessened by Exercise, as it may not in Persons who by Drink supply the Loss which is made by Perspiration, then will the Proportion of Perspiration to Urine be 6 to 1, in Persons who walk at such a Rate as to give a glowing Warmth to their Skins but not to cause Sweat, and 16 to 1 in Persons who walk at such a Rate as to sweat profusely, on Supposition that the Proportion of Perspiration to Urine is 2 to 1 in the Heat of Summer. (64; pp. 281–283)

Also, in relation to this experiment, he noted that "The Exercise of Riding increases Perspiration, but neither so suddenly, or in so great a Degree, as the Exercise of Walking." For example, Robinson said, "A healthful Man upwards of ninety Years of Age, who commonly without Exercise discharged four or five times as much by Urine as he did by Perspiration, observed that in the Night, after riding several Hours the Day before, he always perspired as much as he discharged by Urine." Accordingly, Robinson concluded that in this case, "Perspiration to Urine was increased by Riding in the Proportion of 4 or 5" (64; p. 283).

The earlier work of Hales helped to establish the groundwork for the more exact calculations of total cardiac output as described by the Dutch mathematician and physician Daniel Bernoulli (1700–1782). As noted by Rothschuh, Bernoulli was "the first to make an approximately correct calculation of the cardiac action by considering it the product of force (weight of the blood) and the length of the impulse (height of the column of blood expulsed by the contracting heart)" (65). Bernoulli's major work, *Hydrodynamica*, was written as early as 1734 but was not published until 1738. Earlier in his career, he did experiments and published in the general area of the mechanical aspects of physiology. As a typical iatrophysicist, Bernoulli provided a review of the mechanics of breathing in 1721. Later, in 1728, he published a mechanical theory of muscular contraction. In fact, as his career unfolded, Bernoulli was more a physicist than physician, as he used his mathematics training to assist him in experimentation with physical apparatus and the application of physics and other sciences to man's capacity to perform maximum physical work over a period of sustained time (80).

In *Hydrodynamica*, Bernoulli discussed the physiology of man's use of pumps, levers of windlasses, and oars for rowing, and how they "should be constituted in order that for the minimum fatiguing of the men at the same time the product of their effort by the velocity of all be a maximum." He also wondered "how large the radius should be made in wheels or rollers for treadmills" (2; pp. 185–186). Later, Bernoulli speculated upon the efficiency of a man walking on an incline on a "treadmill machine":

> I would believe that a man of ordinary stature, healthy and robust, marching on a path inclined at 30 degrees will accomplish 3600 feet in a single hour without difficulty, and therefore he will elevate to a vertical height of 1800 feet the weight of his body, which I may assume [to be] 144 pounds or two cubic feet of water. Such a man, therefore, could by means of a treadmill machine, acting in a circle and being most perfect (in which of course nothing of the absolute potential is wasted) elevate in a single hour two cubic feet of water to a vertical height of 1800 feet, or, which is the same, in a single second one cubic foot to the height of one foot. (2; pp. 204–205)

Toward the conclusion of *Hydrodynamica*, Bernoulli provided a detailed explanation of the principles of rowing and noted that the force propelling a boat "is not to be estimated from the pressure of the rowers against the oars, but from the pressure which the extremities of the oars submerged in the water exert against the water" (2; p. 338).

As we move into the 1760s, the experiments on muscular strength and the beginning of quantitative dynamometry by John Theophilus Desaguliers (1683–1744) become known. Desaguliers was born in France but was brought to London in his infancy. He was educated at Oxford, was an ordained priest, and served as curator of experiments for the Royal Society. Desaguliers was particularly gifted at making apparatus and models to demonstrate the principles of applied physics. In this regard, in conjunction with his great interest in the physics of human muscle action and feats of strength performed by strongmen, he invented the Graham-Desaguliers dynamometer. As the historian Robert Schofield so aptly noted about the career of Desaguliers, "it would be hard to find a man more involved in all aspects of British natural philosophy, theoretical, experimental, and practical, for the years from 1714 through 1744" (58, 68).

Desaguliers's major work, *A Course of Experimental Philosophy*, was published in two volumes between 1734 and 1744. A third and revised edition was published in London in 1763 (68). For exercise physiology, it is important to note that he devoted considerable attention to the physicality and movement of both men and animals, with discussions of the mechanics of walking, rope dancing, carrying burdens, lifting barrels, rowing a boat, and lifting weights with the hands and feet (19). Much of his work was presented in the form of "propositions" and "corollaries." For example, in Proposition XXII, he presented his research and experimentation on "the absolute apparent Force, which can be exerted by the two Muscles, the Biceps and Brachiaeus, bending the Cubit (or lower Bone of the Arm) when the whole Arm is in a Supine and horizontal Situation; which is greater than twenty times the Weight that is sustained by them, and exceeds the Force of 560 Pound Weight" (19).

In a later corollary, Desaguliers described the muscles employed "when a man carries a Weight or a Burden upon his Back." As noted, he was intrigued by the lifting feats of the performing strongmen of his era and showed that most of their accomplishments were more attributable to technique and the principles of physics than unusual strength. Yet, he was so impressed with England's William Joy and Thomas Topham, along with the German Johann Karl Von Eckenberg, that he discussed their feats before the Royal Society. In so doing, he made reference to De la Hire's lecture given before the Royal Academy of Sciences in 1699 on "An Examination of the Force of Men to Move Weights, Whether by Lifting, Carrying, or Drawing, Considered as Well Absolutely as When Compared with That of Other Animals Which Carry and Draw, as Horses, & C" (19). Finally, realizing that "all men are not proportionably strong in every Part, but some are strongest in the Arms, some in the Legs, and others in the Back, according to the Work and Exercise which they use," Desaguliers claimed he could not "judge of a Man's strength by lifting only" and offered "a Method ... to compare together the Strength of different Men in the same Parts, and that too without straining the Persons who try the Experiment." Here, he introduced the Graham-Desaguliers dynamometer, provided a detailed description, and published a drawing. With the goal of trying "one's Strength by Means of the Machine," Desaguliers measured lifting power using a yoke-type harness, arm strength, gripping power, and strength of the fingers (19). Pearn suggests that Desaguliers wanted to measure muscular force "in such a way that synergistic muscles could not impart a false mechanical advantage to the test," and that his contributions to dynamometry were: (1) the establishment of the importance of a standard position for testing a particular muscle, (2) making quantitative dynamometry practical for the first time, and (3) noting the variation from person to person of the strength of an individual muscle in comparison to overall body stamina (58).

Adair Crawford (1748–1795), a Scottish physician and chemist, was best known for his work in animal calorimetry. In his major publication, *Experiments and Observations on Animal Heat, and the Inflammation of Combustible Bodies: Being an Attempt to Resolve These Phaenomena into a General Law of Nature* (1779), Crawford acknowledged his indebtedness to his two teachers, William Irvine and Joseph Black, as he set out to determine experimentally which state of air had the greater quantity of absolute heat. Since it was already known that animals with lungs and able to inspire large amounts of air were warm, and that the warmest animals were those with the largest proportional respiratory organs, Crawford added the observation "that in any one animal the degree of heat is proportional to the quantity of air inspired in a given time and that consequently animal heat is increased by exercise or anything else that accelerates the respiration." In this regard, Mendelsohn suggested that Crawford took "the common assertion that heat increases with exercise and related it to the quantity of air respired rather than to increased attrition of the blood in which case the inspired air had been seen as a cooling agent." Accordingly, by 1779, Crawford had shown experimentally that pure or dephlogisticated air contained more heat than fixed air and that arterial blood had a greater capacity for heat than the venous blood and consequently could absorb heat while passing through the lungs (56).

An exceptional work on the therapeutical value of exercise was published in Paris in 1780 by the physician Joseph-Clément Tissot (1747–1826). Similar to the earlier treatise of Fuller (1705), Tissot's *Medicinal and Surgical Gymnastics or Essay on the Usefulness of Movement, or Different Exercises of the Body, and of Rest, in the Treatment of Diseases* was so valued that it was soon translated into German, Italian, and Swedish. Particularly enlightening and informative was Part One, where he discussed the effects of exercise on the body in relation to time, place, intensity, and duration. Similarly, based upon the non-natural tradition, Tissot also devoted his attention to the effects of rest on the human body (81).

For Tissot, "gymnastics" was "that part of medicine which teaches the way to preserve or restore health by means of exercise." Accordingly, he devoted considerable space to describing the physiological benefits of exercise as it related to individual health. Exercise, according to Tissot, included "all the actions which give to the machine a certain motion capable of putting it in play. Exercise is either work or amusement." In a further elaboration of the physiological effects of exercise, Tissot noted that

> All exercise in man is an increased motion, and that motion receives its principal increase in the part which moves most; but soon it extends to all parts following the laws of circulation. Thus through motion the arrested fluids which alkalized themselves flow freely in their channels; those which are too thick are attenuated; those which lack activity have their salts more developed. The fibers enlarge and acquire new strength to push the fluids, to prevent or dissipate swellings; the liquids pushed with more vigor arrive at the secretory tubes of the skin; perspiration becomes more abundant and is transformed into sweat, carries with it the sour salts and a great number of heterogenous parts which would spoil the blood mass. The fluids coming more often to the secretory organs which in turn have received new energy, secretions are made more freely and rid the blood of an infinity of foreign elements. (81; p. 13)

With reference to the duration of exercise, Tissot suggested that "as long as its duration is in proper ratio to the strength of the person exercising ... it always produces most of the good effects of which we have spoken" (81; p. 20). For intensity, Tissot took into consideration a person's age, sex, temperament, and the season of the year and emphasized individual differences related to the ability to perform exercise as well as its physiological effects.

During the last decade of the eighteenth century, several further developments took place in research and experimentation relating to exercise and physiology. Most notable was the work of the French chemist Antoine-Laurent Lavoisier (1743–1794). In 1777, he recognized that both respiration and combustion were linked to what he called "burnable air," or "oxygéne." Subsequent work with the French mathematician Pierre de Laplace (1749–1827) interpreted respiration as a slow oxidation or combustion of carbon in the body (38, 65). Around 1789, Lavoisier improved his experimental study of respiration by doing collaborative work with the French chemist Armand Séguin (1767–1835). Their research led to the conclusion that hydrogen combined with oxygen gave off heat in the lungs, and accordingly, the lungs were the producers of animal heat. And, in some of the first physiological studies to use human subjects under controlled conditions, Lavoisier and Séguin determined that during digestion and physical exercise, oxygen consumption was increased in close correlation to the intensity of the organic combustion that resulted in animal heat (59). They reported in 1790 that

> When by exercise and movement one increases the consumption of oxygen gas in the lungs the circulation accelerates, of which one can easily convince oneself by the pulse beat, and, in general, when the person is breathing without hindrance, the quantity of oxygen consumed is proportional to the increase in the number of pulsations multiplied by the number of inspirations. (47; p. 446)

Evidence of these experiments was the drawing by Madame Lavoisier showing Séguin seated, breathing into a face mask and working a foot treadle. Lavoisier was standing talking to an assistant and Madame Lavoisier was recording the results. Besides using the foot treadle as a form of exercise, Séguin also lifted a weight of about 15 pounds for 15 minutes (20). It became clear to Lavoisier and Séguin that the effects of exercise on respiration were far greater than any other, including external temperature, and concluded that respiration was somehow related to work and to animal heat. Respiration not only replaced heat lost to the environment; it also supplied the heat which was converted into work. Accordingly, Lavoisier and his colleagues showed for the first time the basic ideas of energy transformation and the source of animal heat (38, 47, 65). Lavoisier and Séguin, in a paper on the "Transpiration of Animals" (1790), also demonstrated the importance of perspiration in regulating body temperature (62). Additional information on Lavoisier and his experiment can be found in Chapters 4 and 6.

The eighteenth century was a period when some important progress was made in a variety of physiological subjects. Physiology itself had begun to become more independent as a field of study, but for the most part, research and instruction were still connected with anatomy and medicine. Important insights into combustion, metabolism, and respiration took place, as did a better understanding of muscular contraction and animal heat. Instrumentation also evolved throughout the 1700s as did experimental methodology, publishing outlets, and societies for physiologists. As we have seen, exercise was utilized in many of these new discoveries. As a result, we learned more about the physiological effects of exercise on the human organism.

EARLY NINETEENTH CENTURY

In Europe during the early 1800s, especially England, France, and Germany, and in antebellum America, physiology was becoming a more identifiable field and exercise was receiving more attention from physicians because of its traditionally important role in the maintenance of health. A few scientists were also studying how the human body responded to physical exertion and muscular exercise. A sizable popular literature on hygiene, body function and exercise evolved throughout the first half of the nineteenth century. At about the same time in England and the United States, the term "physical education" was coined and popularized as a new and necessary form of educating individuals about their own bodies, or their "animal economy." As physical exercise became more popular in the 1870s, several books on gymnastics and calisthenics were published. And by the 1850s in the United States, the term "physiology of exercise" was being used in the scientific literature to denote a specific field.

The goal of the popular writers was to provide hygienic information or the "laws of life" to a population not necessarily served by physicians and thus to preserve the health of the nation. They believed that the study of the body's structure and function would provide guidance on the best way to live. Central to this literature was information on the physiology of exercise and its role in providing a "vital" tone to the body. A knowledge of the science of physiology was seen as integral to people's health. In the popular mode, physiology and hygiene often became synonyms. It was this way of thinking that led William A. Alcott and Sylvester Graham to found a society in Boston in 1837 known as the American Physiological Society. Its purpose was "to acquire and diffuse knowledge of the laws of life, and of the means of promoting human health and longevity" (44). This, however, was not the same American Physiological Society that was founded in 1887 and exists today. The new physiology's reputation was further enhanced by its practical application to modern medicine. While exercise was central to these books, they also discussed the other non-naturals, such as diet, and tended to place exercise and diet under the rubric of "regimen." Longevity was another common theme since it was argued that obeying the "laws of health" would lead to a longer life (16, 79).

Robley Dunglison (1798–1869), a physician who was born in England and educated in Germany, became one of the first to publish on physiological topics. He taught first at the University of Virginia and later at Jefferson Medical College in Philadelphia and the University of Maryland's Department of Hygiene, where he authored a two-volume text titled *Practice of Medicine* (1832–1842), as well as *Human Physiology* (1832) (60). However, it was his book *On the Influence of Atmosphere and Locality; Change of Air and Climate; Seasons; Food; Clothing; Bathing; Exercise; Sleep ... on Human Health; Constituting Elements of Hygiene* (1835) that contained significant information on exercise and was identified as "the first American textbook on preventive medicine prepared for use by medical students" (5, 65). Chapter 5 was devoted to exercise and, in his opening sentence, Dunglison said, "there is no hygiene agency of more importance than the due exercise of the body" (21). In this chapter, Dunglison provided a detailed survey of the physiological effects of "active exercise." Specifically, he affirmed that exercise gave "firmness and elasticity to the muscles" and explained that since "fat is absorbed between the muscles and their fasciculi, their outlines become well marked." Exercise also increased the action of the heart, causing the blood to more readily reach the capillaries, which encouraged a "free circulation" and prevented obstructions. Since respiration was directly connected with circulation, Dunglison explained that "if the latter be excited inordinately, the former becomes so likewise." He noted that accordingly:

> when, after violent exercise ... the heart beats violently, and the pulse is accelerated, and
> stronger, the respiratory movements participate in the turmoil so as at times to threaten
> suffocation; whilst the necessary aeration of the blood is interfered with ... that the ex-
> pired air contains less oxygen, and more carbonic acid, than during a state of tranquil-
> ity. (21; p. 427)

The capillaries were also augmented by exercise, therefore accelerating secretion and
nutrition. In this way, Dunglison argued, "we account for the greater development,
acquired by parts that are constantly exercised" and noted that "the body consequently acquires bulk, and vigor" as long as the exercise was not excessive. For excessive exercise, be warned that the body could experience "serious dislocations or
other mischief; —as hernia, aneurisms of the large vessels; dilatation of the cavities
of the heart; hemorrhages from the lungs or nose; sprains and lacerations of the
muscles, & c." In walking up a grade or climbing stairs, Dunglison said:

> the muscles of the anterior part of the thigh of the limb carried forward are powerfully
> exerted to draw the body upwards, and fatigue is experienced in them. The effort re-
> quired, too, hurries the circulation, and respiration, —producing anhelation or panting
> in the most vigorous, if the ascent be long, and steep, and developing it, almost instan-
> taneously, in such as labor under asthma, or serious heart disease: indeed, one of the
> earliest evidences of the existence of the latter may be the panting, and sense of im-
> pending suffocation, when the individual ascends even a moderate flight of stairs. (21;
> p. 433)

Dunglison also discussed the various types of exercise and concluded his chapter on
exercise with a warning that "who resigns himself to inglorious inactivity, suffer"
and that "sedentary habits are injurious" (21; pp. 442–443).

Ideas and information about the physiology of exercise appeared in a few books
written in the early nineteenth century relating to training the human body for athletic competition. It had been noted previously that Hippocrates, Galen, and
Philostratos each wrote about "athletes in training," the dangers of athletic competition, training techniques, and impediments to training. Beginning in the early
1800s, athletics and field sports for upper-class "gentlemen" were popularized in pri-
vate schools and colleges, as well as professionally, especially in England, wrestling,
running, horse racing, rowing, and boxing. With the increased interest in competition came not only the desire to excel but also the interest in playing longer, harder,
and faster than one's opponent. From this new perspective on sports then there
evolved a small literature devoted to the best way to train the human body to suc-
ceed in activities demanding levels of strength, speed, and endurance beyond normal
limits. Collegiate sports were underway in England by the 1820s and professional
events in boxing, horse racing, wrestling, running, distance walking, and weight lifting preceded them. In the United States, college sports did not develop to any extent
until the 1870s. It should be remembered that the ancient Olympic Games tradition
had been banned in 393 A.D. and was not revived until 1896.

In his *A Collection of Papers on the Subject of Athletic Exercise*, published in
London in 1806, Sir John Sinclair (1754–1835) was one of the first to write about
training for sporting competition. The book was an accumulation of the best meth-

ods, techniques, and ideas about training for sport that he could find in England at the time. Much of his information came from a questionnaire completed by "those who profess the art of training pugilists [boxers], wrestlers, and runners of foot-races." It contained queries on "training for athletic exercises," "training for jockeys," "training for race horses," and "training for game cocks" (61).

The most complete summary of Sinclair's findings was published in his *The Code of Health and Longevity* (1807). Based upon his data, the best age for training was between 18 and 40, 2 months were needed to "bring a man into good plight," and the morning air was always preferred for exercise. Sinclair determined that the object of exercise in training was "to enlarge the muscular substance, and to reduce the superfluous fat at the same time, by causing free perspiration, it renders more rapid the changes of absorption and deposition, so that both the fluids and solids of the body are purified and improved" (70; p. 33). He also noted that "the training to athletic exercises, has important effects upon various parts of the body ... and also tends materially to improve, and to preserve the shape of the body, and to promote its duration." For example, Sinclair found that training "always appears to improve the state of the lungs, or to *improve the wind*, as it is said; that is to say, it enables a person to draw a larger inspiration, to hold his breath longer, and to recover it sooner, after it is in a manner lost" (70; pp. 34–35). He also saw the "remarkable effect *upon the bones*," and said that exercise makes them "get harder and tougher, and [they] are less liable to be injured by blows or exercise." Finally, using the aforementioned examples of horses and cocks, Sinclair explained that "running-horses, when trained, do not wear out sooner than other horses; on the contrary, they bear fatigue much better." For game cocks, training did not shorten their lives and they tended to "live longer than common poultry" (70; pp. 35–36).

As noted earlier, Daniel Bernoulli had experimented with physiological studies using a "treadmill machine" as early as the 1730s. The idea resurfaced again in the 1770s when the British Parliament amended its hard-labor requirement for prisoners to include "labour of the hardest and most servile kind ... such as treading in a wheel, or drawing in a capstern for turning a mill or other machine or engine" (14). By 1787, one of Britain's leading reformers, Jeremy Bentham, argued in an article, "Of Airing and Exercise," in favor of adopting "walking in a wheel" as a healthy exercise for those in penitentiaries. His rationale included: "It is not in the smallest degree the less healthful for the profit which it brings: walking up hill is not at all a worse exercise, though it will go farther, than walking on plain ground." Bentham also suggested: "It will be less fatiguing, without being less conducive to health, if performed at twice rather than once, and divided between distant parts of the day. Less than a quarter of an hour each time will hardly answer any purpose; but that time may be doubled, trebled, or quadrupled, if economy should require it" (1; pp. 4, 158–159). By the 1820s, a controversy over the safety and usefulness of the "tread-wheel" for both male and female British prisoners had arisen. And some of the debate was published in the *London Medical and Physical Journal* in 1823. For example, articles on "Observations on the Tread-Wheel" and "Remarks on the

Tread-Wheel," appeared that year. It was into this debate during the 1850s that the British physician Edward Smith (1818–1874) emerged.

Edward Smith received his medical degree in 1843 from the Royal Birmingham Medical School and by the early 1850s had established a practice in London. Smith's reputation in physiological circles was based upon his early work in respiratory physiology, metabolism, and nutrition (14, 15). He published "Hourly Pulsation and Respiration in Health" in 1856 (74) and the following year wrote "The Influence of the Labour of the Treadwheel Over Respiration and Pulsation" (75) and "Inquiries into the Quantity of Air Inspired Throughout the Day and Night, and Under the Influence of Exercise, Food, Medicine, Temperature, & C" (77). It was Smith's early interest in measuring the effects of exercise on respiration that led to his interest in treadwheels and their potential deleterious effects on prisoners. In 1859, more results from his research appeared in major publications. He wrote "Inquiries into the Phenomenon of Respiration" (76), "On the Influence of Exercise Over Respiration and Pulsation" (78), and "Experimental Inquiries into the Chemical and Other Phenomena of Respiration, and Their Modifications by Various Physical Agencies" (73). Largely in recognition of his early research in the quantitative study of human respiratory exchange during exercise, Smith was elected to the Royal Society in 1860 (14). His biographer, Carleton Chapman, said Smith was "the first to devise quantitative methods suitable for studies on the human being during exercise" and argued that his "data on inspiratory volume, respiratory and pulse rates, and carbon dioxide production at rest and at various levels of exercise served as the basis for much of the work on muscular exercise in the latter part of the nineteenth century" (15). For additional information, the reader should consult Chapters 4 and 6.

Two additional individuals did pioneering work in the field of physiology and exercise around the middle of the nineteenth century. In his work on "insensible perspiration" in the 1840s, the French physician Fourcault concluded that its suppression by cold and damp conditions or insufficient exercise was a major cause of chronic disease (62). Then, in the 1850s, the German physician Carl Voit (1831–1908) began a series of studies to examine the effects of coffee, salt, and exercise upon the quantity of urea formed by a dog. In order to measure the effect of muscular activity on nitrogen consumption, Voit constructed a treadmill and had the dog run for 10 minutes, six times a day. In keeping with Justus Liebig's (1803–1873) hypothesis that the breakdown of nitrogenous tissue constituted the source of mechanical work, Voit figured the dog would excrete more urea on the days it exercised compared to days of rest. But, to his surprise, he observed no difference. He continued his research and joined with his colleague Theodor Bischoff to investigate further the "laws" of animal nutrition. Voit became a lecturer at the University of Munich in 1859 and was named professor of physiology in 1863. He and Bischoff published much of their research in their book *The Laws of the Nutrition of Carnivorous Animals* (1860). During the 1860s, Voit combined with another colleague, Max Pettenkofer, to do further studies involving exercise and respiration at their Munich Institute of Physiology (46, 48). Voit had a substantial impact on American physiology in general and in exercise physiology in particular through his American students Wilbur Atwater (1844–1907) and Francis G. Benedict (1870–1957) (65). Their contributions can be found in Chapter 6.

The culmination of this analysis of the "ancient and early influences" on exercise physiology is a discussion of the first and early uses of the term "physiology of exercise." As we have seen, scientists largely trained as physicians had used exercise to better understand human physiology. Also, some scientists, physicians, coaches, and trainers studied physiology to further understand exercise. For the former group, the level and degree of physiological response to exercise provided a unique opportunity to study the range and control mechanisms of various systems in non-normal situations. In addition, by studying individuals who were "trained," they could better understand the processes of physiological adaptation. The latter group, however, utilized physiological knowledge and research techniques to analyze human potential and to improve those aspects of performance most cherished in sporting competition, such as strength, speed, power, and endurance. Yet, nowhere in this vast literature were the terms "exercise physiology" or "physiology of exercise" used to identify their practices or research objectives.

As noted previously, this all changed in 1855 when William H. Byford (1817–1890), a physician from Evansville, Indiana, and professor of obstetrics and diseases of women and children at Chicago's Rush Medical College, published "On the Physiology of Exercise" in the July issue of the *American Journal of Medical Sciences*. Byford had read extensively about the healthfulness of exercise and had performed a variety of experiments on humans and dogs dealing with exertion and its influence on heat, circulation, breathing, and secretions. Moreover, he was puzzled that more research was not being done in the field and wondered why, for the most part, both the medical and physical education professions were largely unconcerned with the scientific analysis of exercise. Accordingly, Byford tried to account for this lack of interest and to encourage others to do more research in this area.

> Although the importance of voluntary exercise has been recognized for centuries and prescribed to its most useful extent by many of the profession, its great practical advantages in a large number of diseases have not been appreciated to their full extent by all. The only reason I can ascribe for this is that its effects upon the animal economy have not been thoroughly investigated and understood. It is with a view to draw the attention of the profession to the importance of more research in this direction, that I wish to record my views upon the subject. (9; pp. 32–33)

Three years later, Byford revisited the subject with a lecture on "Physiology, Pathology, and Therapeutics of Muscular Exercise" delivered before the Cook County, Illinois, Medical Society. The entire speech was published in the August 1858 issue of *The Chicago Medical Journal* and later that year printed in book form by the publisher James Barnet of Chicago (10). Here, Byford discussed some of the experiments he and others had done since his 1855 article and articulated their findings:

> Upon closer scrutiny into the manifestations of muscular exercise, we find the heart and arteries beating rapidly and several of the secretions increased in quantity, and their pe-

culiar products enhanced to a considerable extent. The urinary secretion, the secretion of the liver, skin and the pulmonary excretion, are all carried on more actively than in a state of repose, while the secretion and excretion of the alimentary mucous membrane are diminished. (11; p. 1)

But of course, being an inquisitive scientist, Byford, after stating that "these are the more obvious phenomena of exercise," noted: "The interesting question now arises, how are they produced? What are the intimate circumstances transpiring during the exhibitions mentioned" (11; p. 2)? The remainder of his manuscript dealt with both the pathological and therapeutic effects of muscular exercise. He examined the physiological dangers of "violent efforts" as well as "inactivity," or the "want of exercise," and evaluated some of the most popular exercises of the period.

Exercise physiology was further impacted by two mainstream American physiologists during the midnineteenth century. In his well-known *History of Physiology* text (65), Karl Rothschuh identified Robley Dunglison and Austin Flint (1836–1915) as two of the most prominent authors of the period. Dunglison, who was discussed earlier in the section on popular physiology, was also the teacher of S. Weir Mitchell (1829–1914) of Philadelphia, who studied with Claude Bernard of France, and went on to become one of the founders of the American Physiological Society in 1887. Flint, along with John Dalton, was the inaugurator of laboratory research in the United States. He graduated from Jefferson Medical College in 1857, studied with Dalton at Vermont and with Bernard in France, and served as Professor of Physiology at Buffalo for many years before moving to Bellevue Hospital Medical College in New York City (65). Flint will be remembered for his significant contributions to the infant field of exercise physiology.

Flint's most well-known treatise, the five-volume textbook *The Physiology of Man; Designed to Represent the Existing State of Physiological Science, as Applied to the Functions of the Human Body* (1865–1874), included significant information on exercise. For example, in his volume on *The Blood; Circulation; Respiration*, Flint discussed the influence of posture, muscular exertion, and exercise on the frequency of the heart's action, cited the research of Edward Smith, and analyzed the influence of muscular activity on respiration (24). Exercise was also discussed in Flint's other major work, *A Text-Book of Human Physiology: Designed for the Use of Practitioners and Students of Medicine* (1876) in a section on the "Development of Power and Endurance by Exercise and Diet." He explained that: "A fully-grown, well-developed man, in perfect health, may be trained so as to be brought to what is technically called fine condition, and he will present at that time all the animal functions in their perfection" (25).

Beginning in 1870 and concluding in 1878, Flint authored at least six articles and two books pertaining to his research in exercise physiology. His early research dealt with the effects of muscular exercise on the content of urine and resulted in articles in the *New York Medical Journal* (1870 and 1871), *Medical Gazette* (1870), *Practitioner* (1871), and a book, *On the Physiological Effects of Severe and Protracted Muscular Exercise; with Special Reference to Its Influence upon the Excretion of Nitrogen* (1871). He published another similar article in the *Journal of*

Anatomy and Physiology (1876), wrote "On the Source of Muscular Power" for the same journal the following year, and concluded his work in the field of exercise physiology with another book, *On the Source of Muscular Power. Arguments and Conclusions Drawn from Observations upon the Human Subject, under Conditions of Rest and of Muscular Exercise* (1878) (22, 23).

Finally, the decade of the 1880s marks the recognition of exercise physiology by the profession of physical education and its entrance into Europe, most notably France. Edward M. Hartwell, a leader in the American physical education movement and president of the American Association for the Advancement of Physical Education (AAAPE) (1891–1892 and 1895–1899), earned a doctoral degree in animal physiology from Johns Hopkins (1881) and a medical degree from Miami Medical College in Cincinnati (1882). From 1883 to 1890, he served as associate of physical training and director of the gymnasium at Johns Hopkins (34). During this time, in 1886, he addressed the members of the AAAPE at their annual meeting in Brooklyn on the topic, "On the Physiology of Exercise." His speech was published the following year as a two-part article in the *Boston Medical and Surgical Journal*. Hartwell explained to his readers that "The fundamental and essential characteristics of exercise are so generally misstated and its proper effects so frequently overlooked, that I have chosen the physiology of exercise as my theme" (40; p. 297; 41). Lastly, the French physician Fernand Lagrange (1846–1909) published his book *Physiology of Bodily Exercise* in Paris in 1888. It was translated into English 8 years later as part of the International Scientific Series published by the D. Appleton Company of New York. Spanning almost 400 pages, the text was divided into six major parts: muscular work, fatigue, habituation to work, the different exercises, the results of exercise, and the office of the brain in exercise. Each part was further divided into chapters. For example, the part on fatigue featured chapters on breathlessness, stiffness, overwork, and the theory of fatigue. The part on different exercises had chapters devoted to exercises of strength, exercises of speed, and exercises of endurance, among others (53).

REFERENCES

1. Bentham J. *The Works of Jeremy Bentham*. Edinburgh: William Tait, 1843.
2. Bernoulli D. *Hydrodynamics, by Daniel Bernoulli & Hydraulics, by Johann Bernoulli. Translated from the Latin by Thomas Carmody and Helmut Kobus. Prefaced by Hunter Rouse*. New York: Dover Publications, Inc., 1968.
3. Bernoulli J. *Dissertations on the Mechanics of Effervescence and Fermentation and on the Mechanics of the Movement of the Muscles by Johann Bernoulli*. Philadelphia: American Philosophical Society, 1997.
4. Berryman J. W. Exercise and the medical tradition from Hippocrates through Antebellum America: a review essay. In: *Sport and Exercise Science: Essays in the History of Sports Medicine*, edited by J. W. Berryman and R. J. Park. Urbana and Chicago: University of Illinois Press, 1992, pp. 1–56.
5. Berryman J. W. The tradition of the "six things non-natural": exercise and medicine from Hippocrates through Ante-Bellum America. *Exercise and Sport Science Reviews* 17: 515–559, 1989.

6. Birch T. *The history of the Royal society of London for improving of natural knowledge, from its first rise: In which the most considerable of those papers communicated to the society, which have hitherto not been published, are inserted in their proper order, as a supplement to the Philosophical transactions.* London: Printed for A. Millar, 1756–1757.

7. Brown T. M. Lower, Richard. In: *Dictionary of Scientific Biography*, edited by C. C. Gillispie. New York: Charles Scribner's Sons, Vol. VIII, pp. 523–527, 1973.

8. Brown T. M. Mayow, John. In: *Dictionary of Scientific Biography*, edited by C. C. Gillispie. New York: Charles Scribner's Sons, Vol. IX, pp. 242–247, 1974.

9. Byford W. H. On the physiology of exercise. *The American Journal of Medical Sciences* 30 (July):32–42, 1855.

10. Byford W. H. Physiology, pathology and therapeutics of muscular exercise. (Read before Cook Co. Medical Society, and published at their request). *The Chicago Medical Journal* I (August):357–382, 1858.

11. Byford W. H. *Physiology, Pathology and Therapeutics of Muscular Exercise: a paper read before the Cook County Medical Society and published at their request.* Chicago: James Barnet, 1858.

12. Bylebyl J. J. The school of Padua: humanistic medicine in the sixteenth century. In: *Health, Medicine and Mortality in the Sixteenth Century*, edited by C. Webster. Cambridge: Cambridge University Press, 1979, pp. 335–370.

13. Campbell D. *Arabian Medicine and Its Influence on the Middle Ages*. London: Kegan Paul, Trench, Trubner & Co., Ltd, 1926.

14. Chapman C. B. Edward Smith (?1818–1874) physiologist, human ecologist, reformer. *Journal of the History of Medicine and Allied Sciences* 22:1–26, 1967.

15. Chapman C. B. Smith, Edward. In: *Dictionary of Scientific Biography*, edited by C. C. Gillispie. New York: Charles Scribner's Sons, Vol. XII, pp. 465–467, 1975.

16. Cooter R. The power of the body: the early nineteenth century. In: *Natural Order: Historical Studies of Scientific Culture*. London: Sage Publications, 1979, pp. 73–92.

17. Cournand A. Introduction to 1964 reprint. In: *Statical Essays: Containing Haemastaticks*. New York: Hafner Publishing Company, 1964.

18. Crombie A. C. Decartes, René Du Perron. In: *Dictionary of Scientific Biography*, edited by C. C. Gillispie. New York: Charles Scribner's Sons, Vol. IV, pp. 51–55, 1971.

19. Desaguliers J. T. J. T. *A course of experimental philosophy. By J. T. Desaguliers, L. L. D. F. R. S. Chaplain to his Grace the Duke of Chandos. Vol. I. Adorn'd with thirty-two copper-plates. The third edition corrected.* London: Printed for A. Millar, J. Rivington, R. Baldwin, L. Hawes, W. Clarke, R. Collins, J. Richardson, T. Longman, W. Johnston, and C. Rivington, 1763.

20. Dill D. B. Historical review of exercise physiology science. In: *Science and Medicine of Exercise and Sport* (2nd ed.), edited by W. R. Johnson and E. R. Buskirk. New York: Harper & Row, 1974, pp. 37–41.

21. Dunglison R. *On the Influence of atmosphere and locality; change of air and climate; seasons; food; clothing; bathing; exercise; sleep; corporeal and intellectual pursuits, etc., etc. On human health; constituting elements of hygiene.* Philadelphia: Carey, Lea & Blanchard, 1835.

22. Flint A. *Collected Essays and Articles on Physiology and Medicine.* New York: D. Appleton and Company, 1903.

23. Flint A. *On the Source of Muscular Power. Arguments and Conclusions Drawn from Observations Upon the Human Subject, Under Conditions of Rest and of Muscular Exercise.* New York: D. Appleton and Company, 1878.

24. Flint A. *The physiology of man; designed to represent the existing state of physiological science, as applied to the functions of the human body.* New York: D. Appleton and Company, 1868.

25. Flint A. *A text-book of human physiology: designed for the use of practitioners and students of medicine.* New York: D. Appleton and Company, 1876.

26. Floyer S. J. *The physician's pulse-watch; or, an essay to explain the old art of feeling the pulse, and to improve it by the help of a pulse watch.* London: Printed for Sam. Smith and Benj. Walford, 1707.

27. Foster S. M. *Lectures on the History of Physiology During the Sixteenth, Seventeenth, and Eighteenth Centuries.* New York: Dover, 1970.

28. Frank Jr. R. G. *Harvey and the Oxford Physiologists: A Study of Scientific Ideas and Social Interaction.* Berkeley: University of California Press, 1980.

29. Galen. Good condition. In: *Galen: Selected Works.* New York: Oxford University Press, 1997, pp. 296–298.

30. Galen. The pulse for beginners. In: *Galen: Selected Works.* New York: Oxford University Press, 1997, pp. 325–344.

31. Galen. To Thrasyboulos: is healthiness a part of medicine or of gymnastics? In: *Galen: Selected Works.* New York: Oxford University Press, 1997, pp. 53–99.

32. Galen. *A Translation of Galen's Hygiene (De Sanitate Tuenda).* Springfield, IL: Charles C. Thomas, 1951.

33. Garcia-Ballester L. *Artifex factivus sanitatis*: health and medical care in medieval Latin Galenism. In: *Knowledge and the Scholarly Medical Traditions,* edited by D. Bates. Cambridge: Cambridge University Press, 1995, pp. 127–150.

34. Gerber E. W. *Innovators and Institutions in Physical Education.* Philadelphia: Lea & Febiger, 1971.

35. Granit R. Fernel, Jean Francois. In: *Dictionary of Scientific Biography,* edited by C. C. Gillispie. New York: Charles Scribner's Sons, Vol. IV, pp. 584–586, 1971.

36. Gruner O. C. *A Treatise on the Canon of Medicine of Avicenna, Incorporating a Translation of the First Book.* New York: Augustus M. Kelley, 1970.

37. Guerlac H. Hales, Stephen. In: *Dictionary of Scientific Biography,* edited by C. C. Gillispie. New York: Charles Scribner's Sons, Vol. VI, pp. 35–48, 1972.

38. Guerlac H. Lavoisier, Antoine-Laurent. In: *Dictionary of Scientific Biography,* edited by C. C. Gillispie. New York: Charles Scribner's Sons, Vol. VIII, pp. 66–91, 1973.

39. Hales S. *Statical Essays: Containing Haemastaticks.* New York: Hafner Publishing Company, 1964.

40. Hartwell E. M. On the physiology of exercise (part 1). *Boston Medical and Surgical Journal* 116:297–302, 1887.

41. Hartwell E. M. On the physiology of exercise (part 2). *Boston Medical and Surgical Journal* 116:321–324, 1887.

42. Harvey W. *Exercitatio Anatomica, De Motu Cordis et Sanguinis in Animalibus.* Springfield, IL: Charles C. Thomas, 1931.

43. Hippocrates. *Hippocrates: with an English translation by W. H. S. Jones.* London: William Heinemann Ltd., 1953.

44. Hoff H. E., and Fulton, J. F. Centenary of the first American Physiological Society founded at Boston by William A. Alcott and Sylvester Graham. *Bulletin of the History of Medicine* 5:687–734, 1937.

45. Hoffmann F. *Fundamenta Medicinae.* London: MacDonald, 1971.

46. Holmes F. L. The Formation of the Munich School of Metabolism. In: *The Investigative Enterprise: Experimental Physiology in 19th Century Medicine,* edited by W. Coleman and F. L. Holmes. Berkeley: University of California Press, 1988, pp. 179–209.

47. Holmes F. L. *Lavoisier and the Chemistry of Life: An Exploration of Scientific Creativity.* Madison, WI: University of Wisconsin Press, 1985.

48. Holmes F. L. Voit, Carl von. In: *Dictionary of Scientific Biography,* edited by C. C. Gillispie. New York: Charles Scribner's Sons, Vol. XIV, pp. 63–67, 1976.

49. Kardel T. *Steno on Muscles: Containing Stensen's Myology in Historical Perspective.* Philadelphia: The American Philosophical Society, 1994.
50. Keill J. *An account of animal secretion, the quantity of blood in the humane body, and muscular motion.* London: Printed for George Strahan, 1708.
51. King L. S. *The Growth of Medical Thought.* Chicago: University of Chicago Press, 1963.
52. King L. S. Stahl, George Ernst. In: *Dictionary of Scientific Biography*, edited by C. C. Gillispie. New York: Charles Scribner's Sons, Vol. XII, pp. 599–606, 1975.
53. Lagrange F. *Physiology of Bodily Exercise.* New York: D. Appleton and Company, 1896.
54. Lower R. *A facsimile edition of Tractatus De Corde, Item De Motu & Colore Sanguinis et Chyli in eum Tranfitu or, A Treatise on the Heart, On the Movement and Colour of the Blood and on the Passage of the Chyle into the Blood.* Oxford: Oxford University Press, 1932.
55. Mayow J. *Medico-physical works; being a translation of Tractus quinque medico-physici (1674).* Edinburgh: Alembic Club, 1907.
56. Mendelsohn E. *Heat and Life: The Development of the Theory of Animal Heat.* Cambridge: Harvard University Press, 1964.
57. Payne L. M. Croone, William. In: *Dictionary of Scientific Biography*, edited by C. C. Gillispie. New York: Charles Scribner's Sons, Vol. III, pp. 482–483, 1971.
58. Pearn J. Two early dynamometers: an historical account of the earliest measurements to study human muscular strength. *Journal of the Neurobiological Sciences* 37:127–134, 1978.
59. Pierson S. Seguin, Armand. In: *Dictionary of Scientific Biography*, edited by C. C. Gillispie. New York: Charles Scribner's Sons, Vol. XII, 286–287, 1975.
60. Radbill S. X. Dunglison, Robley. In: *Dictionary of Scientific Biography*, edited by C. C. Gillispie. New York: Charles Scribner's Sons, Vol. IV, pp. 251–253, 1971.
61. Radford P. F. From oral tradition to printed record: British sports science in transition, 1805–1807. *Stadion* 12 & 13:295–304, 1986–1987.
62. Renbourn E. T. The natural history of insensible perspiration: a forgotten doctrine of health and disease. *Medical History* 4:135–152, 1960.
63. Risse G. B. Hoffmann, Friedrich. In: *Dictionary of Scientific Biography*, edited by C. C. Gillispie. New York: Charles Scribner's Sons, Vol. VI, pp. 458–461, 1972.
64. Robinson B. *A Treatise of the Animal Oeconomy.* Dublin: Printed by S. Powell for George Ewing and William Smith, 1734.
65. Rothschuh K. E. *History of Physiology.* Huntington, New York: Robert E. Krieger Publishing Company, 1973.
66. Santorio S. *Medicina Statica: or, Rules of Health, In Eight Sections of Aphorisms.* London: Printed for John Starkey, 1676.
67. Scarborough J. Galen and the Gladiators. *Episteme* 5:98–111, 1971.
68. Schofield R. E. *Mechanism and Materialism: British Natural Philosophy in an Age of Reason.* Princeton, New Jersey: Princeton University Press, 1969.
69. Sherrington S. C. *The Endeavor of Jean Fernel: With a List of Editions of His Writings.* Folkestone & London: Dawsons of Pall Mall, 1974.
70. Sinclair S. J. *The code of health and longevity; or, a general view of the rules and principles calculated for the preservation of health, and the attainment of long life.* London: Printed for Sherwood, Gilbert & Piper; and William Tait, 1833.
71. Singer P. N. Introduction. In: *Galen: Selected Works.* New York: Oxford University Press, 1997.
72. Siraisi N. G. *Medieval and Early Renaissance Medicine: An Introduction to Knowledge and Practice.* Chicago: University of Chicago Press, 1990.
73. Smith E. Experimental inquiries into the chemical and other phenomenon of respiration, and their modifications by various physical agencies. *Royal Society of London—Philosophical Transactions* 149:681–714, 1859.

74. Smith E. Hourly pulsation and respiration in health. With two diagrams and tables. *Medico-Chirurgical Transactions* 39:35–58, 1856.

75. Smith E. The influence of the labour of the treadwheel over respiration and pulsation; and its relation to the waste of the system, and the dietary of the prisoners. *British Medical Journal* 1:591–592, 1857.

76. Smith E. Inquiries into the Phenomena of Respiration. *Proceedings of the Royal Society* 9:611–614, 1859.

77. Smith E. Inquiries into the quantity of air inspired throughout the day and night, and under the influence of exercise, food, medicine, and temperature, & c. *Proceedings of the Royal Society* 8:451–454, 1857.

78. Smith E. On the influence of exercise over respiration and pulsation; with comments. *Edinburgh Medical Journal* 4:614–623, 1859.

79. Smith V. Physical Puritanism and sanitary science: Material and immaterial beliefs in popular physiology, 1650–1840. In: *Medical Fringe & Medical Orthodoxy*, edited by W. F. Bynum and R. Porter. London: Croom Helm, 1987, pp. 174–197.

80. Straub H. Bernoulli, Daniel. In: *Dictionary of Scientific Biography*, edited by C. C. Gillispie. New York: Charles Scribner's Sons, Vol. II, pp. 36–46, 1970.

81. Tissot J.-C. *Gymnastique Medicinale et Chirurgicale*. New Haven, CT: Elizabeth Licht, 1964.

82. Townsend G. L. Sir John Floyer (1649–1734) and His Study of Pulse and Respiration. *Journal of the History of Medicine and Allied Sciences* 22:286–316, 1967.

83. Wilson L. G. William Croone's theory of muscular contraction. *Notes and Records, Royal Society of London* 16:158–178, 1961.

84. Winslow C.-E. A. and Bellinger, R. R. Hippocratic and Galenic concepts of metabolism. *Bulletin of the History of Medicine* 17:127–137, 1945.

85. Wolf A. *A History of Science, Technology and Philosophy in the Sixteenth and Seventeenth Centuries*. New York: Harper and Brothers, 1959.

86. Woody T. Philostratus: Concerning Gymnastics. *Research Quarterly* 7:3–26, 1936.

chapter 2

THE NEUROMUSCULAR SYSTEM

Alan J. McComas

....

IN the first section, which considers the muscles as the system responsible for generating force and movement, I am heavily indebted to the writings of Sir Michael Foster (50, 51), the first professor of physiology at the University of Cambridge, and also to Dr. Dorothy Needham, whose classic text, *Machina Carnis* (127), summarizes in a profound and elegant manner a lifetime's study of muscle. For obvious reasons the original writings of the Greek school, and of some of the classical figures subsequently, are not readily available, and there is the further problem of obtaining translations. In later sections of the chapter the emphasis has been placed on events in the muscles and motoneurons, partly because most is known at this level of the nervous system in relation to exercise, and partly because an analysis of exercise effects in the brain would make the chapter unwieldy. Also, there is the danger that too much attention to the brain would allow the chapter to become not so much a review of exercise as an essay on motor control. At this point it should be added that the review is one written by a neurophysiologist and, although muscle metabolism is touched upon, other chapters deal more thoroughly with this topic, as well as with muscle blood flow and oxygen consumption in exercise. On reading the review, some may think that too much attention has been given to the contributions, between the two world wars, of the Oxford, Cambridge and London schools of physiology. It is certainly true that there were excellent centers of exercise physiology elsewhere, notably in the Harvard Fatigue Laboratory and in Copenhagen. However, the main research directions in these laboratories did not include the excitation and contraction of exercising muscles, and it was in these overlapping fields that the supremacy of the British schools was universally recognized. Even with this caveat, any survey of people and ideas is necessarily selective and so reflects personal choices, although I would like to think that most of my choices would enjoy support. Nevertheless, there are many other muscle physiologists, living and dead, whose work might well have been included in a chapter of this kind and to them I can only apologize.

MUSCLES AS THE GENERATORS OF FORCES AND MOVEMENTS

Nowadays, it seems that even young children have an awareness that certain soft swellings in their bodies are muscles and that, in some way, the size of a muscle, most commonly the brachial biceps, is an indicator of strength. Yet such knowledge was not always apparent, for Hippocrates (460–380 B.C.) and the Hellenistic school considered that the tendons, rather than the muscle bellies, had the ability to produce movement. Moreover, because of their often similar gross appearances, the tendons were not distinguished from nerves. This confusion is evident in the following description of the functions of the principal tissues, given at the end of the third century B.C. "The bones give a body support, straightness and form; the nerves [tendons] give the power of bending, contraction and extension; the flesh and the skin bind the whole together and confer arrangement on it; the blood-vessels spread throughout the body, supply breath and flux and initiate movement" (85; bracketed interpolation added). For Aristotle (384–322 B.C.) also, the tendons were the body structures responsible for producing movements since, like an automatic puppet, "Animals have parts of a similar kind, their organs, the sinewy tendons to wit and the bones; the bones are like the wooden levers in the automaton, and the iron; the tendons are like the strings, for when these are tightened or released movement begins" (3). In contrast to the sinewy tendons, the muscles or "flesh" were thought to convey the sense of touch: "chief of all the primary sensibility is that of touch; and it is the flesh, or analogous substance, which is the organ of this sense" (4).

It was Herophilus of the Alexandrian school (early third century B.C.) who appears to have been the first to recognize the involvement of muscles in producing movements and to distinguish between nerves and tendons, as well as between arteries and veins. Erasistratus, a contemporary of Herophilus, thought that the muscles contracted because they filled with *pneuma*, a vital spirit derived from the air. This spirit was conveyed, in an altered form, from the brain to the muscles by means of the hollow tubes of the nerves. It was thought that as the muscles fill with pneuma they "increase in breadth but diminish in length, and for this reason are contracted" (44).

It was Galen (129–199 A.D.), however, who took the understanding of muscles and nerves to entirely new heights. As well as carrying out public dissections of human bodies, Galen experimented on live animals, and on African monkeys in particular. He wrote copiously on structure and function, as well as on the practice of medicine; he was also a noted philosopher. From his anatomical descriptions, it is apparent that Galen must have been extremely skilled in dissecting human and animal bodies and in recognizing, in great detail, the different parts so revealed. In relation to muscles, he clearly saw that these could only be in one of two states, contracted or relaxed, though he viewed relaxation, incorrectly, as resulting from the pull of antagonist muscles: "The natural activity of the muscles consists of contracting and withdrawing upon themselves, and lengthening and relaxation take place when the antagonist muscles pull and draw towards themselves" (52). So formidable was Galen's intellect, and so extensive and complete were his writings, that his views on muscle contraction remained unchallenged until the Renaissance and the appearance of the next major figure, Andreas Vesalius (1514–1564; Fig. 2.1). Born in Brussels, Vesalius dissected animals while still a classics student at the University of Louvain. Given his medical family background and biological curiosity, it was natural that he should become a physician. In pursuit of that goal, he moved to Paris to be taught by Sylvius. Vesalius soon became dissatisfied with the passive recital of Galen's treatises and, in particular, by dissections which were, by his own standards in animals, clumsily undertaken. Vesalius's evident abilities led the rulers of Venice to appoint him to the chair of surgery and anatomy in the new University of Padua, in which position

Fig. 2.1. Andreas Vesalius (1514–1564). A very handsome portrait of the famous anatomist and pioneer physiologist, seen here holding the dissected flexor tendons to the fingers. Reproduced from Foster (51), with permission from Cambridge University Press.

in it was his responsibility to perform public dissections. This was exactly the opportunity which Vesalius was seeking. After 5 years of unrelenting endeavor, he published his observations on the dissected cadavers in a monumental work entitled *De Humani Corporis Fabrica* (*Structure of the Human Body*). The extraordinary detail of the anatomical parts was illustrated by many plates and woodcuts, the creation of another Belgian, John Stephen Calcar.

When it came to neuromuscular function, Vesalius accepted much of Galen's teaching, as is evident from the following quotation:

> Muscle therefore, which is the instrument of voluntary movement as the eye is the instrument of vision and the tongue of taste, is composed of the substance of the ligament or tendon divided into a great number of fibres and of flesh containing and embracing those fibres. It also receives branches of arteries, veins and nerves, and by reason of the presence of the nerves is never destitute of animal spirits so long as the animal is sound and well. ... I am persuaded that the flesh of muscles, which is different from everything else in the whole body, is the chief agent, by aid of which (the nerves, the messengers of the animal spirits not being wanting) the muscle becomes thicker, shortens and gathers itself together, and so draws to itself and moves the part to which it is attached, and by help of which it again relaxes and extends, and so lets go again of the part which it had so drawn. (147)

By reviving Galen's practice of animal experimentation, Vesalius pointed a way for others to follow; in this respect, he and Galen can be considered the founders of physiology. By vivisection Vesalius was able to show directly—by the simple expedient of ligating a nerve and observing the paralyzing effect on the contractions of the struggling animal—that the muscles really did contract under the influence of nerves. When the ligature was released, the contractions resumed. By dissection he was also able to show that the nerve influence was mediated by the interior of the nerve rather than the surrounding sheath, though he refused to speculate on whether the "animal spirit" was transported in hollow channels within the nerve or through the solid material of the nerves—a dilemma which distantly echoes the present debate over the relative importance of impulse activity and axoplasmic transport in maintaining muscle properties (see p. 76).

Vesalius made important observations on other parts of the body and, while he did not challenge some of Galen's doctrines openly, he gave strong hints of his scepticism. This tendency to deviate from the orthodox view did not pass unnoticed by those who had read the *Fabrica* and, perhaps because of envy, there was resentment and political maneuvering against Vesalius. In anger and disappointment Vesalius turned his back on an academic career by accepting a comfortable appointment as court physician to Emperor Charles V. He eventually died during a journey to Venice.

A more exact description of muscle contraction did not appear until 100 years later, in 1664, with the publication of *De Musculis et Glandulis Observationum Specimen* by the Dane Nicholas Stensen (1638–1686). Apart from being a skilled anatomist (the discoverer of the parotid duct), Stensen was an early geologist who, in effect, founded crystallography and was perhaps the first to recognize the significance of fossils as a biological record. In his anatomical studies, Stensen used one of the early microscopes to examine muscle. He described "motor fibers" each of which was composed of the "most minute fibrils" arranged lengthways. From his account it is evident that Stensen's motor fibers were the muscle fascicles and that the most minute fibrils were the muscle fibers themselves. He went on to state that it was only the fleshy parts of the motor fibers which contracted while the tendinous parts, at the ends of the motor fibers, remained unchanged (144).

Francis Glisson (1597–1677), while regius professor of physics at Cambridge, was one of those who disputed the notion that the nerves conveyed some spirit to the muscles which inflated them and produced the contractions. He put the matter to experimental test by having a subject insert his arm in a glass tube and making a watertight seal around the shoulder. He was then able to show that, when the subject made a forceful contraction, no water was displaced from the tube into a vertical sidearm—that is, the volume of the contracting muscle remained constant. Important observations on the nature of muscle contraction were also made by Glisson's contemporary, Giovanni Borelli (1608–1679). An able mathematician, Borelli was born in Naples but subsequently moved to Rome under the patronage of Christina, the former queen of Sweden. A rough man, largely self-taught and possessed of many interests, Borelli investigated by the application of anatomy and physics, the ways in which movements were produced. Through calculations of angles and lever arms, Borelli was able to analyze a variety of movements, including walking and running, in mathematical terms, and much of his work remains valid today.

Like Glisson, he refuted the notion of animal spirits inflating the muscle, pointing out that no air bubbles could ever be seen to come from cut muscles contracting underwater. Instead it was the muscle tissue itself which contained all the material necessary for the contraction. In regard to the action of the nerves in initiating contractions, Borelli thought that possibly "some commotion must be communicated along some substance in the nerve, in such a way that a very powerful inflation can be brought about in the twinkling of an eye" (16)—a statement foreshadowing the discovery of the action potential.

And so, with the publication of Borelli's treatise on animal motion (*De Motu Animalium*), the correct relationship of the roles of nerves, muscles, and tendons in exercise was finally established. Insights into the nature of the contractile mechanism itself would not be advanced significantly until the development of the sliding filament hypothesis 300 years later (see the section on Machina Carnis: the Flesh Machine).

THE MOTOR UNIT

That there are probably many more muscle fibers than motor nerve fibers must have been obvious to those at the end of the nineteenth century who were able to cut and stain thin cross sections of nerve. The implication of this discrepancy in numbers was clear to Charles Sherrington (1857–1952; Fig. 2.2), who realized not only that each motor nerve fiber must supply many muscle fibers but also that each nerve fiber and

Fig. 2.2. John (Jack) Eccles (1903–1997) and Sir Charles Sherrington (1857– 1952). This photograph, a well-known one, was taken on one of Eccles' visits to Sherrington at his Ipswich home after Sherrington's retirement in 1935 and before Eccles' return to Australia in 1937. From their postures, it would appear that Sherrington has just asked Eccles a difficult question. Unfortunately, there are few photographs of Sherrington as a young man but, even in old age, the lively intelligence is still very evident in his face. Reproduced from Granit (68) with permission.

its colony of muscle fibers could be considered a "motor unit." Having shown that, unlike the situation in crustaceans, mammalian muscles did not have a local inhibitory nerve supply, Sherrington recognized that, in exercise, all movements and forces must be graded according to the numbers of motor units participating. The numbers of active units would, in turn, depend on the numbers of motoneurons in the spinal cord or brain stem that were discharging under the antagonistic influences of central excitation and inhibition. More than any other concept except for that of the contractile mechanism, the motor unit has dominated muscle physiology and has spawned a set of subsidiary questions each of which has importance in its own right. But, first, how could the sizes and tensions of the motor units be measured in a mammalian muscle?

At the time he undertook this problem, in 1928, Sir Charles Scott Sherrington had been Wayneflete professor of physiology at the University of Oxford for 14 years. He was then in his 70th year but as full of energy and enthusiasm as ever, and a magnet for aspiring neurophysiologists from all parts of the world. A shy, unassuming man, slight in build but still sharp of eye and mind, Sherrington had enjoyed a life of great academic distinction. Knighted, a former president of the Royal Society, and the recipient of numerous awards and honorary degrees, Sherrington was in the last phase of an extraordinarily successful academic career when he began to explore the motor unit.

To count the number of motor nerve fibers, and hence the number of motor units in a muscle, it was first necessary to remove the sensory fibers from the muscle nerve. Sherrington did this in the cat by allowing the sensory fibers to degenerate after expertly dividing the dorsal nerve roots between their ganglia and their junctions with the ventral roots. Despite his age and exalted position, Sherrington did much of the nerve histology himself, taking the specimens home and often working through the night in his bathroom (39). The sections of nerve, stained with osmic acid, were given to a young doctoral student from Melbourne, John (Jack) Eccles (Fig. 2.2). Eccles discovered that there were two populations of motor nerve fibers, large and small, and the total number of fibers was divided into the tensions that had been developed by the whole muscle, so as to derive the mean motor unit force. Eccles and Sherrington thought, however, that the small fibers were those which, during development, had arrived too late at the muscle to acquire their full complement of muscle fibers. It was only shown later, by Lars Leksell in Stockholm, Sweden (106), that the small nerve fibers supplied the muscle spindles. Ironically it was Sherrington himself who had first described the motor innervation of the spindles many years earlier. Nevertheless, through the work of Sherrington and Eccles, the first estimates of motor unit numbers and forces had been made in mammalian muscles and the results, distorted as they were by the inclusion of spindle efferents, showed that there were major differences between muscles. The cat gastrocnemius, for example, had twice as many motor units as the semitendinosus, and the gastrocnemius motor units developed four times the mean force (138).

In relation to human muscles, the first quantitative studies were those of Bertram Feinstein and his colleagues in Lund, Sweden, some years later. On the basis of a postmortem study of a patient with poliomyelitis, they assumed that 60% of the large-diameter fibers in a muscle nerve were α-motor. In a variety of limb and cranial muscles they found that the mean sizes of the motor units ranged from more than a 1000 muscle fibers in the medial gastrocnemius to 10 or so fibers in the lateral rectus of the eye, while the mean numbers of motor units in the same muscles were 579 and 2970, respectively (45). Later still, by recording mean motor-unit potential sizes or twitch tensions and comparing these with those of the whole muscle, it became possible to estimate the numbers of human motor units during life (118).

FAST AND SLOW MUSCLES AND MOTOR UNITS

Throughout history those who ate meat could not have helped noticing that some muscles or parts of muscles were darker than others, but the possible significance of these variations in color could not be considered until the muscle bellies, and their component fibers, had been identified as the source of contractions (see first section). Perhaps the first serious comment on the color of muscles was that of Nicholas Stensen (1664), who stated: "The middle parts of the motor fibres [muscle fasciculi], wrapped round by the membraneous fibrillae [muscle fibres], constitute together the fleshy part of the muscle, which, soft, broad and thick, differs in colour in different animals, being reddish or pale or even whitish; in the leg of the rabbit you will find some muscles red and others pale" (144). The next step, a major one, came from Louis-Antoine Ranvier (1835–1922), who began his scientific career as an assistant to Claude Bernard before becoming director of the Histology Laboratory and then professor of anatomy at the Collège de France in Paris. While with Claude Bernard in 1867, Ranvier began to study the fine structure of nerve and muscle and combined his histological investigations with recordings of muscle responses to tetanic stimulation. In comparison with pale muscles, Ranvier noted that red muscles contracted slowly, developed fused tetanic contractions at lower rates of stimulation, and were more resistant to fatigue (131). These were remarkable observations for the time and were not equalled until the work of Denny-Brown.

Born and educated in New Zealand, Derek Denny-Brown developed a strong interest in the nervous system while a medical student, and then as a demonstrator in anatomy (Fig. 2.3, top). To further his knowledge, he joined Sherrington at Oxford in 1925. Although Oxford, because of Sherrington's presence, was the world center for neuroscientific research, the physiology laboratories were primitive: the rooms were cold, damp, and unheated, some lacked electrical outlets, and the doctoral (D.Phil.) students were crowded together. Each laboratory, however, was equipped with a falling plate camera and optical myograph to record the contractile responses. In this system, which required the room to be darkened, the fall of a photographic plate coincided with the release of a stimulus from an induction coil. The ensuing contraction was recorded via the attachment of the muscle tendon to a mirror mounted on a torsion rod, and a narrow beam of light, reflected from the mirror on to the falling plate, registered the tension developed by the muscle (Fig. 2.3, bottom). This sensitive but cumbersome system was not free from error, but in the cat hindlimb Denny-Brown (33) was nevertheless able to confirm Ranvier's observations that, in mammals, the (pale) gastrocnemius muscle had a faster twitch than the (red) soleus; moreover he found that this difference was independent of fiber diameter or the amount of fatty granular material in the fibers. The small rectus muscles of the eyeball had twitches which were faster still. Interestingly, the difference between the gastrocnemius and soleus was reversed in the newborn kitten.

After Denny-Brown's work showing differences between fast and slow muscles, it would have been logical to study individual motor units and to determine to what extent their properties differed from one another. It is tantalizing to realize that such a study could so readily have been carried out in Oxford, but it was not to be.

Instead, the next advance in the fast and slow story was to come from a quite different direction—from the histology laboratory. Wachstein and Meisel may have been the first, in 1955, to show that muscle fibers could be differentiated from each

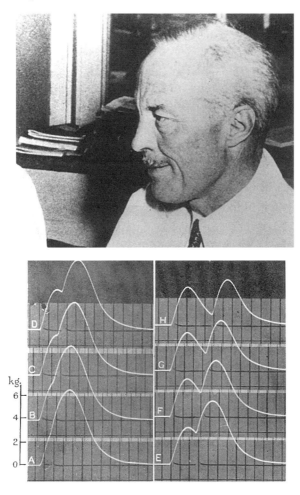

Fig. 2.3. *Top*: Derek Denny-Brown (1901–1981). Almost to the end of his long career, he combined clinical neurology with neuropathology and neurophysiology, with conspicuous success in all three. Reproduced from Gilliatt (59), with permission from *The Canadian Journal of Neurological Sciences* 8:271–273, 1981. *Bottom*: The first satisfactory photographs to be published of isometric muscle twitches, recorded with the Oxford optical myograph and falling-plate camera. The responses, from a cat medial gastrocnemius muscle, are to paired stimuli of increasing separation (A–H). The vertical lines, made by a beam of light emerging from a slit in a rotating drum, represent 20 ms intervals. Reproduced from Cooper and Eccles (30), with permission from The Physiological Society.

other by the avidity with which they stained for certain enzyme reaction products (152). After a stain for succinic dehydrogenase, by Wachstein and Meisel, came one for phosphorylase, by Victor Dubowitz and Everson Pearse (36), and then the most useful stain of all, that for myosin ATPase by King Engel (43). All of these and others, including one for muscle glycogen, revealed the presence of differently staining fibers distributed in an apparently random manner across the muscle belly, giving a checkerboard appearance. With the myosin ATPase reaction, it was possible to show that there were two main types of fiber (I and II) and that the type II fibers could be subdivided by varying the pH of the incubating medium. Thus was born the science of muscle histochemistry. Rather later came the key observation—that all the muscle fibers belonging to a motor unit were of the same histochemical type (40).

It was not until 1965 that the physiology of fast and slow muscle moved forward, and this was through the work of Elwood Henneman (1915–1996; Fig. 2.4, left). At the time Henneman was a professor of physiology at Harvard. He had originally intended to make a career as a neurosurgeon; in preparation he had studied neurophysiology, and it was probably through David Lloyd's influence at the Rockefeller Institute that he had become interested in spinal reflexes. However, unlike the earlier situation in Oxford, muscle seems not to have had any priority at Harvard, and Henneman himself, like Sherrington a quiet and unassuming person, appears as a rather solitary figure. Soon after his appointment to Harvard, Henneman became interested in single motor units from the point of view of their recruitment and, in 1957, had his first publication on this subject (see next section). His technique was to

Fig. 2.4. *Left*: Elwood Henneman (1915–1996), who was the first to record the twitch and tetanic responses of single mammalian motor units, and to propose that the order of motor unit recruitment is governed by the sizes of the parent motoneurons. Courtesy of Dr. Abby Henneman. *Right*: Robert Burke (1934–) whose own motor unit studies were influenced by Henneman's work, and who gave the first comprehensive description of the different types of mammalian motor unit. Courtesy of Dr. Burke.

dissect out sectioned ventral root filaments and to record the centrifugal impulses of two or more fibers during reflex discharges. Later, Henneman took the dissections further by splitting the filaments distal to the section until only a single axon remained on the wire electrode. The isolation of a single fiber was proven by the presence of an all-or-nothing antidromic response when the motor nerve was stimulated in the periphery. The single motor axon could then be stimulated and the contractile responses of the corresponding motor unit could be recorded from the muscle.

In two papers, published back to back in 1965, Henneman, with two younger colleagues, determined the axon conduction velocities, twitch contraction times, and the twitch and tetanic tensions of single motor units in the cat soleus and gastrocnemius muscles (119, 157). In the red soleus muscle, which they found to be largely composed of one histochemical fiber type, the contraction times tended to be long, and most of the tetanic tension was generated at relatively low stimulus frequencies. Even in this uniformly staining muscle, however, the maximum twitch tensions of the units varied more than 10-fold, indicating a similar range in motor unit sizes. In the pale gastrocnemius muscle, the range of tensions was even higher, 100-fold, and there was a greater range of contraction times too, with shorter ones predominating; these units required higher stimulus frequencies to develop maximum force.

These exciting results brought in the modern era of motor unit physiology. Very soon afterward Robert Burke* (an asterisk next to a scientist's name indicates that the scientist is included in the group photograph; see Fig. 2.18) (1934–; Fig. 2.4, right), newly appointed to the National Institutes of Health, Bethesda, carried the physiology one step further by stimulating single motoneurons with intracellular microelectrodes rather than by dissecting out single ventral root fibers (22). In this way he was able to look for any association between the properties of the parent motoneuron and those of the axons and muscle fibers; further, by the technique of glycogen depletion, he was able to count the numbers of muscle fibers in a selected motor unit and to show that the fibers tended to be widely dispersed across the muscle belly. Finally, by combining glycogen depletion with enzyme histochemistry, he could correlate the physiological and histochemical features of the individual motor units. Burke and his colleagues showed that in cat hindlimb muscles there were three main types of motor units—a large, readily fatiguable unit with a fast twitch; a rather smaller unit, which was more fatigue resistant and had a fast twitch; and a considerably smaller unit, which was difficult to fatigue and had a slow twitch. These three types they termed FF (fast contracting, fast fatigue), FR (fast contracting, fatigue resistant), and S (slow contracting), respectively (23). Similar conclusions were reached by V.R. (Reggie) Edgerton* and James Peter, at UCLA (129), on the basis of detailed biochemical studies on rabbit and guinea pig muscles, and by Douglas Stuart and Roger Enoka in Tucson, who, like Burke, employed single motor-unit techniques in the cat. In Edgerton and Peter's classification, the three types of units were termed fast twitch–glycolytic, fast twitch–oxidative glycolytic and slow twitch–oxidative, respectively. In human muscles single motor-unit twitches were first recorded by the technique of threshold motor nerve stimulation and were shown to have a threefold range in contraction times (from 35 to 105 ms; ref. 140; Fig. 2.5, top).

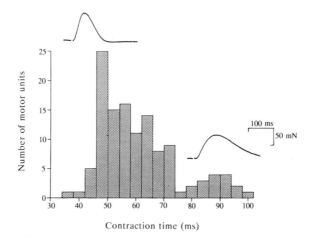

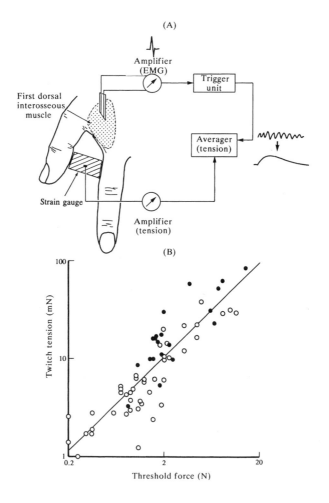

These results were confirmed and extended by intramuscular nerve fiber stimulation in the gastrocnemius by John Stephens and his colleagues in London; as did Burke in the cat, they found evidence of major differences in motor unit type (56).

In more recent studies the main addition to our knowledge of motor units was the recognition that, while certain properties were indeed linked together, as Henneman and Burke had shown, there was actually a continuum of properties among the motor units. Further, the linkage was not inviolable for, at least in some small muscles of the human hand and foot, the relationship between motor unit size and contraction time was absent. This discrepancy introduces a further word of caution. Exercise physiologists talk of "fast" and "slow" muscles, by which they mean fast contracting and slow contracting. Ideally, these terms should refer to the rate of myosin cross-bridge cycling and hence to the myosin–ATPase reaction. At a cellular level, the speed of contraction is best measured by observing the rate of sarcomere shortening in an isotonic contraction. However, muscle twitch recordings in the whole animal are invariably isometric and the durations are determined not so much by the rate of the myosin ATPase reaction as by the amount of time that calcium ions are available to the cross bridges before being taken back into the sarcoplasmic reticulum. The correlations found between contraction time and other motor unit variables actually depend on the differences in calcium pumping that exist between the fiber types.

THE SIZE PRINCIPLE

Having demonstrated the marked variation in motor unit properties, even within the same muscle, it was perhaps inevitable that Henneman should have investigated the

---

Fig. 2.5. *Top*: First records of single human motor unit twitches, obtained from the extensor hallucis brevis muscle using threshold stimulation, transient arterial occlusion, and response averaging. A "fast" unit twitch is shown at the left and a "slow" unit twitch below and to the right. The histogram gives the twitch contraction times of 122 units in 31 individuals. Adapted from Sica and McComas (140), *Journal of Neurology, Neurosurgery and Psychiatry* (1970) 34:113–120 with permission from the BMJ Publishing Group. *Bottom*: Demonstration of the size principle in the human first dorsal interosseous muscle of the hand. The sizes of the motor units, as reflected in their twitch tensions, are given on the vertical axis, while the horizontal axis depicts the muscle force at which each unit was recruited. Filled circles are measurements from single experiments while the open circles are measurements in other experiments with the same subject. The twitch tension of a unit was derived by using the motor unit potential, recorded with an intramuscular needle electrode, to trigger an averager receiving the force output from the whole muscle. With repetition, the force contribution of the triggering motor unit becomes increasingly evident. Reproduced (*lower*) from Milner-Brown et al. (125), with permission from the authors and from The Physiological Society.

possible ways that these differences might be reflected in usage. He had, in fact, considered this problem earlier by recording the discharges of single ventral root fibers in cats during flexion reflexes (75). In that study he had found that the axons of the motoneurons with the lowest thresholds generated the smallest action potentials. With some support from earlier studies by Eccles and Sherrington, Henneman argued that the most readily activated axons belonged to the smallest motoneurons and that these innervated the smallest colonies of muscle fibers. In passing it is instructive to ask why there should be any correlation between the respective sizes of the motor axons and motor units. The answer is that every motor axon necessarily contains enough microtubules and neurofilaments not only for its own upkeep, via axoplasmic transport, but also for the upkeep of the muscle fibers which it innervates. Hence a large motor unit, with many axonal branches, requires a large number of microtubules and neurofilaments, and a thick axon to house them.

But back to Henneman. A problem in his 1957 study was that the motoneuron discharges in the flexor withdrawal reflex were not sustained. There was a burst of impulse activity, lasting perhaps a second, and then silence, so the amplitudes of the nerve fiber action potentials could only be compared over a short period of time. In the later study, with George Somjen and David Carpenter (76), the stretch reflex was employed instead, and in this reflex most motoneurons maintained their discharge for as long as the muscle remained stretched. Using silver wires to record from thin ventral root filaments, the investigators found that motor axons with small action potentials discharged before those with larger potentials, as the reflex became stronger. On the basis of earlier biophysical observations of Herbert Gasser and morphological studies of Ramón y Cajal (26), they argued that motoneuron recruitment was organized according to the sizes of the cell bodies—that is, the larger the cell body, the higher the threshold; this was the *size principle*. It has recently been suggested that priority for formulating the size principle should be given to Denny-Brown on the basis of observations made with Joe Pennybacker during the recruitment of human motor units in voluntary contractions (34). In a study of spontaneous muscle activity in denervated muscle, they wrote:

> a particular voluntary movement appears to begin always with discharge of the same motor unit. More intense contraction is secured by the addition of more and more units added in a particular sequence. This 'recruitment' of motor units into willed contraction is identical to that occurring in certain reflexes. The early motor units in normal graded voluntary contraction are always in our experience small ones. The larger and more powerful units, each controlling many more muscle fibres, enter contraction late.

Interestingly, Denny-Brown and Pennybacker cited Sherrington in relation to the motor unit recruitment that occurs during reflexes and, indeed, Sherrington, in his important 1929 paper summarizing the work of the Oxford school, made it clear that the motoneurons participating in a reflex had differing but reproducible thresholds (138).

We know now that Denny-Brown, as in so much, was correct; larger motor units are called into action as the voluntary effort increases. A beautiful illustration of this was to come from the much later work of Richard Stein* and colleagues in

Edmonton, Canada (125). In the first dorsal interosseous muscle of the human hand, they used spike-triggered averaging to record "twitch" tensions of single motor units and found that, over a 1000-fold range of tensions, there was a linear relationship between motor unit size and threshold (Fig. 2.5, bottom). However, even though Denny-Brown was right, his conclusion rested on flimsy evidence because of the nature of his recording electrode. A coaxial needle electrode such as he used can only record from a very few of the muscle fibers in a motor unit at any given location in the muscle belly. Hence, there is a large chance factor in the determination of the amplitude of the motor unit potential, depending on whether the inner wire of the electrode picks up from one, or more than one, fiber—in any case, probably no more than 1% of the full population. But at the time of Denny-Brown and Pennybacker's paper, the correct motor unit architecture was not known and, indeed, the first concept, that the fibers of a healthy motor unit existed in large clumps or "subunits," proved to be entirely wrong. Although Henneman has been criticized for not citing Denny-Brown and Pennybacker's paper (148)—the omission has been described as "particularly perplexing given the fact that Denny-Brown and Henneman were contemporaries at Harvard"—it is doubtful if Henneman knew of Denny-Brown's work or if the two ever came into meaningful contact in Boston. Further, by the time of Henneman's initial report, in 1957, Denny-Brown was more interested in the rostral end of the central nervous system and, as a practising neurologist, in the solution of clinical problems.

Returning to Henneman's 1965 paper (76), the authors stated that, "we may conclude that there is a general rule or principle applying specifically to motoneurones and perhaps to all neurones, according to which the size of a cell determines its threshold." As to the reason for this, Henneman drew on observations on the amplitudes of miniature end-plate potentials by Bernard Katz and Stephen Thesleff (100) and concluded that "an equal degree of presynaptic activation may generate a larger synaptic potential in small cells due simply to their dimensions and geometry." Once it had been formulated, and so elegantly, the "size principle" was eagerly adopted by muscle physiologists. It did, after all, make good functional sense, in that weak contractions would be smoother and better controlled if they were generated by a number of small units rather than one large unit. However, there was one important dissenting voice—not concerning the experimental observations underlying the size principle, but the principle itself. The voice belonged to Robert Burke, the investigator who, following Henneman, had carried the correlative analysis of motor unit and motoneuron properties to its ultimate level (see previous section). Using intracellular microelectrodes to stimulate cat medial gastrocnemius motoneurons, he was able to measure the input resistances of the motoneurons and to use these as an indication of their relative sizes, the smallest motoneurons having the highest input resistances (21). He found that there was only a weak negative correlation between cell size and the amplitude of the excitatory postsynaptic potential following group Ia (muscle spindle annulospiral afferent) stimulation. Further, he argued, if the densities of the excitatory synaptic contacts on small and large motoneurons were similar, then the synaptic current densities flowing through the membranes of the respective cell bodies (and axon hillocks) would also be similar despite the disparity in cell sizes. A better explanation for the size phenomenon, he suggested, was that the densities of the excitatory synaptic contacts were higher in the smaller motoneurons. Interestingly, Henneman had made the same suggestion, and discarded it, as one of four possible reasons for the size principle: "small cells may receive relatively more synaptic input from stretch receptors and cells connected with them."

There were a number of observations which supported Burke's proposal, beginning with Sherrington's statement that "the very muscles that to the observer are most obviously under excitation by the *tonic* system are those must obviously inhibited by the *phasic* reflex system. (139). While Sherrington's findings showed that the type and/or densities of certain synaptic connections differed among motoneuron pools, it could be argued that they did not necessarily indicate differences in synaptic input between motoneurons innervating the same muscle. Evidence of this kind was to come later from experiments both on animals and on human subjects. In one human study, John Stephens in London showed, with colleagues, that the order of motor unit recruitment during voluntary contractions of the first dorsal interosseous muscle of the hand could be disrupted if weak electrical stimuli were applied continuously to the adjacent index finger (145). Clearly, then the (polysynaptic) synaptic connections from the cutaneous afferent fibers were stronger in the motoneurons with higher thresholds for voluntary activation. In a second, and perhaps more dramatic, example, Lennart Grimby* and his coworkers in Stockholm identified some motor units in the small extensor digitorum brevis muscle on the dorsum of the foot that would only discharge during rapid voluntary contractions or sudden corrective movements and might then do so without activation of the units with the lowest thresholds for slower voluntary contractions (71). At a whole-muscle level this last finding was very much in keeping with the old idea of Sherrington's that different muscles participated preferentially in tonic and phasic contractions (see above).

In summary, then, a size phenomenon was shown to prevail for slow voluntary contractions and certain reflex responses and, from the above considerations, it would have seemed reasonable to accept that its basis was not the size of the motoneuron but rather the type and density of its synaptic connections. This was not the end of the story, however. In 1980 Daniel Kernell and Bert Zwaagstra, in Amsterdam, published the results of a study in which the input resistances of motoneurons were compared with the postmortem dimensions of the same neurons after each cell had been marked with procion dye or horseradish peroxide (101). Kernell and Zwaagstra, quite unexpectedly, found that the smaller motoneurons had higher specific membrane resistances than the larger cells. The specific membrane resistance, it should be explained, is the electrical resistance per unit area of the membrane. As Kernell and Zwaagstra pointed out, this meant that, even if all the motoneurons in a pool were activated by the same densities of excitatory synapses, the smaller motoneurons, because of their higher specific membrane resistances, would automatically develop larger depolarizations and become more easily recruited than the larger motoneurons.

Thus, for a reason which he had not foreseen, Henneman was right after all. Henneman, who died in 1996, was also responsible (with Lorne Mendell) for pioneering spike-triggered averaging, a technique which has had many applications including, appropriately, the demonstration of the size principle in human interosseous muscles by Stein (see above).

THE RECRUITMENT OF DIFFERENT TYPES OF MOTOR UNITS

In the preceding section evidence was given that, in slowly developing contractions, and in certain reflexes, the smallest motor units were the first to be called upon. Since studies in animals, also reviewed earlier, had demonstrated correlations between motor unit size and histochemical type, could differential involvement of muscle fiber types be demonstrated in relatively "natural" activities? In the bushbaby, *Galago senegalensis*, Edgerton and his colleagues at UCLA had approached this problem by causing the animals to either run on a treadmill or to jump. Histochemical analysis of the hindlimb muscles revealed that whereas steady exercise of moderate intensity (running) made greatest use of slow twitch–oxidative (type I) fibers, maximal intermittent activity (jumping) depended heavily on fast twitch–glycolytic (type IIB) fibers (58).

In human subjects, similarly intended experiments could only be performed by analyzing specimens of muscle obtained before, during, and after exercise. The technological advance which made this feasible was the introduction of the muscle biopsy needle by the Swede Jonas Bergström in 1962 (9). Perhaps reintroduction would be a better term, for a biopsy needle had been designed and used 100 years earlier by the remarkable French neurologist Guillaume-Benjamin-Amand Duchenne (1806–1875). Duchenne, yet another solitary figure and one regarded as eccentric by his Parisian contemporaries, not only described the type of muscular dystrophy which today bears his name but also was a pioneer electromyographer, a talented medical artist, an inventor of ingenious orthotic devices, and possibly the world's first clinical photographer. His biopsy needle, or "harpoon," consisted of a hollow needle with a side window set back from the tip. When the needle was inserted into the muscle, any fibers bulging through the side window were cut off and retained by a sharp blade pushed sharply down the inside of the needle (37). Bergström's design was very similar and his development of needle muscle biopsy was an enormous advance, not only in the field of exercise physiology but also in the diagnosis and assessment of neuromuscular disorders.

It was fitting that the first application of the Bergström needle should have been in Sweden, for a strong school of exercise physiology was developing in Scandinavia, one that could trace its roots back to the work of August Krogh (1874–1949) in Copenhagen. Krogh, a Nobel Laureate and inventor, was a man of wide scientific interests but was perhaps best known for his work on the regulation of blood flow in capillaries, including those in contracting frog muscle, and for the diffusion of gases between pulmonary blood and alveolar air. He was also the first to recognize the exponential decline in oxygen consumption after exercise, a phenomenon later termed "oxygen debt" (see Chapter 8). The studies on muscle fiber-type utilization were carried out in Stockholm by Bengt Saltin* and Philip Gollnick,* who stained biopsy sections of muscle fibers for glycogen and for myosin ATPase activity. The depletion of glycogen detected those fibers which had participated in the exercise while the myosin ATPase identified their fiber type. In cycling, Saltin, Gollnick, and colleagues found that the type I (slow twitch) fibers were most heavily used at sub-maximal workloads, whereas type II (fast twitch) fibers also became involved when the contractions were "supramaximal"—that is, beyond the level at which oxygen consumption was maximal (64). In sustained isometric contractions of the same lateral vastus thigh muscles, they found that weak contractions also preferentially employed the type I (slow twitch) fibers and that higher forces made greater use of type II (fast twitch) fibers (63). Thus, both in animals and humans, and in both steady and cyclical exercise, the type I (slow twitch) fibers were the first to be used while type II (fast twitch) fibers contributed increasingly as the effort became greater.

FIRING RATE OR RECRUITMENT?

In Sherrington's work on motor units it was a central theme that contractions, reflex or voluntary, became stronger through the recruitment of additional units. This recruitment came about because more spinal motoneurons had been brought to the threshold for discharging. However, it was also clear to Sherrington, as it had been to Ranvier many years earlier (see p. 46) that force also depended on the frequency of muscle activation as induced by electrical stimulation. To what extent, then, might the frequency of activation vary under natural circumstances?

The key to the understanding of this problem depended on the ability to record impulses in the motor nerve or muscle fibers. To record from either tissue, and particularly from the nerve fibers, was not easy in Sherrington's time, but the breakthrough occurred at the University of Cambridge. In the 1920s the Cambridge Physiology Department, still recovering from the effects of World War I, included, as one of its members, Edgar Adrian (1889–1977; Fig. 2.6, top). A charming and gifted man, Adrian nevertheless preferred to work alone in the laboratory, where he enjoyed a reputation for speed and for the knack of getting experiments to work, often with the help of plasticine and pieces of Meccano. Adrian's first success in recording nerve impulses came by chance when he had a motor nerve over a pair of recording electrodes and noticed some annoyingly persistent electrical "noise"—which he then realized was the discharges of sensory nerve fibers connected to stretch receptors in the dependent muscle belly. Having for the first time "seen" the nerve impulse, Adrian showed that the frequency of the impulse discharge was the information code employed by all the different types of sensory receptor he examined. The next step was to look for a similar code in motor nerve fibers, and he achieved this initially by recording, during reflex contractions in the cat, from fine strands of nerve placed on wire electrodes. For voluntary contractions in human subjects a different approach

Fig. 2.6. *Top*: Lord Adrian (1889–1977) in later life. Adrian became, successively, the head of the Cambridge Physiological Laboratory, president of the Royal Society, and master of Trinity College, Cambridge. He shared the Nobel Prize with Sherrington in 1932. Reproduced from Zotterman (158), *Electroencephalography and Clinical Neurophysiology* 44:137–139, 1978 with permission from Elsevier Science. *Bottom*: Possibly the "cleanest" recording of a single human motor unit discharging during a maximal voluntary contraction. This recording was made with a needle inserted into the adductor pollicis muscle and the unit discharging was one of the few to be innervated by the median nerve, the ulnar nerve having been blocked. The coaxial type of needle electrode was originally designed and used by Adrian and Bronk (1) and is widely used in clinical EMG laboratories. Reproduced from Marsden et al. (112) with permission from the authors and from The Physiological Society.

was needed, and Adrian's solution was simple but effective. By inserting an enamelled copper wire into a fine hypodermic needle, grinding the end obliquely, and using the wire as one electrode and the needle casing as another, the first coaxial needle recording electrode was created (1). In the triceps brachii muscle he and Detlev Bronk found that during slight contractions action currents appeared at frequencies as low as 5 Hz and then, with greater effort, reached 50 Hz or more. Adrian recognized, and confirmed experimentally in the cat by stimulating the motor nerves, that the higher frequencies would be more effective in developing force. He then showed that the range of impulse frequencies encountered among the motor units during reflex and voluntary contractions corresponded to the rising part of the force:frequency curve. Adrian also demonstrated, however, that the recruitment of additional motor units was an important factor in developing more force.

After Adrian, there were many studies of motor unit firing rates in human muscles. In attempts to overcome the problem of deciphering the activities of individual units from the complex multiunit discharge at the onset of a strong contraction, different types of recording electrode were used, including pairs of fine flexible wires and tungsten microelectrodes. Nevertheless, there was still the problem that the electrode tip would be displaced as the muscle fibers in its vicinity shortened. At Cambridge, many years after Adrian's time, David Marsden, John Meadows, and Patrick Merton (112) took advantage of the anomalous innervation of the adductor pollicis muscle in two of their subjects in whom a very few motor units were derived from the median nerve and the remainder from the ulnar nerve. When the ulnar nerve was blocked by local anesthesia, the discharges of the median-innervated units were easily distinguished in recordings with needle electrodes. The maximum firing rates were found to be as high as 100 Hz or more initially and to decline to 20 Hz or so over the next 30 seconds (Fig. 2.6, bottom). An important addendum to the firing rate issue was that the maintained frequency differed between muscles, as Brenda Bigland-Ritchie* was able to show while working in New Haven, Connecticut (8). With the same type of tungsten microelectrode that had been used previously by others for recording from single human peripheral nerve fibers, she found that the maximum maintained rate for the soleus was only about 11 Hz while for the adductor pollicis and the biceps brachii it was around 30 Hz. Bigland-Ritchie and her colleagues pointed out that this difference in rate was appropriate because in humans, as in other mammals, the soleus had a slow (prolonged) twitch and so developed its maximal tension at low excitation frequencies.

But which was the more important strategy for increasing force—rate coding or recruitment of motor units? In going from a series of overlapping twitches at 5 Hz to a fully fused tetanic contraction at 50 Hz, the tension generated by a muscle increases 10-fold while there is a 100-fold range in the forces generated by single motor units. In theory, then, both mechanisms could be extremely effective. In their study of the human first dorsal interosseous muscle, already referred to (p. 53), Richard Stein and his colleagues in Edmonton used spike-triggered averaging to discriminate the contractions of individual motor units (125). Having determined the twitch tensions of a motor unit, they measured the change in firing rate as the muscle contraction became stronger. Using a novel mathematical treatment, they were then able to estimate (124) the relative importance of rate coding and recruitment. In the first dorsal interosseous they found that recruitment was the more important mechanism at low forces, with half of the motor units being engaged when only 10% of the maximum force had been developed. At higher forces, rate coding became increasingly important. In other muscles, though, the situation appeared rather different. Carl Kukulka and Peter Clamann (103), in Richmond, Virginia, showed that while no further motor units were recruited in the adductor pollicis at forces greater than 50% of maximal, in the biceps brachii recruitment continued up to 88% of maximal force. From this and other studies it appeared that the small muscles of the hand depended on rate coding for near-maximal forces, while the larger, more proximal, muscles of the arm employed recruitment as well.

However, the muscle contractions during which such observations were made were necessarily sustained over many seconds. Were there motor units with still higher thresholds that might only be recruited under different circumstances? In the last section it was seen that at least in the extensor digitorum brevis muscle of the foot there were some motor units which only discharged in very rapid voluntary contractions or in sudden corrective movements. Also, there was, as articulated by the Cambridge physiologist, Patrick Merton, "the belief that lunatics, persons suffering from tetanus or convulsions or under hypnosis, and those drowning are exceptionally powerful" (122). Working at the National Hospital for Nervous Diseases in London and usually experimenting on himself with one hand left free to operate the equipment, Merton tested this proposition by the method of twitch occlusion. He designed a ball-race device which enabled the contractions of the adductor pollicis to be isolated when the ulnar nerve was stimulated at the elbow. As the effort to adduct the thumb was increased, the twitch evoked by the interpolated stimulus diminished, almost vanishing when the voluntary contraction was maximal (see p. 58). This diminution indicated that nearly all the motor units in the muscle had already been recruited and were firing impulses at optimal frequencies for force production. Subsequent studies have demonstrated that voluntary motor unit activation is complete, or almost so, in most muscles studied, the most notable exceptions being, in some subjects, the triceps surae and the diaphragm.

MUSCLE WISDOM

In the previous section it was seen that during a maximal voluntary contraction the firing rate of a motoneuron was initially high and then declined to a much lower level. Was there any functional advantage in this? One of the first clues came from a study by David Marsden, John Meadows, and Patrick Merton in Cambridge in which their own adductor pollicis muscles were stimulated tetanically through the ulnar nerves (113). They showed that with a stimulus frequency of 60–80 Hz, fatigue occurred sooner than in a maximal voluntary contraction. However, if the stimulus frequency was progressively lowered from 60 Hz to 20 Hz over the first minute, the rate of fatigue became similar to that in the voluntary contraction. The implication was that in the naturally occurring contraction there was also a progressive reduction in the rate at which the muscle fibers were excited by the motoneurons and that this was optimal for delaying fatigue. Rather provocatively, the Merton group termed this lowering of the muscle firing rate "muscle wisdom." In the course of some subsequent nerve block experiments (see previous section), they were able to follow motor unit firing rates in a maximal voluntary contraction and found that the rates did indeed fall from 100 Hz or higher to around 20 Hz or so by the end of the first minute, just as would have been predicted from the observations on electrical stimulation (112). The relationship between firing rate and fatigue was further explored by David Jones, Brenda Bigland-Ritchie, and Richard Edwards in London (98). Like the Merton group in Cambridge, they compared electrically induced and voluntary contractions. They pointed out that even though the muscle excitation rate fell in a maximal voluntary contraction, the instantaneous rate was still optimal for generating force. The reason for this was that each new excitation was associated with further slowing of the relaxation phase of the contraction; hence less frequent excitations were required for tension to be maintained.

But how did the decline in motoneuron firing rate come about? Part of it was almost certainly due to the phenomenon of "accommodation" by the motoneuron membrane, since a progressive reduction could be seen during the intracellular injection of current pulses in mammalian motoneurons, as Ragnar Granit and David Kernell had observed in Stockholm (69). This was but one of many wide-ranging explorations of sensory and motor function by Granit, himself yet another of Sherrington's former students at Oxford to achieve distinction subsequently and, in this case, a Nobel Prize. However, another explanation for the falling motoneuron firing rate, a particularly intriguing one, was to come from an unexpected source. While working in New Haven, Brenda Bigland-Ritchie (Fig. 2.7, left) proposed that there might be an inhibitory reflex. With Olof Lippold in London, she had, in the early 1950s, studied motoneuron firing rates with fine wire electrodes and also the relationship between rectified EMG activity, recorded with surface electrodes, and voluntary force (11). In 1986, having returned to exercise physiology after a 20 year interval, she suggested, on the basis of experiments with Noel Dawson, Roland Johansson, and her former supervisor, Olof Lippold, that changes in a fatiguing muscle would stimulate afferent fibers which would, through a reflex circuit in the spinal cord, exert inhibition on the discharging motoneurons (12). A critical observation, which supported a concept of this kind, was that occlusion of the blood supply to an exercising muscle would preserve the mechanical fatigue as well as the depression of voluntary EMG activity; both recovered when the cuff was released. Such an observation would have been compatible with an inhibitory stimulus in the muscle belly due to persisting anoxia or the retention of metabolites (H^+, K^+, kinins, etc.). The initiation of afferent discharges by such metabolites in small-diameter fibers had been documented by others (e.g., 120).

Another relevant experiment, both ingenious and technically challenging, was devised by Vaughan Macefield and Simon Gandevia in Sydney, Australia (110). These authors made use of the technique of recording from single human nerve fibers with tungsten microelectrodes pioneered by Karl-Erik Hagbarth in Uppsala, Sweden. With colleagues, Macefield and Gandevia functionally deafferented a human muscle by blocking the motor nerve with local anesthetic. They then recorded from single motor axons with tungsten microelectrodes inserted above the

Fig. 2.7. *Left*: Brenda Bigland-Ritchie, one of the leading workers on muscle fatigue and a former student of A.V. Hill, shown here during her second, "American" career. Courtesy of Dr. Bigland-Ritchie. *Right*: Sir Michael Foster (1836–1907), the first professor of physiology at Cambridge. A great educator, Foster did little experimental work himself but was extremely influential and could spot talent in others, a gift which enabled him to build a strong department. He used his beard to good effect when emphasizing a point. Reproduced from Langley (104), with permission from The Physiological Society.

block. They found that during maximal effort there was no longer any sign of the reduction in motoneuron firing rate which is normally seen. This absence was consistent with interruption of the afferent arm of an inhibitory reflex.

At present it is true to say that there is still no definitive proof of such a reflex: perhaps such a proof will only come from animal experiments. Further, there clearly are other factors that can reduce motoneuron firing in an exercising muscle. In addition to accommodation of the motoneuron membrane (see above), there is the reduced excitatory drive from the muscle spindles to consider, as Hagbarth has suggested (15). Nevertheless, there is something very appealing about the idea of an inhibitory reflex, especially since the inhibition would always be appropriate for the state of the muscle.

Finally, it may be noted that the reduced motoneuron firing rate is not the only example of muscle "wisdom" in exercise. The powerful early boost of the electrogenic Na^+,K^+ pump, maintaining the excitability of the muscle fiber membrane in the face of a rising extracellular K^+ concentration, and the inclusion of noncontracting fibers in such pumping, is an equally impressive example (117). The size principle, discussed in an earlier section (see The Size Principle), is another, and there are more.

CENTRAL FATIGUE

Other than an inhibitory reflex, accommodation, and reduced muscle spindle support, all discussed in the previous section, are there other influences during exercise which ultimately diminish motoneuron excitation? In particular, what about mental factors? After all, to continue exercising in spite of breathlessness, chest discomfort, and a sensation of increasing heaviness in the limbs surely requires an effort of will. This point was nicely made by Michael Foster (1836–1907; Fig. 2.7, right). Foster, later Sir Michael, was the first professor of physiology at the University of Cambridge, holding the chair for 20 years. A learned medical historian, he introduced modern methods of teaching biology, physiology, and embryology, with an emphasis on laboratory work. A keen observer and the author of standard textbooks on embryology and physiology, he wrote in relation to fatigue (50):

> The sense of fatigue of which, after prolonged or unusual exertion, we are conscious in our own bodies, is probably of complex origin, and its nature, like that of the normal muscular sense of which we shall have to speak hereafter, is at present not thoroughly understood. It seems to be in the first place the result of changes in the muscles themselves, but is possibly also caused by changes in nervous apparatus concerned in muscular action, and especially in those parts of the central nervous system which are concerned in the production of voluntary impulses. In any case it cannot be taken as an adequate measure of the actual fatigue of the muscles; for a man who says he is absolutely exhausted may under excitement perform a very large amount of work with his already weary muscles. The will in fact rarely if ever calls for the greatest contractions of which the muscles are capable.

Augustus Waller (1816–1870), a London physician and physiologist, whose name is forever associated with degenerative changes in peripheral nerve after section of the nerve, was another interested in fatigue. He also recognized that there could be causes within the central nervous system as well as in the muscles themselves and, in 1891, wrote:

> The question is: does normal voluntary fatigue depend upon central expenditure of energy, or upon peripheral expenditure of energy, or upon both factors conjointly; and if upon both, in what proportion upon each? The form of this question is justified as follows; a maximum voluntary effort may decline: 1) by decline of cerebral motility; 2) by sub-central or spinal block; *Central*; 3) by motor end-plate block; 4) by decline of muscular energy; *Peripheral*. On the human subject it is impossible to separate factors 1 and 2, we must therefore embrace them under the term central. (153)

Waller went on to conclude that the rostral elements, in the chain of motor command, were more fatiguable than the lower ones, so that a "motor center" in the brain would be more readily exhausted than the muscles and their innervation.

In Italy, the same type of thinking was evident in the writings of Angelo Mosso (1846–1910). Like his contemporaries in the United Kingdom, Mosso was widely read and a good observer with many scientific interests. He was also the inventor of the *ergograph* (Fig. 2.8, top), a device which enabled the contractions of the long fin-

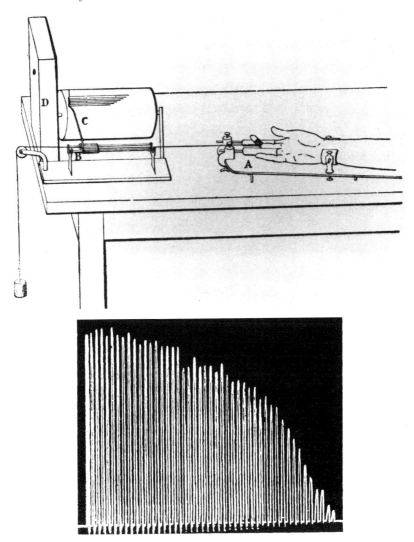

Fig. 2.8. *Top*: One of Mosso's ergographs. In this model a weight is raised by flexing the middle finger; the weight is connected to the finger by a cord which runs over a pulley. The cord is interrupted by a metal "runner," *B*, which slides along two parallel metal rods. As the runner slides it moves a lever, *C*, which writes on smoked paper wrapped around a slowly rotating drum. *Bottom*: Fatigue of maximal voluntary contractions, recorded from Professor Aducco in 1884, a colleague of Mosso's at the University of Turin. (*Top* and *bottom* figures reproduced from Mosso, 126.)

ger flexors (flexor digitorum profundus and sublimis) to be registered on either a horizontal plate or a slowly revolving cylinder. The contractions were confined to movements of the middle finger, which was required to lift a given weight by pulling on a cord running over a pulley. Mosso was able to study his colleagues in the Physiology Department in Turin over many years and to make repeated comparisons of their fatigue records (Fig. 2.8, bottom). Clearly aware of central factors in fatigue, Mosso stated (126): "The way of living, the night's rest, the emotions, mental fatigue, exert an obvious influence upon the curve of fatigue." A little later in the same chapter, however, he appeared to dismiss central fatigue. "However complex may be the mental act which gives rise to a voluntary contraction, we must, after these experiments, recognize that the function of the muscle itself is not less complicated. . . . The most novel and interesting result of these experiments with the ergograph is that we must attribute to the periphery and to the muscles certain phenomena of fatigue which were believed to be of central origin."

An obvious way of establishing whether or not central fatigue was present was to see whether or not any additional force could be developed by electrical stimulation of the exercising muscles, and this was already being done in the latter half of the 19th century (cf. Waller, 153). In more recent times, the investigator who reintroduced muscle stimulation during voluntary contractions was Patrick Merton, in Cambridge, whose classic 1954 paper "Voluntary Strength and Fatigue" has already been referred to. In a review of muscle fatigue 2 years later (121), Merton concluded, "anyone who possesses a sphygmomanometer and an open mind can readily convince himself that the site of fatigue is in the muscles themselves."

Merton's conclusion, if it were true, would contradict the observations of Brenda Bigland-Ritchie and her colleagues, who found in some subjects that either electrical stimulation or a supreme effort added increasing amounts of tension as fatigue progressed during voluntary contractions of the adductor pollicis and quadriceps muscles. Moreover findings similar to Bigland-Ritchie's were obtained by Lennart Grimby and his colleagues in Stockholm, in this instance using tetanic stimulation of the peroneal nerve to detect central fatigue in extension of the great toe (70). And there are other workers who have found evidence of central fatigue. Returning to Merton, though, it is a fact that he tended to make observations on himself and seemed to have a high tolerance for pain so that his results would not necessarily have applied to others. Further, as Simon Gandevia, Gabrielle Allen, and David McKenzie pointed out, close inspection of Merton's twitch interpolation results indicated that even he may not have fully activated the motor units in his adductor pollicis muscle (54). These authors, working in Sydney, Australia, drew attention to the variability in maximal voluntary activations between subjects, even when rested and fresh, and on different occasions in the same subjects. They also commented on the decline in voluntary activation if the mental effort was spread over several muscle groups rather than being concentrated on a single group. Gandevia took the analysis of central fatigue further by looking at the effects of stimulating the human motor cortex (53). Such stimulation can be done using electric shocks applied through scalp electrodes, but it is much more tolerable to use magnetic stimulation; in the latter case, the collapsing magnetic field at the end of a magnetic pulse induces a current in critically oriented axons. Gandevia observed that stimulation of the motor cortex added tension as fatigue developed, a finding which might suggest that the motor cortex itself was not being optimally driven by neural elements "upstream." Eric Newsholme,* at the University of Oxford, has gone one step further by speculating that on the basis of perturbations in plasma amino acid levels during exercise one factor in central fatigue may be excessive firing of sleep-inducing serotonin (5HT) neurons (14).

PERIPHERAL FATIGUE

It is now time to look again at Merton's dictum that "anyone who possesses a sphyg-momanometer and an open mind can readily convince himself that the site of fatigue is in the muscles themselves." In fact, the sphygmomanometer is not really necessary, for physiologists in the nineteenth century had convincingly demonstrated the development of fatigue in isolated frog sciatic nerve–gastrocnemius preparations by using an induction coil to stimulate the nerve and smoked paper on a revolving drum kymograph to record the movements of a lever attached to the tendon. Figure 2.9 (bottom) is an excellent recording of one such experiment, showing the prolongation of the twitch with fatigue. So effective was this kind of experiment that it was to be repeated, using exactly the same type of apparatus, in physiology classrooms for the next hundred years. Perhaps the foremost worker in these early experiments was Hermann von Helmholtz (1821–1894; Fig. 2.9, top), a true genius and, as an inventor, biologist and mathematician, only to be approached by Andrew Huxley in the next century (see the section Machina Carnis: the Flesh Machine). Born in Potsdam, Helmholtz learned seven languages in addition to his native German and became an accomplished pianist. He began his first experiments in a makeshift laboratory as an army doctor. His extraordinary abilities were soon recognized and he became a professor successively in Berlin, Konigsberg, Heidelberg, and finally Berlin again. Among his many achievements he developed the ophthalmoscope, composed a theory of color vision, discovered the principle of the conservation of energy, and performed much work on electricity and magnetism. Using the kymograph to record muscle responses, he was the first, in 1850, to measure the impulse conduction velocities of motor nerve fibers in the human arm (74). The recording devices for muscle were made more sensitive by Étienne-Jules Marey (1830–1904), in Paris, who also devised tambours for recording the arterial and venous pulses. Marey's other contributions included inventing the first ciné camera for recording and analyzing rapid movements and investigating the discharge from the electric eel. With the capillary electrometer, Marey was also the first to record electrical activity in contracting human muscles. Using the Helmholtz apparatus, with Marey's modifications, Hugo Kronecker, in Leipzig, made a special study of fatigue in excised frog muscle and formulated two laws, the first of which stated: "The curve of fatigue of a muscle which contracts at regular intervals, and with equally strong induction shocks, is represented by a straight line." Kronecker's second law, about which more will be said later, was: "The difference in the height of the contractions is less when the intervals of time are greater. In other words, the height of the contractions diminishes the more rapidly, the more rapid is the rhythm in which they are produced, and *vice versa*" (102).

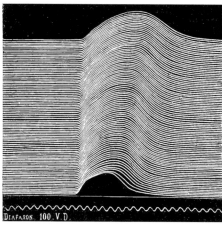

Fig. 2.9. *Top*: Hermann von Helmholtz (1821–1894). Not only was Helmholtz a gifted linguist and musician, but he was also a leading inventor, physiologist, mathematician, and physicist. Helmholtz helped to refine the induction coil–kymograph drum system for recording muscle twitches. Reproduced from Blasius et al. (13), with permission of J.F. Lehmanns Verlag. In the *bottom* figure, made with such an apparatus, the slowing of the twitch during fatigue is nicely shown in successive recordings, from below upward (Mosso, 126). Helmholtz also used this type of equipment to measure the maximum impulse conduction velocities in human motor nerves in 1850.

While on the subject of devices for measuring muscle contractions, it should be noted that a dynamometer had been invented for voluntary efforts in human subjects rather earlier, by George Graham (1675–1751), a master clock maker. The Graham instrument was then improved by John Desaguliers (1683–1744), a fellow Londoner with a flair for invention, who also wrote on optics, steam energy, electricity, and magnetism. The Graham–Desaguliers dynamometer was designed to determine maximum force, as in elbow flexion, and this was measured by a lever arm to which a weight was attached. Unlike the frog muscle experiments, there was no means to obtain a permanent record and, in those early days, no nerve stimulation was performed. The dynamometer was, however, adequate for comparisons of strength in different types of laborers and in men and women (128).

The studies on the isolated frog nerve–muscle preparation, discussed above, demonstrated that muscles would fatigue when stimulated repetitively and that their contractions would become slower, but they did not enable the mechanism of the fatigue to be determined. In the first place, was fatigue due to failure of electrical excitation of the muscle fibers? This is where Patrick Merton made the critical observations with his sphygmomanometer. Merton, already mentioned (Fig. 2.10, top), exhibited in his experiments the simplicity and elegance so prized by Cambridge scientists before and after World War II. In the fatigue experiments (122), Merton inflated the sphygmomanometer cuff around the upper arm to occlude the arterial circulation and proceeded to make repeated maximal voluntary contractions of his adductor pollicis muscle against a strain gauge, interrupting the contractions with single stimuli to the ulnar nerve so as to assess the twitch and muscle compound action potential. Merton's important finding was that when the twitch had become very small, there was still a good-sized action potential (Fig. 2.10, bottom). Although Merton's conclusion, that fatigue is not "electrical," has been questioned—for example, by John Stephens and Anthony Taylor, in London (146)—it is still regarded as essentially correct. Thus, any reduction in the compound action potential of the whole muscle is modest in comparison with the loss of twitch force and, in some human muscles, such as the biceps brachii, there is actually an early potentiation of the action potential due to the hyperpolarizing effect of the electrogenic sodium pump (117). At the single motor-unit level, however, the situation is more complicated, as Peter Clamann, in Richmond, Virginia, has shown (28), and as is evident from the rather earlier recordings of Robert Burke (see p. 49). Thus, while force totally disappears in the FF (fast-contracting, fast-fatigue) units with little change in the action potential, there is more proportionality in the FR (fast-contracting, fatigue-resistant) units, while in the S (slow contracting) units the action potential declines more rapidly than force.

In subsequent years, in collaboration with David Marsden and John Meadows, Merton took his analysis of fatigue further (114). It was already known from Kronecker's work (see above) that the more rapidly a muscle was stimulated the more rapidly it would fatigue. However, Merton went on to assert that the amount of fatigue depended only on the number of impulses, regardless of how they were distributed in time. It might be, he suggested, that in the muscle fiber there was a finite

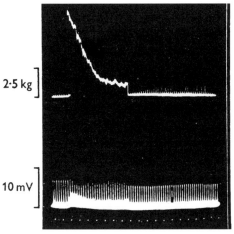

Fig. 2.10. *Top*: Patrick Merton (nearest camera) at the 1969 Hering-Breuer Symposium in London. Merton made important observations on motor unit firing, long loop reflexes, and muscle fatigue, and also helped to introduce magnetic stimulation of the brain. Merton had a particular talent for human experimentation and often used himself as a subject. Courtesy of Dr. Moran Campbell. *Bottom*: Merton's experiment, a classic, which showed that as isometric voluntary force fell (*upper* trace), the muscle compound action potential elicited by single shocks was maintained (*lower* trace). This result pointed to a failure of excitation–contraction coupling and/or the contractile process itself in fatigue. Time pips represent 30 seconds. Reproduced from Merton (122) with permission from the authors and from The Physiological Society.

store of an agent which coupled the action potential to the contractile mechanism, and that a certain amount was used up with each excitation. Merton's conclusion regarding the number of impulses was surprising since it allowed no place for recovery processes in the muscle, which would be more evident with the lower rates of stimulation. However, in isometric contractions, more "work-equivalent" (force × time) would be done by the muscles at the lower rates of stimulation and one might therefore assume that more muscle energy had been consumed. Perhaps the two factors would cancel each other out? There was, however, a flaw in the later Merton experiments, in that fatigue was only measured at the stimulating frequency, rather than at a standard testing frequency. When that was done, it could be seen that fatigue developed with significantly fewer stimuli when the latter were delivered more slowly (55).

Nevertheless, Merton's important observations drew attention to a failure of the excitation–contraction coupling process or of the contractile machinery in fatigue. Which was it? The answer came from New York, in the laboratory of Alexander Sandow. With a young colleague, Arthur Eberstein, Sandow stimulated a frog muscle to fatigue and then applied caffeine to the bathing solution (38). They found that in a 5 mM concentration, caffeine caused the twitches to become very much larger (Fig. 2.11), while at the same time the muscle began to develop a more sustained contraction (i.e., a contracture). Caffeine was known to exert its effect through calcium, and intracellular calcium had long been identified as an important mediator of the contractile process, being released from the sarcoplasmic reticulum with each excitation. Here, then, was important evidence to suggest that the coupling process, linking the action potential to the myofilaments, became increasingly ineffective in prolonged contractile activity and was responsible for fatigue. This conclusion was substantiated by David Allen and his colleagues, first in London and then in Sydney,

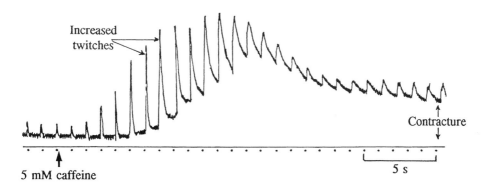

Fig. 2.11. Eberstein and Sandow's experiment showing the dramatic effect of 5 mM caffeine (and calcium mobilization) in temporarily overcoming fatigue in a frog muscle fiber. This important finding drew attention to impaired excitation–contraction coupling in fatigue. Adapted from Eberstein and Sandow (38) with permission from the authors and the Academy of Sciences of the Czechoslovak Republic.

Australia (105). They were able to load single muscle fibers, both amphibian and mammalian, with a dye which fluoresced on exposure to free calcium ions in the cytoplasm. This methodology had been introduced by Colin Ashley and Ellis Ridgway in Bristol (5), using large barnacle muscle fibers. Allen was able to show that fatigue was indeed associated with a smaller release of calcium from the sarcoplasmic reticulum but that there was also reduced sensitivity of the contraction regulating protein, troponin C, to calcium. This last finding raised the next question. What role, if any, had anoxia, depletion of energy, and the accumulation of metabolites in the production of fatigue?

Regarding energy depletion (and the availability of oxygen), crucial information was the concentration of ATP in the muscle fiber and, to a lesser extent, the concentration of phosphocreatine, the ATP-replenishing substrate. With the aid of the muscle biopsy needle, recently introduced by Bergström (see p. 55), this problem was tackled initially by Eric Hultman* and his colleagues in Stockholm, who sampled quadriceps muscles of subjects exercising on the cycle ergometer (87). They found that a decline in phosphocreatine occurred which was related to the work intensity, while the change in ATP concentration was relatively small unless muscles were exercised to fatigue. In the latter condition the mean phosphocreatine concentration was only 10% of the base value, while there was a 40% reduction in the concentration of ATP. These pioneering studies of Hultman were confirmed in a series of investigations by Bengt Saltin and his associates, also in Stockholm. The Scandinavian findings of dramatic declines in phosphocreatine but only modest changes in ATP were confirmed through a methodological advance, the application of nuclear magnetic resonance (NMR) spectroscopy to exercising muscle. Douglas Wilkie* (1922–1998) had started to study exercising muscle under A.V. Hill (see below) in the immediate post–World War II period in London. Like other neurophysiologists at that time, he had built his laboratory equipment from war surplus and he had put it to good use in a meticulous investigation of the force–velocity relationship in the human elbow flexor muscles (156). Never afraid of new ideas, Wilkie had, for example, calculated, on the basis of the power-duration curve, that sustained man-powered flight should be possible. This was a bold prediction, but one which was triumphantly fulfilled on a calm summer day in 1979 when the Gossamer Albatross was slowly but surely pedalled over the English Channel. Wilkie, with Joan Dawson and David Gadian, was fortunate in having access to one of the early NMR machines built in Oxford (the magnets for the first machines were actually made in a garage; Fig. 2.12, left). Having used the methodology on isolated frog muscle initially (31; Fig. 2.12, right), the Wilkie group then tried it on human forearm muscle (154), with highly satisfactory results. Not only could repeated measurements be made in the intact muscle but the results were immediately available. It was found that phosphocreatine decreased, inorganic phosphate rose, pH declined, and, in keeping with the muscle biopsy studies (see above), the ATP concentration remained close to normal until fatigue was far advanced (Fig. 2.12, right). Wilkie argued, however, that it was not the concentration of ATP which was critical, but the rate at which ATP could be hydrolyzed by the myosin cross bridges (and ion pumps), and that the rate would in-

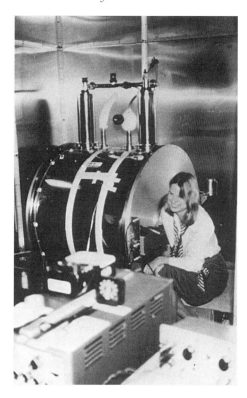

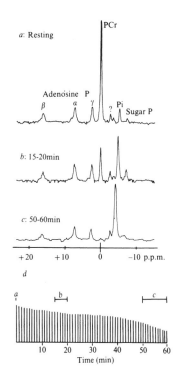

Fig. 2.12. *Left*: The nuclear magnetic resonance (NMR) spectroscopy machine at Oxford used by Douglas Wilkie, David Gadian, and Joan Dawson to study metabolic changes in fatigue, first in frog muscle and then in human forearm muscles. Joan Dawson is shown beside the machine. Reproduced from Wilkie (155) with permission from the Journal of Experimental Biology and from the Company of Biologists Ltd. *Right*: The first published recording of changes in phosphorus metabolites and H^+ concentration, in a stimulated frog gastrocnemius muscle, made by NMR spectroscopy. Comparison of the *upper* (resting) and *lower* (fatigued) recordings shows that inorganic phosphate (Pi) has increased, phosphocreatine (PCr) has disappeared, while much ATP (α, β, γ) remains. The shift in the inorganic phosphate peak reflects the rising H^+ concentration. Reproduced from Dawson et al. (31) with permission from the authors and from *Nature* 274:861–866, 1978, Macmillan Magazines Limited.

evitably be lowered by the rises in concentration of the reaction products, ADP, and inorganic phosphate. Following this pioneer work, NMR spectroscopy has been employed in fatigue research by other investigators, notably Robert Miller and his colleagues in San Francisco.

So far as the role of metabolite accumulation in fatigue was concerned, the starting point was the discovery of lactic acid in the muscles of hunted stags by Berzelius in 1807 (10). Jons Jacob Berzelius (1779–1848) was a largely self-taught Swedish chemist who, among many other achievements, singlehandedly discovered and purified several elements, determined numerous atomic weights, and pioneered electrolysis. Very much later lactic acid was to form an essential part of an erroneous theory of muscle contraction (see p. 85). The possibility that lactic acid might be the direct cause of fatigue had been considered many times but was largely invalidated by the findings of Richard Edwards, who, in London, investigated patients with myophosphorylase deficiency (29). This hereditary disorder had been reported in 1951 (116); affected individuals were unable to degrade muscle glycogen, with the result that no lactic acid was produced in their muscles or appeared in their venous blood. Despite the absence of lactic acid, Edwards and his colleagues found that these patients fatigued even sooner than normal individuals and that their muscle compound action potentials also declined more rapidly.

There were also the puzzling observations of Vøllestad in Oslo to consider (149). By instructing her subjects to make intermittent voluntary contractions which were only 30% of maximal, fatigue occurred slowly. However, when exhaustion was finally reached, there was little change in the concentration of lactic acid—or of inorganic phosphate and hydrogen ions, both of which depress force generation in "skinned" muscle fibers.

So the cause of excitation–contraction coupling failure, and perhaps, to some extent, of an impaired contractile mechanism remains a mystery. Except that there was one very important clue in the old literature, for Angelo Mosso wrote in 1906, "after the frog's leg has been fatigued by prolonged excitation, we can restore its contractility and render it capable of a new series of contractions, simply by washing it. Of course we do not wash the outer surface, but having found the artery which carries blood to the muscle, we pass through it water in place of blood. But not pure water, which is a poison, ... a little kitchen salt is added to the water (seven grammes to a litre), and this solution very closely resembles blood serum. Upon the passage of a current of the liquid through the muscle the fatigue disappears, and the contractions return as vigorously as at the beginning (126)."

So without further oxygen, and hence in the absence of oxidative metabolism to restore ATP and phosphocreatine, fatigue disappears. May there yet be a "toxic" metabolite that induces fatigue?

Lactic Acid and the Oxygen Debt

The formation of lactic acid in exercise, at least in normal subjects under natural conditions, is such a striking event that it is hardly surprising that many experimental studies should have been directed to its mechanism and physiological significance. A significant development in the lactic acid story came from Walter Fletcher and Frederick Gowland Hopkins at Cambridge in 1907 (49). Making use of equipment designed for the study of respiration in plants, they demonstrated that, in isolated frog muscle, lactic acid was indeed produced by contractions. Further, during the recovery period lactic acid was removed by the muscle in an atmosphere of oxygen. The authors may have suspected that glycogen was the precursor of lactic acid, as had others before them, but it was Otto Meyerhof (1884–1951), a medical graduate of the

University of Heidelberg, who convincingly showed that this was so, and that, at least in frog muscle, three-quarters of the lactic acid was converted back to glycogen. Meyerhof suggested that the energy for this synthesis of glycogen came from oxidation of the remaining lactic acid (123). The experimental findings on the recovery period in frog muscle were then found to have parallels in intact human subjects since both Douglas, in Oxford, and Krogh, in Copenhagen, discovered that oxygen consumption remained elevated for some minutes following exercise on a cycle ergometer. A.V. Hill (1886–1977), who had been influenced by Fletcher, his tutor at Cambridge, to study muscle, made a further observation on the recovery period— that the muscle continued to generate heat. This "recovery heat," measured in frog muscle with a specially constructed thermopile and galvanometer, was approximately equal to the initial heat liberated during the contractions, and it required the presence of oxygen for its appearance. Hill combined these various observations by proposing that the formation of lactic acid was necessary for the contractile process itself (see p. 83) and that oxygen was required only for the recovery process, which involved the conversion of lactic acid back to glycogen. The consumption of extra oxygen in the recovery period was a measure of the "oxygen debt" which had accumulated during the muscle contractions. Hill agreed with Meyerhof that the energy for the synthesis of glycogen came from oxidation of a fraction of the lactic acid; it was this metabolism which was also important for the recovery heat. These proposals appeared in a pivotal paper in 1923 (84), which was both profound in terms of the concepts espoused and yet easy to understand. It is interesting that Hill, like Fletcher, was a fine runner and had made many of the human observations on the oxygen debt on himself; indeed, the combination of athleticism and self-experimentation has continued to be a characteristic of many prominent exercise physiologists.

So plausible and ingenious was Hill's synthesis of the known data, and so great his stature in the exercise field—recognized by a Nobel prize in 1922—that several years were to pass before major challenges appeared. One of these came from D.B. (Bruce) Dill (1892–1986) and his colleagues at the Harvard Fatigue Laboratory, who proceeded to make simultaneous measurements of blood lactate and oxygen consumption in human volunteers following running and walking on a treadmill. Because of the huge impact of the Harvard Fatigue Laboratory on the development of exercise physiology in the United States, its history and accomplishments have been described in detail elsewhere (24). In relation to lactate metabolism after exercise, Dill, with H. Edwards and Rodolfo Margaria, showed that in their experiments the additional oxygen consumption was largely over before the blood lactate level began to fall (111). Therefore, they reasoned, only a small part of the oxygen debt repayment could be attributed to lactic acid oxidation; instead the bulk of the oxygen consumption must have been used for the resynthesis of "phosphagen" (phosphocreatine) by the oxidation of "ordinary fuels." Such a conclusion was in keeping with Lundsgaard's (1930) finding that phosphocreatine, which had been discovered in 1927 (see p. 85), could still be partially reformed in muscles poisoned by iodoacetate, which were able to contract without forming lactic acid (108). Dill's findings were extended by Ole Bang, in Copenhagen, who showed, by using different combinations of exercise intensity and duration, that in some circumstances it was possible to return blood lactate to resting levels while exercise was still proceeding even though additional oxygen was still consumed when the exercise was eventually stopped (6).

What did happen to the muscle lactate after exercise, then? It was already known from work in several laboratories, including that of Lovatt Evans in London, that lactic acid, in its buffered state, diffused rapidly from muscle into venous blood, and then from arterial blood into inactive tissue, including muscle. However, it was not until much later that radioisotopes were used to explore the fate of lactate. In rats which had been brought to fatigue, George Brooks and his colleagues at Berkeley, California, injected C^{14}-labeled lactate and then, at intervals afterward, sampled blood, liver, kidney, heart, and skeletal muscle for lactate and its derivatives (17). Using two-dimensional chromatography they found, in conflict with the suppositions of Hill and Meyerhof, that the bulk of the lactate was oxidized to CO_2 and water (presumably through the citric acid cycle) and only a small proportion, less than 20%, was converted to glycogen; a still smaller fraction took part in protein synthesis. In fairness to Hill and Meyerhof, however, there are species differences, such that the formation of glycogen from lactic acid is more pronounced in amphibian than in mammalian muscle. More recently the Scandinavian school has taken the matter of lactate removal further, for Lars Hermansen, in Oslo, has shown that the rate rises as the workload increases, up to a certain point. Largely on the basis of his own work, Hermansen, with Gollnick, has argued that the main fate of lactate is oxidation in the exercising muscles (62).

LONG-TERM EFFECTS OF EXERCISE

It is a commonplace observation that inactivity causes muscles to atrophy, and that an even greater loss of bulk may follow nerve damage or disuse. In this context it is useful to think of the muscle states as a continuum, ranging from denervation, through disuse, to the normal situation, and beyond that, to trained muscle. In this section, however, only the roles of normal daily exercise in maintaining muscle properties and of increased exercise in transforming muscle will be considered.

In 1960, a paper appeared in the *Journal of Physiology* which became a classic (19). It was entitled "Interactions Between Motoneurones and Muscles in Respect of the Characteristic Speeds of Their Responses." In 1957, when this work began, John (Jack) Eccles (1903–1997) was at the Australian National University in Canberra; 20 years earlier he had left Oxford after a highly productive period of muscle research with Sherrington. Now, with his daughter Rosamond (Rose) and Arthur Buller, a newly arrived scientist from the U.K., Eccles was attempting to discover whether a motoneuron might alter its biophysical properties following a change in its peripheral connections. To this end, the nerve supply of the pale gracilis muscle had been cross-united with that of the red crureus in a kitten. On recording the monosynaptic responses of the two populations of motoneurons, it was clear that no change had

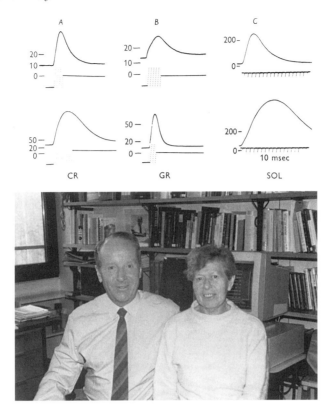

Fig. 2.13. *Top*: Effects of muscle cross innervation. This figure, taken from the classic 1960 paper of Buller, Eccles, and Eccles, shows the twitches of control cat crureus (CR), gracilis (GR) and soleus (SOL) muscles along the bottom row. The top row shows the changes in twitch duration following cross innervation (gracilis nerve to crureus muscle, crureus nerve to gracilis muscle, and peroneus nerve to soleus muscle). The originally "slow" crureus and soleus muscles now have "fast" twitches, while the opposite is true for gracilis. The sum of the near-vertical dots in the time calibrations show the respective contraction times. Reproduced from Buller et al. (19) with permission from the authors and from The Physiological Society. *Bottom*: Gerta Vrbová and Dirk Pette in Pette's laboratory in Konstanz, Germany. Vrbová was the first to show that prolonged low-frequency stimulation made muscles "slow," and Pette exploited this finding in a very comprehensive way. Courtesy of Dr. Vrbová.

taken place in the action potential and after-hyperpolarization. However, on exposing the muscles there was a surprise, for the two muscles had changed their appearances; the crureus had become pale while the gracilis was bright red. Further, the normally "slow" crureus now gave a "fast" twitch, and the "fast" gracilis a "slow" twitch (Fig. 2.13, top). These results were the first evidence that motoneurons regulated the contractile (and biochemical) properties of muscle. Subsequent experiments showed the same phenomenon to exist in cross-innervated soleus and flexor digitorum longus muscles, which were much more suited to this type of experiment.

The paper by Buller, Eccles, and Eccles in 1960 created enormous interest. The results were soon confirmed and it was also shown—for example, by Victor Dubowitz, then working in New York—that the histochemistry of the muscle fibers was transformed by cross innervation (35). But how were the changes brought about? When the Australian work had been presented as a communication to the Physiological Society (U.K.), Andrew Huxley (see p. 88) had made a novel suggestion—that it was the pattern of impulse activity which was the critical factor. This proposal was rejected by Eccles in favor of a neurotrophic (chemical) mechanism. However, the impulse idea attracted the attention of a young woman in the audience. Gerta Vrbová (Fig. 2.13, bottom) had grown up in Czechoslovakia but was forced to flee during the Nazi occupation in World War II and then a second time, during the postwar communist regime, eventually settling in the U.K. In the mysterious world of "trophic" (sustaining) influences between nerve and muscle, Vrbová proceeded to display an intuition which rarely let her down. In her first experiments, reported in 1963, she showed that the rabbit soleus muscle, when deprived of motoneuron excitation by a combination of spinal section and tenotomy, changed from slow twitch to fast twitch (151). Vrbová's results and conclusions were attacked by Arthur Buller (20), by then back in the U.K. and working with Donald Lewis in Bristol, and not merely out of loyalty to Eccles. Buller did have one very valid point: after tenotomy, the rabbit soleus muscle shrank to a fraction of its former size and it could be claimed that it had become a degenerating muscle. Undeterred, Vrbová went a step further, restoring impulse activity in the previously silent, tenotomized soleus muscle by stimulating it for 8–10 hours a day through implanted electrodes (150). She was able to show that if 10 Hz was used as the stimulating frequency, the soleus was prevented from becoming a fast-twitch muscle, while stimulation at a higher frequency (40 Hz) was ineffective; similar results were obtained using an implanted stimulator (134). The role of impulse activity was taken up by other workers, a particularly comprehensive study of the effects of 10 Hz stimulation being carried out by Dirk Pette in Konstanz, Germany (130; Fig. 2.13, bottom). By combining measurements of muscle enzyme activity with those of messenger RNA, one of Pette's interesting findings was that continuous muscle stimulation could elevate enzyme activity so rapidly that increased translation of messenger RNA must have been involved, with the effects of altered gene expression appearing later.

Another laboratory very much involved in this type of work was that of Terje Lømo in Oslo. Lømo had previously published an influential paper with Jean Rosenthal in 1972 (107), in which they reported that some of the features of denervation, such as the spread of acetylcholine sensitivity, fall in resting membrane potential, and fiber atrophy, could be prevented by direct stimulation of the muscle fibers. Lømo, with colleagues, now used this approach to show that, regardless of its prior history, a denervated muscle could be made fast twitch or slow twitch by imposing high-frequency or low-frequency stimulation, respectively. However, the fatiguability of the muscle depended on the number of excitations received in the course of the day, irrespective of their frequency (67). Translating this into the intact animal (or human), it could be said that the soleus was a red, fatigue-resistant, slow twitch, "tonic" muscle because it was in constant use in maintaining posture, being continually excited at low frequencies (ca. 10 Hz) by its motoneurons. In contrast, a muscle such as the medial gastrocnemius was a pale, fatigue-susceptible, fast-twitch "phasic" muscle because it was used only intermittently, as in running and jumping, and was then activated by short bursts of high-frequency impulses (e.g., 40–80 Hz).

While the importance of impulse activity in regulating muscle fiber properties is now universally accepted, it is still not the complete story. For example, Jane Butler, Ethel Cosmos, and Joyce Brierley, in Hamilton, Canada, showed that in the chick embryo muscle fibers were already histochemically differentiated before the motor nerve fibers reached them (25). Further, in human subjects at least, there are some motor units which have such high thresholds that they are rarely used, and yet show no features of denervation. Finally, there are certain membrane properties of muscle, such as the type of sodium channels and the distribution of acetylcholine receptors, which clearly *are* governed by a nonimpulse, trophic influence.

Regardless of the mechanisms involved, it was apparent to Geoffrey Goldspink from the cross innervation and other experiments described above, that the motoneuron must be regulating gene expression in the muscle fibers. First in Hull and then in London, Goldspink had spent much of his career carrying out detailed morphometric analyses of mammalian muscles during normal growth and aging, and after various exercise regimens. He had earlier reported splitting of the myofibrils as fibers hypertrophied and also the ability of sustained muscle lengthening to add new sarcomeres at the ends of the fibers. Now, in 1991, he made use of the last type of experiment by lengthening the rabbit tibialis anterior in a plaster cast and stimulating it electrically (60); within 4 days the muscle mass and RNA had increased significantly, with expression of the gene for slow MHC (myosin heavy chain). Conversely, within 2 days of fixing the rabbit soleus in a shortened position, there was expression of the gene for fast MHC. Frank Booth, in Houston, was another exercise physiologist who, in later years, made an intense exploration of the effects of muscle stretch on gene transcription and messenger RNA translation.

In summary, then, the research since 1951 showed that the pattern and amount of exercise in the normal activities of daily living are largely responsible, through gene expression in the muscle fiber nuclei, for conferring upon muscles their characteristic coloration and their histochemical and contractile properties.

STRENGTH TRAINING

Heavy manual workers, weightlifters, wrestlers, and body builders are remarkable for their bulging muscles, while marathon runners have a lean, greyhound look. How are the contrasting effects of strength and endurance training achieved?

Perhaps the first person to put strength training on a scientific footing was Dr. Thomas DeLorme, a U.S. army captain during the World War II who was responsible for the rehabilitation of injured servicemen (32). One of DeLorme's concerns was the long time required for often-incomplete recovery of quadriceps muscle bulk following injuries or surgery to the thigh and knee. DeLorme reasoned that rather than have his patients undertake a large number of relatively weak contractions involving only a small fraction of the muscle fibers it would be more effective to recruit as much of the muscle as possible through the use of maximal effort. Accordingly, he devised a metal boot to which weights could be attached and determined for each patient the heaviest load which could be raised by extending the knee no more than 10 times (Fig. 2.14, top). The sequence of 10 contractions was repeated seven to 10 times

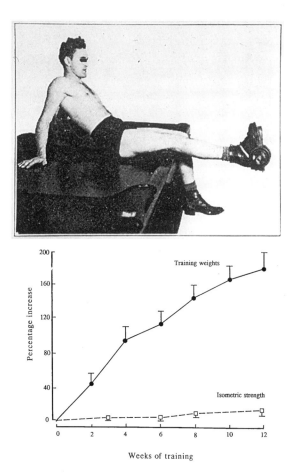

Fig. 2.14. *Top*: DeLorme's system for strengthening quadriceps muscles, which required the subject to lift a metal boot to which weights could be attached. This system was the prototype for modern progressive resistance exercise programs. Reproduced from DeLorme (32) with permission from the *Journal of Bone and Joint Surgery*. *Bottom*: Marked improvement in the ability of the knee extensors to lift weights following the onset of a training program (upper curve). The lower curve shows the relatively small increases in isometric strength, measured with the knee at a right angle, in the same subjects. Adapted from Rutherford and Jones (133), with permission from the authors and from Springer-Verlag New York, Inc.

in the course of a daily workout. As the muscle became stronger, and more easily able to lift the load, the weight on the boot was increased. This method of training was subsequently termed the progressive resistance program, and machines based on this principle are now to be found in many gymnasia. In one study of their effectiveness, David Jones, a former colleague of Richard Edwards (see p. 60) in London, showed that as few as 10 repetitions a day, using loads from 60% to 90% of maximal, increased quadriceps strength by 0.5%–1% per day over several weeks. There were, however, two surprising features about such gains. One was the remarkable specificity of the improvement. By exercising the quadriceps through the lifting of weights, and then testing the strength of the same muscles by isometric contractions against a strain gauge, the knee being held at a right angle, Olga Rutherford and David Jones found that after 12 weeks the mean lifting strength had almost doubled while the improvement in isometric strength was only 15% (133; Fig. 2.14, bottom). They suggested that the difference in performance might lie in some mechanical adaptations during the training exercise, such as a change in the angle of muscle fiber pinnation at different muscle lengths.

The other puzzling feature of such gains in strength was that they far exceeded the increases in muscle cross-sectional area, as had been well documented by Digby Sale,* Duncan MacDougall,* and their colleagues in Hamilton, Canada, in the case of the biceps brachii muscle (109). One possible reason for the discrepancy, as well as for the training specificity, was offered by Enzo Cafarelli, in Toronto; with Barry Carolan, he showed that, as quadriceps training proceeded, there was progressively less coactivation of the antagonistic hamstring muscles (27). Perhaps such a mechanism could be responsible, after all, for extraordinary feats of strength by drowning men and lunatics, as Merton had wondered (see p. 59). The truth of the matter is that this is one of several areas of exercise physiology where uncertainties remain and more study is needed.

When the muscles do hypertrophy, and in body builders and professional wrestlers the increased bulk is particularly impressive, how much is due to enlargement of the muscle fibers and how much to an increase in the numbers of fibers? The possibility of new muscle fibers being formed (hyperplasia) emerged from the chicken experiments of Olav Sola and her associates in Seattle (142). Their experimental design was a simple one and consisted of attaching a weight to one wing while leaving the other wing to serve as a control. Within 5 weeks the weight of the anterior latissimus dorsi muscle had increased by 180% and a significant part of the increase was due to fiber hyperplasia. Similar changes were observed by Steven Alway and his colleagues later (2). Could the same phenomenon occur in overloaded mammalian muscles? William Gonyea,* in Dallas, attempted to answer this question by counting every fiber in strength-trained cat hamstring muscles (57). This was a huge and extremely tedious task, and unfortunately the answer remained elusive because of the normal variations in muscle fiber numbers between animals. The same difficulty was encountered in human subjects by MacDougall, Sale, and their colleagues (109). In brachial biceps muscles the estimated numbers of fibers were found to range more than twofold between subjects. Nevertheless, by comparing muscle fiber diameters with the cross-sectional areas of the biceps muscles, MacDougall and Sale were able to conclude that the muscle enlargement in body builders was largely, if not entirely, due to muscle fiber hypertrophy. The finding of considerable variation in muscle fiber populations, even in untrained subjects, was important for another reason, since it demonstrated that genetic endowment might be an important factor in the gain of strength.

The last aspect of heavy resistance training to have been considered was whether there were changes in the biochemistry of the muscle fibers other than the addition of actin and myosin filaments which was partly responsible for the gain in strength. There have been many studies in this area, but some of the most careful work has been carried out by Robert Staron and his colleagues in Athens, Ohio. In biopsied samples of human lateral vastus muscle, they showed that heavy resistance training is associated with a conversion of histochemical type IIB fibers to type IIA fibers, and with a corresponding change in the myosin heavy-chain composition (143). As in most studies of this kind, particularly when previously untrained subjects were involved, it was found that the type II fibers exhibited the greatest hypertrophy. Staron also demonstrated that the adaptive changes were reversible if training was discontinued, but returned more rapidly than before if training was resumed. Neither in Staron's work nor in that of other investigators was interconversion of type I and type II fibers observed. Further, any changes in glycogen content and in glycolytic and oxidative enzyme activity, although looked for, appeared to be small, while mitochondrial density was found to decrease.

ENDURANCE TRAINING

Is there something special about the muscle fibers in leg muscles of athletes used to long-distance running? This question was tackled in Sweden by Eva Jansson and Lennart Kaijser in a report published in 1977 (96). The authors took needle biopsy samples of quadriceps muscles from orienteers and compared them with a large number of biopsies from control subjects. They found that the proportions of muscle fiber types were indeed different in the two groups. The orienteers had significantly higher percentages of type I (slow contracting, type S) fibers and smaller incidences of type IIB (fast contracting, fast fatigue, type FF) fibers. Similar differences were found by other workers between controls and long-distance runners, cross-country runners, and professional cyclists. These differences were clearly beneficial in that the type I (type S) motor units would be more resistant to fatigue than the type IIB (type FF) units, as described in the section Fast and Slow Muscles and Motor Units.

However, a new question arose. Were the differences in fiber-type proportions between the endurance athletes and controls due to differences in genetic endowment, or did transformation of the fibers occur as a result of the exercise stimuli? This, of course, was a neuromuscular form of the nature vs. nurture issue so common in other fields. In the case of muscle, though, there were ways of resolving the issue. For example, Jean-Aimé Simoneau and his associates in Quebec took 24 untrained subjects and made them perform a program of high-intensity continuous/interval work over a period of 15 weeks (141). A comparison of quadriceps muscle biopsies obtained at the beginning and at the end of the period revealed that there had been a small but significant gain in the percentage of type I fibers and a reduction in type IIB fibers. A decreased proportion of type IIB fibers was, in fact, the most common change in this type of experiment, while some groups also found increases in type I and type IIA (fast contracting, fatigue resistant, type FR) fibers. In general, however, the results in human muscle fiber types were much less impressive than those obtained in animal experiments. Peter Schantz and Gurtej Dhoot, in Stockholm, took the matter further by employing more sensitive tests than routine histochemical staining (137). They made use of recent advances in immunocytochemistry and applied antibody staining and gel electrophoresis to examine the isoforms of the MHCs and troponins. They found that although histochemistry might identify a fiber as belonging to one type or another, immunocytochemistry revealed that many fibers were in a transitional state in that they contained both fast and slow isoforms of the contractile proteins. Similar conclusions were reached by Henrik Klitgaard in Denmark and Hugo Baumann in Switzerland, and their respective collaborators.

In another prospective study, Frank Ingjer, in Oslo, looked at some other features of endurance-trained muscle (95). He took seven young women who underwent cross-country running for 24 weeks. In biopsy samples of quadriceps muscle, he observed that the numbers of capillaries had increased around the muscle fibers regardless of their histochemical type. Further, there were larger numbers of mitochondria, especially in the type I fibers. Both of these adaptations were clearly beneficial since they would facilitate aerobic metabolism in the exercising fibers.

Although the histochemical studies in human muscles were important in showing that muscles could adapt to endurance exercise, by themselves they could only give a limited picture of the changes possible. For a more complete picture it was necessary to subject the muscles to quantitative biochemical techniques, and this was done not only in human volunteers but also in experimental animals. In the latter the first investigations were carried out by John Holloszy, in St. Louis, and were published in 1967 (86). Holloszy trained rats to run continuously on a treadmill for 2 hours each day for 3 months and reported that the exercised leg muscles were a deeper red color, in part due to an increase in the pigment myoglobin. In addition the activities of some mitochondrial enzymes were increased, doubling in the case of the respiratory chain enzymes; there were also increases in fatty acid metabolism and in mitochondrial oxygen uptake. Not only fatty acids but also pyruvate could be oxidized at approximately twice the rates in untrained muscles; in contrast, adaptive changes in glycolytic enzymes were less conspicuous.

Among the many other examinations of muscle metabolism in endurance-trained animals, those of Reggie Edgerton also deserve mention as one of the earliest. Edgerton, then a graduate student at UCLA, showed, with James Barnard and James Peter, that guinea pigs, after 18 weeks of treadmill running, had significant increases in mitochondrial content and in the oxidative capacity of the exercised muscles (7).

Fig. 2.15. Bengt Saltin (*left*) and Paul Gollnick (*right*) have contributed greatly to our understanding of glycolysis in exercising muscles. Courtesy of Dr. Charles Tipton.

Holloszy's detailed biochemical studies in animals also had their human counterpart in a rapidly expanding literature from other laboratories, including those of Michael Houston and Howard Green* in Waterloo, Canada. However, the most complete and wide-ranging investigations were those of Bengt Saltin, Philip Gollnick (1934–1991) (Fig. 2.15, left, right) and their colleagues in Stockholm, most of which made use of the needle muscle biopsy technique reintroduced by Saltin's colleague, Bergström (see p. 55). In relation to glycolysis, Saltin and Karlsson (135) established that, for the same workload, endurance training decreased the rates of glycogen depletion and lactate production as well as the declines in phosphocreatine and ATP. Other studies demonstrated an increase in mitochondrial enzyme activity and a greater capacity to oxidize fats. It was already known, from work on experimental diabetes, that a rise in free fatty acid concentration could depress glucose consumption, and Saltin and Gollnick proposed that the same principle applied in endurance training (65). Thus, in submaximal exercise, an increase in fatty acid metabolism in adapted fibers would have a "sparing" effect on the glycolytic pathway, and so promote prolonged effort. The elevated mitochondrial enzyme capacities demonstrated in human and animal training studies were shown by electron microscopy to be associated with increases in both the numbers and the sizes of the mitochondria.

In a different approach to the effects of endurance training, Karl Hainaut and Jacques Duchateau, in Brussels, decided to study contractile activity, making use of the small muscle brought into prominence by Patrick Merton many years earlier (see p. 58)—the adductor pollicis. They persuaded each of their eight subjects to perform 200 submaximal voluntary adductions of the thumb every day for 3 months (72). Hainaut and Duchateau tested the muscles, both at the beginning and at the end of the training period, by subjecting them to repeated 1-second tetani at 30 Hz. They found that there was a highly significant increase in resistance to fatigue in the trained muscles.

There is, in addition, another major alteration in the muscles of endurance-trained athletes, alluded to at the start of this section. The muscles, rather than hypertrophying, actually become smaller, as do the myofibrils inside them. Geoffrey Goldspink, following up on his extensive studies of exercise effects on muscle, has reasoned that the myofibrils, in becoming thinner, increase their surface-to-volume ratios and would, with their enveloping mitochondria, thereby achieve better perfusion, within the fibers, of nutrients and contraction products (61).

MACHINA CARNIS: THE FLESH MACHINE

It would be reprehensible to conclude a review of the neuromuscular system in exercise without considering the nature of the mechanism which makes exercise possible—that is, the contractile apparatus itself. To cover this subject fully would be too great a task, and fortunately there are excellent reviews, including two by Andrew Huxley (88, 89), dealing with early theories of muscle contraction based largely on observations of the muscle fiber striations. Instead, this section will highlight the work of three persons who made very major contributions to this subject and who dominated the field to an extraordinary extent.

A.V. Hill

Archibald Vivian Hill (1886–1977; Fig. 2.16, left) went to Cambridge to study mathematics but was subsequently persuaded to take up physiology instead. With a strong background in mathematics, physics, and chemistry, Hill was well equipped to investigate muscle as a thermodynamic machine, and this he did for the next 60 years. Hill began by measuring heat production in muscle, a task which had been started in 1848 by Helmholtz (see p. 65). The technical challenge was considerable, for in a single twitch the temperature of a frog sartorius muscle increases by no more than 0.003°C. Hill's solution was to employ the principle of the thermocouple; current will flow in a circuit containing two or more junctions of dissimilar metals (e.g., constantin and iron), if there is a temperature difference between them. By combining many junctions to form a thermopile, and by connecting the latter to a galvanometer, Hill and William Hartree were able to measure astonishingly small changes in temperature (83). Hill was able to show that in a contraction heat was generated in proportion to the lactic acid formed and that in addition to the heat produced during muscle shortening, there was an oxidative delayed heat which persisted for many minutes (see p. 73). Hill had also demonstrated that, during a contraction, the muscle appeared to become more viscous. Combining his observations on lactic acid appearance, heat production, and mechanical changes in the muscle, Hill put for-

Fig. 2.16. *Left*: A.V. Hill (1886–1977), the outstanding figure in muscle physiology for the first half of the twentieth century. Much of Hill's work involved measuring heat production in contracting muscle using an extraordinarily sensitive system made in his laboratory. Hill shared a Nobel Prize with Otto Meyerhof in 1922, though his theory of muscle contraction was subsequently shown to be wrong. Hill gave distinguished service in both world wars as a scientist and scientific advisor, and was also, for a few years, a member of Parliament. This photograph shows Hill in a jovial mood, and indeed his autobiographical writings attest to a excellent sense of humor; but these qualities were not always obvious to those around him! Reproduced from Hill (78). Courtesy of Mrs. Phillippa Hill. *Right*: Wallace Fenn (1893–1971) was one of Hill's earliest overseas research fellows. However, Fenn's finding, that muscles produced more heat if allowed to do work rather than to contract isometrically, was the first indication that Hill's viscoelastic theory of contraction was wrong. Fenn was like Hill in having a keen but subtle sense of humor, while appearing rather intimidating. Fenn had many physiological interests and remained a major figure until the end of his career. Courtesy of Dr. Ethel Cosmos.

ward a viscoelastic theory of muscle contraction (79). The essence of the theory was that the primary biochemical event in muscle contraction is the sudden release of lactic acid from a large precursor molecule, by then known to be glycogen. The hydrogen ions, so liberated, would then neutralize negative charges on the contractile proteins; freed of their negativity, the elastic proteins would fold and shorten.

For his work on muscle, Hill shared the Nobel Prize with Otto Meyerhof in 1922. This was rapid success indeed, for Hill had only started working on muscle in 1909 and had spent most of World War I (1914–1918) developing antiaircraft gunnery. Perhaps it was too rapid for, in 1921, the year before the Nobel award, Embden and Grafe, in Germany, had demonstrated, inexplicably, that phosphate was also produced in contracting muscle (42). Then, in 1923, Wallace Fenn, working in Hill's own laboratory, showed that more heat was generated if the muscle performed work than if it contracted isometrically (46), a fact difficult to reconcile with the idea of a fixed amount of lactic acid being released regardless of the nature of the contraction. The following year, Hill was obliged to write (see ref. 82), "Fenn has provided us with some very difficult problems." Next, in 1927, was the independent discovery of phosphocreatine breakdown in contraction by Philip and Grace Eggleton in London and by Cyrus Fiske and Yellapragada Subbarow at Harvard (41, 48). Finally, in 1930, came the death blow to the Hill–Meyerhof theory with the demonstration by Einar Lundsgaard that a muscle poisoned with iodoacetate was able to contract without producing any lactic acid at all (108). This critical finding, like some of the others considered in this review, was a chance discovery, in this case part of a study of amino acid effects on basal metabolic rate.

Yet, if Hill was wrong with his lactic acid/viscoelastic theory, his approach was perfectly valid, for muscle *is* a machine, albeit a rather inefficient one, and by regarding it as such and asking engineering questions of it, insights can be gained into the nature of the biological mechanisms involved. For example, by recognizing the series-elastic component, Hill deduced that the active contractile process in the muscle fiber must be a briefer event than the recorded twitch (81). In our current understanding, the "active state," in fact, corresponds to the time that calcium ions are available to the contractile filaments. Hill also discovered that when the muscle did work—for example, by lifting a load—the rate of energy liberation (work + shortening heat) decreased as the load increased. From this he derived the "characteristic equation" (80):

$$(P + a)(v + b) = \text{Constant}$$

where P = load, a = shortening heat/cm, v = velocity of shortening in cm/s, and b = increase in total energy rate per g decrease in load.

When tested experimentally, the velocity increased as the load decreased and the points fell on a hyperbolic curve as the equation predicted. This relationship suggested to Hill that the slower shortening with increased loads was due to control of the energy-yielding reactions in the contractile proteins. This idea, that mechanical conditions could control chemical reaction rates, was novel and of fundamental importance. Ultimately, through later discoveries (see below), Hill's results would be applicable in terms of the rate of ATP hydrolysis by the myosin cross bridges.

As noted earlier, Hill dominated muscle physiology in his long career and, indeed, was referred to as the "father" of exercise physiology (82). A tall, distinguished man, but with an intimidating air, he was a keen athlete and always eager to transfer results from isolated muscle to the whole organism. Indeed, some of his earliest research had been on the human elbow flexors, in which an inertia flywheel was used to test the relationship between maximum realizable work and the duration of contraction—a forerunner of the force–velocity curve (Fig. 2.17, bottom). Hill was fascinated by such issues as the progressive reductions in speed with increasing distances run, the fact that small mammals had faster movements than large ones, and that athletic performance was better in men than women. The striking effect of body

Fig. 2.17. *Top*: Hugh Huxley (*left*) and Sir Andrew Huxley photographed during a moment of relaxation on their way to Mount Fuji while on a scientific visit to Japan. Courtesy of Mrs. Fumiko Ebashi (see ref. 115). *Bottom*: An early version of the force–velocity relationship, but using a whole subject rather than a single muscle. The sprinter has to pull against a rotating drum with a variable resistance, and is timed at different points along the track. The sprinter in this case is Charles Best, who had helped Banting to isolate insulin in Toronto. Although Best, perhaps unfairly, did not share in the Nobel Prize with Banting and MacLeod, he nevertheless enjoyed a life of great distinction. Reproduced from Hill (77) with permission from Cambridge University Press.

temperature on movement in a cold-blooded animal, such as a turtle, also intrigued him, prompting him to suggest that raising human body temperature by 1°C, using electric induction or diathermy, would increase running speed by 10%.

Wallace Fenn (1893–1971; Fig. 2.16, right), whose studies on heat production in A.V. Hill's laboratory helped to dismantle the viscoelastic/lactic acid theory of muscle contraction (see above), was another who enjoyed combining observations in vitro with those on intact human subjects. Appointed professor of physiology at the University of Rochester at the early age of 31, Fenn continued to make important contributions in several branches of physiology and, in athletics, investigated the different types of mechanical work performed by sprinters (e.g., 47).

Muscle Power

Hill's interest in athletics, and in the different conditions in which muscles were called upon to contract raised the issue of muscle power—that is, the speed with which a force could be delivered (i.e., force × velocity). With the exception of the tug of war, arm wrestling, and to some extent, weight lifting, it is difficult to think of sports in which muscles are required to contract isometrically. In all others—running, jumping, throwing, cycling, swimming—it is power that determines success or failure. In recent years the ability of muscles to develop power has been examined by Anthony Sargeant (formerly in Amsterdam), David Jones (formerly in London), and John Faulkner* (Ann Arbor, Michigan) with their respective colleagues (136, 97, 18). Such studies have shown that if the maximal rate of shortening is reduced, as in fatiguing muscle, then power must necessarily decline. Also, at progressively higher rates of limb movement, the contributions of the fast-contracting muscle fibers (type II; FF and FR) to power output become increasingly important. Such studies have shown, in addition, that shortening contractions, in which power is developed, are more fatiguable than isometric contractions, and that this is reflected in greater biochemical changes in the fibers. In Philadelphia, Lawrence Rome (132) has argued that in real life, evolutionary forces have ensured that muscles are used in such a way that the shortening velocity is a fairly constant fraction (approximately 0.25 of the maximal velocity) and is then optimal for power production. At the same time muscles are so structured that even when the most extreme movements take place, at some stage in each cycle there is maximal overlap of the myofilaments for force generation. Rome also compared maximal shortening rates in large and small animals, a subject in which A.V. Hill was interested. By studying single fibers Rome was able to demonstrate that the maximal shortening velocity was 1000 times greater in the rat than the horse; this difference was appropriate in that the stepping frequency was much higher in the smaller animal.

Andrew Fielding Huxley and Hugh Esmor Huxley

It was an extraordinary coincidence that the two people who, quite independently, discovered the structural basis of muscle contraction should share a rather uncommon surname and yet be unrelated (see Fig. 2.17, top). Extraordinary too, that after all the complex thermodynamic experiments of Hill and his followers, the secret of muscle contraction should be found by looking at the fine structure of fibers with a microscope.

Hugh Huxley had studied physics at university and after World War II joined the Cavendish Laboratory in Cambridge. There he was one of a group who were employing X-ray diffraction techniques in an attempt to resolve the structure of proteins. While Perutz was concerned with hemoglobin, and Francis Crick and James Watson were to have their spectacular success with DNA, Hugh Huxley chose to work with the contractile proteins of muscle. In a short paper entitled "X-ray Analysis and the Problem of Muscle" (93) Huxley reported in 1953 that the low-power X-ray diffraction patterns indicated the presence of two types of overlapping filaments, actin and myosin, and that these were both arranged hexagonally. By then, however, he had come to the conclusion that electron microscopy might be better for solving the gross structure of actin and myosin, and he elected to follow this direction at the Massachusetts Institute of Technology. Electron microscopy had been applied to muscle before, but Hugh Huxley's approach was to cut thinner sections of myofibrils than had previously been possible using an ultramicrotome which he helped develop, and it was then apparent that the myofilaments were indeed composed of overlapping thick and thin filaments (92). In collaboration with Jean Hanson, the next step was to dissolve myosin and examine the myofibers with the light microscope; in this way the thick filaments could be shown to be formed of myosin while the thin filaments could be concluded to have contained actin. In a letter to *Nature* published in 1953, Hugh Huxley and Jean Hanson made the first proposal that "the two sets of filaments slide past each other" (73). Hugh Huxley subsequently showed, in his ultrathin sections, the presence of projections from the surface of the thick filaments; these, he suggested, were the cross bridges responsible for engaging the actin filaments.

At the very time that Hugh Huxley was developing the sliding filament hypothesis, Andrew Huxley was arriving at rather similar conclusions from a different direction. Born in 1917, Andrew Fielding Huxley is a surviving grandson of Thomas Henry Huxley, the eminent Victorian scientist, educational reformer, and defender of Charles Darwin.

Andrew Huxley's first scientific success, and the one for which he was to share the Nobel Prize in 1963 with Alan Hodgkin (1914–1998) and John Eccles, was the elucidation of the ionic mechanisms underlying the resting and action potentials of nerve fibers. In a dazzling series of experiments, utilizing an electronic circuit to clamp the voltage across the squid membrane, Hodgkin and Huxley showed that the action potential was due to a rapid rise in sodium permeability and was terminated by a slower increase in potassium permeability.

After the voltage-clamp experiments, Andrew Huxley recognized that the analysis of excitation could not be taken any further for the time being and so instead turned his attention to muscle contraction. Andrew Huxley's first step was to design and build a sufficiently high-powered interference microscope for observing the striations of single living amphibian muscle fibers during local changes in fiber

1. Neil McCartney
2. Lennart Grimby
3. Norman L. Jones
4. V. Reggie Edgerton
5. Paavo V. Komi
6. Robert E. Burke
7. Eric A. Newsholme
8. Britton Chance
9. Eric Hultman
10. Howard J. Green
11. Richard B. Stein
12. Alan J. McComas
13. Bengt Saltin
14. Philip D. Gollnick
15. Brenda Bigland-Ritchie
16. Digby G. Sale
17. J. Duncan MacDougall
18. David A. Winter
19. Hans Howald
20. C. David Ianuzzo
21. John A. Faulkner
22. James J. Perrine
23. Kate Bárány
24. Michael Bárány
25. William Gonyea
26. Douglas R. Wilkie

Fig. 2.18. Group photograph, taken at a human muscle power conference at Mc-Master University, Canada, in 1984, includes many of the scientists (indicated by an asterisk in the text) discussed in this chapter. Reproduced from N.L. Jones, N. McCartney, and A.J. McComas, *Human Muscle Power*, 1986, Champaign, IL (99), with permission from Human Kinetics.

length. With Rolf Niedergerke, in Cambridge, he then made the key observation that during passive stretch the dark A-band stayed the same length while the light I-band elongated. In electrically elicited contractions, however, the A-band once more remained the same length but the I-band shortened. This contrasting behavior of the dark and light striations immediately suggested a sliding filament system. The two Huxleys, Andrew and Hugh, eventually met at Woods Hole in the summer of 1953 and compared notes; in the following year, by mutual agreement, they had letters published side by side in *Nature* (94, 90). Andrew Huxley went on to test the sliding filament hypothesis in various ways; one was to show that, in an isometric contraction of a single sarcomere, the force developed was related to the amount of overlap of the thick and thin filaments, and hence to the number of myosin cross bridges that could engage the actin molecules (66).

Hardly had the sliding filament hypothesis been published, than Andrew Huxley had another success. This time he showed, with R.E. Taylor, that there were certain focal sites along the muscle fiber at which excitation could be conducted into the interior so as to activate the myofibrils. When a ciné-film of the experiments was shown to the Physiological Society in 1955, the microscopic nature of these focal sites became evident. They were probably not the Z-lines, as Huxley and Taylor thought, but rather the transverse tubules, a feature of the muscle fiber described and forgotten long before but rediscovered by an electron microscopist in the audience (91).

There are many other examples of Andrew Huxley's brilliance. Many feel that he should have had a second Nobel Prize, to be shared with Hugh Huxley. After all, if A.V. Hill and Otto Meyerhof had been given the prize in 1922, in part for what proved to be an erroneous theory of muscle contraction, should not the codiscovers of the correct mechanism have received one?

The group photograph in Figure 2.18, taken at a human muscle power conference at McMaster University, Canada, in 1984, includes many of the scientists discussed in this chapter.

I am grateful to the Natural Sciences and Engineering Research Council for financial support. I wish to thank Jane Butler for secretarial and technical assistance.

REFERENCES

1. Adrian E.D. and D.W. Bronk. The discharge of impulses in motor nerve fibres. II. The frequency of discharge in reflex and voluntary contractions. *J. Physiol. (Lond.)* 67: 119–151, 1929.
2. Alway S.E., P.K. Winchester, M.E. Davis, and W.J. Gonyea. Regionalized adaptations and muscle fiber proliferation in stretch-induced enlargement. *J. Appl. Physiol.* 66: 771–781, 1989.
3. Aristotle. *De Motu Animalium*. Translation by A.S.L. Farquharson in *The Works of Aristotle*, translated edition by J. A. Smith and W.D. Ross, Vol. 5. Oxford: Clarendon Press, 1912.
4. Aristotle. *De Partibus Animalium*. Translation by W. Ogle in *The Works of Aristotle*, translated edition by J.A. Smith and W.D. Ross, Vol. 5. Oxford: Clarendon Press, 1912.

5. Ashley C.C. and E.B. Ridgway. On the relationships between membrane potential, calcium transient and tension in single barnacle muscle fibres. *J. Physiol. (Lond.)* 209:105–130, 1970.

6. Bang O. The lactate content of the blood during and after muscular exercise in man. *Skand. Archiv. Physiol.* 74 (Suppl. 10):49–82, 1936.

7. Barnard R.J., V.R. Edgerton, and J.B. Peter. Effect of exercise on skeletal muscle. I. Biochemical and histochemical properties. *J. Appl. Physiol.* 28:762–766, 1970.

8. Bellemare F., J.J. Woods, R. Johansson, and B. Bigland-Ritchie. Motor-unit discharge rates in maximal voluntary contractions of three human muscles. *J. Neurophysiol.* 50: 1380–1392, 1983.

9. Bergström J. Muscle electrolytes in man. Determined by neutron activation analysis on needle biopsy specimens. A study on normal subjects, kidney patients, and patients with chronic diarrhoea. *Scand. J. Clin. Lab. Invest.* 14(Suppl. 68):1–110, 1962.

10. Berzelius, J., 1807. Cited by D.M. Needham in *Machina Carnis. The Biochemistry of Muscular Contraction in Its Historical Development.* Cambridge: University Press, 1971, p. 41.

11. Bigland B. and O.C.J. Lippold. Motor unit activity in the voluntary contraction of human muscle. *J. Physiol. (Lond.)* 125:322–335, 1954.

12. Bigland-Ritchie B.R., N.J. Dawson, R.S. Johansson, and O.C.J. Lippold. Reflex origin for the slowing of motoneurone firing rates in fatigue of human voluntary contractions. *J. Physiol. (Lond.)* 376:451–459, 1986.

13. Blasius W., J. Boylan, and K. Kramer (eds). *Founders of Experimental Physiology.* Munich: J.F. Lehmanns Verlag, 1971, p. 241.

14. Blomstrand E., F. Celsing, and E. A. Newsholme. Changes in plasma concentrations of aromatic and branched-chain amino acids during sustained exercise in man and their possible role in fatigue. *Acta Physiol. Scand.* 133:115–121, 1988.

15. Bongiovanni L.G. and K.-E. Hagbarth. Tonic vibration reflexes elicited during fatigue from maximal voluntary contractions in man. *J. Physiol. (Lond.)* 423:1–14, 1990.

16. Borelli G. *De Motu Animalium.* Rome: Angelo Bernabo, 1680. Cited by Sir M. Foster, *Lectures on the History of Physiology During the Sixteenth, Seventeenth and Eighteenth Centuries.* Cambridge: University Press, 1901, pp. 73–75.

17. Brooks G.A. and G.A. Gaesser. End points of lactate and glucose metabolism after exhausting exercise. *J. Appl. Physiol.* 49:1057–1069, 1980.

18. Brooks S.V. and J.A. Faulkner. Forces and powers of slow and fast skeletal muscles in mice during repeated contractions. *J. Physiol. (Lond.)* 436:701–710, 1991.

19. Buller A.J., J.C. Eccles, and R.M. Eccles. Interactions between motoneurones and muscles in respect of the characteristic speeds of their responses. *J. Physiol. (Lond.)* 150: 417–439, 1960.

20. Buller A.J. and D.M. Lewis. Some observations on the effects of tenotomy in the rabbit. *J. Physiol. (Lond.)* 178:326–342, 1965.

21. Burke R.E. Group Ia synaptic input to fast and slow twitch motor units of cat triceps surae. *J. Physiol. (Lond.)* 196:605–630, 1968.

22. Burke R.E. Motor unit types of cat triceps surae muscle. *J. Physiol. (Lond.)* 193:141–160, 1967.

23. Burke R.E., R.N. Levine, F.E. Zajac III, P. Tsairis, and W.K. Engel. Mammalian motor units: physiological-histochemical correlation in three types in cat gastrocnemius. *Science* 174:709–712, 1971.

24. Buskirk E.R. and C. M. Tipton. Exercise physiology. In: The History of Exercise and Sport Sciences, edited by J.D. Massengale and R.A. Swanson. Champaign, IL: Human Kinetics, 1997, pp. 367–438.

25. Butler J., E. Cosmos, and J. Brierley. Differentiation of muscle fiber types in aneurogenic brachial muscles of the chick embryo. *J. Exp. Zool.* 224:65–80, 1982.

26. Cajal R.S. *Histologie du Systeme Nerveux de l'Homme et des Vertebré*, Vol 1. Paris: Maloine, 1909, pp. 485–489.

27. Carolan B. and E. Cafarelli. Adaptations in coactivation after isometric resistance training. *J. Appl. Physiol.* 73:911–917, 1992.

28. Clamann H.P. and K.T. Broecker. Relation between force and fatigability of red and pale skeletal muscles in man. *Am. J. Phys. Med.* 58:70–85, 1979.

29. Cooper R.G., M.J. Stokes, and R.H.T. Edwards. Myofibrillar activation failure in McArdle's disease. *J. Neurol. Sci.* 93:1–10, 1989.

30. Cooper S. and J.C. Eccles. The isometric response of mammalian muscles. *J. Physiol. (Lond.)* 69:377–385, 1930.

31. Dawson M.J., D.G. Gadian, and D.R. Wilkie. Muscular fatigue investigated by phosphorus nuclear magnetic resonance. *Nature* 274:861–866, 1978.

32. DeLorme T.L. Restoration of muscle power by heavy resistance exercises. *J. Bone Joint Surg.* 27:645–667, 1945.

33. Denny-Brown D. The histological features of striped muscle in relation to its functional activity. *Proc. R. Soc. Lond. Ser. B* 104:371–411, 1929.

34. Denny-Brown D. and J.B. Pennybacker. Fibrillation and fasciculation in voluntary muscle. *Brain* 61:311–334, 1938.

35. Dubowitz V. Cross-innervated mammalian skeletal muscle: histochemical, physiological and biochemical observations. *J. Physiol. (Lond.)* 193:481–496, 1967.

36. Dubowitz V. and A.G.E. Pearse. Reciprocal relationship of phosphorylase and oxidative enzymes in skeletal muscle. *Nature* 185:701–702, 1960.

37. Duchenne G.B.A. De l'Électrisation Localisée et de son Application à la Physiologie, à la Pathologie et à al Thérapeutique (2nd and 3rd eds.). Paris: J.-B. Baillière et Fils, 1861, 1872.

38. Eberstein A. and A. Sandow. Fatigue mechanisms in muscle fibres. In: *The Effect of Use and Disuse on Neuromuscular Functions*, edited by E. Gutmann and P. Hník. Prague: Publishing House of the Czechoslovak Academy of Sciences, 1963, pp. 515–526.

39. Eccles J.C. and W.C. Gibson. *Sherrington. His Life and Thoughts.* Berlin: Springer-Verlag, 1979, p. 50.

40. Edström L. and E. Kugelberg. Histochemical composition, distribution of fibres and fatigability of single motor units. Anterior tibial muscles of the rat. *J. Neurol. Neurosurg. Psychiatry* 31:424–433, 1968.

41. Eggleton P. and G.P. Eggleton. The inorganic phosphate and a labile form of organic phosphate in the gastrocnemius of the frog. *Biochem. J.* 21:190–195.

42. Embden G. and E. Grafe. Über den einfluss der muskelarbeit auf die phosphorsäureausscheidung. *Z. Physiol. Chem.* 113:108–137, 1921.

43. Engel W.K. The essentiality of histo- and cytochemical studies of skeletal muscle in the investigation of neuromuscular disease. *Neurology* 12:778–794, 1962.

44. Erasistratus, cited by Galen in *De Locis Affectis*. Translation by K. G. Kühn, Vol. 8, p. 429; cited by D.M. Needham in *Machina Carnis. The Biochemistry of Muscular Contraction in Its Historical Development.* Cambridge: University Press, 1971, p. 8.

45. Feinstein B., B. Lindegård, E. Nyman, and G. Wohlfart. Morphologic studies of motor units in normal human muscles. *Acta Anat.* 23:127–142, 1955.

46. Fenn W.O. A quantitative comparison between the energy liberated and the work performed by the isolated sartorius muscle of the frog. *J. Physiol. (Lond.)* 58:175–203, 1923.

47. Fenn W.O. Frictional and kinetic factors in the work of sprint running. *Am. J. Physiol.* 92:583–611, 1930.

48. Fiske C.H. and Y. Subbarow. The nature of the 'inorganic phosphate' in voluntary muscle. *Science* 65:401–403.

49. Fletcher W.M. and F.G. Hopkins. Lactic acid in amphibian muscle. *J. Physiol. (Lond.)* 35:247–309, 1907.

50. Foster M.A. *A Textbook of Physiology* (4th ed.). London: MacMillan & Co., 1883, pp. 96–97.

51. Foster, Sir M. *Lectures on the History of Physiology During the Sixteenth, Seventeenth and Eighteenth Centuries.* Cambridge: University Press, 1901.

52. Galen C. On muscular movements. Translation by C.V. Daremberg in *Oeuvres Anatomiques, Physiologiques et Médicales de Galien,* Vol. 2. Paris: J.B. Baillière, 1870, p. 334.

53. Gandevia S.C., G.M. Allen, J.E. Butler, and J.L. Taylor. Supraspinal factors in human muscle fatigue: evidence for suboptimal output from the motor cortex. *J. Physiol. (Lond.)* 490:529–536, 1996.

54. Gandevia S.C., G.M. Allen, and D.K. McKenzie. Central fatigue. Critical issues, quantification and practical implications. In: *Fatigue. Neural and Muscular Mechanisms,* edited by S.C. Gandevia, R.M. Enoka, A.J. McComas, D.G. Stuart, and C.K. Thomas. New York: Plenum Press, 1995, pp. 281–294.

55. Garland S.J., S.H. Garner, and A.J. McComas. Relationship between numbers and frequencies of stimuli in human fatigue. *J. Appl. Physiol.* 65:89–93, 1988.

56. Garnett R.A., M.J. O'Donovan, J.A. Stephens, and A. Taylor. Motor unit organization of human medial gastrocnemius. *J. Physiol. (Lond.)* 287:33–43, 1979.

57. Giddings C.J. and W.J. Gonyea. Morphological observations supporting muscle fiber hyperplasia following weight-lifting exercise in cats. *Anat. Rec.* 233:178–195, 1992.

58. Gillespie C.A., D.R. Simpson, and V.R. Edgerton. Motor unit recruitment as reflected by muscle fibre glycogen loss in a prosimian (bushbaby) after running and jumping. *J. Neurol. Neurosurg. Psychiatry* 37:817–824, 1974.

59. Gilliatt R.W. Dr. Derek Denny-Brown, OBE, MD, DPHIL, FRCP, 1901–1981. An appreciation. *Canad. J. Neurol. Sci.* 8:271–273, 1981.

60. Goldspink G., A. Scutt, J. Martindale, T. Jaenicke, L. Turay, and G.-F. Gerlach. Stretch and force generation induce rapid hypertrophy and myosin isoform gene switching in adult skeletal muscle. *Biochem. Soc. Trans.* 19:368–373, 1991.

61. Goldspink G. The proliferation of myofibrils during muscle fibre growth. *J. Cell Sci.* 6:593–603, 1970.

62. Gollnick P.D. and L. Hermansen. Biochemical adaption to exercise: anaerobic metabolism. *Exercise Sports Sci. Rev.* 1:1–43, 1973.

63. Gollnick P.D., J. Karlsson, K. Piel, and B. Saltin. Selective glycogen depletion in skeletal muscle of fibres of man following sustained contractions. *J. Physiol. (Lond.)* 241:59–67, 1974.

64. Gollnick P.D., K. Piehl, and B. Saltin. Selective glycogen depletion pattern in human muscle fibres after exercise of varying intensity and at varying pedalling rates. *J. Physiol. (Lond.)* 241:45–57, 1974.

65. Gollnick P.D. and B. Saltin. Significance of skeletal muscle oxidative enzyme enhancement with endurance training. *Clin. Physiol.* 2:1012, 1082.

66. Gordon A.M., A.F. Huxley, and F.J. Julian. The variation in isometric tension with sarcomere length in vertebrate muscle fibres. *J. Physiol. (Lond.)* 184:170–192, 1966.

67. Gorza L., K. Gundersen, T. Lømo, S. Schiaffino, and R.H. Westgaard. Slow-to-fast transformation of denervated soleus muscles by chronic high-frequency stimulation in the rat. *J. Physiol. (Lond.)* 402:627–649, 1988.

68. Granit R. *Charles Scott Sherrington. An Appraisal.* Doubleday & Company, Inc.: Garden City, New York, 1967.

69. Granit R., D. Kernell, and G.K. Shortess. The behaviour of mammalian motoneurones during long-lasting orthodromic, antidromic and transmembrane stimulation. *J. Physiol. (Lond.)* 169:743–754, 1963.

70. Grimby L., J. Hannerz, J. Borg, and B. Hedman. Firing properties of single human motor units on maintained maximal voluntary effort. In: *Human Muscle Fatigue: Physiological Mechanisms,* edited by R. Porter and J. Whelan. London: Pitman Medical, 1981, pp. 157–177.
71. Grimby L., J. Hannerz, and B. Hedman. The fatigue and voluntary discharge properties of single motor units in man. *J. Physiol. (Lond.)* 316:545–554, 1981.
72. Hainaut K. and J. Duchateau. Muscle fatigue, effects of training and disuse. *Muscle Nerve* 12:660–669, 1989.
73. Hanson J. and H.E. Huxley. Structural basis of the cross-striations in muscle. *Nature* 172:530–532, 1953.
74. Helmholtz H., 1848. Cited by D.M. Needham in *Machina Carnis. The Biochemistry of Muscular Contraction in Its Historical Development.* Cambridge: University Press, 1971, p. 40.
75. Henneman E. Relation between size of neurons and their susceptibility to discharge. *Science* 126:1345–1346, 1957.
76. Henneman E., G. Somjen, and D.O. Carpenter. Functional significance of cell size in spinal motoneurons. *J. Neurophysiol.* 28:560–580, 1965.
77. Hill A.V. *First and Last Experiments in Muscle Mechanics.* London: Cambridge University Press, 1970, p. 26 (facing page).
78. Hill A.V. Memories and reflections. Library of Churchill College Cambridge, 1974, pp. 1–25.
79. Hill A.V. The absolute mechanical efficiency of the contraction of an isolated muscle. *J. Physiol. (Lond.)* 46:435–469, 1913.
80. Hill A.V. The heat of shortening and the dynamic constants of muscle. *Proc. R. Soc. Ser. B* 126:136–195, 1938.
81. Hill A.V. The mechanics of active muscle. *Proc. R. Soc. Ser. B* 141:104–117, 1953.
82. Hill A.V. *Trails and Trials in Physiology.* London: Edward Arnold, 1965.
83. Hill A.V. and W. Hartree. The four phases of heat production of muscle. *J. Physiol. (Lond.)* 54:84–128, 1920.
84. Hill A.V. and H. Lupton. Muscular exercise, lactic acid, and the supply utilization of oxygen. *Quart. J. Med.* 16:1335–1371, 1923.
85. Hippocrates [Hippocratic Corpus]. *On the Nature of Bones; On the Sacred Disease; Epidemics; On Maladies.* Translation by E. Littré in *Oeuvres Complètes d'Hippocrates,* 10 vols. Paris: J.B. Baillière, 1839–1861.
86. Holloszy J.O. Biochemical adaptations in muscle. Effects of exercise in mitochondrial oxygen uptake and respiratory enzyme activity in skeletal muscle. *J. Biol. Chem.* 242:2278–2282, 1967.
87. Hultman E., J. Bergström, and N. McLennan Andersen. Breakdown and resynthesis of phosphocreatine and adenosine triphosphate in connection with muscular work in man. *Scand. J. Clin. Lab. Invest.* 19:56–66, 1967.
88. Huxley A.F. Looking back on muscle. In: *The Pursuit of Nature. Informal Essays on the History of Physiology,* edited by A.L. Hodgkin, A.F. Huxley, W. Felding, W.A.H. Rushton, R.A. Gregory, and R.A. McCance. Cambridge: Cambridge University Press, 1977, pp. 23–64.
89. Huxley A.F. Muscular contraction (Review lecture). *J. Physiol. (Lond.)* 243:1–43, 1974.
90. Huxley A.F. and R. Niedergerke. Structural changes in muscle during contraction. Interference microscopy of living muscle fibres. *Nature* 173:971–973, 1954.
91. Huxley A.F. and R.E. Taylor. Local activation of striated muscle fibres. *J. Physiol. (Lond.)* 144:426–441, 1958.
92. Huxley H.E. Electron microscope studies of the vagination of the filaments in striated muscle. *Biochim. Biophys. Acta* 12:387–394, 1953.
93. Huxley H.E. X-ray analysis and the problem of muscle. *Proc. R. Soc. Ser. B* 141:59–62, 1953.

94. Huxley H.E. and J. Hanson. Changes in the cross-striations of muscle during contraction and stretch and their structural interpretation. *Nature* 173:973–976, 1954.

95. Ingjer F. Effects of endurance training on muscle fibre ATP-ase activity, capillary supply and mitochondrial content in man. *J. Physiol. (Lond.)* 294:419–432, 1979.

96. Jansson E. and L. Kaijser. Muscle adaptation to extreme endurance training in man. *Acta Physiol. Scand.* 100:315–324, 1977.

97. Jones D.A. How far can experiments in the laboratory explain the fatigue of athletes in the field? In: *Neuromuscular Fatigue,* edited by A.J. Sargeant and D. Kernell. Amsterdam: North Holland, 1992, pp. 100–108.

98. Jones D.A., B. Bigland-Ritchie, and R.H.T. Edwards. Excitation frequency and muscle fatigue: mechanical responses during voluntary and stimulated contractions. *Exp. Neurol.* 64:401–413, 1979.

99. Jones N.L., N. McCartney, and A.J. McComas (eds). *Human Muscle Power.* Champaign, IL: Human Kinetics Publishers, 1986.

100. Katz B. and S. Thesleff. On the factors which determine the amplitude of the "miniature end-plate potential." *J. Physiol. (Lond.)* 137:267–278, 1957.

101. Kernell D. and B. Zwaagstra. Input conductance, axonal conduction velocity and cell size among hindlimb motoneurones of the cat. *Brain Res.* 204:311–326, 1980.

102. Kronecker H. *Ueber die Ermüdung und Erholung der Quergestreiften Muskeln.* Berichte der Verhandlungen d. h. Sächsischen Gesell. der Wiss. zu Leipzig, 1871, p. 718.

103. Kukulka C.G. and H.P. Clamann. Comparison of the recruitment and discharge properties of motor units in human brachial biceps and adductor pollicis during isometric contractions. *Brain Res.* 219:45–55, 1981.

104. Langley J.N. Sir Michael Foster. In memoriam. *J. Physiol. (Lond.)* 35:233–246, 1906.

105. Lee J.A., H. Westerblad, and D.G. Allen. Changes in tetanic and resting $[Ca^{2+}]_i$ during fatigue and recovery of single muscle fibres from *Xenopus laevis. J. Physiol. (Lond.)* 433:307–326, 1991.

106. Leksell L. The action potential and excitatory effects of the small ventral root fibres to skeletal muscle. *Acta Physiol. Scand.* 10 (Suppl. 31):1–84, 1945.

107. Lømo T. and J. Rosenthal. Control of Ach sensitivity by muscle activity in the rat. *J. Physiol. (Lond.)* 221:493–513, 1972.

108. Lundsgaard E. Untersuchungen über muskelkontraktion ohne milchsäure. *Biochem. Z.* 217:162–177, 1930.

109. MacDougall J.D., D.G. Sale, S.E. Alway, and J.R. Sutton. Muscle fiber number in biceps brachii in body builders and control subjects. *J. Appl. Physiol.* 57:1399–1403, 1984.

110. Macefield V., S.C. Gandevia, B. Bigland-Ritchie, R.B. Gorman, and D. Burke. The firing rates of human motoneurones voluntarily activated in the absence of muscle afferent feedback. *J. Physiol. (Lond.)* 471:429–443, 1993.

111. Margaria R., H.T. Edwards, and D.B. Dill. The possible mechanisms of contracting and paying the oxygen debt and the role of lactic acid in muscular contraction. *Am. J. Physiol.* 106:689–715, 1933.

112. Marsden C.D., J.C. Meadows, and P.A. Merton. Isolated single motor units in human muscle and their rate of discharge during maximal voluntary effort. *J. Physiol. (Lond.)* 217:12–13P, 1971.

113. Marsden C.D., J.C. Meadows, and P.A. Merton. Muscle wisdom. *J. Physiol. (Lond.)* 200:15P, 1969.

114. Marsden C.D., J.C. Meadows, and P.A. Merton. "Muscle wisdom" that minimizes fatigue during prolonged effort in man: peak rates of motoneuron discharge and slowing of discharge during fatigue. *Adv. Neurol.* 39:169–211, 1983.

115. Maruyama K. Birth of the sliding filament concept in muscle contraction. *J. Biochem.* 117:1–6, 1995.

116. McArdle B. Myopathy due to a defect in muscle glycogen breakdown. *Clin. Sci.* 10:13–35, 1951.

117. McComas A.J., V. Galea, and R.W. Einhorn. Pseudofacilitation: a misleading term. *Muscle Nerve* 17:599–607, 1994.

118. McComas A.J., P.R.W. Fawcett, M.J. Campbell, and R.E.P. Sica. Electrophysiological estimation of the number of motor units within a human muscle. *J. Neurol. Neurosurg. Psychiatry* 34:121–131, 1971.

119. McPhedran A.M., R.B. Wuerker, and E. Henneman. Properties of motor units in a homogeneous red muscle (soleus) of the cat. *J. Neurophysiol.* 28:71–84, 1965.

120. Mense S. Nervous outflow from skeletal muscle following chemical noxious stimulation. *J. Physiol. (Lond.)* 267:75–88, 1977.

121. Merton P.A. Problems of muscular fatigue. *Brit. Med. Bull.* 12:219–239, 1956.

122. Merton P.A. Voluntary strength and fatigue. *J. Physiol. (Lond.)* 123:553–564, 1954.

123. Meyerhof O. Die Energiumwandlungen im muskel. I. Uber die beziehungen der milchsäure zur wärmebildung und arbeitsleistung des muskels in der anaerobiose. *Pflugers Archiv ges. Physiol.* 182:232–283, 1920.

124. Milner-Brown H.S., R.B. Stein, and R. Yemm. Changes in firing rate of human motor units during linearly changing voluntary contractions. *J. Physiol. (Lond.)* 230:371–390, 1973.

125. Milner-Brown H.S., R.B. Stein, and R. Yemm. The orderly recruitment of human motor units during voluntary isometric contractions. *J. Physiol. (Lond.)* 230:359–370, 1973.

126. Mosso A. *Fatigue.* Translation by M. Drummond and W.B. Drummond. London: George Allen & Unwin Ltd; New York: G.P. Putnam's Sons, 1915.

127. Needham D.M. *Machina Carnis. The Biochemistry of Muscular Contraction in Its Historical Development.* Cambridge: University Press, 1971.

128. Pearn J. Two early dynamometers. An historical account of the earliest measurements to study human muscular strength. *J. Neurol. Sci.* 37:127–134, 1978.

129. Peter J.B., R.J. Barnard, V.R. Edgerton, C.A. Gillespie, and K.E. Stempel. Metabolic profiles of three fiber types of skeletal muscle in guinea pigs and rabbits. *Biochemistry* 11:2627–2633, 1972.

130. Pette D., B.U. Ramirez, W. Müller, R. Simon, G.U. Exner, and R. Hildebrand. Influence of intermittent long-term stimulation on contractile, histochemical and metabolic properties of fibre populations in fast and slow rabbit muscles. *Pflügers Archiv.* 361:1–7, 1975.

131. Ranvier L. *Leçons d'Anatomie Générale sur le Système Musculaire.* Paris, 1880.

132. Rome L.C. The design of the muscular system. In: *Neuromuscular Fatigue,* edited by A.J. Sargeant and D. Kernell. Amsterdam: North Holland, 1992, pp. 129–136.

133. Rutherford O.M. and D.A. Jones. The role of learning and coordination in strength training. *Eur. J. Appl. Physiol.* 55:100–105, 1986.

134. Salmons S. and G. Vrbová. The influence of activity on some contractile characteristics of mammalian fast and slow muscles. *J. Physiol. (Lond.)* 201:535–549, 1969.

135. Saltin B. and J. Karlsson. Muscle ATP, CP, and lactate during exercise after physical conditioning. In: Muscle Metabolism During Exercise, edited by B. Pernow and B. Saltin. New York: Plenum, 1971, pp. 395–399.

136. Sargeant A.J., E. Hoinville, and A. Young. Maximum leg force and power output during short-term dynamic exercise. *J. Appl. Physiol.* 51:1175–1182, 1981.

137. Schantz P.G. and G. K. Dhoot. Coexistence of slow and fast isoforms of contractile and regulatory proteins in human skeletal muscle fibres induced by endurance training. *Acta Physiol. Scand.* 131:147–154, 1987.

138. Sherrington C.S. Some functional problems attaching to convergence. *Proc. R. Soc. Lond. Ser. B* 105:332–362, 1929.

139. Sherrington C.S. *The Integrative Action of the Nervous System.* New Haven: Yale University Press, 1906 (reprinted 1947).

140. Sica R.E.P. and A.J. McComas. Fast and slow twitch units in a human muscle. *J. Neurol. Neurosurg. Psychiatry* 34:113–120, 1970.

141. Simoneau J.A., G. Lortie, M.R. Bonlay, C.M. Marcotte, M.C. Thibault, and C. Bouchard. Human skeletal muscle fibre type alteration with high-intensity intermittent training. *Eur. J. Appl. Physiol.* 54:250–253, 1985.

142. Sola O.M., D.L. Christensen, and A.W. Martin. Hypertrophy and hyperplasia in adult chicken anterior latissimus dorsi muscle following stretch with and without denervation. *Exp. Neurol.* 41:76–100, 1973.

143. Staron R.S, M.J Leonardi, D.L Karapondt, E.S Malicky, F.C Falkel, F.C Hagerman, and R.S. Hikida. Strength and skeletal muscle adaptations in heavy resistance-trained women after detraining and retraining. *J. Appl. Physiol.* 70:631–640, 1991.

144. Stensen N. *De Musculis et Glandulis Observationum Specimen,* 1664. Cited by Sir M. Foster. *Lectures on the History of Physiology During the Sixteenth, Seventeenth and Eighteenth Centuries.* Cambridge: University Press, 1901, p. 71.

145. Stephens, J.A., R. Garnett, and N.P. Buller. Reversal of recruitment order of single motor units produced by cutaneous stimulation during voluntary muscle contraction in man. *Nature* 272:362–364, 1978.

146. Stephens J.A. and A. Taylor. Fatigue of maintained voluntary contraction in man. *J. Physiol. (Lond.)* 220:1–18, 1972.

147. Vesalius A. *Fabrica Humani Corporis.* Basel, 1543. Cited by Sir M. Foster. *Lectures on the History of Physiology During the Sixteenth, Seventeenth and Eighteenth Centuries.* Cambridge: University Press, 1901, pp. 69–70.

148. Vilensky J.A. and S. Gilman. Renaming the Henneman size principle. *Science* 280:203, 1998.

149. Vøllestad N.K., O.M. Sejersted, R. Bahr, J.J. Woods, and B. Bigland-Ritchie. Motor drive and metabolic responses during repeated submaximal contractions in humans. *J. Appl. Physiol.* 64:1421–1427, 1988.

150. Vrbová G. Factors determining the speed of contraction of striated muscle. *J. Physiol. (Lond.)* 185:17–18P, 1966.

151. Vrbová G. The effect of motoneurone activity on the speed of contraction of striated muscle. *J. Physiol. (Lond.)* 169:513–526, 1963.

152. Wachstein M. and E. Meisel. The distribution of demonstrable succinic dehydrogenase and mitochondria in tongue and skeletal muscle. *J. Biophys. Biochem. Cytol.* 1:483–488, 1955.

153. Waller A. The sense of effort: an objective study. *Brain* 14:179–249, 1891.

154. Wilkie D.R. Muscle function: a historical view. In: *Human Muscle Power,* edited by N.L. Jones, N. McCartney, and A.J. McComas. Champaign, IL: Human Kinetics Publishers, 1986, pp. 3–13.

155. Wilkie D.R. Muscle function: a personal view. *J. Exp. Biol.* 115:1–13, 1985.

156. Wilkie D.R. The relation between force and velocity in human muscle. *J. Physiol. (Lond.)* 110:249–280, 1950.

157. Wuerker R.B., A.M. McPhedran, and E. Henneman. Properties of motor units in a heterogeneous pale muscle (m. gastrocnemius) of the cat. *J. Neurophysiol.* 28:85–99, 1965.

158. Zotterman Y. Obituary. Lord Adrian. *EEG. Clin. Neurophysiol.* 44:137–139, 1978.

chapter 3

THE CARDIOVASCULAR SYSTEM

Loring B. Rowell

It may be said ... that the majority of the adaptations of which the circulation is capable are designed to increase the blood supply to the muscle in exercise.
—R.T. S. McDowell, 1956 (95)

THE history of studies of cardiovascular function during dynamic exercise encompasses virtually every area of circulatory physiology. Investigations extended naturally from rest to exercise because insights into mechanisms of control often require the forcing of functions over a broad range to assure that insights apply generally and not to limited conditions. This thinking was most clearly expressed by Sir Joseph Barcroft (1872–1947) in 1934 in his monograph entitled *Features in the Architecture of Physiological Function* (7). Exercise has provided the ultimate tool for synthesis and integration in cardiovascular and respiratory physiology and in other areas as well. Of course, understanding circulatory control during exercise has intrinsic value as well.

During the last half of the nineteenth century, physiologists developed strong foundations for understanding the mechanisms and control of cardiac pumping, characteristics of cardiac contraction, and the control of blood vessels. In their chapter in *Circulation of Blood: Men and Ideas* (41), William F. Hamilton (1893–1964) and D.W. Richards wondered why, despite this progress, no quantitative relationship among the above areas—that is, no integration or synthesis, had been established (60). They suggested that this was due to the complexity of the system, noting that effects of nerve stimulation were so variable that they offered little hope of insight into integration. One could argue that meaningful integration requires an understanding of the quantitative relationships between input and output variables that are sensed and processed by the major controllers of the system. Inasmuch as exercise makes the study of integration and synthesis more possible, it has become one of the important tools in cardiovascular physiology. The history will speak for itself.

The history of cardiovascular function in exercise cannot be covered in an entire book of this size. Its treatment in one chapter forces the coverage to be highly selective and the choices reflect my own interests. Even so, the number of important references available far exceeds the number that could be cited. Accordingly, I have had to cite reviews and chapters that include references to many of the important works that are discussed. There are also too many heroes to show in photographs and far too many classic illustrations to include, so there are none of either. I apologize to all contributors whose important ideas were not included and whose great value to the field is not covered here, but who are, thankfully, well recognized elsewhere.

This chapter begins with a brief background sketch of some important people and ideas that prevailed before the twentieth century. Their ideas provided the scientific background for what follows. The chapter's emphasis is on ideas of people who worked, or still work, in the twentieth century. During this century, a rapid evolution of new methods permitted important experiments to be done on conscious behaving animals as well as on human subjects.

The objective of this chapter is to present those ideas that initiated a long battle over what determines cardiac performance during exercise. Starling's "law of the heart" was at the center of the dispute, which grew in crescendo from the 1950s through the 1970s. The central theme of this chapter is how a number of investigators revealed that the heart is mechanically coupled to the peripheral vasculature and the function of one dictates that of the other. The history unveils a circulation that must be altered between rest and exercise by both neurally controlled cardiac and vasomotor events and by mechanical ones if cardiac performance is to be increased. Some strong opposition to these ideas by other investigators is presented as well. There were some stirring disputes, and those of us who witnessed them will never forget.

BACKGROUND

The chain of discoveries that elucidate function and control of the cardiovascular system is long and complex. Sudden flashes of genius like those of Sir William Harvey (1578–1657) and Stephen Hales (1677–1761) were rare. Rather, the systematic application of strong intelligence, keen powers of observation, and ability to build on the findings of others gave this science its growth and its continuity.

Some Early Highlights

Sir William Harvey is credited with discovering the circulation of blood and also with a new method of scientific approach to physiological problems that included: (1) careful analysis of observed phenomena, (2) testing of specific hypotheses by experimentation, and (3) "the startling innovation of quantitative reasoning to prove (or disprove) a proposed theory" (63). Harvey did not believe the ventricles could suck blood into themselves during diastole and proposed the following alternative: "It [blood] tends to move from the tiny veins to the intermediate branches and then to the larger branches because of the movements of the extremities and the compression of muscles." Harvey realized that the heart could not put out what it had not received, and that during activity, some other force external to the heart was needed to fill the right ventricle so as to maintain cardiac output.

In the 370 years since *Du Mortu Cordis* was published by Harvey in 1628, this last concept regarding venous return has continued to remain a stumbling block and a point that is recurrent in this chapter.

The remarkable Reverend Stephen Hales, vicar of Teddington, made fundamental contributions. These included inventing a manometer to measure arterial and venous pressures, and the first quantitative estimates of the "circulation rate" in the horse—based on his measurement of heart rate followed by estimation of the dead animal's ventricular volume (from a wax mold). The product yielded his estimate of cardiac output. He made similar calculations based on Harvey's estimates of human ventricular ejection volume: he calculated cardiac output to be approximately 4 liters per minute. Hales noted that systole normally occupied one-third of the cardiac cycle and postulated that diastolic runoff was determined by the elasticity of the large arteries. He introduced the concept of resistance owing to "different Degrees of Viscidity or Fluidity of the Blood, or by several Degrees of Constriction or Relaxation of those fine Vessels" (86). He estimated vascular diameters, total cross-sectional areas, and resistance across the vascular tree. He even estimated the transit time of red blood cells through the pulmonary capillaries. His ingenious animal experiments were published in 1733 in a volume entitled *Haemastatiks* (57). The great contributions of Stephen Hales were broad concepts of blood pressure, blood flow, blood velocity, and resistance and his quantitative measurements and calculations of each. The development of modern cardiovascular physiology is based on these contributions (60, 86). He deserves the title "The father of Hemodynamics" (94).

The mathematician Daniel Bernoulli (1700–1782), a contemporary of Hales, had developed the laws of fluid hydraulics and his work provided the foundation for the contributions of Jean Léonard Marie Poiseuille (1799–1869) (30, 60). In 1842 Poiseuille described the fundamental relationship governing the flow of Newtonian fluids in very narrow tubes. He also discovered the nonlinear relationship between pressure and flow when flow becomes turbulent. The law of Pierre LaPlace (1749–1827) relating vascular wall tension to vascular transmural pressure and diameter (a relationship previously observed by Bernoulli) appeared in 1882. Auguste Chauveau (1827–1917) and Etienne Jules Marey (1830–1904) used air-filled manometers and catheters, including the first double-lumen catheter, to record pressures in the right and left ventricles and pulmonary artery of the horse. They noted the pulmonary arterial pressure was one-fifth of the systemic arterial pressure (25).

In 1896 Ernest Starling (1860–1927) developed a hypothesis to explain transcapillary fluid exchange, a key to which was his evidence that serum proteins and not crystalloids provided the osmotic or reabsorptive force (131). In 1733 Stephen Hales had imposed known pressures and measured flows and observed that increased "insinuation" (i.e., filtration) of water from the capillary vessels caused a counterpressure that reduced flow. From his direct measure of mean arterial and venous pressures Hales calculated capillary pressure. Not until 1926 were capillary pressures measured directly by Eugene Landis (1901–1987) (60, 86).

The Concept of Control

The concept that the heart and blood vessels are actively controlled began to develop in the midnineteenth century, at which time physiology was also emerging as a distinct science.

Extrinsic, Neural Control of the Heart

The cardioinhibitory influence of the vagus nerve was discovered by the brothers Ernst H. Weber (1795–1878) and Eduard Fredrick Weber (1806–1871) in 1845 (148). In 1861, Jules Marey reported that heart rate and blood pressure are related in an inverse manner; this was later called Marey's law. Parenthetically this "law" caused considerable confusion much later when it was shown not to apply to exercise. Albert von Bezold (1838–1868) discovered the cardiac accelerator nerves in 1861 (10). The discovery of the depressor nerve by Elias von (Elie de) Cyon (1849–1912) and Karl Frederick Wilhelm Ludwig (1816–1895) in 1866 led to important advances discussed below (32).

Intrinsic Control of the Heart

The development of a stable heart–lung preparation isolated in dogs by Henry Newell Martin (1848–1896) in 1876 led his students William Henry Howell (1860–1945) and Frank Donaldson in 1884 at The Johns Hopkins University to publish a study entitled *Experiments upon the Heart of the Dog with Reference to the Maximum Amount of Blood Sent Out by the Left Ventricle in a Single Beat and the Influence of Variations in Venous Pressure, Arterial Pressure and Pulse Rate upon the Work Done by the Heart* (76). They describe a law for the work of the left ventricle that enables it to overcome an increase in outflow resistance, increased venous filling pressure, increased left ventricular stroke output, and left ventricular work. From their tables, the ascending portion of what later became known as a "Starling curve" could be constructed.

The next big steps in describing ventricular function were those of Otto Frank (1865–1944), one of Karl Ludwig's last students. Otto Frank's goal was to make rigorous quantitative comparisons of mechanical responses of cardiac muscle with those of skeletal muscle as previously established by Adolph Fick (1829–1901), Johannes von Kriese (1853–1928), and Magnus G. Blix (1849–1904), who described the relationship between skeletal muscle fiber length and force of contraction (45). Frank used frog hearts perfused with diluted ox or sheep blood and altered diastolic filling of the ventricles. He measured resultant changes in intraventricular and atrial pressures and observed stroke volumes. He established a quantitative relationship between initial myocardial wall tension and the isometric pressure developed. He then examined isotonic ventricular contractions in which he determined stroke volume, speed of ventricular pressure development, and end systolic volume. The features that governed strength of contraction were shown to be the same for skeletal and cardiac muscle (45).

Fifteen years later Ernest H. Starling modified and refined Martin's isolated canine heart–lung preparation and performed experiments similar to those made by Howell and Donaldson. An important advance was Starlings' development of the "Starling resistor," which provided a means of controlling peripheral resistance, a variable left uncontrolled and one that clearly limited Howell's and Donaldson's experiments (76). In the latter studies arterial pressure rose in proportion to left ventricular output, thereby making arterial pressure (afterload) a determinant of left ventricular work along with venous filling pressure (preload). Starling expressed the concept of the "law of the heart" in this way: "increased diastolic distention exercises a strong augmenting effect on the left ventricular contraction" (99). The combined insights of Frank and Starling led to naming the cardiac "length–tension relationship" the *Frank–Starling relationship*.

Starling saw the immediate adjustment to exercise in an upright human as follows:

> Thus as a man starts to run, his muscular movements pump more blood into the heart so increasing the venous filling [i.e., and thus diastolic volume and stroke volume] while the central nervous system, by contracting the arteries, raises blood pressure and forces all the available blood through the muscles. (132)

In the decades following Starling's experiments the general applicability of the "law of the heart" to intact animals, including humans, came to be seriously questioned by physiologists in Europe and in the Americas, particularly during exercise. Even today opinion continues to swing between acceptance and rejection. Many seemed unaware that Starling also appreciated the potential importance of neurohumoral actions for cardiac performance in exercise (133).

Neural Control of Blood Vessels

The notion of vascular tone is implied in Stephen Hales writings as he speaks of "Different Degrees of Constriction or Relaxation" in blood vessels, but this observation might have been due to the physical alterations in vascular resistance during his perfusion experiments. The idea of "tone" was first applied to the veins by Richard Lower (1631–1691) in his *Tractatus de Corde* in 1669 (91). This idea was not followed up until 1864 when Frederick Goltz (1834–1902) reported that "venous tone" (he assumed it was venous) depends on the presence of the spinal cord and the medulla oblongata (49).

The discovery of vasomotor nerves is often credited to Francois Pourfois du Petit (1664–1741), who in 1727 reported that section of cervical sympathetic nerves causes conjunctival vessels to dilate (71, 103). However, Jean-Baptiste de Senác (1705–1770) provided the first clear statement that the diameter of arteries is controlled (124). In 1777 he stated: "The arteries which are so active are true hearts in another guise; … [their] movements are alternate dilations and contractions" (translation from Heymans and Folkow [69]).

Although scientists began to realize two centuries ago that blood vessels were controlled by nerves, they did not grasp the functional significance of this control until the mid-nineteenth century when, within a period of months, between 1852 and 1853, Charles E. Brown-Sequard (1818–1894), Claude Bernard (1813–1878), and Augustus V. Waller (1816–1870) all reported the *tonic* vasoconstrictor influence of sympathetic nerves on blood vessels (69). For nearly 10 years, Bernard argued that the sudden rise in surface temperature attending section of the cervical sympathetic fibers joining the sympathetic ganglia was an effect of sympathetic nerves on tissue heat production: vasodilation was seen as a secondary effect (103). Conversely, Brown-Sequard stood by his view that local temperature increased as a consequence of vasodilation and increased blood flow owing to the loss of vasomotor tone. Subsequently, Claude Bernard agreed. Waller added to this story the observation that stimulation of sympathetic nerves caused vasoconstriction (69). The door was open to study vasomotor tone and the regulation of blood pressure, body temperature, and nutrient blood flow by reflexes and local factors. Here, first the local factors.

Local Control of Blood Vessels

Chemical factors, particularly metabolites, were explored as causes of vasodilation late in the nineteenth century. Julius Friedrich Cohnheim (1839–1884) demonstrated reactive hyperemia after stoppage of blood flow and ascribed it as a reaction to tissue metabolism (69). J. Latschenberger and A. Deahna in 1876 (87) and Walter H. Gaskell (1847–1914) (48) in 1878 first proposed that vascular tone is inhibited by metabolic by-products in active muscle. These initial observations were followed in 1887 by those of Auguste Chauveau and M. Kaufmann, who observed marked hyperemia in the lip muscles of a horse during chewing. These were the first clear demonstrations of exercise hyperemia (26).

By the end of the nineteenth century, the metabolic hypothesis of vasodilation in muscle was accepted and the preoccupation with "neurovasodilation" eventually faded (69). Although questions still remain about cholinergic vasodilation of skeletal muscle in some species, humans appear not to have such a mechanism. (In some cases, the vasodilation was probably caused by epinephrine) (109). As we near the end of the twentieth century, there is still no known combination of vasoactive metabolites or of vasodilator drugs that can elicit the degree of hyperemia seen when muscle is heavily exercised (89, 126). The influence of purely mechanical effects associated with muscle contraction on muscle blood flow was not considered in this context (see below).

The concept of myogenic vasomotion was introduced in 1902 by William Maddock Bayliss (1860–1924) (9). This intrinsic reaction of vascular smooth muscle to tension came to be seen as the basic element in vascular tone called "basal tone." Extrinsic neural and chemical influences are superimposed on basal tone, to which they are added or subtracted to cause vasoconstriction and vasodilation (21).

Control of the Cardiovascular System by Reflexes

The idea that reflexes controlled the cardiovascular system apparently sprang from the discovery of the depressor nerve by von Cyon and Karl Ludwig in 1866 (32). Stimulation of the central end of the cut nerve elicited bradycardia and arterial hypotension. The reflex hypotension persisted when bradycardia was prevented by atropine. They thought that the reflex originated in the heart, a view subsequently supported by A. von Bezold's and L. Hirt's observation in 1867 that intracardiac injection of veratrine elicited bradycardia and hypotension (11). The idea, of course, proved to be incorrect.

The arterial baroreflex. What follows is a brief history of the steps that led us to understand the sine qua non of cardiovascular reflexes—the arterial baroreflex. This reflex was once thought to be inactive during exercise but is now considered by many to be essential if not the primary reflex engaged in cardiovascular control during dynamic exercise (for references see 98, 114). This important history is greatly complicated by the existence of and separate discovery of the aortic and carotid sinus baroreceptors and by the persistent idea that cerebral ischemia, or other factors not related to baroreceptors, was at the origin of any responses to experimental manipulation of the carotid sinus and of arterial pressure. The following are some key steps.

In 1836 Sir Ashley Cooper (1768–1841) observed that occlusion of the common carotid arteries precipitated arterial hypertension (29). In 1885 Henry Sewall (1855–1936) and D.W. Steiner (125) found that bilateral section of the aortic depressor nerves caused arterial pressure to rise. They, along with Cooper, ascribed the increased blood pressure to ischemia of the vasomotor center. In 1902 G. Köster and J.N. Tschermak showed that the aortic depressor nerve arose from the aortic arch; aortic distension generated action potentials in this nerve as did the pulsations associated with each heart beat as shown by Willem Einthoven (1860–1920) in 1908 (81). When J.A.E. Eyster and D.R. Hooker (1908) distended the aortic arch they observed reflex bradycardia (39). In 1924 and 1925, Cornelius Heymans (1892–1968) and A. Ladon used a donor dog to perfuse the head of a recipient dog and showed that the responses to increases or decreases in blood pressure of the recipient dog's body (bradycardia or tachycardia, respectively) were abolished by section of the vagus nerves (66). Thus these responses were via a reflex and not due to cerebral ischemia.

The role of the carotid sinus was not appreciated until 1923 when Heinrich Ewald Hering (1866–1936) reported his classic observations of a "carotid sinus reflex." In 1924 Hering proved that stimulation of the carotid sinus wall not only caused the bradycardia but also caused the arterial hypotension. Hering recognized the tonic influence of these nerves on heart rate and blood pressure; in 1927 he observed arterial hypertension when both sinus nerves were cut and he called these nerves *blutdruckzügler* (65). Hering's work was confirmed and extended by many, and very importantly, by Eberhardt Koch (1931) (79) and Cornelius Heymans (68).

The importance of the peripheral circulation in blood pressure control, and particularly the role of the splanchnic region, was appreciated even before the arterial baroreflex was discovered (79). The vascular effector arm of the baroreflex was shown by several groups to be composed entirely of sympathetic adrenergic fibers. Neither sympathetic nor parasympathetic vasodilator fibers (cholinergic) are activated (10, 22). Thus it became clear that blood pressure is controlled by changes in heart rate and cardiac output and by sympathetically mediated adjustments in peripheral vascular resistance.

The expression of baroreceptor properties and particularly lumped properties of the entire reflex in quantitative terms (gains, operating points, thresholds, etc.) was brilliantly pioneered by Eberhart Koch in 1931 (79). This important quantitative analysis of a control system was furthered by the work of D. W. Bronk and D. Stella in 1932 (18) and H.W. Ead, J.H. Green, and Eric Neil in 1952 (37). These investigators recorded impulse traffic in single- and multifiber afferent nerve preparations from the carotid sinus baroreceptor and related this traffic to simultaneously recorded arterial pressure. They showed that these baroreceptors respond to changes in mean pressure, pulse pressure, and rate of change of pressure.

Analysis of the control system was furthered by the work of Landgren 1952 (85) and became even more analytical in the classic studies of the carotid sinus reflex as a servocontrol system by Allen Scher and Allen Young in 1963 (122). Resetting of the arterial *baroreceptors* (not the reflex) by local application of drugs was first demonstrated by Heymans (1950) (67). Mechanical properties of the receptors changed so as to shift the operating range of the receptors toward higher or lower pressures. McCubbin, Green, and Page in 1956 (93) indicated that arterial baroreceptors are "reset" during chronic hypertension.

Kiichi Sagawa (1926–1989) captured the best of these contemporary quantitative analyses in his landmark review of 1983 (119). Sagawa and also Paul Korner in 1979 (80) emphasized the important distinction between adaptation (often called "resetting") of a rapidly adapting mechanoreceptor and the *resetting of the entire baroreflex arc by the central nervous system.* Today these ideas about resetting of this reflex play a major part in our understanding of how the cardiovascular system is controlled during exercise (98, 102). Resetting of the baroreflex is now viewed as a prominent feature of cardiovascular control in exercise (for references see 98, 102, 114).

CONTROL OF CARDIAC PERFORMANCE DURING EXERCISE

Development of Methods to Measure Cardiac Output

The first step in understanding cardiac performance was to be able to measure cardiac output, "but the measurement of flow is difficult while that of pressure is easy so that our knowledge of flow is usually derivatory"—attributed by Kenneth J. Franklin to Karl Ludwig (94).

Sir William Harvey realized that blood either had to circulate or that the human body would have to produce from 20 to 160 pounds of blood each hour. Similarly,

Adolph Fick realized that to consume a given amount of oxygen, any specific difference in the concentration of oxygen leaving and entering the lungs required a specific blood flow (40). Both men clearly understood that mass is conserved.

Much of the experimental work in cardiovascular physiology, especially that involving humans, relied on Ficks' statement of the conservation of mass in terms of the dilution principle. Fick's idea, thereafter known as the Fick principle, was "an evanescent idea that he barely took time to put on paper and to which he never returned" (58), yet the concept underlies most accepted methods for measuring cardiac output and blood flow to organs and tissues. These methods are referred to as either the direct Fick method, which requires direct sampling of mixed venous blood from the pulmonary artery and of arterial blood from a systemic artery, or indirect Fick methods in which cardiac output and organ blood flow are derived from the dilution principle. The latter methods include foreign gas rebreathing, indicator dilution, and isotope clearance techniques. In 1898, Nathan Zuntz (1848–1920) and O. Hagemann meticulously applied the direct Fick method to the exercising horse and saw changes in cardiac output consistent with those we would expect today (153). Y. Henderson felt that the method should be named after Zuntz and Hagemann rather than Fick (58). The direct Fick method was not used on exercising humans for another 50 years until its use by Andre Cournand and his colleagues after introduction of catheterization techniques on himself by Werner Forssman in 1929 and the heroic self-measurements of cardiac output by O. Klein in 1930 (146).

To avoid the use of pulmonary arterial catheterization to acquire mixed venous blood samples, A. Loewy and H. von Schrotter in 1903 and 1905 brilliantly conceived of using the whole lung as a tonometer in humans to obtain mixed venous CO_2 values (thus the term "indirect Fick"). They and J. Plesch in 1909 (Plesch is also credited with this idea) had subjects rebreathe gas mixtures, allowing rapid equilibration between blood entering the lung and a continuous lung-bag system. In 1912 Zuntz's group in Germany and August Krogh (1874–1949) and Johannes Lindhard (1870–1947) in Copenhagen introduced the nitrous oxide rebreathing technique, substantially improving the original rebreathing technique of Bornstein (1910), who used less soluble nitrogen. Apparently the first measurements of cardiac output in humans during exercise were made in 1915 by Lindhard, who found a linear relationship between oxygen uptake and cardiac output. In 1928 Arthur Grollman introduced the acetylene rebreathing method; acetylene was much easier to analyze than nitrous oxide. It became "the gold standard" among rebreathing techniques until later advances were made in CO_2 rebreathing techniques originally used by Douglas and Haldane in England and by Dill and coworkers at the Harvard Fatigue Laboratory 1928–1930 (for references see 4, 59, 146). Despite being the gold standard, the original values for cardiac output based on acetylene rebreathing were low (146).

In addition to the great technical and analytical skill required for these rebreathing techniques, another problem was that the recirculation of blood previously exposed to the lung-rebreathing bag gas mixture occurred much faster than the

10–18 seconds seen at rest. This would cause cardiac output to be underestimated. In 1931, Eric Hohü-Christensen (1904–1996) in Copenhagen proposed that the first blood to return to the circulation would contain little acetylene because most of it would have diffused into tissue fluids initially containing no acetylene (27). Ancel Keys confirmed this in 1941 but later Carlton Chapman, Henry Taylor, Keys and colleagues in Minneapolis showed that early recirculation of acetylene could cause some underestimation of cardiac output (24). Nevertheless, subsequent comparison of this technique with the dye dilution technique (see below) during mild to heavy exercise revealed no systematic differences (4, 5, 120).

The basic notion of the indicator dilution technique was introduced in 1897 by George Neil Stewart, who injected concentrated saline solution in the right atrium of dogs and recorded the change in conductivity (via a telephone circuit) when the solution reached the femoral artery. William F. Hamilton and his colleagues J.M. Kinsman and J.W. Moore at Louisville are credited with development in 1932 of this method, using dye rather than saline, for application to humans (59, 60).

Confidence in cardiac output methods was increased by favorable comparisons with methods based on different principles. Hamilton's dye dilution technique provided values in good agreement with those obtained by the direct Fick method in six resting dogs (58). Twenty years later in 1948, the Hamilton and Cournand groups met in New York and made the first simultaneous comparisons of the dye dilution and direct Fick methods in man. They found no systematic differences (146).

Despite persistent criticism (59, 60, 116, 118) improvements in rebreathing techniques in the second half of this century normally led to good comparisons with measurements by dye dilution and direct Fick methods (4, 120, 146). The point is that arguments about cardiac performance rarely hinged on the variability, unless extreme, of methods for measurement of cardiac output. An exception may be some of the more recent applications of radionuclide angiographic measurements of left ventricular dimensions during exercise. The potentially large errors in absolute volume measurements (25% for end-diastolic volume and 50% for end-systolic volume) can lead to large errors in the calculation of stroke volume and especially cardiac output in exercise (52, 114). Until the mid 1990s, these errors spawned serious disagreement concerning the responses to exercise in the aged for whom the unusually high maximal values for cardiac output and stroke volume disagreed with previous findings based on other accepted methods (for references see 77, 114).

It is important that the early comparisons of these cardiac output techniques were fraught with problems caused by different body postures. The direct Fick procedure was done in supine subjects because of the catheterization, whereas the rebreathing techniques were usually carried out in subjects who were seated. In the sitting position the cardiac output is lower and heart rate is higher than in supine posture. *Failure to recognize and include the large effects of posture on central hemodynamics in humans was, from the start, a stumbling block* and it has continued to cloud interpretation of results well after the midtwentieth century especially in comparing and contrasting responses in bipeds and quadrupeds.

Flow Meters

The development of electromagnetic and Doppler ultrasonic flow meters in the 1950s had a major impact on our understanding of both cardiac and vascular functions in nonhuman species. These methods had the enormous value of providing instantaneous and continuous determinations of blood flow during both dynamic conditions producing large transient changes as well as during steady-state conditions. A. Kolin's and E. Wetterer's electromagnetic flow meter and Rushmer and colleagues' ultrasonic flow meters were two among many comparable devices able to track flow continuously (see references in 59).

Intrinsic Control of Cardiac Performance

At mid-twentieth century, the widely accepted view based mainly on Starling's work was, according to Robert F. Rushmer and others as well, that cardiac output is raised during exercise by increases in stroke volume and heart rate. The rise in stroke volume was thought to be caused by an increased "venous return" (i.e., a rise in end-diastolic volume or myofibril length) via the Frank–Starling mechanism. This rise was attributed to muscle pumping and peripheral vasodilation. Rushmer and his group in Seattle became the major opponents of this view.

The Great Debate

The antagonists spoke strongly and pointedly; Starling's work was at the center of the dispute. Their unbridled arguments at the federation meetings during the 1950s and early 60s became legend. The viewpoints of four leaders in the field are expressed below.

W.F. Hamilton and D.W. Richards stated:

> It is now held that in the normal animal as distinguished from the usual physiologist's preparation, the venous return is always ample to fill the heart and hence is not a key to the regulation of cardiac output. On the other hand the pumping action of the heart is regulated reflexively to maintain arterial pressure within physiological limits despite the changes in flow. The changes in flow are the direct result of peripheral vasomotion in response in humoral stimuli and to locally produced metabolic vasodilator influences. (60)

R.F. Rushmer and colleagues observed reductions in some left ventricular dimensions, thought to reflect end diastolic volume, in exercising dogs and no change in left ventricular end diastolic pressure, which led them to conclude that: "Starling's Law of the Heart is applicable to exposed or isolated hearts but not to the intact hearts of animals and man" (115). The debate therefore centered on whether the heart by itself can provide the cardiac output required for exercise by simply increasing the rate and force of contraction once muscle has vasodilated and vascular conductance has risen.

Arthur C. Guyton and coworkers concluded the heart could not do what Rushmer proposed and he stated:

The normal [i.e., at rest] circulatory system operates near this limit [i.e., collapse of central veins owing to increased cardiac output] so that an increase in efficacy of the heart as a pump cannot by itself increase cardiac output more than a few percent unless some simultaneous effect takes place in the peripheral circulation at the same time to translocate blood from the peripheral vessels to the heart. (55)

Stanley J. Sarnoff (1916–1989): Although Sarnoff and Rushmer "agreed" in general (but not noticeably) on the importance of increased contractility, Sarnoff saw the length–tension relationship as being present but profoundly altered and masked by changes in contractility. In 1954 he and Erik Berglund produced a family of ventricular function curves that included simultaneous actions of length–tension or Starling effects and both the neural and the humoral factors emphasized by Rushmer. The combined actions of Starling effects and neural and humoral factors would, with each altered state, shift control of the heart from one ventricular function curve to another (121).

Conflicting (?) and Competing Ideas

Contractility vs. Length–Tension

Rushmer's group studied chronically instrumented conscious dogs, which permitted continuous on-line measurements of heart rate, left ventricular diameter, left ventricular length and circumference, intraventricular pressures, and intrapleural pressure (to derive effective filling pressure). During voluntary mild exercise they saw no consistent changes that would support a significant role for length–tension (Starling) effects on stroke volume. Rather, left ventricular diameter and circumference often decreased slightly and stroke volume was maintained by greater systolic ejection through increased contractility. Rushmer concluded that increased contractility is the dominant mechanism of increasing cardiac performance along with increased heart rate during exercise (116, 118).

Sarnoff's experiments were on anesthetized, thoracotomized dogs in which he progressively elevated right and left atrial pressures over a wide range. He assumed that diastolic ventricular volume was directly related to filling pressure. When neural and humoral factors shifted the stroke work–pressure relationship to different ventricular function curves (he called them "Starling curves") he concluded that a change in ventricular filling pressure could occur in one direction accompanied by a change in ventricular stroke work in the opposite direction—but still the Frank-Starling relationship applies. His advance did little at the time to assuage the dispute, which rested heavily on responses seen during exercise. Rushmer and Smith objected to the "flexibility" of a concept that could explain "any possible change" and that could not be tested in intact animals (118).

Increased Venous Return

Guyton's famous experiments are depicted in his well-known venous return–cardiac output curves (or vascular function and cardiac function curves). These were the end product of an effort to understand both the peripheral vascular factors controlling venous return to the heart and the left ventricular output (54). These experiments were conducted on anesthetized, thoracotomized dogs with right heart bypass pumps.

Guyton's focus on venous return revealed its finite limits owing to the effects of blood flow on peripheral vs. central venous pressures and volumes. He showed that increasing cardiac output reduced central venous pressure to the point of venous collapse, making further increases in cardiac output impossible unless an added force could provide an increased driving pressure for venous return. He went beyond cardiac determinants of cardiac performance and included as one of the primary determinants the *nature of the mechanical coupling between the heart and the peripheral circulation*. This important contribution was made in spite of widespread misunderstanding and misinterpretation concerning the physiological meaning of his vascular and cardiac function curves and especially of mean circulatory pressure. For example the need, for graphic purposes, to reverse the independent variable (cardiac output or venous return) from the x-axis (the stimulus) to the y-axis (the response) led many to conclude that central venous pressure determines venous return rather than the converse. The problem was put into perspective by Matthew Levy in 1979 (90). In an accompanying editorial Guyton graciously acknowledged the problem; indeed, Guyton had made in 1962 a fundamental and related point (cited above) regarding exercise [55].

Rushmer and Smith questioned the importance of increased venous return to cardiac performance during exercise (115, 118). They compared responses to mild exercise to those obtained when venous return and effective ventricular filling pressure were increased by volume infusions, abdominal compression, and head-down tilt. They concluded that increased venous return is no substitute for exercise and proposed that "the concept of 'increased venous return' be completely abandoned" (118). This misperception was fundamental; these maneuvers simply transiently shifted blood volume centrally as the brief increase in cardiac output quickly redistributed volume into the peripheral vasculature where any effects were further blunted by reflexes.

Did the Data Really Conflict?

The fact that cardiac output can determine central venous pressure through its interaction with the peripheral vasculature (not the converse) provides the key to a major misunderstanding. Guyton's work and subsequently that of others helped us to understand why it is not possible to raise resting cardiac output substantially by ventricular pacing or a variety of other means (77, 90, 141). For example, Rushmer tried unsuccessfully to raise cardiac output by lowering peripheral vascular resistance via drugs, by humoral stimulation of the heart, and by triggering tachycardia by electrical pacing and electrical stimulation of cardiac sympathetic nerves. As with central volume loading, these maneuvers also failed to simulate exercise.

Many had observed much earlier that *the ability to raise resting cardiac output is self-limited by its negative effect on central venous pressure*. Because of the capacitive and resistive properties of the peripheral circulation, whose venous volume increases with increased flow, the central vascular volume is depleted, thus lowering central venous pressure and causing a self-limiting effect on cardiac output. The inverse relationship between cardiac output and central venous pressure was shown in humans in 1944 by J. McMichael and E.P. Sharpey-Shafer in an early application of the direct Fick method (96) and again over the years by many investigators (12). Initially the cause was unknown and attempts to explain the phenomenon preceded our understanding of the mechanical coupling between the heart and peripheral circulation. For example, Hamilton stated: "When the heart accelerates and pumps more blood it becomes smaller; when it slows, it fills more and becomes larger; changes in rate thus work against any application of Starling's Law in the intact animal" (58). Now we know that it is the negative effect of raising peripheral blood flow on ventricular filling pressure that limits the rise in cardiac output through length–tension effects (and vice versa when blood flow falls)—that is, *it is an illustration of Starling's law.*

The crucial point put forth by Guyton and later demonstrated directly by D.D. Sheriff and colleagues (1993) by their control of cardiac output in intact voluntarily exercising dogs is that something must reverse the tendency for any *rise* in cardiac output (at any given work intensity) to *lower* end-diastolic volume and stroke volume (127). The combined actions of skeletal muscle and abdominal and respiratory pumps (primarily skeletal muscles—see below) must reverse the negative relationship between cardiac output and both diastolic and stroke volume so that all rise together. Here the disagreement with Rushmer and Hamilton was fundamental.

The now often-observed constancy of stroke volume and end-diastolic volume in exercising dogs first noted by Rushmer is also an argument *for* Starling's law. This law must be invoked to explain why these variables do not *decrease* with increased cardiac output as they do at rest; namely, the increase in filling pressure owing to muscle pumping maintains stroke volume via length–tension effects. In short, the findings by the four adversaries and their colleagues were not really in conflict nor did they conflict with the mainstream of the contemporary findings of most others from animals and humans; it was their interpretations that were often in striking conflict.

Conflicting Data (?): Humans vs. Quadrupeds

The Issue of Stroke Volume and Posture

This conflict concerning the constancy of stroke volume erupted under the critical eye of Rushmer, who in 1959 and 1962 reviewed the literature on the subject and concluded that, like dogs, humans often showed no consistent rise in stroke volume with increasing severity of exercise up to maximal levels (116, 118)—whereas others found progressive increases. Asmussen and Nielsen in 1952 often saw that cardiac output was increased by increasing heart rate without raising stroke volume during exercise. As techniques for measuring cardiac output progressed, the observation of nearly constant stroke volume across graded intensities of exercise up to severe levels became commonplace (for references see 4, 5, 12, 112, 120, 146). There-fore, as Rushmer argued, the responses to graded exercise were apparently not con-sistent with what many felt Starling would have predicted (cf. 133)—namely, pro-gressive increments in end-diastolic volume and stroke volume.

Again the problem was whether exercise commenced with subjects in upright or supine posture. Wade and Bishop (1961) and Wang, Marshall, and Shepherd (1960) and Bevegard, Holmgren, and Jonsson (1960) studied subjects in both pos-tures and confirmed that substantial increases in stroke volume occurred only when exercise commenced in the upright posture—that is, when stroke volume was low due to gravitational "pooling" of blood in the legs during orthostasis (12, 146). In his typically incisive manner, Erling Asmussen (1907–1991) came directly to the point, which was: "the variations in stroke volume called forth by variations in the filling of the heart and central veins [i.e., owing to change in posture] are large compared to the changes in stroke volume that accompany muscular exercise [per se]." In short, the marked reductions in thoracic blood volume, central venous pressure, end-diastolic volume, and stroke volume attending upright posture are quickly reversed as central volumes and pressures are restored by exercise and the sudden increase in venous return. (This important example of Starling's law was conceded by Rushmer in 1962 [116].) The conclusion that flaws in the rebreathing techniques caused the apparent discrepancy in measurements of stroke volume does not hold up (e.g., 4, 5, 120).

Rushmer observed that the dogs' left ventricle functions near its maximum di-astolic dimensions during exercise, as evidenced by the failure of left ventricular di-ameter to increase when filling pressure was increased by intravenous infusions (116, 118). This observation is consistent with the fact that 75% of the dog's total blood volume is at or above heart level, so filling pressure should normally be opti-mal and unopposed by the large hydrostatic forces as seen in humans.

Preload: Does Exercise Increase Effective Filling Pressure and End-Diastolic Volume?

Filling pressure. The earliest attempts to measure the changes in venous pressure in response to exercise (H.D. Hooker in 1911 was one of the earliest to try) were made in peripheral veins positioned at venostatic levels in which pressure almost invari-ably rose (14). However, these measures were shown not to reflect right atrial pres-sure or central venous pressure (75). In 1951, Louis Dexter made the first measure-ments of left atrial pressure (i.e., pulmonary wedge pressure [PWP]) in humans during supine exercise: this pressure rose in six of seven subjects (35). A major crit-icism of atrial pressure as normally measured—that is, referred to atmospheric pres-sure—has been that it may not reflect an *effective filling pressure.* At constant atrial or ventricular pressures effective filling pressure can increase because intrapleural pressure falls or vice versa. Conversely, recent studies (see references 77 and 106) show that the positive and negative intrathoracic pressure changes with breathing over several cycles average to zero net pressure difference so that measures of cardiac pressure referenced to atmospheric could be valid estimates of the transmural pressure or effective pressure. Nevertheless, a warning from R.F. Rushmer is germane here; it is that a difference of 1 to 2 mmHg in filling pressure can make a great difference in ventricular volume and function in intact animals (especially on the steep portion of the ventricular pressure–volume curve) and this "cannot be resolved in intact exercising animals" (116).

In humans in the last one or two decades, measures of right atrial pressure indicate a sudden rise at the onset of upright exercise (107), little further increase through mild and moderate exercise, and then further increase with severe exercise (105). Measurements of PWP which reflect left atrial filling pressure show a consistent pattern (105, 137). The PWP and end-diastolic pressure in both supine and upright humans rise in response to exercise (138). The PWP rises in proportion to right atrial pressure and both are highly correlated with pulmonary arterial pressure; in endurance athletes with high cardiac output, PWP can rise from a few mmHg at rest to greater than 45 mmHg when cardiac output equals 35 liters per minute. So right and left atrial filling pressures clearly rise (106).

End-diastolic volume. Here there is conflict. A plethora of both two- and three-dimensional radiological studies dating back to 1920 has never revealed any marked or consistent increase in human heart size that is commensurate with that which many would have predicted from Starling's law (4, 118, 146). In 1920 Starling himself, however, commented on the lack of increase in the size of the intact heart detectable by radiological means during exercise. He saw the intact heart as being "reined in and controlled by cardiac centers" of the central nervous system (133). Modern techniques of radionuclide imaging of ventricular volumes have enabled specific measurements of left ventricular volume; nevertheless, errors of measurement can be quite large (52). Clearly, end-diastolic volume decreases during the transition from supine to upright posture and is restored or increased by upright exercise. Results from the 1980s and 90s indicate that end-diastolic volume increases markedly from upright rest to exercise and remains relatively constant up to peak exercise, whereas the decrements in end systolic volume, owing to increased contractility and increased ejection fraction, contribute to further increments in stroke volume (70, 100). The Starling mechanism accounts for two-thirds of the increase in stroke volume and contractility for about one-third. These and other results (106, 137) indicate that despite the continuous rise in PWP with increasing work rate, end diastolic volume and stroke volume stop rising and may even decline slightly.

The latter observations prompted the suggestion by John Reeves (in reference 106) and Joseph Janicki (in reference 77) that this dissociation of end-diastolic volume from filling pressure indicates greater restriction to filling of the left ventricle and that this could be due to the encroachment of pericardial restraint on ventricular filling. Parenthetically, another suggestion is that PWP may rise during heavy exercise because the high heart rate limits diastolic filling time and thereby creates a diastolic pressure gradient between the left atrium and the left ventricle (106); or perhaps more precisely, the high heart rate may displace the relationship between *mean* left atrial pressure (a time average) and the unaveraged end-diastolic pressure.

Afterload

Afterload represents the mechanical load that opposes ventricular ejections; it too can affect stroke volume. Physiologists have traditionally used the aortic pressure that the heart must overcome to define afterload rather than the hydraulic input impedance and ejection dynamics.

Measurement of arterial pressure. Stephen Hales introduced the U-tube manometer (containing water) and Poiseuille introduced the mercury-containing version that enabled more convenient pressure measurements. Karl Ludwig added the float and pointer for inscribing blood pressure continuously on a smoked paper kymograph. The development of manometers that could transduce pulsatile pressures came from Otto Frank in 1903. Carl Wiggers (1883–1963) used such manometers in his exhaustive studies of pressure wave forms presented in his classic monograph: *The Pressure Pulses of the Cardiovascular System 1928* (151). In the 1930s William Hamilton improved the frequency response and sensitivity of Frank's manometer. Earl Wood, at the Mayo Clinic, developed and refined with Lambert the strain gauge manometer, bringing us to the modern state of the art (84).

Blood pressure in exercise. As with heart rate, for nearly a century the measurement and descriptions of arterial blood pressure changes during exercise were a laboratory exercise. In 1892 M.M. Koffman and in 1898 Zuntz and Hagemann measured arterial pressure in exercising horses. Both saw blood pressure fall in response to exercise (95). Later Zuntz studied dogs and saw, with exercise, an immediate rise in blood pressure in 60% and a transient decrease in 40% of the dogs followed by a rise with increasing work rate (95). The rise in blood pressure was measured in humans by Zadek in 1881 and Leonard E. Hill (1866–1952) in 1898 (14). Bowen in 1904 provided the first detailed description of the magnitude and time course of blood pressure changes in humans including its slow decline as exercise is prolonged (14)—a phenomenon studied in detail by Lars Goran Ekelund and Alf Holmgren in the 1960s (38). A.V. Bock, C. van Caulaert, and David Bruce Dill in Boston in 1928 showed the rise in blood pressure in the marathon runner DeMar to be much less at a given work rate than in a sedentary subject—perhaps one of the earliest descriptions of the influence of physical conditioning (14).

The question of a fall in blood pressure at the onset of exercise has persisted from the beginning. Krogh (1912) thought there was no fall in spite of vasodilation in muscle and he concluded that there must be an increase in circulating blood volume due to the delivery of blood from the splanchnic region (83). Holmgren in 1956 reviewed the history of earlier observations and his own detailed measurements (75). He showed that often there is an initial transient dip in blood pressure at the onset of exercise. Sometimes it did not occur at all and in most subjects it disappeared on repeated trials (i.e., the fall in blood pressure is clearly not the stimulus to raise cardiac output—as many had thought).

Apparently the first indication that the site of measurement influenced the amplitude of the pressure pulse and systolic pressure came from K. Hüthle in 1890. Later, Otto Frank considered the increase in pulse and systolic pressures in peripheral vessels relative to those recorded centrally to be a prominent feature of blood pressure. These amplitude differences were also observed by William Hamilton and coworkers (for references see 82). In 1955 Kroeker and Earl Wood observed that at supine rest brachial, femoral, and radial arterial pulse pressures were 131, 139, and 146% of aortic pulse pressure respectively (82). This amplification of pulse pressure was similar during mild supine exercise. In 1968 Rowell and colleagues compared simultaneous measurements of ascending aortic and radial arterial pressures during rest and mild to maximal exercise in upright humans (111). From rest to maximal exercise aortic pulse pressure rose by a factor of two whereas it rose threefold at the radial artery. This pulse wave amplification originates from the summation of the incident pulse wave with waves reflected from the periphery and from resonance effects in peripheral arteries (94).

Aortic pressure, afterload, and stroke volume. Janicki concluded from his experimental results that a rise in aortic blood pressure from 112/68 to 154/70 mmHg at peak exercise would lower stroke volume by about 10 mL if filling volume and contractility remained the same (77). In general, the normal left ventricle appears to be remarkably insensitive to a wide range of acute changes in afterload. This is in contrast to earlier findings in the anesthetized dog with external manipulation of aortic pressure (for reference see 77). Clearly the rise in afterload calculated from aortic systolic pressure is far less than that calculated from peripheral arterial systolic pressure—a point of clinical importance.

Pericardial Constraints

The notion that ventricular function might be constrained under normal conditions by the pericardium has a history of controversy. In 1895 H.L. Barnard, a colleague of Leonard Hill, observed that the pericardium could withstand more strain than the myocardium and he proposed that the pericardium "limits passive dilation of the heart" (cited by Reeves and Taylor [106]). In 1915 Y. Kuno observed that "When the pericardium is intact, the heart, in order to perform a certain amount of work, requires a higher venous pressure than when the pericardium is open" (cited in ref. 106). Barnard's suggestion was reinforced by Yandell Henderson in 1920 when he maintained that the rise in human stroke volume above resting was limited to 20%–30% by the pericardium. This drew a "pungent" rebuke from Zuntz, as well as disapproval from Starling (146). (Note: The dog hearts in Starling's experiments had no pericardium; also his disapproval is surprising in light of his comments in 1920 and statements above concerning constancy of heart size [see ref. 133].)

Rushmer's observation that the dog's left ventricle functions near its maximal diastolic dimensions supports the view that the ventricles are normally operating at the limit of a mechanical constraint imposed by the pericardium during exercise (115, 116, 118). In a classic experiment, David Donald and John Shepherd of the

Mayo Clinic (1963) showed that in dogs after cardiac denervation, which necessitated cutting the pericardium, a normal rise in cardiac output with exercise up to heavy levels was achieved mainly by raising stroke volume with relatively small increments in heart rate—that is, a reversal of the normal response (36). Recent experiments on dogs by Jere Mitchell's group in Dallas in 1986 (136) and on pigs in 1992 by H.K. Hammond and colleagues (61) emphasize the significance of pericardial restraint. Pericardiectomy enabled these animals to increase both stroke volume and cardiac output during voluntary exercise and especially at maximal exercise. Thus the dissociation between end-diastolic volume and pulmonary wedge pressure could be caused by even small increases in heart volume that cause the pericardium to reach its limits of compliance rapidly as proposed by Reeves and Taylor (106).

Despite decades of debate, the history of cardiac performance in intact animals and humans during exercise appears not to be in conflict with the principles put forward by Howell and Donaldson (1894) and by Starling (1918). Furthermore, the potential of the Frank–Starling mechanism was also unveiled by pericardiectomy.

Ventricular Filling by Respiratory and Skeletal Muscle Pumps

The Respiratory Pump

The mechanical effect of breathing in augmenting venous return received its initial attention from Antonio Valsalva (1666–1723). He observed collapse of the jugular vein during inspiration and filling of the vein during expiration (16). The venerable Rev. Hales attributed the fluctuations he observed in his pioneering measurements of arterial and central venous pressure to be the consequence of blood being expressed from the lungs by respiratory movement. Albrecht von Haller (1708–1777) perfused frog lungs with dye and saw pulmonary blood flow increase with lung inflation and decrease with deflation (16, 30). John Hunter in 1794 referred to a "degree of stagnation" during expiration whereas "during inspiration the veins readily empty themselves." In 1830 Poiseuille demonstrated the effect of negative intrathoracic pressure on venous return. Von Haller's work, which initiated a 100-year controversy, was upheld in the closed-chested animal in 1881 by P. Héger and E. Spehl in Belgium, who also made the first estimate of pulmonary blood volumes during inspiration and expiration (16). In 1888 J. Pal proposed that hepatic venous outflow is dominated by respiratory movement of the diaphragm (95). Leonard Hill and H.L. Barnard in 1897 concluded that respiratory pumping of blood is sufficient to maintain arterial pressure in an animal placed in tail-down vertical posture (95). In 1937 K.J. Franklin further emphasized the action of breathing on hepatic venous outflow. He saw it acting mainly through compression of the liver and its subsequent release from the liver by the fall and rise of the diaphragm during inspiration and expiration (47). These observations were skillfully advanced in technically demanding studies in the 1960s by A.H. Moreno and colleagues (97).

An argument that opposed the respiratory pump concept maintained that the negative pressure in the thorax would collapse the veins during inspiration so that respiration could not increase venous return (16). Gerhard Brecher (1956) employed the unusual "bristle flowmeter" in dogs and demonstrated the effectiveness of the respiratory pump, showing the collapse to be a time-dependent event that would restrict caval flow only if the fall in downstream intravascular pressure was too prolonged or unusually deep (16). Wexler and coworkers in 1968 directly measured the acceleration of caval blood flow with inspiration in normal humans and noted its greater increase with inspiration during exercise (149).

The Skeletal Muscle Pump

The phenomenon of "muscle pumping" first received attention from William Harvey in *Du Mortu Cordis* (1628) and was subsequently referred to by R. Lower (1669), von Haller (1763), and others. For example, John Hunter (1794) and Francoise Magendie (1830) called attention to the importance of muscle contraction in reducing fluid accumulation in the feet in upright posture (95). More precise measurements were made in the later nineteenth century—for example, by W.H. Gaskell (1877) (95). Leonard Hill (1898) emphasized the importance of mechanically squeezing blood out of the veins of active muscle to facilitate the return of blood to the heart. In 1902 Burton-Opitz used a "Stromuhr" (flowmeter) and noted three phases of flow with muscle contraction: (1) high flow during shortening, (2) low flow during sustained contraction, and (3) increased flow during relaxation (95). Tigersted (1909) emphasized the mechanical nature of the muscle pump when he observed that contraction of the hindlimbs after lumbar section of the spinal cord still raises venous return and cardiac output (83). Then several observed that when upright humans began to walk, pressure in superficial veins of the ankle suddenly fell below the expected hydrostatic pressure (D.R. Hooker 1911, Leonard Hill 1909, August Krogh 1929) (95). In 1936 Beecher, Field, and Krogh speculated that there was enough compression of the superficial veins when underlying muscles contracted and pressed against the skin to drive blood into the deep veins (i.e., via the small perforating veins) (95).

During the 1930s, G.V. Anrep and his colleagues made detailed studies of the muscle pump, measuring venous outflow from the muscle by the heated wire method. They observed four distinct phases of inflow and outflow and emphasized the large venous outflow during shortening and the second flow increase during relaxation. They also showed that the magnitude and duration of hyperemia during and after contraction varied with duration and strength of each contraction (1, 2).

In 1949 Henry Barcroft (1904–1998) and A.C. Dornhorst demonstrated that contracting muscles could increase venous outflow from the muscle against venous back pressures as high as 90–100 mmHg (raised by partial venous occlusion) (8). In 1949 A.A. Pollack and E.H. Wood showed in a classic study that within six to seven walking steps starting from upright rest, pressure in a superficial ankle vein fell from 90 to ~25 mmHg. Like Krogh and colleagues, (1936) they thought this decline was slowed because volume from these cutaneous veins must be forced (sucked?) into deep muscle veins through the perforating veins (101).

The volume of blood mobilized by the muscle pump cannot be easily separated from the contribution of the respiratory and abdominal pumps. In 1962 Guyton and collaborators demonstrated in anesthetized and areflexic dogs on right heart bypass that static contractions of hindlimb and abdominal muscles immediately increased central venous pressure by 7.5 mmHg (55). This rise represents such a large volume of blood that some of it must have also been expelled from the splanchnic organs by contraction of the abdominal muscles. This abdominal contribution may occur in intact animals and humans as well because H.F. Stegal (1966) (134) showed that abdominal pressure rises 25 mmHg or more during exercise in humans. S.D. Flamm and colleagues (1990) used radionuclide imaging to estimate regional changes in the volume of technetium-^{99}m-labeled red blood cells in humans during mild to severe leg exercise (cycling) (42). Owing to muscle pumping, leg blood volume fell by 22%–30% whereas total thoracic blood volume and individual volumes of the heart and lung all rose by 20%–30%. Calculations from Flamm's data and those of John Ludbrook (1966) of Australia and D.D. Sheriff (1993) indicate that approximately 450 ml of blood is displaced from leg veins by rhythmic contractions under zero loading conditions (92, 128).

In addition to the effect of contraction on muscle blood volume, the muscle pump also modifies fluid filtration from muscle capillaries and prevents edema as noted by John Hunter in 1794 and recently quantified in exercising humans by J. Stick and colleagues (1992) in Germany. The edema-preventing function is only partly due to reduced venous and capillary back pressures. The main factors appear to be the mechanical expulsion of lymph from the lymphatics (the "muscle lymph pump") and the rise in muscle tissue pressure which opposes capillary filtration and reduces lymph formation (62, 135).

Does the muscle pump actually increase muscle blood flow? Bjorn Folkow and his colleagues (1970) proposed that the pumping action of a muscle could increase its own rate of perfusion (43). They saw muscle contraction as widening the arterial–venous pressure difference across the muscle by its momentary lowering of muscle venous hydrostatic pressure to zero immediately after contraction; at the same time the height-related hydrostatic effect on arterial driving pressure persisted so that, momentarily, driving pressure was increased by approximately 90 mmHg (depending on height). Folkow and colleagues (1971) further demonstrated that this effect of gravity on driving pressure could explain why peak blood flow to the exercising human leg is greater in upright than in supine posture (44).

In 1987 Harold Laughlin put his own research and ideas together in an important review article (88). His findings indicated that muscle pumping action must be critically dependent on the type of contraction with normal locomotory exercise being far more effective than various modes of electrically induced contraction. He concluded that the efficacy of the muscle pump is determined by the spatial and temporal sequence of the pressure developed as well as its magnitude. Rhythmic lengthening and shortening are the key factors in effective pumping. Laughlin observed that peak vascular conductance in resting muscle perfused with potent vasodilator drugs to elicit "maximal" vasodilation was far below peak "conductance" during voluntary contractions. However in the latter case the term "conductance" is not applicable. Laughlin described what Sheriff and colleagues (1993) subsequently called "virtual conductance," signifying that a simple ohmic conductance does not exist across a pump that directly imparts energy to the system (127). Laughlin proposed that the restorative forces within the muscle immediately after the contraction could pull open veins that are tethered to the muscle, causing their lumen pressure to become transiently negative. Something like this must happen because the small animals studied by Laughlin and also by Sheriff lack the hydrostatic influence on veins seen in humans and emphasized by Folkow, yet the pump is still highly effective.

In 1993, D.D. Sheriff and colleagues in Seattle showed that as dogs begin voluntary treadmill exercise, terminal aortic (hindlimb) and "virtual conductance" rose immediately and was followed in about 5 seconds by a secondary rise in "conductance," which was probably due to vasodilation (128). Approximately 5 seconds was suggested by R.J. Gorcynski, B. Keltzman, and B. Duling (1978) (51) (isolated muscle) and by K. Toska and M. Ericksen (1994) (140) (human quadriceps) to be the minimum time necessary for vasodilation to occur. Sheriff experimentally ruled out a myogenic relaxation of vasculature following compression by contracting muscles, and autonomic blockade ruled out either activation of sympathetic cholinergic nerves or inhibition of sympathetic noradrenergic nerves as causes of the sudden rise in "conductance" (128). Further, when they held cardiac output constant by ventricular pacing, arterial blood pressure fell 40 mmHg in 1 second after the onset of exercise—far too rapidly for vasodilation. Thus the muscle pump must have withdrawn a volume of blood from its stiff arterial supply, thereby causing a large drop in pressure before any vasodilation. The negative pressure in small muscle veins proposed by Folkow and by Laughlin would be necessary for this to occur. The proposal was supported by Sheriff and Van Bibber, who demonstrated in 1998 the flow-generating capability of the hindlimb skeletal muscle pump when it was surgically isolated from the cardiac pump in anesthetized pigs (129). Hindlimb muscles perfused themselves for brief periods by driving their blood supply around in a surgically created short circuit that isolated them from the remainder of the circulation.

The importance of the muscle pump to venous return and ventricular filling was questioned in the 1950s and 60s and is again accepted. Increases in heart rate and cardiac contractile force cannot by themselves (cf. 115) maintain adequate cardiac filling pressure—in fact, by themselves they would reduce it.

Matching Right and Left Ventricular Output: The Fundamental Importance of the Frank–Starling Relationship

"The Starling mechanism also plays an important role in equalizing the output of the right and left ventricles which have no intercommunicating reflexes nor separate hormonal control" (60). This quotation applies to what Janicki and colleagues called "ventricular interdependence" (77). That is, the two ventricles are functionally and inextricably related by the common muscular wall surrounding them, by the pericardium, and by the intraventricular septum. Contraction of one ventricle affects the other; thus the phase relationship between right and left ventricular contraction is important.

Breathing, standing up, and exercise cause transient imbalances between right and left ventricular output. Julian Hoffman and colleagues (1965) in San Francisco concluded that due to the Starling mechanism, maximum right ventricular stroke volume during inspiration could be from 10% to 50% greater than minimal stroke volume on expiration (72). Left ventricular stroke volume was less variable owing to the capacitive properties of the pulmonary circulation, which buffered the changes in left ventricular filling. Abraham Guz (1983) observed that the inspiratory fall in left ventricular stroke volume is more pronounced in humans because as the right ventricle expands, left ventricular compliance is reduced by rising pericardial restraint; this restricts left ventricular filling and thus the two ventricles must compete for intrapericardial space (see ref. 77). Guz showed this effect of inspiration on left ventricular stroke volume was minimized in patients after pericardiectomy.

Robotham and Wise (in ref. 77) also pointed in addition to the potential effect of falling intrathoracic pressure on left ventricular stroke volume. This would be mechanically equivalent to raising left ventricular afterload. They also pointed to similar forces that undoubtedly influence left and right ventricular performance during exercise. These included the pronounced rise in abdominal pressure and a possibly larger influence of the pericardium. This is important unfinished business as pointed out in 1996 in a superb *Handbook* chapter by Janicki, Sheriff, Robotham, and Wise (77).

It is on this matter of matching ventricular outputs that the persistent opposition of Rushmer's group to the applicability of the Starling mechanism ran aground. D.L. Franklin and coworkers concluded that (in dogs) neural control kept both ventricles in phase and in balance (46). However, the 10 seconds or so required for a sympathetic response to affect heart rate and contractile force is far too slow to balance ventricular outputs in response to breathing or standing up. In fact, the right and left ventricular effects of breathing are 180° out of phase, which also refutes the conclusion of Franklin and colleagues (46).

THE DISTRIBUTION OF CARDIAC OUTPUT AND BLOOD VOLUME— ANOTHER DETERMINANT OF END-DIASTOLIC VOLUME

The previous section revealed an apparent paradox, which is that when cardiac output was increased during rest, the consequent decrease in central venous pressure and in end-diastolic volume reduced stroke volume and thereby limited the rise in cardiac output. Conversely, cardiac output cannot be substantially increased without raising or at least maintaining central venous pressure and end-diastolic volume (55). We have seen that these latter conditions could be achieved by (1) the skeletal muscle pump and (2) by the less powerful respiratory and abdominal pumps. The third idea, which is the subject of this section, is that active vasoconstriction of nonexercising regions and the resultant passive release of blood volume back to the heart raise end-diastolic volume.

Some Early Ideas and Models

In 1884 A.F. Dastre and J.T. Morat proposed that a large increase in blood flow to one region, such as the cutaneous circulation, is compensated for by a reduction in blood flow to visceral organs. They comment "Ce balancement entre la circulation cutanée et la circulation viscerale est un fait remarquable" (This balance between the cutaneous circulation and the visceral circulation is a remarkable fact) (33).

In 1886 S. DeJager suggested that a change in blood flow through distensible vessels would cause their blood volume to change proportionally (34). He described the factors thought to be important in determining the distribution of blood flow and blood volume, in particular the effects of changes in resistance on downstream pressure and volume. In 1912 August Krogh introduced a simple model of the circulation which contained a pump that supplied two parallel circuits—one compliant (the portal system) and one noncompliant (e.g., skeletal muscle) (83). He proposed that the increased resistance to flow into the portal system would, as a consequence of reducing its blood flow and thus its transmural pressure, passively (by venous collapse) expel its large blood volume back to the central veins. This would raise ventricular filling pressure and cardiac output. Krogh considered the portal system to be a regulator of both pressure in the central veins and the supply of blood to the right heart. In short, the volume of blood available to fill the heart in rest and exercise depends on the fraction of cardiac output that perfuses the compliant vs. the noncompliant circuits. This important insight is still central to our understanding of how blood volume is distributed during rest and exercise.

Krogh's ideas were verified (apparently unknowingly) by Henry Barcroft and A. Samaan in 1935 when they observed that occlusion of the dog's aorta above the mesenteric arteries caused a sudden increase in cardiac output, whereas occlusion below these arteries reduced cardiac output (6). The former maneuver reduced vascular transmural pressure and blood volume in the capacious splanchnic region (the compliant region in Krogh's model) causing a passive expulsion of its blood volume back to the heart. In contrast, the latter maneuver increased the pressure, flow, and volume in that region and thereby reduced ventricular filling volume and pressure. The less compliant regions below the occlusion contained too little volume to affect cardiac output. In 1983, O. Skokland provided a more detailed analysis of these types of experiments and a confirmation of the major effects of passively releasing blood from the splanchnic region on cardiac filling pressure and volume (130).

In 1974 P. Caldini, S. Permutt, J.A. Waddell, and R.L. Riley at Johns Hopkins offered a more complex version of Krogh's model (19). Their model was analogous to an electrical network of parallel circuits containing arterial and venous resistances and venous capacitances and a single arterial capacitance. By analogy, a "time constant" for venous return was derived for the major compliant and noncompliant circuits. For example, venous return from compliant skin and splanchnic regions would have a long time constant whereas the return from noncompliant skeletal muscles would have a short time constant. The model was tested on dogs by infusion of drugs with effects on venous return to the heart that were in a broad sense consistent with the model and Krogh's original concepts.

As adequate measures of regional blood flow became available in animals and in human subjects the old disputes stemming from differences in species and methods and in posture emerged again. It was assumed by many that if such regional vascular adjustments are essential to cardiac function in exercise, they should be present in most terrestrial mammals that are exposed to dynamic exercise. Initial emphasis was on renal blood flow because it could be measured—not because of its influence on end diastolic volume (i.e., the renal vasculature has high blood flow but low compliance and low vascular volume). Presumably renal blood flow responses reflected those of other major vascular beds. Later, measurement of splanchnic blood flow and splanchnic blood volume became feasible. This region's high compliance and large blood volume (~25% of total), which is blood flow dependent, makes its perfusion an important determinant of end-diastolic volume.

Regional Blood Flow during Exercise

Early Observations for and against Redistribution of Blood Flow

A.V. Bock and D.B. Dill (1931) cite observations by Weber (1907) that even the thought of exercise expanded leg blood volume, which he presumed was a consequence of constricting splanchnic vessels (14). Joseph Barcroft (1872–1947) and H. Florey in 1929 observed in a dog with exteriorized spleen and an exteriorized patch of colon that short bursts of running elicited marked pallor of the colon and splenic contractions, but during more prolonged exertions the spleen remained contracted but the colonic pallor was not maintained (14, 30). The inability to measure regional blood flows forced early experimenters to focus on blood volume rather than blood flow. Nevertheless it appears to have been generally assumed that exercise was accompanied by redistribution of blood flow to maximize the blood supply to active muscle. The splanchnic circulation was predicted to be a major site for vasoconstriction (14).

In 1936 Francisco Grande (Covian) worked with Paul Brandt Rehberg and August Krogh on human renal function during exercise (50). One study by Grande and Rehberg revealed that urine formation was reduced not only by greater tubular reabsorption of water (increased ADH?) but also by a 65% reduction in glomerular filtration rate. This and a much reduced creatinine clearance suggested that exercise reduced renal blood flow.

In 1940, Herrick, Grindlay, Baldes, and Mann implanted thermostromuhrs in the renal, superior mesenteric, and iliac arteries in dogs (64). During exercise blood flow to the hindlimbs rose markedly but blood flow in the mesenteric and renal arteries was not reduced. Soon afterward the thermostromuhr technique became obsolete.

Plasma clearance techniques for determination of renal function and blood flow, developed by Paul Rehberg and Homer Smith and colleagues were applied to the investigation of renal blood flow during exercise. In 1947 Barclay and colleagues noted that water diuresis was inhibited by exercise. Further, brief strenuous exercise decreased renal plasma flow by 40% as estimated after exercise by clearance of diodone. Subsequently most investigators chose the more reliable technique of para-amino hippurate (PAH) clearance (123). In 1948 Carlton Chapman and Ancel Keys and their colleagues in Minnesota estimated renal blood flow in human subjects by renal clearance of PAH and found it to decrease in proportion to exercise intensity (23). Also in 1948 H.L. White and D. Rolf observed that during severe exercise (running) renal blood flow fell 80% (150). Sid Robinson (1949) and colleagues at the University of Indiana noted that moderate exercise combined with heat stress reduced renal blood flow to a greater degree at a given work rate than in cool conditions; dehydration further augmented the reduction (104, see also 108).

Regional Blood Flow—1950 on

The Newer Methods

Another era began with the gradual advent of central venous catheterization, made possible in human studies by the heroic efforts of Werner Forssman (1929) and O. Klein (1930). Andre Cournand and Helmut Ranges in 1941 implemented central catheterization for use in humans (31). This permitted catheters to be placed under fluoroscopic guidance into renal and hepatic veins to measure extraction and clearance of indicators (i.e., direct Fick method). In the 1950s the noncannulating electromagnetic and ultrasonic flowmeters were introduced. These instruments could be surgically placed around the superior mesenteric and renal arteries to provide continuous estimates of blood flow in experimental animals. During the 1970s injections of radioactively labeled microspheres became another accepted method. The microspheres are mixed in the bloodstream and are distributed to all organs (and lodged in their microvessels) in proportion to each organ's blood flow.

Renal Blood Flow in Exercising Dogs and a New Debate

None of the measurements of PAH clearance in exercising dogs during the 1950s revealed any significant changes in renal blood flow or in clearances of either inulin or creatinine (123). Ewald E. Selkurt (1962) concluded that "the dog and man differ significantly in the response of renal blood flow to exercise" (123). Application of more invasive techniques to dogs yielded the same results as the renal clearance methods. In 1961 Rushmer, Franklin, Van Citters, and Smith implanted pulsed ultrasonic flowmeters on one renal artery and saw no exercise-induced fall in blood flow (117). It was argued by some that the different responses of dogs vs. humans existed because the level of treadmill exercise was too mild for dogs. Rushmer's colleagues, however, argued that the different responses of the two species were not attributable to "species differences" but rather to basic flaws in the clearance methods. The most serious alleged flaw was that they were *"indirect"* measures of blood flow whereas the flowmeters were perceived as providing *direct* measurements (certainly a contested point) (144). A series of studies by Stephen Vatner and his colleagues, first in Seattle and then in Boston, in 1971 and 1972, put to rest the argument concerning severity of exercise. They observed no significant changes in renal blood flow and only small increments (28%–50%) in renal and mesenteric (see below) vascular resistance in dogs during moderate to severe exercise (for refs. see 144). As revealed in the following sections, the dispute over methods was sorted out gradually.

Renal Blood Flow in Humans and Primates

The evidence, based on PAH clearance, for marked decrements in renal blood flow in upright exercising man is long-standing (12, 146). A group of important studies from Sweden by Gunnar Grimby (53) and Jan Castenfors (20) between 1965 and 1967 provided multiple measurements of renal blood flow in their subjects over a wide range of exercise intensities. Their results revealed that renal blood flow falls in proportion to the severity of exercise and declines by as much as 80% at near-maximal levels. Both Grimby and Castenfors found a high negative correlation between renal blood flow and heart rate during exercise. The negative correlation coefficients, slopes, and intercepts of their regression lines were virtually the same (108). Both investigators also measured PAH extraction by renal venous catheterization and found no effect of exercise and decreased blood flow on renal extraction. Thus clearance was directly related to renal blood flow, refuting one earlier criticism of the method.

In 1991 B. Tidgren and coworkers measured renal venous outflow from one kidney by thermal dilution technique in humans during graded exercise (139). They observed the same graded decrements in flow as Grimby and Castenfors and also saw increments in the overflow of norepinephrine, renin, and immunoreactive neuropeptide Y into the renal vein. All these experiments reflected the graded renal responses to progressive increases in renal sympathetic nerve activity during graded exercise.

Surprisingly, in 1978 Vatner and coworkers found no significant decrease in renal arterial blood flow in free-ranging baboons when they exercised (but renal vascular resistance did rise significantly) (145). Because of the phylogenetic closeness of baboons to humans, these results cast some doubts on human findings. Conversely, Roger Hohimer and Orville Smith (1979), in Seattle, found that renal arterial blood flow (electromagnetic flowmeter) fell in proportion to the severity of leg cycling exercise and was closely related to the rise in heart rate in baboons, as in humans (74). Blood flow to the contralateral denervated kidney transiently rose with exercise and then returned to resting baseline as exercise continued. Thus the renal vasoconstriction was caused by sympathetic nerves. As expected and revealed below, renal responses were paralleled by those subsequently found for the splanchnic region (and other vascular beds) in humans and primates. In this one case the conflict with Vatner remains unresolved (see below).

Splanchnic Blood Flow

Studies on Dogs

As with renal blood flow, little or no decrease in splanchnic blood flow was observed as measured by superior mesenteric arterial flowmeters (117, 144). Van Citters and Franklin (1969) found no decrease in mesenteric blood flow (nor in renal blood flow) in Alaskan sled dogs during severe exercise in their natural environment (142). Also the studies employing radioactive microspheres revealed no decrements in renal blood flow and either no change or small but significant decreases in visceral organ blood flow (for refs. see 73).

Splanchnic Blood Flow in Humans and Primates

Humans. In 1945 Stanley Bradley and his colleagues at Columbia University established a method for determining total hepatic–splanchnic blood flow in dogs and humans by measuring hepatic clearance rate and extraction of sulfobromophathalein sodium (BSP). Measurement of extraction required catheterization of an hepatic vein. In 1948 Bradley and his colleagues made the first measurements of splanchnic blood flow in humans during supine exercise and they observed a significant decrease (15). Later, Bradley's group (1955) devised a two-indicator method for simultaneously determining hepatic–splanchnic mean transit time and splanchnic blood flow in order to derive total splanchnic blood volume. In 1956 O.L. Wade and his colleagues at Birmingham, England, together with A. Cournand and S. Bradley applied these techniques to normal subjects and found that splanchnic blood flow fell 350 ml min^{-1} or by only 20% but splanchnic blood volume fell by 35% during moderate supine exercise (147).

In John M. Bishop's thesis for doctor of medicine at the University of Birmingham in 1956, he described the widening A–V oxygen difference of several organs and the limbs during mild-to-heavy supine exercise in normal subjects and in cardiac patients with different degrees of functional limitation. The marked widening of the hepatic–splanchnic A–V oxygen difference (the assumption was that organ's oxygen consumption [$\dot{V}O_2$] did not change) led Bishop to conclude that splanchnic blood flow normally decreased in proportion to the severity of supine exercise. He calculated a reduction of 60% during moderately heavy exercise (146).

In 1964, Rowell, Robert A. Bruce, and John R. Blackmon in Seattle employed a bolus dye (indocyanine green) injection technique and measured its hepatic clearance at rest and at multiple levels of exercise up to those requiring a maximal oxygen consumption ($\dot{V}O_2$max) (110). Hepatic extraction of both dye and oxygen increased and dye clearance rate fell in proportion to work intensity. These changes all paralleled the reductions in blood flow. The widening of hepatic A–V oxygen difference also paralleled the decrements in splanchnic blood flow as had been assumed earlier by J.M. Bishop. The slopes and intercepts of the regression lines and the correlation coefficients relating the decrements in splanchnic blood flow to heart rate were virtually identical to those for renal blood flow found by Grimby (1965), Castenfors (1967), and later by Tidgren (1993). As with renal blood flow, decrements in splanchnic blood flow reached 70%–80% at or near $\dot{V}O_2$max. Also, relative to both heart rate and to the percent of $\dot{V}O_2$max required, decrements in both renal and splanchnic blood flows were the same for sedentary and physically well-conditioned individuals but were *very different relative to absolute* $\dot{V}O_2$ (28, 110). Taken together, this group of studies from Sweden, Copenhagen, and Seattle revealed that sympathetic neural outflow to the heart and to the splanchnic organs and kidneys was increased by the same relative degree and in a highly predictable manner (28, 108, 110, 114).

Primates. Hohimer, Hales, Rowell, and Smith in 1983 repeatedly measured the distribution of radioactive microspheres containing multiple nuclide labels in baboons during rest and voluntary exercise. Blood flows to the kidneys, all splanchnic organs, body skin, nonworking muscles, and adipose tissue all fell by 11%–31%. This revealed an active redistribution of blood flow like that seen in humans. Cerebral blood flow did not change (73).

Effects of Circulatory Impairment on Redistribution of Cardiac Output

In their 1962 monograph Owen Wade and J.M. Bishop from Birmingham, England, reviewed their extensive measurements of regional A–V oxygen differences along with the related findings of others from cardiac patients and normal subjects (146). Their research was stimulated by their surprising observation that the oxygen saturation of mixed venous blood dropped to less than 10% in heart patients during mild exercise. They then observed that superior vena caval blood returning from the inactive upper one-half of the body was as low in oxygen saturation as blood returning from the exercising legs. Thus a pronounced redistribution of cardiac output must have reduced blood flow to inactive regions to a point at which the oxygen extraction of these regions equalled that of active muscle. The marked widening of A–V oxygen difference suggested that organ blood flow reached the same low levels during mild supine exercise in patients as seen subsequently in normal individuals, including athletes, during maximal upright exercise. In 1967, John Blackmon and colleagues in Seattle measured hepatic clearance of indocyanine green during rest and upright exercise that required 30%–100% of $\dot{V}O_2$max in patients with pure mitral stenosis (13). They found the slope relating indocyanine green clearance to *percent of* $\dot{V}O_2$ max required was normal but relative *to absolute* $\dot{V}O_2$, the decline was abnormally steep, thus confirming Bishop's estimates based on oxygen extraction.

The studies of Robert Zelis from Pennsylvania State University in 1969 also underlined the intense sympathetic vasoconstriction seen in cardiac patients during exercise. His group saw intense vasoconstriction of *active muscle* which limited muscle vasodilation and prevented hypotension, as well as vasoconstriction of skin which impeded temperature regulation (152). The vasoconstrictor responses in the resting forearm were far greater than those established for normal subjects by S. Bevegard and John Shepherd from the Mayo Clinic in the 1960s (12).

During 1970–1974, Stephen Vatner and his colleagues in Boston measured renal arterial and superior mesenteric arterial blood flow in dogs during severe voluntary exercise after oxygen delivery to the working muscle had been impaired by (1) surgically induced heart block, (2) congestive right heart failure, (3) severe chronic anemia, and (4) splenectomy. In contrast to the normal dog in which visceral organ blood flow is maintained during severe exercise, these impaired dogs sustained marked decreases (up to 90%) in renal and mesenteric blood flows during severe exercise (144). Again, the marked decrements in regional blood flows that accompany severe exercise in normal humans also occur in patients with circulatory impairment, but at much lower levels of exercise in the patients (146).

Resolution of the Dispute

It eventually became clear that the different degrees of regional vasoconstriction in dogs as opposed to humans during exercise could not be attributed to the methods used to measure blood flow. Things began to come together when Vatner and colleagues revealed that impairment of oxygen delivery in the dog produces a human-like redistribution of blood flow during exercise. Humans normally have only one-half to one-third of the cardiac pumping capacity per kilogram of body weight of a dog. Thus when the capacities to transport oxygen became similar in the two species so did their regional vasomotor responses.

The Importance of Regional Vasoconstriction in Humans

The early ideas of Dastre and Morat (1884), of DeJager (1886), and especially of Krogh (1912) regarding the distribution of blood flow had gradually come to life. In addition to the importance of this regional vasoconstriction to both the redistribution of oxygen from less active to more active regions and to the maintenance of arterial pressure, there is the greater importance (in the context of this chapter) to the maintenance of end-diastolic volume. The opponents of Starling's concept did not accept the need for any redistribution of blood volume nor did they find evidence that blood volume is redistributed during exercise by active regional vasoconstriction in accordance with Krogh's model (1912) so that end-diastolic volume could be increased or maintained in accordance with Starling's law.

When, in 1973, E. Ashkar from Argentina surgically eliminated neural control of splanchnic blood vessels and the adrenal medulla by thoracic sympathectomy, his dogs could no longer maintain cardiac output in exercise owing to the fall in stroke volume (3). Stroke volume fell because without vasoconstriction splanchnic blood volume must have risen and depleted central blood volume and lowered ventricular filling pressure. Thus there had to be enough vasoconstriction to at least prevent a rise in splanchnic blood flow and blood volume. In humans loss of vasoconstrictor control of splanchnic organs by surgical damage to the celiac and superior mesenteric ganglia had similar deleterious consequences (143). Flamm's and colleagues' findings from humans, mentioned earlier, showed that in addition to the blood volume driven centrally by the muscle pump, reductions in hepatic, splanchnic, and total abdominal blood volume (about 20%) contributed significantly to the rise in thoracic blood volume (42). When regional vasoconstriction is lost, end-diastolic volume can no longer be maintained, just as Krogh would have predicted.

The Problem of Cutaneous Vasodilation

The combination of exercise and hyperthermia probably imposes the most severe regulatory problems ever placed on the human cardiovascular system, except possibly for severe hemorrhage. There are two problems:

1. Cutaneous vasodilation translocates blood volume into the uniquely large and capacious cutaneous circulation of humans. This lowers ventricular filling pressure and stroke volume and consequently reduces cardiac pumping capacity at a time when demands for blood flow are the greatest ever experienced.
2. Skeletal muscle and skin circulations must compete for blood flow because their combined needs can exceed cardiac pumping capacity.

Demands imposed on the circulations of other terrestrial mammals by heat do not parallel those in humans, as revealed, for example, by the studies of J.R.S. Hales and his colleagues (56). The situation in humans who, unlike other mammals require high levels of skin blood flow, brought to life the ideas of Krogh, Guyton, and Caldini and coworkers. The predictions based on Krogh's two-compartment model are realized when an increasingly large fraction of cardiac output is distributed to a compliant region like skin—namely end-diastolic volume should fall. An old idea which was, in general, counter to Krogh's thinking was that the demands for skin and muscle blood flow were simply summed during exercise so that a high cardiac output fully supplied both demands until a normal maximal cardiac output and stroke volume were achieved. During heat stress, maximal cardiac output would be achieved but at a much lower $\dot{V}O_2$ because a smaller fraction of that total would supply active muscle, and skin extracts so little oxygen. However, in some earlier studies by Dill and colleagues (1931), Asmussen (1940), Williams and Windham and colleagues in South Africa (1962) in which cardiac output was measured by acetylene and CO_2 rebreathing techniques, heat stress had no significant effect on cardiac output during moderate exercise but stroke volume was always much lower in the heat (for refs. see 108).

During 1965–1970, Rowell and colleagues examined the central circulatory consequences of cutaneous vasodilation during exercise (108). Exercise was either mild and prolonged or graded from moderate to near maximal in both neutral and in hot, dry environments (43°–49°C). In mild exercise, heat stress raised cardiac output by an additional 2–3 l min^{-1} by markedly elevating heart rate, but it lowered stroke volume and central (thoracic) blood volume (CBV) by 15%–20% (e.g., 112). In moderate-to-severe graded exercise owing to the lower stroke volume cardiac output could only be maintained at its normothermic levels during moderate exercise plus heat by abnormally high heart rates. But as work intensity rose toward maximal levels in the heat, heart rate prematurely reached maximum values so that the low stroke volumes actually reduced cardiac output below the normal values seen for this work intensity in cooler conditions (112).

In additional studies at the same ambient temperatures, hepatic splanchnic clearance and extraction of indocyanine green were determined at the same levels of exercise (108). At any given $\dot{V}O_2$, splanchnic blood flow was reduced by an additional

20% with heat stress. But this vasoconstriction clearly did not passively displace a sufficient quantity of blood volume from the splanchnic veins to prevent the large decline in CBV and stroke volume (108, 112).

A closer look at the hemodynamics was enabled by the use of water-perfused suits (designed for heating or cooling Apollo astronauts) in order to control body skin temperature (T_s) and skin blood flow and to switch these variables from normal to high and to low values during prolonged continuous exercise (113). When T_s and skin blood flow were raised, right atrial pressure fell and CBV dropped suddenly; this revealed a transient period in which left ventricular output momentarily exceeded right ventricular output owing to rising cutaneous venous volume and a momentary reduction in venous return. Also cardiac output rose steeply at first but gradually leveled off when right atrial pressure approached zero. Forearm skin blood flow was later shown by Brengelmann and colleagues (17) and John Johnson and coworkers (78) to respond in a pattern similar to that of cardiac output.

One question was why cardiac output and skin blood flow leveled off despite a continuously increasing body temperature and demand for blood flow. In 1993, D.D. Sheriff and colleagues controlled the cardiac outputs of voluntarily exercising dogs to examine the interaction between cardiac output, stroke volume, central venous pressure, and heart rate (127). They saw that lowering heart rate and cardiac output (by ventricular pacing) raised central venous pressure and stroke volume whereas raising heart rate and cardiac output lowered central venous pressure and stroke volume just as this occurs at rest. Cardiac output could only be raised until central venous pressure approached 0 mmHg and at that point further rise in cardiac output would be prevented by collapse of the cavae (54, 55, 127). These experiments and those that manipulated skin blood flow in humans illustrated the concepts of Krogh and Guyton concerning the need for peripheral adjustments to maintain central venous pressure and stroke volume. In short, when cardiac output was raised, either by heat stress or by pacing, a portion of the rise went to a compliant region and thereby lowered central venous pressure and stroke volume; the rise in cardiac output during heating became limited when central venous pressure approached zero.

When the skin was suddenly cooled after the heating in the exercising humans (113), stroke volume, central venous pressure, and CBV immediately rose owing both to the sudden cutaneous venoconstriction, which actively expelled cutaneous blood volume back to the heart, and to the rapid fall in skin blood flow, which passively expelled its volume and raised venous return momentarily. So for a moment right ventricular output exceeded left ventricular output, causing sudden restoration of CBV and also central venous pressure and stroke volume. In short, heating lowered end-diastolic volume and stroke volume whereas cooling raised end-diastolic volume and stroke volume (113).

Experiments that have, for example, experimentally manipulated the skin circulation in humans or cardiac output in other animals have provided a unique view of the interdependence of cardiac and peripheral vascular functions that was the prediction of August Krogh and demonstrated in Guyton's experiments on venous return. In short, the cardiac *muscle length–tension relationship* and the peripheral vascular *pressure–volume–flow relationship* represent the key functional links in the mechanical coupling between the heart and peripheral circulation.

The final decades of the twentieth century leave us little doubt about the importance of the Frank–Starling mechanism as one determinant of cardiac performance. However, those who opposed its applicability were nevertheless justified in their support for neural and humoral control, which are also important but not exclusively so. Today we are better able to assess quantitatively how all of these mechanisms act in concert to regulate cardiac performance. Dynamic exercise has provided both a crucial proving ground for these ideas and a powerful tool for integrating them into whole-body function.

Some of the investigators who contributed to a portion of this history have been good friends and colleagues and mentors. Their contributions to science and to me personally are acknowledged with deep gratitude. I am also grateful to my colleagues, Drs. George Brengelmann, Jerry Dempsey, Eric Feigl, and Allen Scher, who provided information and advice during the preparation of this chapter; to Mrs. Judy Schroeder, who prepared the manuscript; and to Mrs. Pam Stevens Campbell, who prepared the bibliography.

REFERENCES

1. Anrep C.V. The circulation in striated and plain muscles in relation to their activity. In: *The Harvey Lectures*. Baltimore: Williams and Wilkins, 1936, pp. 146–169.
2. Anrep C.V. and E. von Saalfeld. The blood flow through the skeletal muscle in relation to its contraction. *J. Physiol. (Lond.)* 85:375–399, 1935.
3. Ashkar E. Effects of bilateral splanchnicectomy on circulation during exercise in dogs. *Acta Physiol. Lat. Am.* 23:171–177, 1973.
4. Asmussen E. Muscular exercise. In: *Handbook of Physiology. Respiration*, edited by W.O. Fenn and H. Rahn. Washington, DC: American Physiological Society, 1965, Sect. 3, Vol. II, Chapt. 36, pp. 939–978.
5. Asmussen E. and M. Nielsen. Cardiac output during muscular work and its regulation. *Physiol. Rev.* 35:778–800, 1955.
6. Barcroft H. and S. Samaan. Explanation of the increase in systemic flow caused by occluding the descending thoracic aorta. *J. Physiol. (Lond.)* 85:47–61, 1935.
7. Barcroft J. *Features in the Architecture of Physiological Function*. Cambridge: Cambridge University Press, 1934.
8. Barcroft H. and A.C. Dornhorst. The blood flow through the human calf during rhythmic exercise. *J. Physiol. (Lond.)* 109:402–410, 1949.
9. Bayliss W.M. On the local reactions of the arterial wall to changes of internal pressure. *J. Physiol. (Lond.)* 28:220–231, 1902.
10. Bezold A. von. *Untersuchungen uber die Innervation des Herzens*. Leipzig: W. Englemann, 1863.
11. Bezold A. von and L. Hirt. Ueber die physiologischen Wirkungen des essigsauren Veratrin's. *Untersuch. Physiol. Lab. Würzburg* 1:75, 1867.
12. Bevegård B.S. and J.T. Shepherd. Regulation of the circulation during exercise in man. *Physiol. Rev.* 47:178–213, 1967.
13. Blackmon J.R., L.B. Rowell, J.W. Kennedy, R.D. Twiss, and R.D. Conn. Physiological significance of maximal oxygen intake in pure mitral stenosis. *Circulation* 36:497–510, 1967.

14. Bock A.V. and D.B. Dill. *The Physiology of Muscular Exercise* (3rd ed.). London: Longmans, Green, and Co., 1931.

15. Bradley S.E. Hepatic blood flow. Effect of posture and exercise upon blood flow through the liver. In: *Transactions of the Seventh Conference on Liver Injury*, edited by F.W. Hoffbauer. New York: Josiah Macy, Jr. Found., 1948, pp. 53–56.

16. Brecher G.A. *Venous Return*. New York: Grune and Stratton, 1956.

17. Brengelmann G.L., J.M. Johnson, L. Hermansen, and L.B. Rowell. Altered control of skin blood flow during exercise at high internal temperature. *J. Appl. Physiol.: Respir. Environ. Exerc. Physiol.* 43:790–794, 1977.

18. Bronk D.W. and G. Stella. Afferent impulses in the carotid nerve. I. The relation of the discharge from single end organs to arterial blood pressure. *J. Cell. Comp. Physiol.* 1:113–130, 1932.

19. Caldini P., S. Permutt, J.A. Waddell, and R.L. Riley. Effect of epinephrine on pressure, flow, and volume relationships in the systemic circulation of dogs. *Circ. Res.* 34:606–623, 1974.

20. Castenfors J. Renal clearances and urinary sodium and potassium excretion during supine exercise in normal subjects. *Acta Physiol. Scand.* 70:207–214, 1967.

21. Celander O. The range of control exercised by the "sympathico-adrenal system." *Acta Physiol. Scand. Suppl.* 116:1–132, 1954.

22. Celander O. and B. Folkow. Are parasympathetic vasodilator fibres involved in depressor reflexes elicited from the barorecepror regions? *Acta Physiol. Scand.* 23:64–77, 1951.

23. Chapman C.B., A. Henschel, J. Minckler, A. Forsgren, and A. Keys. The effect of exercise on renal plasma flow in normal male subjects. *J. Clin. Invest.* 27:639–644, 1948.

24. Chapman C.B., H.L. Taylor, C. Borden, R.V. Ebert, A. Keys, and W.S. Carlson. Simultaneous determination of the resting arteriovenous difference by the acetylene and direct Fick methods. *J. Clin. Invest.* 29:651–659, 1950.

25. Chauveau A. and J. Marey. Détermination graphique des rapports du choc du coeur avec les mouvements des oreillettes et des ventricules: expérience faite à l'aide d'un appareil enregistreur (sphygmographe). *C. R. Acad. Sci. (Paris)* 53:622, 1861.

26. Chauveau A. and M. Kaufmann. Expériences pour la détermination du coefficient de l'activité nutritive et respiratoirs des muscles en repos et en travail. *C.R. Acad. Sci. (Paris)* 104:1126, 1887.

27. Christensen E.H. Beitrage zur Physiologie schwerer körperlicher Arbeit. Gas analytische Methoden zur Bestimmung des Herzminutenvolumens in Ruhe und während körperlicher Arbeit. *Arbeitsphysiologie* 4:175–202, 1931.

28. Clausen J.P. Effect of physical training on cardiovascular adjustments to exercise in man. *Physiol. Rev.* 57:779–815, 1977.

29. Cooper A. Some experiments and observations on tying the cartoid and vertebral arteries, and the pneumo-gastric, phrenic, and sympathetic nerves. *Guy's Hosp. Rep.* 1:457–475, 1836.

30. Cournand A. Air and blood. In: *Circulation of Blood: Men and Ideas*, edited by A.P. Fishman and D.W. Richards. Bethesda, MD: American Physiological Society, 1982, Chapt. I.

31. Cournand A. and H.A. Ranges. Catheterization of the right auricle in man. *Proc. Soc. Exp. Biol.* 46:462–466, 1941.

32. Cyon E. de and C. Ludwig. Die Reflexe eines der sensiblen Nerven des Herzens auf die motorischen-nerven der Blutgefässe. *Arb. Physiol. Anstalt Leipzig* 128–149, 1867.

33. Dastre A. and J.P. Morat. *Recherches expérimentales sur le système nerveax vasomoteur*. Paris: Masson, 1884.

34. DeJager S. Experiments and considerations of haemodynamics. *J. Physiol. (Lond.)* 7:130–215, 1886.

35. Dexter L., J.L. Whittenberger, F.W. Haynes, W.T. Goodale, R. Gorlin, and C.G. Sawyer. Effect of exercise on circulatory dynamics of normal individuals. *J. Appl. Physiol.* 3:439–453, 1951.

36. Donald D.E. and J.T. Shepherd. Response to exercise in dogs with cardiac denervation. *Am. J. Physiol.* 205:393–400, 1963.

37. Ead H.W., J.H. Green, and E. Neil. A comparison of the effects of pulsatile and non-pulsatile blood flow through the carotid sinus on the reflexogenic activity of the sinus baroreceptors in the cat. *J. Physiol. (Lond.)* 118:509–519, 1952.

38. Ekelund L.-G. Circulatory and respiratory adaptation during prolonged exercise. *Acta Physiol. Scand.* 70(*Suppl.*):292, 1967.

39. Eyster J.A.E. and D.R. Hooker. Direct and reflex response of the cardio-inhibitory centre to increased blood pressure. *Am. J. Physiol.* 21:373–399, 1908.

40. Fick A. Ueber die Messung des Blutquantums in den Herzventrikeln. *S.B. phys.-med. Ges. Würzburg*, July 9, 1870.

41. Fishman A.P. and D.W. Richards (eds). *Circulation of Blood. Men and Ideas.* Bethesda, MD: American Physiological Society, 1982.

42. Flamm S.D., J. Taki, R. Moore, S.F. Lewis, F. Keech, F. Maltais, M. Ahmad, R. Callahan, S. Dragotakes, N. Alpert, and H.W. Strauss. Redistribution of regional and organ blood volume and effect on cardiac function in relation to upright exercise intensity in healthy human subjects. *Circulation* 81:1550–1559, 1990.

43. Folkow B., P. Gaskell, and B.A. Waller. Blood flow through limb muscles during heavy rhythmic exercise. *Acta Physiol. Scand.* 80:61–72, 1970.

44. Folkow B., U. Haglund, M. Joday, and O. Lundgren. Blood flow in the calf muscle of man during heavy rhythmic exercise. *Acta Physiol. Scand.* 81:157–163, 1971.

45. Frank O. (1895) On the dynamics of cardiac muscle. English translation by C.B. Chapman and E. Wasserman. *Am. Heart J.* 58:282–317, 467–478, 1959.

46. Franklin D.L., R.L. Van Citters, and R.F. Rushmer. Balance between right and left ventricular output. *Circ. Res.* 10:17–26, 1962.

47. Franklin K.J. *A Monograph on Veins.* Springfield, IL: C.C. Thomas, 1937.

48. Gaskell W.H. Further researches on the vasomsotor nerves of ordinary muscles. *J. Physiol. (Lond.)* 1:262–302, 1878–1879.

49. Goltz F. Reflexlähmung des Tonus der Gefässe. *Zbl. Med. Wiss* 2:625, 1864.

50. Grande Covian F. and P.B. Rehberg. Über die Nierenfunktion während schwerer Muskelarbeit. *Skand. Arch. f. Physiol.* 75:21–37, 1936.

51. Gorczynski R.J., B. Klitzman, and B.R. Duling. Interrelations between contracting striated muscle and precapillary microvessels. *Am. J. Physiol.* 235 (*Heart Circ. Physiol.* 4): H494–H504, 1978.

52. Gould K.L. Quantitative imaging in nuclear cardiology. *Circulation* 66:1141–1146, 1982.

53. Grimby G. Renal clearances during prolonged supine exercise at different loads. *J. Appl. Physiol.* 20:1294–1298, 1965.

54. Guyton A.C. Venous return. In: *Handbook of Physiology. Circulation,* edited by W.F. Hamilton and P. Dow. Washington, DC: American Physiological Society, 1963, Sect. 2, Vol. II, pp. 1099–1133.

55. Guyton A.C., B.H. Douglas, J.B. Langston, and T.Q. Richardson. Instantaneous increase in mean circulatory pressure and cardiac output at onset of muscular activity. *Circ. Res.* 11:431–441, 1962.

56. Hales J.R.S. Thermoregulatory requirements for circulatory adjustments to promote heat loss in aminals: a review. *J. Thermal Biol.* 8:219–224, 1983.

57. Hales S. *Statistical Essays. Vol. II. Haemastatiks.* London: Innys and Manby, 1733.

58. Hamilton W.F. The physiology of cardiac output. *Circulation* 8:527–543, 1953.

59. Hamilton W.F. Measurement of cardiac output. In: *Handbook of Physiology. Circulation,* edited by W.F. Hamilton and P. Dow. Washington, DC: American Physiological Society, 1962, Sect. 2, Vol. I, pp. 551–584.

60. Hamilton W.F. and D.W. Richards. The output of the heart. In: *Circulation of Blood: Men and Ideas,* edited by A.P. Fishman and D.W. Richards. Bethesda, MD: American Physiological Society, 1982, Chapt. II.

61. Hammond H.K., F.C. White, V. Bhargava, and R. Shabetai. Heart size and maximal cardiac output are limited by the pericardium. *Am. J. Physiol.* 263 (*Heart Circ. Physiol.* 32): H1675–H1681, 1992.

62. Hargens A.R., R.W. Millard, K. Pettersson, and K. Johansen. Gravitational haemodynamics and oedema prevention in the giraffe. *Nature* 329:59–60, 1987.

63. Harvey W. *Exercitatio anatomica de motu cordis et sanguinis in animalibus,* with an English translation by C.D. Leake. Springfield, IL: Charles C. Thomas, 1928.

64. Herrick J.F., J.H. Grindlay, E. J. Baldes, and F.C. Mann. Effect of exercise on the blood flow in the superior mesenteric, renal, and common iliac arteries. *Am. J. Physiol.* 128: 338–344, 1940.

65. Hering H.E. *Karotissinusreflexe auf Herz und Gefässe, vom normalphysiologischen, pathologische-physiologischen und klinischen Standpunkt.* Dresden: Steinkopff, 1927.

66. Heymans C. and A. Ladon. Sur le mécanisms de la bradycardie hypertensive et adrénalinique. *C.R. Soc. Biol.* (*Paris*) 90:966, 1924.

67. Heymans C. and G. Van Den Heuvel-Heymans. Action of drugs on arterial wall of carotid sinus and blood pressure. *Arch Int. Pharmacodyn.* 83:520–528, 1950.

68. Heymans C. and E. Neil. *Reflexogenic Areas of the Cardiovascular System.* London: Churchill, 1958.

69. Heymans C.J.F. and B. Folkow. Vasomotor control and the regulation of blood pressure. In: *Circulation of Blood: Men and Ideas,* edited by A.P. Fishman and D.W. Richards. Bethesda, MD: American Physiological Society, 1982, Chapt. VI.

70. Higginbotham M.B., K.G. Morris, R.S. Williams, P.A. McHale, R.E. Coleman, and F.R. Cobb. Regulation of stoke volume during submaximal and maximal upright exercise in normal man. *Circ. Res.* 58:281–291, 1986.

71. Hoff H.H. and R. Guillermin. Claude Bernard and the vasomotor system. In: *Claude Bernard and Experimental Medicine,* edited by F. Grande and M.B. Visscher. Cambridge, MA: Schenkman Publishing, 1967.

72. Hoffman J.I.E., A. Guz, A.A. Charlier, and D.E.L. Wilcken. Stroke volume in conscious dogs: effect on respiration, posture, and vascular occlusion. *J. Appl. Physiol.* 20: 865–877, 1965.

73. Hohimer A.R., J.R.S. Hales, L.B. Rowell, and O.A. Smith. Regional distribution of blood flow during mild dynamic leg exercise in the baboon. *J. Appl. Physiol.: Respirat. Environ. Exerc. Physiol.* 55:1173–1177, 1983.

74. Hohimer A.R. and O.A. Smith. Decreased renal blood flow in the baboon during mild dynamic leg exercise. *Am. J. Physiol.* 236 (*Heart Circ. Physiol.* 5):H141–H150, 1989.

75. Holmgren A. Circulatory changes during muscular work in man: with special reference to arterial and central venous pressures in the systemic circulation. *Scand. J. Clin. Lab. Invest.* 8 (*Suppl.* 24):1–97, 1956.

76. Howell W.H. and F. Donaldson, Jr. Experiments upon the heart of the dog with reference to the maximum volume of blood sent out by the left ventricle in a single beat, and the influence of variations in venous pressure, arterial pressure, and pulse-rate upon the work done by the heart. *Philos. Trans. Pt.* I:139–160, 1884.

77. Janicki J.S., D.D. Sheriff, J.L. Robotham, and R.A. Wise. Cardiac output during exercise: contributions of the cardiac, circulatory, and respiratory systems. In: *Handbook of Physiology: Exercise, Regulation and Integration of Multiple Systems,* edited by L.B. Rowell and J.T. Shepherd. Bethesda, MD: American Physiological Society, 1996, pp. 649–704.

78. Johnson J.M. Nonthermoregulatory control of human skin blood flow. *J. Appl. Physiol.* 61:1613–1622, 1986.

79. Koch E. *Die reflektorische Selbststeuerung des Kreislaufes.* Dresden: Steinkopff, 1931.

80. Korner P.I. Central nervous control of autonomic cardiovascular function. In: *Handbook of Physiology, The Cardiovascular System, The Heart,* edited by R.M. Berne and N. Sperelakis. Bethesda, MD: American Physiological Society, 1979, pp. 691–739.

81. Köster G. and A. Tschermak. Ueber den N. depressor als Reflexner der Aorta. *Pflügers Arch. ges Physiol.* 93:24, 1903.

82. Krocker E.J. and E.H. Wood. Comparison of simultaneously recorded central and peripheral arterial pressure pulses during rest, exercise and tilted position in man. *Circ. Res.* 3:623–632, 1955.

83. Krogh A. Regulation of the supply of blood to the right heart (with a description of a new circulation model). *Scand. Arch. Physiol.* 27:227–248, 1912.

84. Lambert E.H. and E.H. Wood. The use of a resistance wire, strain gauge manometer to measure intraarterial pressure. *Proc. Soc. Exp. Biol. Med.* 64:186–190, 1947.

85. Landgren S. On the excitation mechanism of the carotid baroreceptors. *Acta Physiol. Scand.* 26:1–34, 1952.

86. Landis E.M. The capillary circulation. In: *Circulation of Blood: Men and Ideas,* edited by A.P. Fishman and D.W. Richards. Bethesda, MD: American Physiological Society, 1982, Chapt. IV.

87. Latschenberger J. and A. Deahna. Beiträge zur Lehre von der reflectorischen Erregung der Gefässmuskeln. *Pflügers Arch. ges Physiol.* 12:157–204, 1876.

88. Laughlin M.H. Skeletal muscle blood flow capacity: role of muscle pump in exercise hyperemia. *Am. J. Physiol.* 253 (*Heart Circ. Physiol.* 22):H993–H1004, 1987.

89. Laughlin M.H., R. Korthuis, D. Duncker, and R.J. Bache. Control of blood flow to cardiac and skeletal muscle during exercise. In: *Handbook of Physiology: Exercise, Regulation and Integration of Multiple Systems,* edited by L.B. Rowell and J.T. Shepherd. Bethesda, MD: American Physiological Society, 1996, pp. 705–769.

90. Levy M.N. The cardiac and vascular factors that determine systemic blood flow. *Circ. Res.* 44:739–746, 1979.

91. Lower R. *Tractatus de Corde.* Translated by K.J. Franklin. In: R.T. Gunther, *Early Science in Oxford.* Oxford: Clarendon Press, 1932, Vol. IX.

92. Ludbrook J. *Aspects of Venous Function in the Lower Limbs.* Springfield, IL: Charles C. Thomas, 1966.

93. McCubbin J.W., J.H. Green, and I.H. Page. Baroreceptor function in chronic renal hypertension. *Circ. Res.* 4:205–210, 1956.

94. McDonald D.A. *Blood Flow in Arteries.* Baltimore: Williams and Wilkins Co., 1974.

95. McDowall R.J.S. *The Control of the Circulation of the Blood.* London: Longmans, Green, 1938.

96. McMichael J. and E.P. Sharpey-Shafer. Cardiac output in man by a direct Fick method. *Brit. Heart J.* 6:33–40, 1944.

97. Moreno A.H., A.R. Burchell, R. van der Woude, and J.H. Burke. Respiratory regulation of splanchnic and systemic venous return. *Am. J. Physiol.* 213:455–465, 1967.

98. Papelier Y., P. Escourrou, J.P. Gauthier, and L.B. Rowell. Carotid baroreflex control of blood pressure and heart rate in man during dynamic exercise. *J. Appl. Physiol.* 77:502–506, 1994.

99. Patterson S.W., H. Piper, and E.H. Starling. The regulation of the heart beat. *J. Physiol.* (*Lond.*) 48:465–513, 1914.

100. Poliner L.R., G.J. Dehmer, S.E. Lewis, R.W. Parkey, C.G. Blomqvist, and J.T. Willerson. Left ventricular performance in normal subjects: a comparison of the responses to exercise in the upright and supine positions. *Circulation* 62:528–534, 1980.

101. Pollack A.A. and E.H. Wood. Venous pressure in the saphenous vein at the ankle in man during exercise and changes in posture. *J. Appl. Physiol.* 1:649–662, 1949.

102. Potts J.T., S.R. Shi, and P.B. Raven. Carotid baroreflex responsiveness during dynamic exercise in humans. *Am. J. Physiol.* 265 (*Heart Circ. Physiol.* 34):H1928–H1938, 1993.

103. Pourfois du Petit F. Mémoire dans lequel il est demonstré que les nerfs intercostaux fournissent des rameaux qui portent des esprits dans les nerfs. *Hist. Acad. Roy. Sc. Paris*, 1–19, 1727.

104. Radigan L.R. and R. Robinson. Effects of environmental heat stress and exercise on renal blood flow and filtration rate. *J. Appl. Physiol.* 2:185–191, 1949.

105. Reeves J.T., B.M. Groves, A. Cymerman, J.T. Sutton, P.D. Wagner, D. Turkevich, and C.S. Houston. Operation Everest II: cardiac filling pressures during cycle exercise at sea level. *Respir. Physiol.* 80:147–154, 1990.

106. Reeves J.T. and A.E. Taylor. Pulmonary hemodynamics and fluid exchange in the lungs during exercise. In: *Handbook of Physiology: Exercise, Regulation and Integration of Multiple Systems*, edited by L.B. Rowell and J.T. Shepherd. Bethesda, MD: American Physiological Society, 1996, pp. 558–613.

107. Robinson B.R., S.E. Epstein, R.L. Kahler, and E. Braunwald. Circulatory effects of acute expansion of blood volume: studies during maximal exercise and at rest. *Circ. Res.* 19:26–32, 1966.

108. Rowell L.B. Human cardiovascular adjustments to exercise and thermal stress. *Physiol. Rev.* 54:75–159, 1974.

109. Rowell L.B. Active neurogenic vasodilatation in man. In: *Vasodilatation*, edited by P.M. Vanhoutte and I. Leusen. New York: Raven Press, 1981, pp. 1–17.

110. Rowell L.B., J.R. Blackmon, and R.A. Bruce. Indocyanine green clearance and estimated hepatic blood flow during mild to maximal exercise in upright man. *J. Clin. Invest.* 43:1677–1690, 1964.

111. Rowell L.B., G.L. Brengelmann, J.R. Blackmon, R.A. Bruce, and J.A. Murray. Desparities between aortic and peripheral pulse pressures induced by upright exercise and vasomotor changes in man. *Circulation* 37:954–964, 1968.

112. Rowell L.B., H.J. Marx, R.A. Bruce, R.D. Conn, and F. Kusimi. Reductions in cardiac output, central blood volume, and stroke volume with thermal stress in normal men during exercise. *J. Clin. Invest.* 45:1801–1816, 1966.

113. Rowell L.B., J.A. Murray, G.L. Brengelmann, and K.K. Kraning, II. Human cardiovascular adjustments to rapid changes in skin temperature during exercise. *Circ. Res.* 24:711–724, 1969.

114. Rowell L.B., D.S. O'Leary, and D.L. Kellogg, Jr. Integration of cardiovascular control systems in dynamic exercise. In: *Handbook of Physiology: Exercise, Regulation and Integration of Multiple Systems*, edited by L.B. Rowell and J.T. Shepherd. Bethesda, MD: American Physiological Society, 1996, Chapt. 17, pp. 770–838.

115. Rushmer R.F., O. Smith, and D. Franklin. Mechanisms of cardiac control in exercise. *Circ. Res.* 7:602–627, 1959.

116. Rushmer R.F. Effects of nerve stimulation and hormones on the heart; the role of the heart in general circulatory regulation. In: *Handbook of Physiology. Circulation*, edited by W.F. Hamilton and P. Dow. Washington, DC: American Physiological Society, 1962, Sect. 2, Vol. I, pp. 533–550.

117. Rushmer R.F., D.L. Franklin, R.L. Van Citters, and O.A. Smith. Changes in peripheral blood flow distribution in healthy dogs. *Circ. Res.* 9:675–687, 1961.

118. Rushmer R.F. and A.O. Smith, Jr. Cardiac control. *Physiol. Rev.* 39:41–68, 1959.

119. Sagawa K. Baroreflex control of systemic arterial pressure and vascular bed. In: *Handbook of Physiology, The Cardiovascular System, Peripheral Circulation and Organ Blood Flow*, edited by J.T. Shepherd and F.M. Abboud. Bethesda, MD: American Physiological Society, 1983, pp. 453–496.

120. Saltin B. Aerobic work capacity and circulation at exercise in man. *Acta Physiol. Scand.* 62(*Suppl.*) 230:1–52, 1964.

121. Sarnoff S.J. and E. Berglund. Ventricular function. I. Starling's law of the heart studied by means of simultaneous right and left function curves in the dog. *Circulation* 9:706–718, 1954.

122. Scher A.M. and A.C. Young. Servoanalysis of carotid sinus reflex effects on peripheral resistance. *Circ. Res.* 12:152–162, 1963.

123. Selkurt E.E. The renal circulation. In: *Handbook of Physiology. Circulation*, edited by W.F. Hamilton and P. Dow. Washington, DC: American Physiological Society, 1963, Sect. 2, Vol. II, pp. 1457–1516.

124. Sénac J.B. DE *Traité de la structure du couer*. 2. éd. Paris: Méquignon l'aîné, 1783.

125. Sewall H. and D.W. Steiner. A study of the action of the depressor nerve, and a consideration of the effect of blood-pressure upon the heart regarded as a sensory organ. *J. Physiol. (Lond.)* 6:162–176, 1885.

126. Shepherd J.T. Circulation to skeletal muscle in man. In: *Handbook of Physiology, The Cardiovascular System, Peripheral Circulation and Organ Blood Flow*, edited by J.T. Shepherd and F.M. Abboud. Bethesda, MD: American Physiological Society, 1983, pp. 319–370.

127. Sheriff D.D., X.P. Zhou, A.M. Scher, and L.B. Rowell. Dependence of cardiac filling pressure on cardiac output during rest and dynamic exercise in dogs. *Am. J. Physiol.* 265 (*Heart Circ. Physiol.* 34):H316–H322, 1993.

128. Sheriff D.D., L.B. Rowell, and A.M. Scher. Is rapid rise in vascular conductance at onset of dynamic exercise due to muscle pump? *Am. J. Physiol.* 265 (*Heart Circ. Physiol.* 34): H1227–H1234, 1993.

129. Sheriff D.D. and R. Van Bibber. Flow-generating capability of the isolated skeletal muscle pump. *Am. J. Physiol.* 274 (*Heart Circ. Physiol.* 43):H1502–H1508, 1998.

130. Skokland O. Factors contributing to acute blood pressure elevation. *J. Oslo City Hosp.* 33:81–95, 1983.

131. Starling E.H. On the absorption of fluids from the connective tissue spaces. *J. Physiol. (Lond.)* 19:312–326, 1896.

132. Starling E.H. *The Linacre Lecture on the Law of the Heart*. London: Longmans, Green, 1918.

133. Starling E.H. On the circulatory changes associated with exercise. *J. Roy. Army Med. Corps* 34:258, 1920.

134. Stegall H.F. Muscle pumping in the dependent leg. *Circ. Res.* 19:180–190, 1966.

135. Stick C., H. Jaeger, and E. Witzleb. Measurements of volume changes and venous pressure in the human lower leg during walking and running. *J. Appl. Physiol.* 72: 2063–2068, 1992.

136. Stray-Gunderson J., T.I. Musch, G.C. Haidet, D.P. Swain, G.A. Ordway, and J.H. Mitchell. The effect of pericardiectomy on maximal oxygen consumption and maximal cardiac output in untrained dogs. *Circ. Res.* 58:523–530, 1986.

137. Sullivan M.J., F.R. Cobb, and M.D. Higgenbotham. Stroke volume increases by similar mechanisms during upright exercise in normal men and women. *Am. J. Cardiol.* 67: 1405–1412, 1991.

138. Thadani U. and J.O. Parker. Hemodynamics at rest and during supine and sitting bicycle exercise in normal subjects. *Am. J. Cardiol.* 41:52–59, 1978.

139. Tidgren B., P. Hjemdahl, E. Theodorsson, and J. Nussberger. Renal neurohormonal and vascular responses to dynamic exercise in humans. *J. Appl. Physiol.* 70:2279–2286, 1991.

140. Toska K. and M. Ericksen. Peripheral vasoconstriction shortly after onset of moderate exercise in humans. *J. Appl. Physiol.* 77:1519–1525, 1994.

141. Tyberg J.V. Venous modulation of ventricular preload. *Am. Heart J.* 123:1098–1104, 1992.

142. Van Citters R.L. and D.L. Franklin. Cardiovascular performance of Alaska sled dogs during exercise. *Circ. Res.* 24:33–42, 1969.

143. van Lieshout J.J., W. Wieling, K.H. Wesseling, E. Endert, and J.M. Karemaker. Orthostatic hypotension caused by sympathectomies performed for hyperhidrosis. *Neth. J. Med.* 36:53–57, 1990.

144. Vatner S.F. Effects of exercise on distribution of regional blood flows and resistances. In: *The Peripheral Circulations,* edited by R. Zelis. New York: Grune and Stratton, 1975, pp. 211–233.

145. Vatner S.F. Effects of exercise and excitement on mesenteric and renal dynamics in conscious, unrestrained baboons. *Am. J. Physiol.* 234 (*Heart Circ. Physiol.* 3):H210–H214, 1978.

146. Wade O.L. and J.M. Bishop. *Cardiac Output and Regional Blood Flow.* Oxford: Blackwell, 1962.

147. Wade O.L., B. Combes, A.W. Childs, H.W. Wheeler, A. Cournand, and S.E. Bradley. The effects of exercise on the splanchnic blood flow and splanchnic blood volume in normal man. *Clin. Sci.* 15:457–463, 1956.

148. Weber E.F.W. and E.H. Weber. Experimenta, quibus probatur nervos vagos rotatione machinae galvanomagneticae irritatos, motum cordis retardare et adeo intercipare. *Ann. Univ. Med. (Milano)* 20:227–233, 1845. Partial translation in Fulton J.F. *Selected Readings in the History of Physiology* (2nd ed.). Springfield, IL: Charles C. Thomas, 1966, p. 296.

149. Wexler L., D.H. Bergel, I.T. Gabe, G.S. Makin, and C.J. Mills. Velocity of blood flow in normal human venae cavae. *Circ. Res.* 23:349–359, 1968.

150. White H.L. and D. Rolf. Effects of exercise and of some other influences on the renal circulation in man. *Am. J. Physiol.* 152:505–516, 1948.

151. Wiggers C.J. *The Pressure Pulses in the Cardiovascular System.* London: Longmans, Green, and Co., 1928.

152. Zelis R., D.T. Mason, and E. Braunwald. Partition of blood flow to the cutaneous and muscular beds of the forearm at rest and during leg exercise in normal subjects and in patients with heart failure. *Circ. Res.* 24:799–806, 1969.

153. Zuntz N. and O. Hagemann. Untersuchungen über den Stoffwechsel des Pferdes bei Ruhe und Arbeit. *Landw. Jb.* 27 (*Ergänz. Bd.* 3), 371–412, 1898.

chapter 4

THE RESPIRATORY SYSTEM

Jerome A. Dempsey and Brian J. Whipp

T HE history of science deals with events and concepts—but above all, with people. In our treatment of the history of respiratory physiology as applied to exercise, we have chosen to follow the lead of J.B. West and his contributors (155) in emphasizing the people and ideas theme used in their recent history of respiratory physiology. To this end, we have corresponded with several of the major scientists in this area over the past 50 years who have shared their personal motivation to study respiration and exercise, their own perception of their contributions, and finally the important questions they feel remain unanswered in the field. We have also chosen to subdivide our topic into studies of (1) the mechanical aspects of the lung and chest wall, (2) alveolar to arterial gas exchange, (3) the control of exercise hyperpnea, and (4) the adaptability of the pulmonary system and its role in the limitation of maximal O_2 transport and exercise performance.

BREATHING MECHANICS

The *Physiological Reviews* of 1954 article by Arthur Otis on the work of breathing (115) and of 1961 by Jere Mead on the mechanical properties of the lungs (107) are classical summaries of the field halfway through the twentieth century. Before the 1940s, several concepts were already well established (for reviews and refs. 115 and 107). For example, Galen (ca. 129–200) had already recognized the importance of the respiratory muscles in moving the lung (rather than vice versa) and Leonardo da Vinci (1452–1519) described the bellows-like action of the rib cage. Andreas Vesalius (1514–1564) in the mid-sixteenth century performed what might have been the first experiment in respiratory mechanics by visualizing the retraction of the lung on a living dog upon puncturing the pleura. The Englishman John Mayow (1645–1679) in the seventeenth century modeled the lung and thorax mechanical interactions and recognized the principle of elastic lung recoil. The Scottish physician Johannas Carson made the first measurements of lung elasticity in the early nineteenth century and described in colorful prose the battle between muscle and elastic forces: "Breathing is in a great measure the effect of this interminable contest between the elasticity of lungs and the irritability of the diaphragm!" It was not really until the 1920s that the first comprehensive conceptual analysis of pleural pressure and its significance to the mechanics of breathing was formulated in Basel, Switzerland, by Fritz Rohrer and his associates VonNeergard and Carl Wirtz (a physiology research fellow). These investigators simplified the complex motions and forces of respiration to single variables—volume and pressure. They also introduced the important concept of surface tension as a critical retractive force in the lung.

So the stage was set for the golden age of quantitation of breathing mechanics, which spanned the 1940s, 50s, and 60s. The scientific giants of this era in breathing mechanics research—most of them physicians—included Wallace Fenn (1893–1971), Herman Rahn (1912–1990), Jere Mead, and Arthur Otis, and a little later, Richard Riley, Moran Campbell, Joseph Milic-Emili, Solbert Permutt, Peter Macklem, and Robert Hyatt. Mead pointed out in 1961 that up to the 1940s ideas and techniques were in place, but it was in the next two decades that *measurements* were first made. These measurements were greatly assisted by electrical recording devices and the use of the esophageal balloon technique, the latter being invented in the 40s and refined in the late 50s and 1960s as a simple indirect method for measuring pleural pressure. Thus, the static compliance of the human lung and chest wall and the components of airway resistance were quantified, and the work of breathing was approximated from the Campbell pressure:volume diagram. The dynamic compression of airways on forced expiration leading to flow limitation was discovered, as was the reactivity of bronchiolar smooth muscle. The other fundamental advance was the detailed morphology of the lung and airways and diffusion surface provided by Ewald Weibel (151).

The first published works applying these principles and measurements to exercise appeared in the 1950s and were largely motivated by the desire to better understand the dyspnea of exercise in patients with pulmonary and cardiovascular diseases. We now outline several of the landmark studies and concepts.

Work and the Oxygen and Circulatory Cost of Breathing

Joseph Milic-Emili (Fig. 4.1) was mentored from 1957 to 1960 by a famous exercise physiologist, Rudolpho Margaria, in Milan. Together with Jean Marie-Petit, they produced a series of papers between 1958 and 1960 that described their refinement of the esophageal balloon technique in order to measure the work of breathing during muscular exercise and in order to estimate the mechanical efficiency of breathing, which they found to be on the order of 20%–25% (i.e., similar to limb muscle) (105). A major problem in these estimates, which was suspected but did not surface

Fig. 4.1. Professor Josef Milic-Emili (center) of McGill University with Professor Peter Macklem (left) discussing the subtleties of breathing mechanics with one of their admiring fans. Photograph provided to the authors by Professor Milic-Emili upon request.

until the 1970s, was that the pressure:volume diagram alone does not account for the work done to overcome chest wall distortion and the work by the abdominal muscles done on the abdominal wall during exercise (64). Another confounder in these measurements was that the several estimates of the O_2 cost of breathing obtained from stimulated breathing at rest did not mimic the respiratory muscle actions required during exercise. The true mechanical efficiency of the respiratory muscles during exercise continues to remain elusive, and estimates ranged all the way from 5% to 30%. In reality, heavy exercise presents such a complex array of muscle actions and shifts between abdominal and rib cage contributions to the breath and chest wall distortion that the mechanical efficiency of breathing must be constantly changing and highly unpredictable.

In the past decade, some progress has been made in more precisely estimating the oxygen and blood flow required by the respiratory muscles for exercise hyperpnea. Estimates at max exercise in highly trained humans and animals are in the range of 14%–16% percent of both total $\dot{V}_{O_2}$ (1) and of cardiac output. These blood flow estimates were obtained directly by measuring the deposition of infused microspheres in the respiratory muscles of the exercising pony (104); and in the human, estimates were extrapolated from the reduction in total cardiac output observed when the respiratory muscles were mechanically unloaded using a positive-pressure mechanical ventilator at max exercise (70). Murli Manohar, a veterinary scientist at Southern Illinois University, carried out a unique series of studies in the 1980s in a pony model in which he measured microsphere deposition in muscle and sampled the phrenic venous effluent for the first time, thereby producing the first direct measures of arterial-to-venous O_2 content difference across the diaphragm and diaphragm $\dot{V}O_2$ during exercise. These data showed that the diaphragm reached blood flows which (per gram of tissue) were comparable to those of limb locomotor muscles during max exercise (104).

The early series of studies on the work of breathing by Milic-Emili, Otis, and colleagues also addressed in an elegant and straightforward manner several additional problems, which continue to be investigated today. Most notably, these included the principle that the minimal or optimal work of breathing during exercise (especially heavy exercise where tidal volume encroaches on both expiratory and inspiratory reserve volumes) occurs when subjects are allowed to spontaneously choose their frequency and tidal volume. Otis was a key member of the famous Fenn and Rahn mechanics group at Rochester in the 1940s and 1950s and was responsible for much of the theory underlying the concept of minimization of the work of breathing. In view of these original findings and those that followed, it is surprising that many popular theories still expound the value of *voluntary* control over one's breathing pattern during exercise and during athletic endeavors.

Expiratory Flow Limitation in Exercise

Robert Hyatt (Fig. 4.2), currently at the Mayo Clinic, was taught in Rochester's medical school by Fenn, Rahn, and their colleagues, but it was not until he was a research fellow at the National Heart Institute in the mid-1950s and met Donald Fry that he became intrigued by lung mechanics. In a paper with Fry in 1958 (79) and several that followed, the maximum flow:volume envelope and the concept of an effort-independent expiratory flow rate was described. They also presented several techniques for quantifying the limits of maximum effective expiratory pressure. Then in the late 1960s, Olaffson and Hyatt presented the first application of these principles to exercise, showing that acute exercise had little effect on the lung's mechanical properties. This confirmed the pioneering work of two great Danish physiologists, Erling Asmussen and Marius Neilsen, from the late 1930s (4). Hyatt also demonstrated that tidal volume increased during exercise because of encroachment on both the inspiratory and expiratory reserve volumes and that the tidal flow:volume loop encroached on the expiratory limb of the maximum flow volume loop during heavy exercise but that tidal expiratory pressures did not exceed maximum effective pressure. Hence, expiration remained efficient.

The question of expiratory flow limitation and its importance to exercise hyperpnea began in the 1960s and had a rebirth in the 1980s and 90s. First, Gunnar Grimby (Fig. 4.3) in the late 1960s (63) showed that many athletes required sufficiently high flow rates in heavy exercise to encroach on their maximum flow:volume envelope and to increase their end-expiratory lung volume. Bruce Johnson and associates in the late 1980s and early 1990s observed substantial flow limitation in young and especially older athletes, the latter having experienced substantial loss of elastic

Fig. 4.2. Robert Hyatt (center), with Bruce Johnson (left) and Kenneth Beck, all of the Mayo Clinic in 1999. Dr. Hyatt began research into expiratory flow limitation in the late 1950s and continues today advising his younger Mayo colleagues in their studies of breathing mechanics during exercise in asthma (Beck) and heart failure (Johnson). Photograph provided to the authors by Dr. Robert Hyatt upon request.

lung recoil with aging despite their continued training regimens (82, 84). These data, along with the effects of He:O_2 breathing which increased the maximum flow volume envelope and resulted in greater increases in $\dot{V}E$ during heavy exercise in flow-limited subjects also emphasized that $\dot{V}E$ at heavy and max exercise was indeed mechanically constrained in highly fit subjects. Hyatt recently suggested that even approaching flow limitation might trigger a reflex cessation of expiration and initiation of inspiration so that end-expiratory lung volume would increase. Unfortunately, we are unable to more directly address many of these questions of neural reflex control of respiratory motor output during exercise because we still lack an accurate means of measuring neural respiratory motor output under these dynamic conditions in the whole animal. Perhaps some of the newer methods of quantifying diaphragmatic function (e.g. EMG) in the human will prove helpful in this regard.

Hyatt's fundamental concepts of flow limitation revealed the complexity of the nature of the mechanical loads that occur once expiratory flow limitation begins during exercise. Because of hyperinflation to avoid expiratory flow limitation, this causes the subject to breathe on the upper, stiffer portion of the pressure:volume relationship. If the hyperinflation is sufficient (such as in the asthmatic subject) an inspiratory threshold load is incurred whereby the inspiratory muscles must overcome the combined inward recoil of the lung and chest wall at the onset of inspiration before any inspiratory flow is initiated. What begins then as an expiratory resistive

Fig. 4.3. Professor Gunnar Grimby of Sahlgrenska University, Goteberg, Sweden. Photograph provided to the authors by Professor Grimby upon request.

load ends up producing nearly intolerable elastic loads as the end-expiratory lung volume is elevated (112).

Milic-Emili, who was missing from exercise science for over 30 years, again recently reemerged to inform us that the maximum expiratory flow:volume curve depends critically on the time course of the preceding inspiration. Accordingly, the max flow:volume envelope may not be accurately compared to the tidal loop. He and his colleagues suggest that true flow limitation is best evaluated in a more straightforward manner simply by applying a negative pressure during expiration and observing whether flow rate increases (95).

Control of Airway Caliber during Exercise

Several investigators, beginning with Asmussen and Nielson in the 1930s (4), recognized how closely airway caliber was protected during exercise. Airway resistance during inspiration remains constant throughout exercise despite up to 10-fold increases in flow rate. Two types of mechanisms appear to be operative. First, caliber of the extrathoracic upper airway is maximized by activation of dilator muscles of the oral pharynx and laryngeal abductors and by sympathetically induced vasoconstriction of arterioles of the nasal pharynx. The maximal abduction of the laryngeal folds throughout the respiratory cycle during exercise was beautifully demonstrated by Sandra England and Donald Bartlett of Dartmouth University in the early 1980s using a bronchoscope to visualize the dynamic changes in airway caliber throughout the respiratory cycle in exercising humans (51). The role of feedback and feedforward mechanisms controlling motor output to the upper airway muscles, and also regulating the use of predominantly oral vs. nasal breathing routes during exercise are unknown—although strong cases can certainly be made for both mechanisms (43). Dilation of bronchiolar smooth muscle also occurs in normal subjects during exercise and has been attributed primarily to removal of cholinergic tone to the airways, analogous to the marked reduction in parasympathetic tone to the heart.

A major advance in understanding airway function during exercise with substantial implications for the common clinical problem of exercise-induced asthma occurred in the late 1970s with the discovery that at high flow rates with dry air, substantial water loss and change in osmolarity occurred in airway lung fluid, even at the bronchiolar level (38). In turn, this drying effect set up the airway for potential bronchoconstriction by causing release of inflammatory mediators. However, that powerful bronchodilatory forces must be also present during exercise is evident from the observation that bronchoconstrictive effects of histamine are greatly reduced during exercise and the consistent finding that even asthmatic patients only very rarely show bronchoconstriction *during* exercise. Most often, bronchoconstriction is unmasked in recovery only when the exercise influences are removed. Several potential mediators of this bronchodilation are under active investigation, ranging from mediator release (such as nitric oxide) to the mechanical dilator effects of lung expansion acting via parenchymal-airway attachments (19).

Respiratory Muscles and Exercise

Peter Macklem trained at McGill University under David Bates and at Harvard with Jere Mead. His early research, in the 1960s, was the first to explore the mechanics of the peripheral airways. In the late 1970s, driven by the clinical problem of respiratory failure in chronic obstructive pulmonary disorder (COPD), Macklem and his colleague Charis Roussos were the first to address the fatigability of the diaphragm by showing that sustained voluntary hyperpnea at high transdiaphragmatic pressures resulted in respiratory task failure in healthy volunteers (129). Over the next decade, several studies centered on this hot topic, including methods for testing diaphragm fatigue, diaphragm fatigue-resistant characteristics, multiple actions of the costal and crural diaphragm, acute and chronic effects of diaphragm length changes, reflex effects emanating from the diaphragm, etc. Alejandro Grassino from Argentina conducted a series of studies in humans and animals at McGill which very carefully documented the conditions of force output and duty cycle which would lead to diaphragm fatigue (20). In the early 1980s, Roussos and colleagues showed for the first time that the physiologic stress of heavy endurance exercise in humans caused changes in EMG frequency spectrum in the diaphragm which were consistent with fatigue (27). Several studies in the rat also observed significant reductions in glycogen content in the diaphragm following exhaustive endurance exercise.

Dempsey was skeptical of the indirect EMG findings claiming exercise-induced diaphragm fatigue. He set out with Bruce Johnson and Mark Babcock at Wisconsin to apply the Bellemare, Bigland-Ritchie bilateral phrenic nerve stimulation technique to exercising subjects. Their rationale was that the highly oxidative, richly perfused diaphragm muscle would not be susceptible to fatigue during even exhausting whole-body exercise. In fact, just the opposite was shown—that is, the transdiaphragmatic pressure (P_{di}) with supramaximal phrenic nerve stimulation did indeed fall significantly following sustained exercise to exhaustion at greater than 85% $\dot{V}O_2$max and did not recover until more than 1 hour of rest in the postexercise period (83). Further studies showed that exercise-induced diaphragm fatigue occurred in both inactive and highly trained subjects and it occurred more readily and took longer to recover when hypoxia was present. Furthermore, for a given magnitude and duration of increased force output by the diaphragm, fatigue occurred much more readily during heavy treadmill running exercise than it did at rest (8). This effect of exercise on precipitating diaphragm fatigue was attributed to the competition between limb and respiratory muscles for a share of the limited cardiac output during heavy exercise. Alternatively, the metabolic acidosis created by whole-body exercise could also contribute to this "sensitizing" effect of exercise on diaphragm fatigue.

Macklem and his Italian colleagues have recently applied new imaging techniques in an attempt to understand respiratory muscle recruitment and the interaction among respiratory muscles during exercise (92). Earlier estimates of rib cage vs. abdominal contributions had been made in the 1960s and 1970s by Grimby, Goldman, and Mead using pressure measurements together with surface displacements of the chest and abdominal wall—the latter measurement often fraught with the uncertainty of changing calibrations during hyperpneic states (60). The recent imaging studies by Macklem and colleagues now suggest that during exercise, the expiratory muscles relax slowly on inspiration (92). This action concomitantly minimizes rib cage distortion, thereby sparing the diaphragm the relatively large amount of force required to produce distortion and allowing its work to be devoted to inflating the rib cage. These recent findings also revealed how very little we know about the mechanisms which influence the strategy of how the diaphragm and accessory respiratory muscles are recruited during exercise, especially when fatigue of the primary pump muscle begins to occur. Again, these problems cry out for newer approaches to measuring neural respiratory motor output and its distribution in the human.

Exertional Dyspnea

In health, breathing is a beautifully orchestrated, low-cost, automatic activity of which we are rarely ever aware. Thus, when breathing efforts are consciously perceived during exercise, this represents a serious breach of the automatic, homeostatic response. These sensations have many important implications. They are the most important cause of exercise limitation in lung disease and in heart failure. That is, exercise becomes symptom-limited rather than oxygen transport-limited—and under some circumstances dyspnea probably plays a major role in regulating breathing and even exercise performance in the healthy subject. The ability to perceive breathing efforts in response to mechanical loads has also recently been linked to asthmatic

Fig. 4.4. E.J. Moran Campbell (left) with the late John R. Sutton (second from left) of McMaster University together with the mayor of Hamilton, Canada and a medical school colleague, shown at the dedication of a plaque in honor of the nineteenth century arctic explorer Dr. John Rae. Photograph provided to the authors by Dr. Campbell upon request.

patients' propensity to seek and/or administer the appropriate medications in order to manage their bronchoconstriction/inflammation. The search for the mechanisms of this perceived effort and discomfort has as long a history as that of exercise hyperpnea—but the science has been difficult to carry out because there are no animal models in whom perception can be quantified.

E.J. Moran Campbell (Fig. 4.4), a British-trained chest physician/respiratory physiologist, was driven by the dyspnea problem to study the mechanics of breathing and respiratory muscles during exercise and published one of the first books on this topic in the mid-1950s in which he describes his EMG measurements on expiratory muscles in exercising humans (28). Campbell emigrated to Canada in the 1960s, eventually recruiting Norman Jones from the United Kingdom and then Kieran Killian from Ireland to Hamilton, Ontario, and McMaster University. This group has studied several hundred patients over the past 30 years using standardized exercise protocols and induced symptoms of respiratory distress in several ways, including chemostimulation, respiratory loads, and neuromuscular blockade. It is their view that dyspneic sensations are not uniquely related to muscle tension, displacement (ventilation), work, pressure time:product, or any of the other obvious compound factors. Rather, the severity of these sensations is related to the magnitude of motor command or effort. They propose further that motor commands to both inspiratory and locomotor muscles occur in parallel and that effort perception is the cornerstone of fatigue and exercise limitation (93).

Campbell for the first time provided a conceptual framework for the term "dyspnea" by labeling its cause as an "inappropriate" relationship between length and tension developed in the respiratory system. Others have similarly used such phrases as "unsatisfied inspiratory efforts" to describe the sensation when the effort exerted by the respiratory muscles to produce a breath greatly exceeds the resulting ventilatory response because of mechanical (lung, muscle, airway) deficits. The occurrence of an inspiratory threshold load when hyperinflation results from expiratory flow limitation (see above) is a good example of this unsatisfied effort (112). In the 1980s and 1990s, studies on selectively denervated human patients and in acutely paralyzed respiratory physiologists have shown that sensory input to produce dyspnea may arise from many receptor sites sensitive to chemical and mechanical perturbations (12). These latter studies, especially those by Robert Banzett and associates at Harvard, have also recently questioned the concept of an obligatory role for respiratory muscle afferents in giving rise to dyspneic sensations.

A major hindrance to understanding mechanisms of dyspnea has been the dearth of knowledge of the function and even existence of supramedullary neural pathways and receptor sites which might serve to bring sensory inputs to the level of cortical perception. An exciting beginning in this direction includes the recent discovery in Frederick Eldridge's laboratory (see below) of neural pathways in the decorticate cat ascending from the brain stem to mesencephalic and thalamic neurons that develop rhythmic activity in response to chemostimulation (30). Equally insightful is evidence from John Orem's conscious cat model which was conditioned (via negative feedback) to voluntarily alter its breathing pattern in response to an auditory signal while recordings were made from inspiratory medullary neurons (114). The findings clearly demonstrated that a voluntary effort to breathe is, in fact, sensed by medullary inspiratory neurons, thereby dispelling the notion that cortical spinal tracts are the only means of affecting breathing via (higher) central command.

PULMONARY GAS EXCHANGE

It is clear that scientists as early as several centuries before the Christian era were already wondering about the relationship between the air being breathed—specifically, its contribution to the blood through the *pneuma*—and what we would today (thanks to Schwann's mid-nineteenth century definition) term "metabolism." Erasistratus of Alexandria (~340–280 BC) went beyond supposition, however. He enclosed a fowl in a jar in order to establish the variation in its weight as a function of the weights of its ingesta and excreta. He concluded that the animal was consuming itself and, with remarkable prescience, that the function of the pneuma was to "transmute nourishment into a form suitable for supplying the place of the matter carried off." These energy-balance studies predate the more widely acknowledged "weighing chair" experiments of Santorio by almost two millennia—although in Santorio's case the experiments were performed on humans.

Metabolic Rate

It would not be until the late eighteenth century, however, that it would be shown that O_2 uptake and CO_2 output in humans actually increased with muscular exercise. This was demonstrated by the great Antoine-Laurent Lavoisier (1743–1794)—who also coined the name "oxygen" for the "eminently respirable air." He also, remarkably, determined that what we now term the "respiratory quotient" was normally approximately 0.8. However, there was then, and for the best part of the next century, great uncertainty regarding the site of the increased metabolic rate—although even at this time the Italian physiologist Lazaro Spallanzani (1729–1799) argued "that combustion took place neither in the lungs nor the blood but in the muscles and organs of the body." But it was not until the 1860s and 1870s that the German physiologists Carl von Voit (1831–1908) and Eduard Pflüger (1829–1910) separately established that the oxygen consumption of living cells is determined by their metabolic needs rather than the oxygen content of the arterial blood.

This period also provided important advances in the quantification of O_2 uptake during exercise. Voit, for example, performed important experiments as early as 1857 that firmly established the concept of steady-state "nitrogen equilibrium" in the body and hence of N_2 conservation during the breath—although the notion itself can be traced back to a letter from Lavoisier to Joseph Black in 1790 (98). This, of course, was an essential development for measuring oxygen consumption during exercise using expiratory (or inspiratory) volume measurements only. This was subsequently formalized by Geppert and Zuntz (60), in 1888, into the now-familiar equation that forms the basis of the "true O_2" calculation:

$$\dot{V}_I = \dot{V}_{Ex}(F_EN_2/F_IN_2)$$

Metabolic rate could now be accurately quantified, allowing the relationship between ventilation and O_2 uptake, for example, to be established as a basis for understanding pulmonary gas exchange and the regulation of blood-gas and acid–base status.

The subsequent development of techniques for sampling "alveolar" and "arterial" blood-gas tensions made it clear that the relationship between ventilation and the rates of pulmonary gas exchange, at least for steady-state or quasi–steady-state exercise, was sufficiently close to maintain alveolar and arterial P_{CO_2} and P_{O_2} relatively constant in normal subjects—at sea level. The later development of rapidly responding gas analyzers and the use of digital computers allowed the details of the pulmonary gas exchange transients to be determined breath-by-breath by J. Howland Auchincloss and his associates (7) and by William Beaver, Karlman Wasserman, and Brian Whipp (18), an issue previously addressed.

The resulting studies on the non–steady state of exercise made it clear that the challenges to the exchange of pulmonary gases are different for O_2 uptake and CO_2 output during the phase in which the mixed-venous gas tensions change as a result of increased muscle gas exchange. This is not the case for the initial "cardiodynamic" phase, as termed by Wasserman and his associates (150), in which the O_2 uptake is blood flow dependent—as originally proposed by August Krogh (1874–1949) and Joseph Lindhard (97) in 1913. The subsequent transient decrease of the respiratory exchange ratio (R) during the on-transient non–steady-state (a result of the higher capacitance of the tissues for CO_2 than that for O_2) was characterized by Rahn and Fenn (123) in their classic monograph on the "O_2–CO_2" diagram in 1955. Ventilation, however, was shown to remain closely correlated with the exchange rate of CO_2 at the lung throughout the transient phase of the exercise. This was demonstrated using a range of different dynamic forcing strategies but most strikingly perhaps by Casaburi and colleagues using sinusoidal work-rate forcings (29). Consequently, alveolar and arterial P_{CO_2} were shown to be regulated close to the control level—although the small but significant difference in the time constants for ventilation and CO_2 results in a small but significant fluctuation of P_{CO_2} (1–2 mmHg) during the transient period. As the time course of the ventilation response is slow relative to that of O_2 uptake, transient reductions of alveolar and arterial P_{O_2} were characteristic.

Alveolar-to-Arterial O_2 Exchange

In order to gain insight into the adequacy of the pulmonary gas-exchange mechanisms, determinations of the difference between alveolar and arterial gas tensions were required. Establishing the arterial blood-gas tensions was conceptually straightforward even though initially technically challenging. The measurement of alveolar gas tension, however, was conceptually much more demanding: what is "alveolar" P_{O_2}? The concern about what actually constituted an "alveolar" gas concentration was expressed by Geppert and Zuntz more than 100 years ago. They also stated that (60) "even if we knew the exact composition of the alveolar air, it remains to be proven that an absolute equilibrium in tension between the arterial blood and the alveolar air is established."

The notion that arterial P_{O_2} could differ from that of alveolar P_{O_2} because of regional variations in alveolar ventilation to perfusion was clearly recognized by both John Scott Haldane (1860–1936) in England and Krogh and Lindhard in Denmark. Haldane had also recognized, however, that while such inhomogeneities of ventilation and perfusion would necessarily result in a difference between alveolar and arterial P_{O_2}, there would be relatively little effect on the alveolar-to-arterial CO_2 difference as a result of the relatively linear slope of the CO_2 dissociation curve in the physiological range.

Two papers, interestingly both published in 1949, provided a conceptual solution for understanding pulmonary gas exchange: Riley and Cournand's concept of the "ideal" alveolar air (126) and Rahn's concept of "mean alveolar air" (122). In essence, this allows pulmonary gas exchange to be considered with respect to exchange of an "ideal" lung compartment and two others that lead to inefficiencies of gas exchange—as a result of their dead-space-like and shunt-like effects, respectively. And so, based upon the useful, but erroneous, assumption that mean alveolar and arterial P_{CO_2}s are identical, an "ideal" alveolar P_{O_2} could be established to serve as a frame of reference for the measured arterial value—the difference being an index of the "in-efficiency" of the gas exchange. It also provided the basis for the development of the O_2–CO_2 diagram" which incorporated influences of variations of both R and $\dot{V}_A/\dot{Q}$.

Direct measurements of the effects of muscular exercise on this alveolar-to-arterial Po_2 difference revealed little or no widening of the gradient at low work rates but significant and systematic increases at higher work rates: that is, there is a decrease in the efficiency of pulmonary gas exchange during muscular exercise even in normal subjects, especially at high work rates. The increase in the (A–a) Po_2 difference was shown to be a result of an increased alveolar O_2 tension rather than a decrease in arterial Po_2. The increased P_AO_2 at high work rates was explicable by the increase in the ventilatory equivalent for oxygen ($\dot{V}E/\dot{V}O_2$) that is associated with the development of a metabolic acidosis: the failure of PaO_2 to increase in concert was the concern. The physiological basis for this decreased pulmonary gas-exchange efficiency could not be discerned from these measurements alone. The initial studies on the distribution of alveolar ventilation to perfusion during exercise, using radioactive tracer gases, were consistent with the ratio becoming more uniform (10). This was supported by the demonstration that the alveolar-to-arterial nitrogen difference decreased "to practically zero" during exercise (9). Other mechanisms were therefore sought for the decreased gas-exchange efficiency. As the widening of the A–a gradient for Po_2 began at metabolic rates too low for diffusion impairment to be thought to contribute, the effect of right-to-left shunt (especially associated with the low mixed venous Po_2) was considered to be the major determinant.

While it was a common assumption up until the 1980's that arterial Po_2 was maintained at, or close to, resting levels during even exhaustive exercise in normal subjects, Lillienthal, working with Riley (100), had shown as early as 1946 that arterial Po_2 could decrease in normal subjects at high work rates. The finding was supported by the studies of Holmgren and Linderholm (75) and also Doll and his associates (44), among others, in the late 1950s and mid-1960s, respectively. Further along this line, Shepherd (133) carried out important computer modeling studies in 1959 on the effect of various levels of pulmonary diffusing capacity for O_2 and of physical fitness (as expressed by the maximum $\dot{V}O_2$) on arterial O_2 saturation during exercise. He demonstrated that an inadequate diffusing capacity for O_2 would predispose to arterial oxyhemoglobin desaturation at high work rates. This was subsequently decisively demonstrated by Dempsey and his associates (41).

Ventilation-to-Perfusion Distribution and Diffusion Limitation

Further development in this area, however, demanded more refined techniques. The seminal advance was supplied, in the early 1970s by the multiple inert gas elimination technique (MIGET) of Wagner and West (130). They used an elegant extension of the concept, formalized by Leon Farhi and his associates (52), that the ratio of alveolar to mixed-venous gas tensions of an inert gas (x) is an exclusive function of its partition coefficient (l) and its ventilation-to-perfusion ratio; that is:

$$\frac{P_AX}{P_VX} = \frac{\lambda}{\lambda + (\dot{V}_A/\dot{Q})}$$

Infusing a mixture of inert gases, with a suitably large range of partition coefficients, into the mixed-venous blood allows a functionally continuous distribution of ventilation-to-perfusion ratios to be established in the lung, extending to a $\dot{V}_A/\dot{Q}$ ratio of zero (i.e., a shunt). Consequently, if the measured arterial P_{O_2} is found to be less than that predicted from the sum of the effects of the ventilation-to-perfusion ratio distribution and right-to-left shunt (determined by this technique) and any hypoventilation (assessed by measuring arterial P_{CO_2}), then the only remaining contributory mechanism will be the "diffusion" component. When the MIGET technique was applied to exercising normal subjects, Gledhill and colleagues (62) were able to demonstrate that approximately 40% of the (A–a) P_{O_2} difference, both at rest and during muscular exercise, was attributable to anatomically shunted blood and the remainder to non-uniformity of the ventilation-to-perfusion ratios. That is, in these studies there was no significant component of diffusion impairment. However, the surprising finding in this study, and the subsequent studies of Wagner and his associates (57, 67), was that the distribution of ventilation-to-perfusion ratios in the lung actually became greater rather than less with muscular exercise. This finding was clearly notionally at odds with the previous results of less sophisticated methods. The greater sensitivity of the MIGET technique in detecting intraregional variations of $\dot{V}_A$ and $\dot{Q}$ was proposed to account for the different perspectives.

In the mid-1980s however, Wagner and his associates demonstrated that diffusion "impairment" contributed significantly to the widening of the (A–a) P_{O_2} at high work rates in healthy subjects. This provided experimental support for the earlier suggestion of Dempsey and coworkers (41) that "high-enough" levels of pulmonary blood flow would predispose to hypoxemia as a result of the pulmonary–capillary transit time being too short to allow equilibration of the alveolar and pulmonary end-capillary blood. Piiper and Scheid (108) extended these considerations by stressing the importance of the $D/\dot{Q}\beta$ ratio in determining such diffusion "limitation" during exercise. D in this ratio is the O_2 diffusing capacity, $\dot{Q}$ is the pulmonary blood flow, and β is the effective slope of the O_2 dissociation curve in that region.

More recently the question has been raised as to what extent pulmonary interstitial edema, and in extreme cases even disruption of the pulmonary capillary interface, may contribute to the diffusion component of the markedly widened (A–a) P_{O_2} gradient in athletes? That is, could this be a necessary consequence of the high pulmonary vascular pressures associated with the high levels of pulmonary blood flow achieved by athletes at high work rates? West and his associates (141) have reported that such disruption of the capillary interface can indeed occur in humans, but only, it seems, in highly fit subjects, and then only at extremely high power outputs. However, establishing the magnitude, threshold, and even occurrence, of interstitial edema as a determinant of impaired gas exchange at high work rates remains a significant technical challenge.

That regional variations in ventilation-to-perfusion ratios in the lung result in differences between the alveolar and arterial gas partial pressures, particularly with respect to P_{O_2}, was apparent both to Krogh and Lindhard and to Haldane in the second decade of the twentieth century. It was therefore a natural extension subsequently to move from the notion that such spatial and also temporal variations in pulmonary gas exchange could influence the profile of alveolar gas partial pressure during an exhalation (although DuBois and his associates [47] made important calculations regarding the time course of gas exchange during both inspiratory and expiratory phases of the breath) to the notion that analysis of the alveolar gas exchange throughout an exhalation could also provide information regarding pulmonary function. For example, the analysis of the alveolar profile following the inhalation of a tracer gas (including O_2), such as the "single-breath N_2 test" or the "closing volume," provided important information with respect to clinical pulmonary physiology.

CO₂ Differences across the Lung

The major use of single-breath analysis during exercise has been to estimate the anatomical or series dead space. This has been most commonly done using the Fowler technique, although an ingenious extension of this method was devised by Cumming (35) in the mid-1980s. In this, he plotted the cumulative volume of CO_2 expired as a function of the volume of the expirate throughout the exhalation. This is equivalent to integrating the expired P_{CO_2} profile with respect to expired volume. He showed that this produced, to a high level of approximation, a linear relationship between the variables, the slope of which provided the flow-weighted mean alveolar P_{CO_2} and with an intercept on the volume axis equal to the volume of the series dead space. This technique was applied to exercise by Lamarra and his associates (156) in the mid-1980s to determine the breath-by-breath profiles of the mean alveolar P_{CO_2} and the series dead space. However, as mentioned above, this technique provides a flow-weighted average of the alveolar P_{CO_2} rather than the time average, which is more appropriate for estimating arterial P_{CO_2}; for example, from the midexpiratory alveolar value. As the end-tidal P_{CO_2} exceeds the mean arterial P_{CO_2} during exercise (by an amount that depends upon the CO_2 output and the breathing pattern) by 6 mmHg or more during heavy exercise, end-tidal P_{CO_2} is an entirely inappropriate index either of mean alveolar P_{CO_2} or of arterial P_{CO_2} during exercise. To overcome this concern, Jones and his associates (85) established an empirical relationship to predict the arterial P_{CO_2} in normal subjects based upon the tidal volume and end-tidal P_{CO_2}. Although differences of up to ±4 mmHg were evident between the predicted and the measured arterial P_{CO_2} in individual subjects, the group-mean response has been shown to be very good.

Alveolar P_{CO_2} has more commonly been used to estimate mixed-venous P_{CO_2}. These single-breath estimates of P_{CO_2} were based upon the recognition that the rate at which CO_2 is evolved into the alveolar gas during a prolonged exhalation decreases progressively as the alveolar P_{CO_2} increases toward the mixed-venous level.

This also results in a progressive decrease of the respiratory exchange ratio (R) during the expirate as oxygen continues to be taken up into the blood at an effectively constant rate. The alveolar P_{CO_2} at which pulmonary CO_2 output ceases and hence at which R has fallen to zero is the oxygenated mixed-venous value rather than the "true" mixed-venous value. Kim, Rahn, and Farhi (94), however, developed an ingenious single-breath technique that could be used to establish the "true" mixed-venous P_{CO_2}. This was based on the known magnitude of the Haldane shift, which unloads CO_2 from hemoglobin in the pulmonary capillaries without change in P_{CO_2}—that is, as the blood is oxygenated. This is equivalent to the "vertical shift" in the CO_2 dissociation curve. This was shown to occur at an intrabreath R of 0.32. Kim and colleagues therefore had subjects perform prolonged exhalations during which the time course of intrabreath R was computed for successive samples of the expirate. The P_{CO_2} at which R had decreased to the required value of 0.32 was taken as the true mixed-venous P_{CO_2}. Unfortunately the ingenuity of this technique is matched by its demands for a high degree of technical rigor. Consequently, it has not been widely applied during exercise: most investigators still prefer to use the rebreathing method to estimate the oxygenated mixed-venous P_{CO_2} as an expedient means of computing cardiac output during exercise.

CONTROL OF EXERCISE HYPERPNEA

The raised metabolic demands caused by muscular exercise present the greatest physiologic challenge to the ventilatory control system in its role as the first line of defense of systemic acid–base status and oxygenation. All would agree that the control system is suitably built for its task and is near perfect in terms of the high degree of precision with which it matches the increase in alveolar ventilation to metabolic rate and the minimum of effort that is put forward by the respiratory muscles to generate the manyfold increases in alveolar ventilation above resting levels during exercise. Most investigators would also agree that more than one mechanism is responsible for exercise hyperpnea as espoused in the mathematical models from the 1950s of Frederick Grodins (Fig. 4.5) (1915–1989) (65) and Pierre Dejours (1922–) (40). The controversy, then, is over which is the primary, dominant mechanism for the hyperpnea. The foundation was already laid by Geppert and Zuntz before the turn of the twentieth century (60) and Krogh and Lindhard early in the twentieth century (97), who spelled out the options of a predominantly "feed-forward" mechanism in the form of "cortical irradiation" *or* predominantly feedback stimuli transmitted either by neural reflex mechanisms from the working limbs or by humoral stimuli initiated by the increase in muscle metabolism. In the ensuing century, several ingenious approaches have been offered in human and animal models in an attempt to isolate *the stimulus* in question. These approaches have included multiple levels of denervation of selected sensory inputs including chordotomy, limb ischemia—to occlude and/or enhance the humoral stimulus-passive exercise—cross-circulation or extracorpeal circulation, and electrical or pharmacologic stimulation of

Fig. 4.5. Frederick Grodins (left) with William Yamamoto (center) and Robert Gelfand (right) in a discussion during a coffee break at the second of the Oxford Conferences on Modeling and Control of Breathing at Lake Arrowhead, CA, in 1982. Photograph provided by Brian J. Whipp.

motor areas in the central nervous system. We now summarize selective contributions to this rich history of research.

"CO_2" and the Exercise Hyperpnea

CO_2-Linked Control

The close correlation between pulmonary CO_2 output ($\dot{V}CO_2$) and ventilation ($\dot{V}E$) during exercise has stimulated, and continues to stimulate, investigators to search for a CO_2-related stimulus to breathing. This notion can be traced back to the studies of Haldane and his associates at Oxford, who determined that the "respiratory center" was exquisitely sensitive to small changes in PCO_2. And although not stated explicitly in these terms, the idea was that respiratory control during exercise involved a high-gain negative feedback loop for PCO_2. The alveolar gas-sampling techniques at that time, however, made it difficult to establish a precise level for mean alveolar PCO_2 (a residual concern even today), as the procedures predominantly sampled end-expiratory gas partial pressures which, especially during exercise, will systematically exceed the mean level.

Further support for a control process somehow sensitive to changes in metabolic CO_2 output came from the experiment of C.G. Douglas (1882–1963) (45) in which he ingested spoonfuls of sugar while sitting quietly at rest. He found that alveolar ventilation increased in proportion to the change in pulmonary CO_2 output as the substrate mixture undergoing oxidation shifted R toward one involving greater carbohydrate utilization. A similar close correlation between CO_2 output and ventilation, with shifts of respiratory quotient (in this case consequent to endurance training) was subsequently demonstrated during the steady state of exercise by Jones and his associates (139).

Further correlative support for a control link between ventilation and "CO_2" came in 1932 when Herxheimer and Kost (72) published an important paper on the response dynamics. In this, they addressed the interrelationships among ventilation, CO_2 output, and O_2 uptake throughout the transient phase of recovery from muscular exercise. They demonstrated that $\dot{V}E$ changed as a linear function of $\dot{V}CO_2$ but not of $\dot{V}O_2$. This important observation has been repeatedly confirmed subsequently using the more sophisticated techniques of dynamic systems analysis to study the response transients. Casaburi and his associates (29), for example, used sinusoidal work-rate protocols and showed that the tight coupling between the $\dot{V}E$ and $\dot{V}CO_2$ dynamics operated a wide range of sinusoidal frequencies. This was not the case for the $\dot{V}E-\dot{V}O_2$ relationship. Consequently, clues to the control of the hyperpnea seemed to reside in features of the pulmonary CO_2 output and not the metabolic CO_2 production: that is, the component of the CO_2 produced metabolically that does not reach the lungs for exchange (being stored, predominantly in the muscles) did not seem to "influence" $\dot{V}E$. These authors were also able to determine, from considerations of the relative phase lags, that the closeness of the coupling was not a consequence of a primary stimulus to the hyperpnea simply washing out CO_2 proportionally as a result. But while such correlations have proven enticing they have not, to date, been supported by convincing control mechanisms.

A disquieting feature of the theory that the exercise hyperpnea was controlled, in large part, by a CO_2-linked feedback system was the absence of a consistent increase in arterial PCO_2. This led to a succession of theories being proposed attempting to reconcile the correlative attractions of CO_2-linked control for the exercise hyperpnea of moderate exercise and the apparent lack of a discernible CO_2-related signal (at least as reflected by mean arterial PCO_2 or pH) at known sites of chemosensitivity.

Mixed Venous Chemoreceptors

In the early 1960s, Armstrong and coworkers (2) and Riley (128) resurrected a theory initially proposed by Zuntz and Geppert in the late nineteenth century—that the CO_2-linked signal may trigger intrathoracic chemoreceptors sensing mixed-venous blood. The suggestion has obvious a priori attractions as the mixed-venous PCO_2 changes in proportion to metabolic rate and hence, in theory at least, could trigger a hyperpneic response that would clear the additional CO_2 from the blood, thereby regulating arterial PCO_2 and pH. There would consequently be no downstream error. The suggestion proved short-lived, however, as both Cropp and Comroe (34) and Sylvester and colleagues (137) demonstrated that hypercapnic blood infused into the pulmonary artery or right atrium of dogs did not stimulate ventilation until the blood, with its increased CO_2 load, had traversed the entire pulmonary circulation to stimulate arterial chemoreceptors. Pierre Dejours, in his classic chapter on the exercise hyperpnea in the 1964 *Handbook of Physiology* (40), concluded that if venous chemoreceptors were not contributory to the exercise hyperpnea then "the slow component of the ventilatory reaction to exercise is under the control of some chemical properties of the arterial blood that influence the respiratory centers, either directly or through arterial chemoreceptors." That is, the notion of a downstream error signal mediating proportional control still prevailed.

Oscillations of Arterial CO_2–H^+

In 1961, Yamamoto and Edwards (159) published a paper that was startling in its conceptual ingenuity. They proposed that a CO_2-linked error signal in the arterial blood might be manifest during exercise but operates independent of the mean arterial P_{CO_2}. They recognized, as had Dubois, Britt, and Fenn (among others) earlier, that the phasic ventilation of the alveoli, operating on a functionally constant mixed venous P_{CO_2} input to the lung, would result in intrabreath fluctuations or oscillations of alveolar P_{CO_2} that would also be manifest in the arterial blood. As the characteristics of this intrabreath oscillation, especially its rate of change, would be a function of the mixed-venous P_{CO_2}, which in turn would be a function of the metabolic rate, the attraction of the theory was immediately obvious. The demonstration by Yamamoto and Edwards that CO_2 infused into the venous system of the rat produced ventilatory changes consistent with their theory took the suggestion beyond mere theoretical possibility. An obvious concern with the theory, however, was the degree of fidelity with which the oscillations of alveolar P_{CO_2} would be reflected in those of the arterial blood.

The development of a rapidly responding intravascular pH electrode by Semple and Band (11) put this issue beyond serious doubt. The pH oscillations in the arterial blood (there being no sufficiently rapid P_{CO_2} electrode) were both evident and consistent with those expected from the alveolar fluctuations.

Furthermore, they were able to demonstrate a close correlation between the rate of change of the downstroke of the arterial pH oscillation and the magnitude of the hyperpneic response under a range of conditions including intravenous CO_2 loading. These oscillatory signals in the arterial blood were considered to provide the stimulus for the oscillations in carotid chemoreceptor discharge; these also had the period of the respiratory cycle, as first demonstrated by Hornbein and coworkers (76).

Black and Torrance (22) subsequently established a phenomenon which had particular relevance to the issue of arterial P_{CO_2}-pH oscillations as potential respiratory stimuli during exercise. They were able to demonstrate that the functional sensitivity of the pontomedullary respiratory complex to afferent signals from, for example, the peripheral chemoreceptors was not constant throughout the breathing cycle. Rather, it was shown to vary phasically such that signals arriving during the ongoing inspiratory phase of the breath stimulated ventilation, whereas signals arriving during the expiratory phase of the ongoing breath provided little or no additional stimulation. Consequently, the peak or maximum rate of change of the pH oscillation stimulating the carotid bodies—for example, during inspiration—would stimulate ventilation whereas the same signal shifted by $180°$ would have little or no effect.

Dan Cunningham and his associates then utilized procedures that would change the shape and/or arrival phase of the oscillatory signal reaching the peripheral chemoreceptors in humans (37). While they were able to provide evidence in favor of such a phase-coupling mechanism in the control of ventilation (especially when peripheral chemoreceptor sensitivity was increased—by hypoxia, for example), they could find no consistent relationship between the arrival time of the peak of the P_{CO_2} oscillation at the carotid bodies and the magnitude of the exercise hyperpnea. This finding, coupled with the fact that subjects who had undergone bilateral carotid body resection (103) did not have a reduced level of ventilation in the steady state of moderate intensity exercise, apparently ruled out an obligatory role for the "oscillation mechanism."

Disequilibrium Theory

Crandall and Forster (46) proposed a mechanism involving the intriguing possibility that a change in mean arterial pH would, in fact, be operative at sites of chemoreception downstream of the lung but would not be apparent in the measurement as standardly analyzed from a syringe sample of blood by a blood-gas analyzer. This was based upon the notion that if carbonic anhydrase were restricted to the interior of the erythrocyte, then the dehydration reaction would continue in the functionally carbonic anhydrase free plasma phase during the transit from the pulmonary capillary to the chemoreceptors. Consequently, as the time course of the uncatalyzed reaction is long (90% response of approximately 1 minute) relative to the transit time to the chemoreceptors (approximately 4–6 seconds during exercise), then the $[H^+]$ would continue to fall in the blood collected for subsequent analysis. Thus the chemoreceptors would "see" a $[H^+]$ that was higher and a pH that was lower than that measured in the standard blood-gas analyzer. This difference would be significantly greater during exercise due to both the increased blood flow and the mixed-venous–to–arterial differences in both $[H^+]$ and P_{CO_2}. These slow changes in plasma pH and P_{CO_2} will only occur, however, if the plasma does not have access to a source of carbonic anhydrase.

Interest in this mechanism quickly dissipated, however, when Lewis and Hill (99) injected bovine carbonic anhydrase into the plasma of dogs and were unable to document any reduction in ventilation during steady-state exercise. This would be expected if the disequilibrium mechanism were operative. Furthermore, Crandall, using a carbonic anhydrase inhibitor which is restricted largely to the plasma (benzolamide), demonstrated a marked slowing of the dehydration reaction in the plasma—suggestive of carbonic anhydrase being accessible to the plasma phase from sources other than the erythrocyte. Effros and his associates (48) demonstrated such a source of accessible carbonic anhydrase at the pulmonary capillary endothelial surface. This was subsequently confirmed specifically by Linnerholm (101) and by Ryan and colleagues (131). The disequilibrium mechanism for a CO_2-linked component of the exercise hyperpnea was therefore seen to be untenable.

Cardiodynamic Control

Karlman Wasserman (1927–) advanced a theory in the mid-1970s that provoked widespread interest. This was based on the observation that although the onset of exercise hyperpnea is rapid, often doubling within a breath or two when the exercise starts from rest, this hyperpnea is commonly not associated with evidence of frank hyperventilation—i.e., end-tidal P_{CO_2} falling and the gas-exchange ratio increasing, although these changes naturally can be demonstrated in some subjects. This meant that there were proportional increases in ventilation and pulmonary blood flow at the onset of exercise. The increase in oxygen uptake and CO_2 output during the first 15 seconds or so of exercise would be largely independent of changes in mixed-venous gas concentration as a result of altered tissue metabolic rate; rather it would be a consequence of the increase in pulmonary blood flow, per se, a phenomenon recognized by Krogh and Lindhard as early as 1913. Wasserman, Whipp (Fig. 4.6), and their associates therefore questioned whether all or part of this early hyperpneic response might be secondary to primary changes in pulmonary blood flow, a mechanism that they termed "cardiodynamic hyperpnea." Such a mechanism had also seemed attractive to Bock and Dill (23), who wrote in 1931 that "reflex stimulation

Fig. 4.6. Karlman Wasserman (right) and Brian Whipp (left), about to depart from their hotel during a visit to Sendai, Japan. Photograph provided by Brian J. Whipp.

of the respiratory center from the heart itself would offer a more logical means of correlating the intake of oxygen through the lungs and the oxygen needs of the organism."

Several studies, mainly from the Wasserman group, gave results which were consistent with such a mechanism. When the β-adrenergic-blocking drug propranolol, for example, was intravenously infused into exercising humans (23), a transient fall of ventilation could be demonstrated during the phase in which the cardiac output decreased. This was followed by an increase of ventilation back to the control level, thought to be consequent to the subsequent increase in mixed-venous CO_2 partial pressure. These changes occurred with little or no alterations in alveolar partial pressure of CO_2. Wasserman and associates also stimulated cardiac output with intravenous infusions of the β-agonist isoproterenol in cats (148). Their experiment confirmed a close correlation between the cardiac output increase and that of ventilation. That this somehow involved a CO_2-related signal was suggested by the absence of an hyperpneic response to the drug following acute hyperventilation (of the cat) despite a marked increase in cardiac output. Eldridge, however, subsequently showed that a component (at least) of the hyperpneic response to the infused isoproterenol was through direct stimulation of the carotid bodies (49). This, however, could not explain the hyperpnea that resulted when cardiac output was increased by electrical stimulation of the heart. What was the stimulus, however?

The original hypothesis was that it was a signal downstream of the lung proportional to the "CO_2 flux" to the lung. This stimulated an active period of research on the topic that had investigators performing experiments, often of a high degree of technical complexity, that loaded CO_2 into or out of the systemic venous inflow to the lung in several animal species. All found that such loading induced hyperpnea and unloading to hypopnea; some proposed that these ventilatory changes were isocapnic with respect to the arterial blood; some said that the response, while not isocapnic, was greater than that induced by inhaled CO_2, and others asserted that the response was nothing more than could be accounted for by the animals' inhaled CO_2-response curve. The question, to uninvolved reviewers such as Cunningham, Robbins, and Woolf, has not, it seems, been decisively resolved (37).

The Wasserman group subsequently proposed a different mechanism for the "cardiodynamic" hyperpnea (87): that the hyperpnea was mechanically mediated, transmitted via cardiac afferents in response to what they termed "right ventricular strain." And although several groups of investigators had demonstrated that hyperpnea could be evoked by stimulating particular cardiac afferents, the lack of reduction in the exercise hyperpnea in humans who had undergone cardiac transplantation seems to rule out an obligatory role for a cardiodynamic mechanism in the exercise hyperpnea. Furthermore, Huszczuk and associates (77) could demonstrate no reduction in either the magnitude or the rapidity of the onset of the hyperpneic response to treadmill exercise in their studies of calves with artificial hearts (Jarvik-type).

The quest for a cardiovascular source of cardiopulmonary coupling to the exercise hyperpnea continues, however. Haouzi and his associates (68), for example, have recently provided experimental evidence for a ventilatory stimulus consequent to vascular distension in the exercising muscles, including neurophysiological evidence of the source of the afferent drive. How important this mechanism is as a proportional contributor to the exercise hyperpnea remains to be resolved (see also next section).

Peripheral Work Factor Contribution to Exercise Hyperpnea

E.A. Asmussen (1907–1991), commonly labeled as one of three Danish musketeers along with E.H. Christensen and Marius Nielson, was instrumental in entrenching Scandinavian scientists at the forefront of research into respiratory and circulatory system regulation during exercise. The Scandinavian presence in these fields was subsequently strengthened by Hermansen, Holtman, Astrand, Bergstrom, Grimby, and several others throughout the 1960s and 1970s and continues today through the research of Bengt Saltin and his colleagues at the Muscle Research Institute in Copenhagen. Asmussen began as an unpaid assistant in the 1920s to J. Lindhard—of Krogh and Lindhard fame—conducting his initial studies in muscle physiology concerning the classic question of the all-or-none response of skeletal muscle to motor nerve stimulation, followed later by studies showing the reduced energy expenditure of eccentric vs. concentric work in humans. Asmussen and Nielson carried out over 40 years of collaborative work as faculty members, and eventually professors, in the Laboratory for the Theory of Gymnastics in Copenhagen.

The series of 25–30 published studies conducted by Asmussen and Nielsen, which spanned almost three decades from the late 1930s to the 60s, set the standard for investigation of the peripheral neurogenic work-factor mechanism of exercise hyperpnea in humans. These are beautifully summarized in Asmussen's 1963 American Physiological Society (APS) handbook chapter (4) and a few years later with a slightly different perspective in a concise summary of their work (5). Their 1943 study showed that limb muscle electrical stimulation produced an isocapnic hyperpnea which was identical to that obtained with whole-body exercise; it was especially important that $\dot{V}CO_2$ (and $\dot{V}E$) were increased to about four tmes rest with the electrical stimulation so that small systematic changes in $PaCO_2$ within the random error of measurement could not explain this substantial hyperpnea. Variations of these studies have been repeated several times in the 1970s and 80s with similar results.

Asmussen also used several insightful approaches to heighten the peripheral work factor, including arm vs. leg and negative vs. positive work. They also exercised tabetic patients with lesions of the dorsal spinal columns and loss of sensory myotatic reflexes, and they all showed a discernable but small hyperventilatory response. Increased $\dot{V}E/\dot{V}O_2$ was especially evident with two additional models—blocking the circulation to the working limb, and exercise with varying degrees of paralysis achieved via curare in healthy subjects. Asmussen and Nielson interpreted the resultant increase in $\dot{V}E/\dot{V}O_2$ to mean that fatigue of the individual motor units had occurred. In turn, fatigue would require that the innervation be extended to a larger portion of the motor pool, thereby demanding a larger voluntary effort. However, Asmussen favored the interpretation that this work factor was of "peripheral nervous origin, closely correlated with aerobic metabolism and somehow dependent on the mechanical conditions of the muscle which, via afferent nerves, were able to alter the state of the reticular formation surrounding the respiratory centers." In the 1960s Asmussen acknowledged that his curare-and-ischemia-induced increases in $\dot{V}E/\dot{V}O_2$ during exercise could equally as well be explained by Krogh's cortical irradiation (see below). However, he also pointed to his electrical muscle stimulation results and the then-completed Frederick F. Kao's classic cross-circulation experiments in the anesthetized dog (88), which both supported the sensory input from exercising muscle as a major source of the hyperpnea.

In the proud tradition of this elusive problem, subsequent studies in both the human and animal models have shown that spinal cord denervation did *not* prevent the exercise hyperpnea secondary to muscle electrical stimulation (90). The "CO_2 flow stimulus" concept (see preceding section) appeared as a viable alternative. Thus, as with all other proposed "primary" stimuli to exercise hyperpnea, the peripheral "work factor" was also *not obligatory* for the normal ventilatory response.

Beginning in the early 1970s, neurophysiologists and cardiovascular physiologists intensified the search for this "work factor" in the form of muscle metabo- and mechanoreceptors. Beginning with the classic studies of Ian McCloskey and Jere Mitchell in anesthetized cats (106), through the microneurographic studies of muscle sympathetic nerve activity in humans over the past decade (142), it is now clear that type III–IV receptors in skeletal muscle play a substantial role in the sympathetically mediated pressor response to exercise. They also play a significant, although less well defined, role in the hyperpnea of exercise.

Marc Kaufman has been one of the major neuroscientists working in this field over the past decade. He is a native New Yorker who studied physiological psychology as a graduate student at the University of Miami. He learned to record from type III–IV afferents innervating the heart and lungs under Hazel and John Colleridge at the Cardiovascular Research Institute in San Francisco and transferred this expertise to contracting skeletal muscle under the urging of Jere Mitchell at Southwestern Medical School in Dallas. Kaufman revealed a significant role for electrical stimulation of the muscle afferents in the bronchodilation and pressor response accompanying muscular contraction. However, the afferent recordings were all in response to electrical muscle stimulation or mechanical or pharmacologic manipulation of the muscle or muscle vasculature. Kaufman's most important work was only very recently completed in a special cat preparation made to exercise naturally via stimulation of CNS motor areas (also see Eldridge below). This physiologically induced contraction of locomotor muscles was shown to stimulate type III–IV muscle afferents from which he was recording (90). These physiologic preparations are very difficult to prepare and to use but are absolutely essential in demonstrating the physiologic relevance of these muscle receptors.

Evidence is also accumulating to show that many of these muscle afferents are stimulated by changes in the contracting muscles chemical environment ($[K^+]$, $[H^+]$) and even by venous distension secondary to changes in vascular conductance (69).

These latter findings, together with previous work showing a ventilatory response to increases in muscle blood flow (68), provide a strong case in favor of both vascular and intramuscular stimulation of muscle receptors and a clear link of cardiovascular to ventilatory regulation during exercise by these peripheral "work factors."

Central Command Contributions to Exercise Hyperpnea

Frederick Eldridge (Fig. 4.7) has used his anesthetized cat model for over 30 years from the 1960s to the 90s to investigate neural mechanisms underlying the control of breathing. He and his associates completed a variety of studies ranging from mechanisms of memory-like phenomenon such as short-term potentiation of respiratory motor output *to* central serotonergic actions *to* medullary chemoreception *to* mathematical modeling *to* suprapontine mechanisms of breath perception. Eldridge was attracted by the neurohumoral (fast–slow components) control theory of Pierre Dejours, but was not convinced that a peripheral mechano-sensitive stimulus was *the* neural mechanism involved. Kao's findings using electrical stimulation of locomotor muscles in anesthetized dogs were commonly cited as the primary proof for the pe-

Fig. 4.7. Frederick Eldridge and the North Carolina Central Command Research Group. Clockwise from top left: Tony Waldrop, Missy Klinger, David Millhorn, and Eldridge, 1982. Photograph provided to the authors by Frederick Eldridge upon request.

ripheral neural mechanism (see Asmussen above) but Eldridge was not at all convinced that these controversial cross-circulation findings were the answer to the neural-control of exercise hyperpnea: (1) exercise was nonphysiologic as it was caused by electrically induced contraction of the limb muscles; (2) there were serious questions of whether the neural dogs' chemoreceptors in Kao's scheme were, in fact, truly separated from the donor (humoral) dogs' effluent blood; and (3) others were having difficulty replicating the finding that spinal cord denervation eliminated the hyperpneic response to muscle stimulation.

Even at this early stage, Eldridge favored the central mechanism proposed by Krogh in 1913—a mechanism which was often referred to but never studied, apparently because of the difficulty in isolating the stimulus in a physiologic preparation. Eldridge chose to pursue this mechanism decerebrate and then a decorticate diencephalic (thalamic) cat preparation which he discovered through the Russian literature (50). The preparation appeared ideal because: (1) locomotion could be elicited spontaneously or by electrical or pharmacologic stimulation of the hypothalamus; (2) phrenic nerve activity and quadriceps EMG activities were used to represent breathing and locomotor activity, respectively; and (3) fictive locomotion could be provided by skeletal muscle paralysis. The key finding was that locomotion—whether active or fictive—that is, with or without feedback or a changing metabolic rate, produced identical increases in phrenic nerve activity and blood pressure (50). This was powerful science, which clearly identified a strong feedforward mechanism. Recently, in his mathematical model, Eldridge acknowledged the contributions of input from receptors in the peripheral musculature, of short-term potentiation of central respiratory neurons, and of a slow rise in the humoral component, K^+ (acting on the peripheral chemoreceptors) as part of the neurohumoral axis for exercise hyperpnea. He also claimed from his modeling approach that the gain of the central neural mechanism is such that it was required "for the rapid ventilatory response at exercise onset and necessary for the maintenance of isocapnia in the steady-state."

A graduate student co-investigator with Eldridge in these central command studies was Tony Waldrop, a native of North Carolina who received all of his formal education at Chapel Hill and then completed postdoctoral study with Jere Mitchell in Dallas. Waldrop continues to conduct in vivo and in vitro studies on the central command question at the University of Illinois at Urbana–Champaign. He has documented an association between so-called locomotor sites in the brain with loci that control cardiorespiratory function. His work has also shown that the caudal hypothalamus receives input from skeletal muscle receptors, thereby providing the circuitry for interaction between central command and muscle feedback (144, 145). This type of information means that it is highly unlikely that the feed-forward control mechanism (once initiated) would ever act in isolation in effecting exercise hyperpnea—that is, in the absence of feedback influence from the contracting limb muscle. Accordingly, one might theorize that the (descending) feed-forward stimuli of Krogh and Eldridge act in concert with the (ascending) feedback influence of Asmussen and Kao in producing a significant share of the hyperpnea.

Waldrop and Eldridge both believe a major need in this field is to identify the full neuronal circuitry involved in mediating central command. This is a fascinating, challenging problem. Of course, many others involved with this nearly unfathomable problem of exercise hyperpnea do not fully accept that the induced locomotion of a decorticate cat with its relatively small increases in metabolic rate represents the real world of physiologic exercise in the intact state. Certainly, these design limitations are worth heeding in limiting extrapolations from reduced preparations. Nevertheless, it is equally important to recognize that these types of findings in the cat point strongly to a significant feed-forward contribution to the hyperpnea of exercise that must be incorporated into our current view of this difficult problem. This mechanism for hyperpnea also parallels the better-established central command mechanism for exercise tachycardia.

A footnote to this brief synopsis of central neural control of exercise hyperpnea is to acknowledge the important and varied contributions of Jere Mitchell, of Southwestern Medical School in Dallas. A cardiologist by training, Mitchell directed a landmark study in the late 1950s concerning the physiological meaning of maximal oxygen uptake in healthy humans, directed key animal and human studies on central and peripheral neural mechanisms of cardiorespiratory control in exercise (106, 108), and mentored several important scientists at both ends of the feed-forward/feedback spectrum, including both Kaufman and Waldrop.

Hyperventilation of Heavy Exercise

The mechanisms that normally result in even-greater responses to exercise associated with a metabolic acidosis also continue to provide a challenge to investigators of exercise hyperpnea. As early as 1888, however, Geppert and Zuntz (60) had proposed that ventilation during exercise was controlled, at least in part, by metabolites formed in the exercising limbs and transported to the "respiratory center." The subsequent demonstration by Fletcher and Hopkins (53) in 1907 that lactic acid was a normal end product of metabolism and, as originally shown by Ryffel (132), that its concentration in the blood was increased in humans consequent to muscular exercise provided strong suggestive evidence that the lactic acidosis of exercise contributed to the hyperpnea. In 1914 Christensen, Douglas, and Haldane (of "CHD" or "Haldane" effect fame) at Oxford showed that there was a marked reduction in the CO_2 combining capacity of venous blood following exercise in humans, a finding supported by several groups over the next decade.

In 1923 Barr, Himwich, and Green (14) sampled arterial and venous blood simultaneously and demonstrated a break point in the CO_2 combining power of the blood with increasing rates of work. In 1927 C.G. Douglas (45) and his student, W.H. Owles (116), were able to demonstrate (45) what was already implicit in the work of Barr, Himwich, and Green: that there was a range of work rates at which there was little or no change in blood lactate, but after which further increases in work rate resulted in systematic increases—that is, there appeared to be a "threshold" of response to the profile of metabolic acidemia during progressive exercise.

In 1936, Neilsen (110) showed that ventilation increased more rapidly during exercise at work rates associated with a metabolic acidosis, with a consequent reduction in alveolar P_{CO_2}. He noted further that an "exceptionally well-trained subject could attain a pulmonary ventilation of more than 70 l/min without any increase in the H^+ concentration of the arterial blood." Considerable support for the relationships among these variables emerged subsequently from numerous laboratories. That this additional component of ventilatory response was associated with the lactic acidosis subsequently formed the basis for actually using the ventilatory response to estimate the onset of the metabolic (chiefly lactic) acidemia.

Although the mechanisms of a threshold of ventilatory drive consequent to lactic acidemia remain controversial, the vast weight of evidence supports the view that at the work rate associated with the development of metabolic acidosis there is a more rapid increase in ventilation which serves to constrain the fall of pH: that is, the subject with the largest compensatory increase in ventilation has the smallest decrease in pH—suggesting that the hydrogen ion, per se, is unlikely to be the stimulus. Numerous other potential stimuli for this additional component of ventilatory drive during exercise have been demonstrated, many of which seem to change relatively abruptly in concert with the developing acidemia. These include epinephrine and norepinephrine, potassium, osmolarity, and even body temperature. Despite this, however, many investigators continue to focus their studies on a single mechanism.

Isocapnic Buffering

The development of rapidly responding gas analyzers and the advent of digital computers in the 1960s and 70s allowed the profiles of these ventilatory and pulmonary gas-exchange responses during exercise to be investigated in greater detail. An initially surprising finding of rapid incremental exercise tests was that although ventilation in normal subjects could be shown to increase disproportionately with respect to metabolic rate as the lactic acidosis developed, frank hyperventilation (as manifested by a decrease in alveolar and arterial P_{CO_2}) was delayed for a metabolic rate range of 0.5–1.0 l/min of oxygen uptake—that is, there appeared to be an "isocapnic buffering" region with the metabolic acidosis (149). This region was characterized by the ventilatory equivalent for O_2 ($\dot{V}_E/\dot{V}_{O_2}$) increasing, having initially decreased during the moderate intensity domain, but with the ventilatory equivalent for CO_2 ($\dot{V}_E/\dot{V}_{CO_2}$) being constant. The metabolic rate at which $\dot{V}_E/\dot{V}_{CO_2}$ began to increase and hyperventilation became manifest was termed the "respiratory compensation point" by the Wasserman group (149) and the "threshold of decompensated metabolic acidosis" by Reinhard and his associates (125). It is still not clear why hyperventilation in response to the metabolic acidosis of exercise is so delayed, as the time course of ventilatory response attributable to the carotid bodies (at least with respect to hypoxia) is known to be fast, with a time constant in humans of the order of 5–10 seconds (44).

An important clue to the site of action of this metabolic-acidosis-related increment of ventilation at high work rates was provided by Asmussen and Neilsen (3) in 1946. They showed that the abrupt administration of 100% oxygen to subjects exercising at high work rates resulted in a rapid and sustained decline of ventilation—but only at high work rates. They concluded that "during heavy work, some substance which increases the ventilation by the arterial chemoreceptors is produced in the working muscles." They also concluded, however, that this "substance" was not lactic acid itself.

Pierre Dejours and his colleagues (39), recognizing that hyperoxia exerted its effect on ventilation during exercise predominantly by suppressing peripheral chemosensitivity, developed what has subsequently come to be known as the "Dejours test." With this procedure, "slugs" of oxygen lasting as briefly as a few breaths are surreptitiously administered to the inspirate during exercise. The higher time resolution of Dejours's measurements allowed these investigators to demonstrate a transient reduction in ventilation at both moderate and heavy exercise. The amount of the decrement consequent to the oxygen breathing was considered to represent the ongoing component of the ventilatory drive attributable to the peripheral chemoreceptors. This, and other similar tests of hypoxic responsiveness, have subsequently been used over a range of work rates with surprisingly consistent results among different investigators—the peripheral chemoreceptors account for some 15%–20% of the exercise hyperpnea during both moderate and heavy exercise.

The availability of subjects (CBR subjects) who had both carotid bodies surgically resected (predominantly for the relief of dyspnea in patients with obstructive lung disease) provided a further level of insight with respect to factors controlling the compensatory hyperventilation. What emerged from a large body of work on these patients was that there was no demonstrable role for the aortic bodies as ventilatory chemosensors in humans and that, despite the absence of peripheral chemosensitivity, the ventilatory response to moderate exercise was not appreciably different during air breathing compared with either normal controls or an appropriate group of asthmatic controls. However, unlike the latter control groups, the steady-state level of ventilation during moderate exercise was no different in the CBR subjects with air breathing or breathing 12% oxygen (103). However, what was clearly apparent during heavy exercise, whether at the higher reaches of an incremental test or during sustained heavy exercise, was the marked reduction or even absence of the normal compensatory hyperventilation. That is, from these results, the carotid bodies appeared to be the dominant mediators of the compensatory hyperventilation for an acute metabolic acidosis.

Further studies on this mechanism were performed in normal subjects by Rausch and coworkers (124), who demonstrated that, in normal subjects, the degree of acidemia associated with a given degree of metabolic acidosis is highly dependent on the sensitivity of the peripheral chemoreceptors. A surprising additional finding in this experiment, however, was that with peripheral chemosensitivity being presumably abolished by hyperoxia there was still a small compensatory hyperventilation: the mechanism for this component of the response, however, still awaits elucidation.

As neither hydrogen nor potassium ions, for example, cross the blood–brain barrier (at least as reflected by measurements in bulk CSF), central chemoreceptors are thought not to mediate a significant component of the compensatory hyperventilation. They may, however, constrain its magnitude. Bisgard and his associates (21) demonstrated that the cerebrospinal fluid actually becomes alkaline at high work rates in ponies consequent to the reduced arterial and CSF P_{CO_2}—that is, conditions consistent with constraint of the ventilatory response deriving from other stimuli to the hyperpnea.

Alternative Extra Chemoreceptor Mechanisms

Until the early 1980s it was pretty well accepted that while the exercise hyperpnea question was still a true dilemma, the hyperventilation of heavy exercise was relatively straightforward. As outlined above, a known stimulus (metabolic acidosis in arterial blood) acting at a well-established receptor site (carotid chemoreceptors) would produce the compensatory hyperventilatory response. However, as one might have expected, several findings have arisen since this time which question whether this model is sufficient to explain the hyperventilatory response.

First, increased circulating lactic acid was shown *not* to be obligatory to the hyperventilatory response to heavy exercise in humans because when dietary-induced glycogen depletion was used to greatly reduce exercise-induced lactic acidosis in normal subjects, the hyperventilatory response to exercise still occurred and was unaltered from normal (26, 90). Similar findings in the patient with McArdle's syndrome are often cited as evidence against an obligatory role for metabolic acidosis (66); but it is also clear that these patients hyperventilate even at rest and in mild exercise, so it is doubtful that this complex clinical entity is simply a pure model of normal exercise without metabolic acidosis.

Second, it was pointed out that the human carotid-body-denervated patients who lacked hyperventilation in heavy exercise were asthmatics. Thus, the fact that they actually hypoventilated in heavy exercise (rather than just failed to hyperventilate) was thought to be as much attributable to abnormal mechanical constraint as it was to their lack of carotid chemoreceptors (55)—although the resting lung function in these patients was only mildly abnormal. Furthermore, when Larry Pan and Bert Forster in Milwaukee denervated the carotid bodies in otherwise-normal ponies, these denervated animals actually hyperventilated *more*—rather than less—given the same levels of lactic acidosis in heavy exercise (90).

Third, more complex models of carotid chemoreceptor function in heavy exercise have been proposed. These include David Patterson's (of Oxford University) suggestion that carotid chemoreceptor stimulation in heavy exercise by multiple interactive stimuli—that is, potassium, norepinephrine, metabolic H^+—may, in turn, interact with a descending central command mechanism (118). It has also been shown recently, however, in awake dogs that when isolated carotid chemoreceptors were made selectively hypocapnic via extracorporeal circulation, this caused a marked depression of tidal volume and a substantial inhibition of the ventilatory response to superimposed carotid body hypoxia (134). Accordingly, during heavy exercise the slowly developing systemic hypocapnia which occurs would be expected to exert at least some braking effect on the ventilatory response to the circulating stimuli present.

This series of findings does not rule out significant contributions from the carotid bodies in the intact, heavily exercising acidotic human. Indeed, systemic metabolic acidosis, per se, does increase sensory input from the carotid chemoreceptors and it is conventional to view the hyperventilation of heavy exercise as a respiratory compensation for the accompanying metabolic acidosis (see above). However, these findings do open the door for serious consideration of alternative mechanisms and a more complex model beyond just circulating humoral stimuli. For example, James Duffin of the University of Toronto presents locomotor muscle EMG data in humans which suggest that as limb muscle fatigue sets in during heavy exercise, more motor units are recruited to maintain a given muscle tension (80). Accordingly, the postulate is that this extra motor output would also augment in parallel the descending central command mechanism for respiratory motor output—or, equally well, the ascending muscle afferent mechanism—and cause an extra increase in ventilation.

So again, many of the similar controversies and inconsistencies that have plagued the search for the exercise hyperpnea mechanism also confront the question concerning hyperventilation mechanism in heavy exercise. Indeed, because of the complexities added by heavier exercise, this problem may prove to be even harder to solve or even to investigate. Certainly carotid body denervation is also not the pure intervention one would desire because secondary adaptations are likely to occur—possibly centrally—in response to the removal of this important sensory input. The question is further complicated by the fact that at maximum exercise and very high ventilations the higher airway flow rates and force outputs required by the respiratory muscles may impose the added confounder of mechanical constraints on ventilatory output during heavy exercise—at least in fit subjects (see below).

The Dilemma of Exercise Hyperpnea: What Next?

Research on the primary underlying mechanism of the exercise hyperpnea remains at a standstill. Evidence supporting three schools of thought (as outlined above) was established more than a decade ago. Certainly, this body of knowledge is extremely important in demonstrating beyond a reasonable doubt that descending neural influences from supramedullary motor areas and ascending influences from contracting limb muscles are both extremely important in the regulation of breathing and of hyperpnea. A great deal of evidence has also been forthcoming over the past two decades detailing the nature of the muscle receptors, their stimuli, and their sensory neural pathways. Furthermore, while some lingering mysteries need to be clarified (see above), it seems likely that the humoral control of exercise hyperpnea via a changing CO_2 flow to the lung and acting via established or presumed chemoreceptors plays a minor (if any) role, except perhaps during heavy-intensity exercise.

So, research over the past century has clearly established these mechanisms in isolation but their relative importance and their mutual dependencies remain unresolved. Several carefully conducted studies in physiologic animal and human models have demonstrated that one or more of those mechanisms are *not* obligatory for the hyperpnea. The engineer and modeler William Yamamoto (Fig. 4.5) (see CO_2 and the exercise hyperpnea section above) analyzed this dilemma in a symposium over two decades ago by invoking Sherrington's concept of occlusion as it applies to interpretation of conflicting experimental data. "The appearance of any surface activity, like the adjustment of pulmonary ventilation, may in any given instant depend upon the merging of alternative equivalently powerful mechanisms by 'policy maker' mechanisms that adapt the output ventilation by adding, compromising, or ignoring signals pertinent to several conflicting goals of the organism" (159). In response to further questioning, Yamamoto explained: "You may have many sufficient mechanisms, each of which in a given isolated circumstance explains the whole phenomenon. When they act simultaneously, they mask each other. Unfortunately, at the present time, and I've tried for a long time to write a mathematical subsystem showing occlusion and I cannot do so."

At least two attempts have been made in the 1980s to test this occlusion hypothesis. Waldrop found in anesthetized cats that the sum of the phrenic nerve responses to activation of central command and to the peripheral neurogenic stimuli, when given separately, exceeded the response when activated simultaneously (144). Hubert Forster developed the awake, exercising pony as an experimental model to study exercise hyperpnea. A long series of single-sensory-pathway denervation studies conducted throughout the 1980s and 90s culminated in the investigation of the effects of simultaneous lesioning of three sensory pathways—carotid sinus nerves, hilar nerves, and (partially) spinal afferents (117). Unpredictably, they found that the effect of three simultaneous lesions on the exercise hyperpnea exceeded the sum of the effects of the three individual lesions. Clearly, these results from the multiple denervation experiments emphasize the amazing robustness of the exercise hyperpnea stimulus but are difficult to interpret because there must certainly exist secondary adaptations to the permanent removal of sensory pathways.

This is obviously a tough problem to crack. While, as Yamamoto suggested, mathematical modeling is welcomed in this multifaceted field, it is also clear the modelers need much more information, and the source of this could come from the burgeoning field of neurobiology—although much caution must be exercised in translating findings in experimental animals to humans. For example, we need to know much more about the structure and interconnections between locomotor and respiratory neuronal networks in the CNS. This is not a simple task given the current controversies concerning the precise location and operation of the respiratory pattern generator itself. Nonetheless, some progress is being made as indicated by recent findings demonstrating an important role for the fastigial nucleus in the cerebellum in the control of certain types of hyperpnea (156).

Finally, if we are ever going to solve this huge question of the *relative contributions* of the several potential mechanisms to the total hyperpnea, we must develop unanesthetized chronically instrumented animal models in whom specific stimuli and receptor sites can be manipulated or controlled without the need for denervation of sensory pathways or lesioning of central circuits. It does not seem feasible to remove important sensory inputs without causing substantial adaptations by other mechanisms—perhaps even by the medullary pattern generator itself. It is also absolutely essential for future experiments to use exercise modes and models which truly mimic physiologic exercise. Certainly the modern approach to manipulating these proposed mechanisms would be through the use of molecular, genetic techniques which would start by identifying differences among strains in the hyperpneic response, then identify genomic regions and ultimately the genes involved, and finally study the effects of the isolated expressed proteins on the hyperpnea (56). These approaches are already being applied in mice to questions concerning interindividual differences in chemosensitivity and propensity toward sleep apnea.

EXERCISE AND ACID–BASE REGULATION

Evaluation of the extra- and intracellular acid–base status during exercise has a long history and remains an exciting and evolving field today. Indeed, our understanding of acid–base regulation is an integral part of many facets of exercise physiology including regulation of O_2 and CO_2 transport, control of breathing, regulation of muscle metabolism, and even muscle fatigue. The fundamental methods for compartmentalizing and assessing acid–base balance were developed in the early twentieth century, emphasizing the so-called Henderson–Hassellblach relationships, with changes in P_{CO_2} being used to characterize the respiratory contributions and changes in HCO_3 (at a given P_{CO_2}) being classified as the metabolic contribution to any observed changes in hydrogen ion. These traditional concepts continue to be used widely in research, in teaching, and in clinical practice.

Peter Stewart (1921–1993) was educated in physics and chemistry at the University of Manitoba and was a long-time faculty member at Brown University. In the late 1970s and early 1980s he proposed a physicochemical approach which purports to quantify the contribution of various ionic changes to the hydrogen ion concentration (135). This concept utilizes such independent variables as the strong ion difference (SID^+) or the net charge balance of fully dissociated or "strong" ions; the total concentration of metabolic acids and bases $[A_{tot}]$; and the P_{CO_2}. The equations developed utilizing these parameters allows one to apportion changes in each variable to the observed change in $[H^+]$ (or even to calculate the $[H^+]$).

In the past decade Stewart's approaches have acquired a number of followers who study exercise effects on intra- and extracellular acid–base regulation and their consequences (78, 86, 96). These scientists favor the Stewart approach especially for its distinction between independent and dependent variables and his treatment of $[H^+]$ and $[HCO_3^-]$ as being dependent on the interaction between several acid-base systems. Accordingly, several long-held concepts have recently been challenged by claims that lactic acid is not buffered by bicarbonate; that pH is not "controlled" by the P_{CO_2}/balance; that bicarbonate is not conserved or excreted by the kidneys; and that $[SID^+]$ and changes in osmolality are key regulators of breathing. In fact, a major purpose of ventilatory control may be to preserve protein function. Thus, proponents of this approach do not accept that [H$^+$] is "controlled" by changes in [HCO_3^-] or that movements of H$^+$ or HCO_3^- ions, per se, are of any relevance to acid–base regulation. Furthermore, it is claimed that the Stewart approach has "opened the door to understanding the close links between fluid and electrolyte control and the physiological and biochemical events occurring during exercise in muscle, the circulation and respiration" (86).

Not unexpectedly, many scientists have not embraced the Stewart approach. Rather, many view these concepts as unnecessary and confusing rather than illuminating or insightful and still others even question the validity of the chemistry. Undoubtedly, these disagreements will continue and it is too early to determine if indeed this approach will enhance our understanding of extracellular and especially intracellular acid–base status in muscle during exercise, its regulation, and its physiologic consequences. In the meantime, at the very least, this new approach and the controversy it has created have precipitated exciting new research into this important problem of acid–base regulation.

ARE THE LUNG AND CHEST WALL BUILT FOR EXERCISE?

What links in the O$_2$ transport/utilization scheme are most important in limiting V̇O$_2$max? In 1923 the Nobel laureate A.V. Hill (1886–1977) and his colleague Lupton represented a common view among physiologists in speculating that either the heart or the lung would be prime candidates as major limiting factors: "At higher speeds of blood flow, the blood is imperfectly oxygenated in its rapid passage through the lung" (73). The resulting exercise-induced arterial hypoxemia (EIAH) would, of course, reduce systemic O$_2$ transport and limit the ability to widen O$_2$ content difference across the working muscle. Theoretically, the ventilatory response to maximum exercise may also limit performance if: (1) it is inadequate relative to metabolic demand (therefore EIAH and respiratory acidosis will develop), or if (2) achieving adequate alveolar ventilation for gas exchange purposes may require such high levels of respiratory muscle work that blood flow may be diverted from locomotor to respiratory muscles, or intolerable dyspneic sensations may develop and/or respiratory muscle fatigue may occur.

Overbuilt?

Of course, this was all theoretical speculation and once measurements of arterial blood gases and ventilation gradually became plentiful throughout the 1950s and 60s it was obvious that arterial PO$_2$ (and O$_2$ content) was held constant even during maximal exercise; furthermore, the ventilation achieved in max exercise was always considerably less than that achievable voluntarily at rest (see the earlier section titled Pulmonary Gas Exchange). Robert Forster and Sir Francis Roughton's (1899–1972) methods for measuring pulmonary diffusion capacity and capillary blood volume were developed in the 1950s (54) based in part on the earlier pioneering work of Marie Krogh (1874–1943), wife of the Danish Nobel laureate August Krogh (1874–1949). Measurements of $D_{L_{CO}}$ during exercise followed, most notably by Robert L. Johnson, a protégé of Forster and a major contributor to exercise gas-exchange research for over 40 years (81). In turn, theoretical modeling of the lung's diffusion capacity confirmed the concept of a truly overbuilt lung by predicting that regardless of the $\dot{V}_{O_2}$ demanded by the muscle and the degree of O_2 desaturation in mixed-venous blood, end-capillary arterial O_2 saturation would be maintained in the healthy lung (133). It was further predicted that only exercise in the hypoxic environment of high altitude would result in diffusion limitation. Similarly, the work and O_2 cost of breathing during even max exercise was dismissed as a limiting factor. Richard L. Riley, the pioneer of the bubble electrode method for blood-gas measurements, summarized contemporary thinking in 1960 in a chapter of a book entitled *Science & Medicine of Exercise & Sports* (127). Riley used the few estimates of the O_2 cost of breathing available at the time to claim that the O_2 cost of ventilation at the presumed maximum $\dot{V}_E$ achievable in exercise would not exceed the benefit of the increased ventilation to transporting O_2.

With all this considerable evidence mounting to support the concept of an extremely overbuilt lung and chest wall whose capacity greatly exceeded demand for gas transport, the debate regarding limiting factors shifted to the relative roles of cardiac output (and O_2 delivery) vs. muscle metabolic capacity. This debate continues today with the bulk of evidence favoring cardiac output—specifically max stroke volume—and O_2 transport combined with capillary-to-mitochondrial diffusion as the major limitations to $\dot{V}_{O_2}$max, at least in the normally fit human subject (109). Most of this evidence is based on cross-sectional data and shows a strong correlation between $\dot{V}_{O_2}$max and the product of blood flow and arterial O_2 content across a wide range of fitness levels. Furthermore, experimental manipulation of inspired O_2 levels, hemoglobin concentration, circulating blood volume, and cardiac output usually shows the appropriate increase or decrease in $\dot{V}_{O_2}$max and sometimes parallel changes in max work rate. Some exceptions to the rule are in extremely sedentary subjects and animals in whom $\dot{V}_{O_2}$max remains unaffected by experimental changes in arterial O_2 content.

Underbuilt? Costly?

Beginning in the early 1960s, through the present, several studies began to more closely examine this long-held concept of an overbuilt healthy lung and chest wall. The question was asked whether the pulmonary system in different groups of healthy humans (and other mammals) is actually susceptible to failure, especially in the face of the extraordinary maximal metabolic demands required by highly fit humans or other athletic species. The evidence is certainly not all in at this point, nor conclusive, but we can list the following sets of findings which build a case for a limit to the healthy pulmonary system's physiologic capabilities.

Alveolar-to-Arterial Gas Exchange

Up through the early 1960s there were isolated reports of exercise-induced arterial hypoxemia in individual subjects. For example, Richard Riley in his chapter (see above) mentioned that his own arterial PO_2 had fallen more than 20 mmHg to the low 70s during stressful exercise in one experiment. However, he also stated, "Riley appears to be unusual in this respect because the reduction in PaO_2 is not the expected response in normal people" (127). Loring Rowell, who was training with Henry Longstreet Taylor in Minneapolis in the early 1960s, spent over 35 years contributing importantly to our understanding of cardiovascular hemodynamics and its regulation during exercise in humans (see Chapter 3 of this book). He also made an important observation of substantial arterial HbO_2 desaturation in a group of athletes at $\dot{V}O_2$max using the VanSlyke technique to measure arterial O_2 content (130). Twenty years passed before others confirmed and extended these findings by measuring arterial blood gases in many highly fit subjects in heavy exercise (see Pulmonary Gas Exchange section, above). These studies produced three major findings.

First, EIAH was found to occur in a significant number of fit subjects including young adult men, women, and older fit subjects (71). O_2 desaturations in the 85%–92% range were achieved by reductions in PaO_2 of 20–30 mmHg combined with a rightward shift in the O_2 dissociation curve induced by coincident increases in metabolic acidity and blood temperature.

Second, a widened alveolar-to-arterial O_2 difference correlated best with the degree of EIAH. The causes of this excessive A–aDO_2 remain unresolved—although some indirect evidence has been mounted in favor of a diffusion limitation (see Pulmonary Gas Exchange section above). A major technical problem facing the testing of these hypotheses is our inability to quantify changes in extravascular lung water in vivo.

Finally, absolute CO_2 retention rarely occurred in the face of EIAH, but subjects with the greatest hypoxemia often showed minimal hyperventilation. In other words, the hyperventilatory response was inadequate to compensate for an excessively widened alveolar-to-arterial O_2 difference. Thoroughbred horses represent an extreme in this regard as they all begin to experience EIAH even in submaximal exercise and this is accompanied by marked CO_2 retention (17).

Airway Limitation to Expiratory Flow

Expiratory flow limitation, like EIAH, also occurs only in the highly fit at or with heavy or near-maximum exercise, simply because the maximum flow:volume envelope is sufficient to accommodate exercise-induced demands on tidal volume and flow rate achieved at normal $\dot{V}O_2$max (see Breathing Mechanics section above for descriptive observations). The increased end-expiratory lung volume and elastic inspiratory work associated with expiratory flow limitation have two types of implications for performance limitation: (1) the hyperventilatory response to heavy exercise will be mechanically constrained and (2) the increased elastic work of the respiratory muscles associated with hyperinflation will require a greater $\dot{V}O_2$ and, therefore, a greater share of the cardiac output.

Exercise-Induced Diaphragmatic Fatigue

Diaphragm fatigue occurs in very heavy prolonged exercise to exhaustion (see above Breathing Mechanics section). Unlike exercise-induced arterial hypoxemia or expiratory flow limitation, exercise-induced diaphragm fatigue occurs across a wide range of fitness levels from the near sedentary to the highly fit. Diaphragm fatigue does not compromise ventilation in long-term heavy constant-load exercise because accessory muscles are recruited to provide most of the required ventilation. Indeed, even unloading the respiratory muscles using a proportional assist ventilator has no consistent effect on the ventilatory response to heavy exercise (58). However, the increased use of accessory muscles may mean that breathing becomes mechanically inefficient at a time when the diaphragm's contribution to pressure generation is decreasing with exercise duration, thereby increasing the oxygen and circulatory cost of breathing.

Competition for the Available Blood Flow

In the late 1970s, spurred by the respiratory muscle fatigue studies of Macklem, Roussos, Grassino, and Bellmare at McGill University (see Breathing Mechanics section above), much interest was shown in respiratory muscle function—including muscle properties, metabolic capacities, susceptibility to acid–base disturbances, and blood flow regulation. Furthermore, it was also shown by Craig Harms and colleagues that in humans the work of breathing normally incurred at $\dot{V}O_2$max caused vasoconstriction of working limb muscle so that when the work of breathing was reduced at max exercise, limb vascular resistance fell and limb flow increased (70). Conversely, adding to the work of breathing at $\dot{V}O_2$max increased limb vascular resistance and decreased blood flow. The reflex mechanisms causing this sympathetically mediated vasoconstriction in limb muscle arterioles remain unknown. It was speculated that mechanoreceptor and/or chemoreceptor reflexes originating in respiratory muscles when undergoing heavy, fatiguing work may be important. Type III–IV reflex receptors have been located in the diaphragm, and when thin fiber phrenic afferents are stimulated electrically systemic vasoconstriction occurs in several vascular beds (74); but whether they explain the effects of changing the work of breathing on vascular resistance in exercising limbs remains untested.

Demand vs. Capacity: Are Lungs Adaptable?

So the lung and chest wall are not immune from failure or from incurring extremely high levels of ventilatory work, both of which compromise O_2 transport to working limbs and increase the perception of effort. In most cases, with the exception of diaphragmatic fatigue, these failures and high costs are reserved mainly for the highly trained human. This implies that the increased demand for oxygen transport in the trained is greater than the capacity to meet this demand of the lung, the airways, and the chest wall. In turn, this imbalance must mean *that the structures serving pulmonary gas transport are less malleable (or adaptable) in response to physical training* and to the trained state than are other steps in the scheme of O_2 transport and utilization.

The history of research into the question of malleability of organ system structure is fascinating. The lung's premier morphologist, Ewald Weibel, chairman of the Anatomy Department at the University of Berne, Switzerland, pioneered in this area. Professor Weibel graduated from medical school in Zurich in the mid-1950s and received postdoctoral training at Columbia University under Andre Cournand and Dickinson Richards and also with Domingo Gomez. Weibel and Gomez collaborated to develop ingenious morphometric methods for estimating the lung's diffusion surface area, and their first application of this method was to establish that the human lung contains about 300,000,000 alveoli. Weibel's 1963 book *Morphometry of the Human Lung* (151) and his 1973 *Physiological Reviews* on the same subject continue to be the ultimate references on lung microstructure. Much of Weibel's work from 1959 to the present was based on his strong conviction that animals are built "reasonably," in the sense that they build enough but not too much structure to support the functional demands on the system (152). Accordingly, all parts should be adjusted to the maximal functional capacity (or $\dot{V}O_2$max)—a notion he termed "symmorphosis," or well-balanced design.

This concept demands comparable adaptability by all organs involved in gas transport—including the lung—and indeed, this concept was beautifully demonstrated by Marsh Tenney and his protégé, John Remmers, at Dartmouth in the early 1960s. These authors had shown that alveolar surface area changed with resting $\dot{V}O_2$ across mammals of widely varying size ranging from the Etruscan shrew to the whale (140). The key follow-up question here was whether the regulation of lung compartmentalization was genetically determined by evolutionary pressures on species or whether it represented a response of the individual organ system to functional demands imposed by metabolic requirements. The idea of adaptive growth is particularly attractive in the case of the lung since so much of the pulmonary development occurs after birth when environmental and behavioral factors will vary greatly. Accordingly, many investigators have shown diffusion surface area to increase substantially when rats or dogs or humans were exposed to long-term hypoxia at high altitude during development (81). Similarly, even in the mature lung, alveolar capillary surface area in remaining lung tissue was shown to grow and expand following lung resection. Increased mechanical strain via alveolar inflation, release of growth factors, and activation of transcription factor have all been implicated as signals for gene induction in the remaining lung tissue (61). So, lung structure was clearly adaptable, but does this also apply to a changing metabolic demand?

Adaptation within a Species

Professor Weibel and colleagues discovered the ideal animal model in which to study the influence of elevated metabolic rate in the form of the Japanese waltzing mouse, an animal that is constantly in motion and whose metabolic rate is 70% greater than that of normal mice because of a genetically determined vestibular defect. In line with their hypotheses, the Swiss scientists found a substantial increase in alveolar surface area in the waltzing mouse compared to age-matched albino mice (59). Furthermore, they later showed a similar enhancement of alveolar surface area in response to pharmacologically induced long-term hyperactivity and hypermetabolism in otherwise-normal albino mice (25).

Donald Bartlett, another protégé of Marsh Tenney at Dartmouth, had also been influenced by Tenney and Remmers' cross-species correlation of metabolic rate to lung surface area. However, his early longitudinal studies using chronic exercise training and thyroid hormone administration to chronically elevate $\dot{V}O_2$ in rats yielded no measurable changes in lung structure (15). Bartlett then reasoned that the enhanced alveolar surface area observed in the Japanese waltzing mouse may have been due to genetic mechanisms rather than due to adaptive responses to chronic exercise stress and reinvestigated the question by using phenotypically normal littermates heterozygous for the waltzing trait. Their analysis showed no difference between the paired animals' lung surface area despite the severalfold difference in chronic metabolic rate (16). Similar negative findings have recently been reported by the Canadian pathologist Thurlbeck, who subjected very young mice shortly after weaning to chronic physical training (141). These reports in rodents are consistent with mostly cross-sectional studies of pulmonary function tests and diffusion capacity in human athletes vs. nonathletes, as well as with limited numbers of longitudinal studies over several weeks of physical training. Longitudinal studies of highly active aging athletes in the sixth and seventh decades also reported a normal age-dependent decline in lung elastic recoil and diffusion capacity despite their continued high level of training and high $\dot{V}O_2$max.

Certainly there are some reports of exceptional groups of athletes with very large diffusion capacities or maximum flow:volume envelopes, especially in swimmers. For example, P.O. Astrand, a charter member of Sweden's famous group of exercise physiologists reported very large lung volumes in a large group of girl swimmers in the early 1960s (6). There is also some evidence that swim training during growth in humans will cause some increase in lung volume and diffusion capacity (31).

Are Respiratory Muscles Adaptable to the Training Stimulus?
Although the oxidative capacity and fatigue resistance of the diaphragm exceeds that of limb musculature, specific overload training of respiratory muscles has substantial effects on their strength or endurance in humans and animals. With whole-body endurance training, several studies conducted throughout the 1980s in rodents reported widely different effects on the diaphragm's enzymatic capacities. In the 1990s, Scott Powers of the University of Florida completed a comprehensive analysis of this question in rats by systematically altering training duration and intensity and by distinguishing between influences on the costal and crural diaphragm and even other respiratory muscles (121). His findings are clear that training does, indeed, produce relatively small (compared to limb musculature) but consistent increases in mito-

chondrial and antioxidant enzymatic activities in the costal diaphragm. Much greater training intensity and duration were required to elicit comparable significant changes in the crural diaphragm. Whether these enzymatic changes actually enhance diaphragmatic contractile performance remains unknown. However, it has been shown in humans that fitter, trained subjects do experience levels of exercise-induced diaphragmatic fatigue similar to those in less fit subjects; but at similar relative exercise intensities, the trained subject incurs substantially higher ventilatory loads and force outputs by the diaphragm (43).

In summary, these data show that the lung and airways, and to a lesser extent the respiratory muscles, do not adhere strictly to the concept of symmorphosis; rather, most of these important links in the gas transport chain lack the capability for adaptation—at least within a given animal or human in response to the stimulus of physical training. Accordingly, although training—or the trained state—evokes substantial adaptive changes in locomotor skeletal muscle metabolic capacity and in capillary density and in circulating blood volume, the lung structure eventually becomes underbuilt with respect to an increasing maximal demand for gas transport. This absence of adaptation to the training stimulus is most striking in the lung's gas-exchange apparatus and in its airway dimensions and elastic properties. Functionally, the result of this acquired imbalance between demand (for O_2 transport) and capacity (of the lung and chest wall) in many highly trained subjects is exercise-induced hypoxemia, expiratory flow limitation, high levels of ventilatory work, and diaphragmatic fatigue (42).

Interspecies Adaptation

Does this imbalance between gas transport systems also apply *across species*, especially under conditions where marked differences exist in maximal metabolic requirements? Ewald Weibel addressed this question in collaboration with Richard Taylor (1939–1995) (Fig. 4.8), director of Harvard's Concord Field Station. Their goal was to combine Weibel's expertise in lung morphology with Taylor's substantial experience in comparative physiology of the energetics of locomotion. Together, they traveled throughout the world and produced a remarkable series of studies of several athletic and sedentary animal species with the emphasis on comparing animals of similar body mass but markedly different athletic prowess and $\dot{V}O_2$max. These pairings included dog vs. goat, pony vs. calf and racehorse vs. steer. The differences in $\dot{V}O_2$max averaged 2.5-fold and the volume of mitochondria in skeletal muscle, as well as various indices of muscle capillarity, red cell mass, and ventricular volume all showed up-regulation in the athletic species which were proportional, or nearly proportional, to $\dot{V}O_2$max. The lung's morphometric diffusion capacity was an exception. Although it was significantly higher in athletic species, differences were 30%–40% less than the differences in $\dot{V}O_2$max or mitochondrial volume. Nonetheless, based on his estimates of transit time and alveolar capillary equilibrium, Weibel reasoned that both the athletic and sedentary animals and humans would still show equilibrium of alveolar air with capillary blood before the blood left the capillary bed, even at $\dot{V}O_2$max (89, 152). The athletes would simply use more of their redundancy (or

Fig. 4.8. Professors Ewald R. Weibel (right) of the University of Berne, Switzerland and C. Richard Taylor (left) of Harvard University. Photograph provided to the authors by Ewald R. Weibel upon request.

reserve) in diffusion capability than would the sedentary subjects. However, as outlined above, these claims of merely a reduced redundancy in diffusion capacity at high $\dot{V}_{O_2}$max clearly do *not* apply to many fit humans of varying age and gender with a $\dot{V}_{O_2}$max up to two times their sedentary contemporaries or to the thoroughbred horse with a $\dot{V}_{O_2}$max twice the fittest humans. The evidence against these claims is simply that many of these athletes *do* experience markedly widened A–aDO2 and EIAH—along with expiratory flow limitation and high levels of ventilatory work! In brief, the limited malleability in the lung's airways and diffusion surface meant that *all* of its reserve was utilized in order to meet the high metabolic demand required by the more malleable downstream structures subserving O_2 transport.

There are several species such as the fox and the prong-horned antelope, however, with a $\dot{V}_{O_2}$max severalfold that of the fittest humans (when expressed with reference to body mass) who do *not* experience EIAH during max exercise and who hyperventilate substantially in heavy exercise (102). Accordingly, morphometric diffusion capacity of the lung was shown to be enhanced *in proportion to* $\dot{V}_{O_2}$max in these species. We speculate that these animals would also have appropriate structural adaptations in their airways and respiratory muscles because they showed a significant hyperventilatory response to maximum exercise. We do not, however, know the O_2 or circulatory cost of breathing at max exercise in these remarkable natural athletes.

In summary, the question of lung structure adaptability appropriate to increased metabolic demands defies generalization. Within a species (including the human), little or no adaptability occurs in response to the training stimulus. Across species, cursorial animals may or may not show significant pulmonary adaptations to their markedly elevated $\dot{V}O_2$max, ranging from the woefully inadequate gas exchange shown in the human-engineered equine to the near-optimal proportional adaptations in the lungs of several extraordinarily fit athletes in the wild. These widely diverse degrees of structural adaptability coincide with the changing role of the lung's contribution to limitation of $\dot{V}O_2$max.

Traditional Ideas Regarding O_2 Transport and Exercise Limitation Challenged

While we have given a popular, simple view of O_2 transport to working muscle as a key determinant of $\dot{V}O_2$max, this topic is much more complex and has become even more controversial in recent years. Since this topic is outside the scope of our chapter, we offer only a few brief comments on controversial points and their spokespersons.

Hugh Welch of the University of Tennessee has published several studies throughout the 1970s and 80s which question the simple view of O_2 transport effects as influenced by increases and decreases in PaO_2, on $\dot{V}O_2$max and performance (155). Welch stresses the huge technical problems encountered even in measuring $\dot{V}O_2$max when FiO_2 is raised. He has also documented that while changing FiO_2 and PaO_2 affects the arterial O_2 content, its effects on O_2 delivery to the working muscle may well be completely negated by reciprocal changes in local blood flow. He joins with his mentor, Wendell Stainsby, who contributed the isolated perfused hindlimb preparation to the study of gas transport and muscle physiology and colleague Jack Barclay in proposing that it is not really O_2 transport but blood flow to the working muscle that is the key determinant of $\dot{V}O_2$max and muscle performance. Muscle acid–base regulation may be the more important controlled variable in this scheme.

Consistent with these theories of Welch and colleagues over the past decade or so, others have proposed that the changes in peak performance caused by changing PaO_2 (and CaO_2) may not be attributable *directly* to O_2 transport to the working muscle. Exercise performance may be curtailed—for example, in hypoxia—because of the need to spare the function of other, more vital organ systems. John Sutton (1941–1996), a native Australian who spent his career in sports sciences at McMaster University (Canada) and Sydney University (Australia), researched the physiologic effects of high-altitude hypoxia as much for the adventure (and risk) as for the physiologic insights to be gained. His most notable scientific contribution was his direction of a distinguished group of investigators in a simulated climb of Mount Everest in a low-pressure chamber in the mid-late 1980s. Among the many seminal discoveries that came out of this epic study was the concept that the marked reductions in $\dot{V}O_2$max seen with increasing altitude may have been attributable, in part, to the influence of brain hypoxia on locomotor areas of the CNS. That result, in turn, may have led to the curtailment of activation of force output by the limb muscles (136).

Other suggestions for indirect effects of arterial hypoxemia on locomotor muscle performance have included the threat of an inadequate myocardial O_2 supply, which would limit ventricular function; thus, limb muscle force output would be reflexively inhibited to spare myocardial function (111). Similarly, others have noted that limb fatigue did not occur during exhaustive exercise at high altitude and suggested that a reflex inhibition of force output of limb muscles might have occurred in order to spare fatigue of the diaphragm (91). Finally, recent observations suggest that high levels of respiratory muscle work during max exercise might increase sympathetic efferent activity and cause vasoconstriction of limb muscle arterioles and thereby limit their blood flow (70). A.S. Paintal suggested that a vagally mediated reflex inhibition of limb skeletal muscle contractibility may originate in the lungs' C-fibers (116). Pulmonary C-fibers are stimulated by lung stretch, distention of the great vessels or heart or interstitial pulmonary edema—all of which may occur in heavy exercise. More recently, this viscerosomatic reflex effect was beautifully demonstrated in Kaufman's mesencephalic cats (see above), which showed marked inhibition of locomotor muscle intensity when pulmonary congestion was produced transiently via raising left atrial pressure (119). If indeed these inhibitory reflexes from the lungs are also sensitive in humans, then they might make a significant contribution to the limitation of limb muscle performance in health and especially in patients with congestive heart failure.

It is extremely healthy for the field that such time-honored concepts as oxygen transport to muscle limits $\dot{V}O_2max$ (dating back to the time of A.V. Hill) are being questioned several decades later. It's also disconcerting that as a new century begins we are still so far from the evidence necessary to test these hypotheses one way or another. Hopefully, new methods of exploring and even quantifying levels of mitochondrial oxygenation in working skeletal muscle, in vivo, will make some of these goals more attainable.

REFERENCES

1. Aaron E.A., J.C. Seow, B.D. Johnson, and J.A. Dempsey. Oxygen cost of exercise hyperpnea: implications for performance. *J. Appl. Physiol.* 72:1818–1825, 1992.
2. Armstrong W., H.H. Hurt, R.W. Blide, and J.M. Workman. The humoral regulation of breathing. *Science* 133:1897–1906, 1961.
3. Asmussen E. and M. Neilsen. Studies on the regulation of respiration in heavy work. *Acta Physiol. Scand.* 12:171–188, 1946.
4. Asmussen E. Muscular exercise. In: *Handbook of Physiology, Respiration,* edited by W.O. Fenn and H. Rahn. Washington, DC: American Physiological Society, 1964, pp. 939–978.
5. Asmussen E. Exercise and the regulation of ventilation. In: *Physiology of Muscular Exercise,* edited by C.B. Chapman. New York: American Heart Association, 1967, pp. I-132–I-145.

6. Astrand P.O., L. Engstrom, B.O. Eriksson, P. Karlberg, I. Nylander, B. Saltin, and C. Thoren. Girl swimmers. *Acta Paediat. Scand.* (Suppl) 147:1–75, 1963.

7. Auchincloss J.H., R. Gilbert, and G.H. Baule. Effect of ventilation on oxygen transfer during early exercise. *J. Appl. Physiol.* 21:810–818, 1966.

8. Babcock M.A., D.F. Pegelow, S.R. McClaran, O.E. Suman, and J.A. Dempsey. Contribution of diaphragmatic force output to exercise-induced diaphragm fatigue. *J. Appl. Physiol.* 78:1710–1719, 1995.

9. Bachofen H., H.J. Hobi, and M. Scherrer. Alveolar-arterial N_2 gradients at rest and during exercise in healthy men of different ages. *J. Appl. Physiol.* 34:137–142, 1973.

10. Bake B., J. Bjure, and J. Widimsky. The effect of sitting and graded exercise on the distribution of pulmonary blood flow in healthy subjects studied with the 133xenon technique. *Scand. J. Clin. Lab. Invest.* 22:99–106, 1968.

11. Band D.M. and S.J.G. Semple. Continuous measurement of blood pH with an indwelling arterial glass electrode. *J. Appl. Physiol.* 22:854–857, 1967.

12. Banzett R.B., R.W. Lansing, R. Brown, G.P. Topulos, D. Yager, S.M. Steele, B. Londono, S.H. Loring, M.B. Reid, L. Adams, and C.S. Nations. "Air hunger" from increased PCO_2 persists after complete neuromuscular block in humans. *Respir. Physiol.* 81:1–17, 1990.

13. Barcroft J. *Features in the Architecture of Physiological Function.* Cambridge: Cambridge University Press, 1934.

14. Barr D.P., H.E. Himwich, and R.P. Green. Studies in the physiology of muscular exercise. *J. Biol. Chem.* 55:495–537, 1923.

15. Bartlett D. Postnatal growth of the mammalian lung: influence of exercise and thyroid activity. *Respir. Physiol.* 9:50–57, 1970.

16. Bartlett D. and J.G. Areson. Quantitative lung morphology in Japanese waltzing mice. *J. Appl. Physiol.* 44:446–449, 1978.

17. Bayly W.M., D.R. Hodgson, D.A. Schultz, J.A. Dempsey, and P.D. Gollnick. Exercise-induced hypercapnia in the horse. *J. Appl. Physiol.* 67:1958–1966, 1989.

18. Beaver W.L., K. Wasserman, and B.J. Whipp. On-line computer analysis and breath-by-breath graphical display of exercise function tests. *J. Appl. Physiol.* 34:128–132, 1973.

19. Beck K.C. Control of airway function during and after exercise in asthmatics. *Med. Sci. Sports Exerc.* 31(Suppl):54–511, 1999.

20. Bellemare F., D. Wight, C.M. Lavigne, and A. Grassino. Effect of tension and timing of contraction on the blood flow of the diaphragm. *J. Appl. Physiol.* 54:1597–1606, 1983.

21. Bisgard G.E., H.V. Forster, B. Byrnes, K. Stanek, J. Klein, and M. Manohar. Cerebrospinal fluid acid-base balance during muscular exercise. *J. Appl. Physiol.* 45:94–101, 1978.

22. Black A.M.S. and R.W. Torrance. Respiratory oscillations in chemoreceptor discharge in control of breathing. *Respir. Physiol.* 13:221–237, 1971.

23. Bock A.V. and D.B. Dill. The physiology of muscular exercise. In: *F.A. Bainbridge*, 3rd ed., rewritten by Bock A.V. and D.B. Dill. London: Longmans, Green and Co., 1931, p. 157.

24. Brown H.V., K. Wasserman, and B.J. Whipp. Effects of beta-adrenergic blockade during exercise on ventilation and gas exchange. *J. Appl. Physiol.* 41:886–892, 1976.

25. Burri P.H. and E.R. Weibel. Morphometric estimation of pulmonary diffusion capacity. II. Effect of PO_2 on the growing lung: adaptation of the growing rat lung to hypoxia and hyperoxia. *Respir. Physiol.* 11:247–264, 1971.

26. Busse M.W., N. Maassen, and H. Konrad. "Relation between plasma K^+ and after glycogen depletion and repletion in man. *J. Physiol.* 443:469–476, 1991.

27. Bye P.T.P., S.A. Esau, K.R. Walley, P.T. Macklem, and R.L. Pardy. Ventilatory muscles during exercise in air and oxygen in normal men. *J. Appl. Physiol. Respir. Environ. Exerc. Physiol.* 56:464–471, 1984.

28. Campbell E.J.M. *The Respiratory Muscles and the Mechanics of Breathing.* Chicago: Yearbook Publishers, 1958.

29. Casaburi R., B.J. Whipp, K. Wasserman, W.L. Beaver, and S.N. Koyal. Ventilatory and gas exchange dynamics in response to sinusoidal work. *J. Appl. Physiol.* 42:300–311, 1977.

30. Chen Z., F.L. Edridge, and P.G. Wagner. Respiratory-associated thalamic activity is related to level of respiratory drive. *Respir. Physiol.* 90:99–113, 1992.

31. Clanton T., G.F. Dixon, J. Drake, and J. Gadek. Effects of swim training on lung volumes and inspiratory muscle conditioning. *J. Appl. Physiol.* 62:39–46, 1987.

32. Cotes J.E. *Lung Function and Assessment and Application in Medicine* (3rd ed). Oxford: Blackwell, 1975.

31. Crandall E.D. and R.E. Forster. Rapid ion exchange across the red cell membrane. *Adv. Chem. Ser.* 118:65–87, 1973.

34. Cropp G.J.A. and J.H. Comroe, Jr. Role of mixed venous CO_2 in respiratory control. *J. Appl. Physiol.* 16:1029–1033, 1961.

35. Cumming G. Gas mixing in disease. In: *Scientific Foundations of Medicine*, edited by J.G. Scalding and G. Cumming. London: Heinemann, 1981.

36. Cunningham D.J.C. Integrative aspects of the regulation of breathing: a personal view. In: *MTP International Review of Science: Ser 1, Physiology*, edited by J.G. Widdicombe. Baltimore: University Park Press, 1974, pp. 303–369.

37. Cunningham D.J.C., P.A. Robbins, and C.B. Wolff. Integration of respiratory responses to changes in alveolar partial pressures of CO_2 and O_2 and in arterial pH. In: *Handbook of Physiology: The Respiratory System, Vol. II, Control of Breathing*, edited by N.S. Cherniack and J.G. Widdecombe. Bethesda: American Physiological Society, 1986, pp. 475–528.

38. Deal E.C., E.R. McFadden, R.H. Ingram, R.H. Strauss, and J.J. Jaeger. Role of respiratory heat exchange in exercise-induced asthma. *J. Appl. Physiol.* 46:467–475, 1979.

39. Dejours P., Y. Labrousse, J. Raynaud, F. Girard, and A. Teillac. Stimulus oxygne de la ventilation au repos at au cours de l'exercise musculaire, a basse altitude (50 m) chez l'Homme. *Rev. Franc. Etudes Clin. Biol.* 3:105–123, 1958.

40. Dejours P. Control of respiration in muscular exercise. In: *Handbook of Physiology*, edited by W.O. Fenn and H. Rahn. Washington, DC: American Physiological Society, 1964, pp. 631–648.

41. Dempsey J., P. Hansen, and K. Henderson. Exercise-induced hypoxemia in healthy persons at sea level. *J. Physiol. (Lond.)* 355:161–175, 1984.

42. Dempsey J. A. Is the lung built for exercise? *Med. Sci. Sports* 18:143–155, 1986.

43. Dempsey J. A., L. Adams, D. Ainsworth, R. Fregosi, C. Gallagher, A. Guz, B. Johnson, and S. Powers. Airway, lung and respiratory muscle function during exercise. In: *Handbook of Physiology—XII—Exercise*, edited by L.B. and J.T.S. Rowell. Oxford: Oxford Press, 1996, pp. 448–515.

44. Doll E., J. Keul, A. Brechtel, R. Limon-Lason, and H. Reindell. Der Einfluss krplicher Arbeit auf die arteriellen Blutgase in Freibur und in Mexico City. *Sportarzt Sportsmedizin* 8:317–325, 1967.

45. Douglas C.G. Co-ordination of the respiration and circulation with variations in bodily activity. *Lancet* 2:213–218, 1927.

46. Downes J.J. and C.J. Lambertsen. Dynamic characteristics of ventilatory depression in man on abrupt administration of O_2. *J. Appl. Physiol.* 21:447–453, 1966.

47. DuBois A.B., A.G. Britt, and W.O. Fenn. Alveolar CO_2 during the respiratory cycle. *J. Appl. Physiol.* 4:535–548, 1952.

48. Effros R.M., R.S.Y. Chang, and P. Silverman. Acceleration of plasma bicarbonate conversion to carbon dioxide by pulmonary carbonic anhydrase. *Science* 199:427–429, 1978.

49. Eldridge F.L. and P. Gill-Kumar. Mechanisms of hyperpnea induced by isoproterenol. *Respir. Physiol.* 40:349–363, 1980.

50. Eldridge F., D.E. Millhorn, and T.G. Waldrop. Exercise hyperpnea and locomotion: parallel activation from the hypothalamus. *Science* 211:844–846, 1981.

51. England S.J. and D. Bartlett, Jr. Changes in respiratory movements of the human vocal cords during hyperpnea. *J. Appl. Physiol.* 52:780–785, 1982.

52. Farhi L.E. Ventilation–perfusion relationships. In: *Handbook of Physiology, Sect. 3, The Respiratory System*, edited by L.E. Farhi and S.M. Tenney. Bethesda: American Physiological Society, 1987, pp. 199–215.

53. Fletcher W.M. and F.G. Hopkins. Lactic acid in amphibian muscle. *J. Physiol. (Lond.)* 35:247, 1907.

54. Forster R.E. Exchange of gases between alveolar air and pulmonary capillary blood: pulmonary diffusing capacity. *Physiol. Rev.* 37:391–452, 1957.

55. Forster H.V. and L.G. Pan. Exercise hyperpnea. In: *The Lung: Scientific Foundations*, edited by R.G. Crystal and J.B. West. New York: Raven Press, pp. 1553–1564, 1991.

56. Forster H.V. Exercise hyperpnea: where do we go from here? *Exerc. Sport Sci. Rev.* 28:133–137, 2000.

57. Gale G.E., J. Torre-Bueno, R. Moon, H.A. Saltzman, and P.D. Wagner. $\dot{V}A/\dot{Q}$ inequality in normal man during exercise at sea level and simulated altitude. *J. Appl. Physiol.* 58:978–988, 1985.

58. Gallagher C.G. and M. Younes. Effect of pressure assist on ventilation and respiratory mechanics in heavy exercise. *J. Appl. Physiol.* 66:1824–1837, 1989.

59. Geelhaar A. and E.R. Weibel. Morphometric estimation of pulmonary diffusing capacity. III. The effect of increased oxygen consumption in Japanese waltzing mice. *Respir. Physiol.* 11:354–366, 1971.

60. Geppert J. and N. Zuntz. Ueber die Regulation der Atmung. *Arch. Ges. Physiol.* 42: 189–244, 1888.

61. Gilbert K.A. and D.E. Rannels. From limbs to lungs: a newt perspective on compensatory lung growth. *News Physiol. Sci.* 14:260–267, 1999.

62. Gledhill N., A.B. Froese, and J.A. Dempsey. Ventilation to perfusion distribution during exercise in health. In: *Muscular Exercise and the Lung*, edited by J.A. Dempsey and C.E. Reed. Madison: University of Wisconsin Press, 1977, pp. 325–343.

63. Grimby G., B. Saltin, and L. Wilhemson. Pulmonary flow, volume and pressure volume relationships during submaximal and maximal exercise in young well-trained men. *Physiolpathol. Respir.* 7:57–168, 1971.

64. Grimby G., M. Goldman, and J. Mead. Respiratory muscle action inferred from rib cage and abdominal V-P partitioning. *J. Appl. Physiol.* 41:739–751, 1976.

65. Grodins F. Analysis of factors concerned with the regulation of breathing in exercise. *Physiol. Rev.* 30:220–239, 1950.

66. Hagberg J.M., E.F. Coyle, J.E. Carroll, J.M. Miller, W.H. Martin, and M.H. Brooke. Exercise hyperventilation in patients with McArdle's disease. *J. Appl. Physiol.* 52:991–994, 1982.

67. Hammond M.D., G.E. Gale, K.S. Kapitan, A. Ries, and P.D. Wagner. Pulmonary gas exchange in humans during exercise at sea level. *J. Appl. Physiol.* 60:1590–1598, 1986.

68. Hanouzi P., A. Huszczuk, J.P. Gilles, B. Chalon, F. Marchal, J.P. Crance, and B.J. Whipp. Vascular distension in muscles contributes to respiratory control in sheep. *Respir. Physiol.* 99:41–50, 1995.

69. Hanouzi P., J.M. Hill, B.K. Lewis, and M.P. Kaufman. Responses of group III and IV muscle afferents to distension of the peripheral vascular bed. *J. Appl. Physiol.* 87:545–553, 1999.

70. Harms C.A., T.J. Wetter, S.R. McClaran, D.F. Pegelow, G.A. Nickele, W.B. Nelson, P. Hanson, and J.A. Dempsey. Effects of respiratory muscle work on cardiac output and its distribution during maximal exercise. *J. Appl. Physiol.* 85:609–618, 1998.

71. Harms C.A., S.R. McClaran, D.F. Pegelow, G. A. Nickele, W.B. Nelson, and J.A. Dempsey. Exercise-induced arterial hypoxemia in healthy young women. *J. Physiol.* 507: 619–628, 1998.

72. Herxheimer H. and R. Kost. Das verhaltnis von sauerstoffaufnahme und kohlensaurausscheidung zur ventilation bei harter muskelarbeit. *Z. Klin. Med.* 108:240–247, 1932.

73. Hill A.V. and H. Lupton. Muscular exercise, lactic acid, and the supply and utilization of oxygen. *Q. J. Med.* 16:135–171, 1923.

74. Hill J.M. Discharge of group IV phrenic afferent fibres increases during diaphragmatic fatigue. *Brain Res.* 856:240–244, 2000.

75. Holmgren A. and H. Linderholm. Oxygen and carbon dioxide tensions of arterial blood during heavy and exhaustive exercise. *Acta Physiol. Scand.* 44:203–215, 1958.

76. Hornbein T.F., Z.J. Griffo, and A. Roos. Quantitation of chemoreceptor activity: interrelation of hypoxia and hypercapnia. *J. Neurophysiol.* 24:561–568, 1961.

77. Huszczuk A., B.J. Whipp, T.D. Adams, A.G. Fisher, R.O. Crapo, G.C. Elliott, K. Wasserman, and D.B. Olsen. Ventilatory control during exercise in calves with artificial hearts. *J. Appl. Physiol.* 68:2604–2611, 1990.

78. Jennings D.B. Respiratory control during exercise: hormones, osmolality, strong ions and PaCO$_2$. *Can. J. Appl. Physiol.* 19:334–349, 1994.

79. Hyatt R. E., D.P. Schilder, and D.L. Fry. Relationship between maximum expiratory flow and degree of lung inflation. *J. Appl. Physiol.* 13:331–338, 1958.

80. Jeyaranjian R., R. Goode, and J. Duffin. Role of lactic acidosis on the ventilatory response to heavy exercise. *Respiration* 55:202–209, 1989.

81. Johnson R.L., S.C. Cassidy, R. Grover, J. Schutte, and R.H. Epstein. Functional capacities of lungs and thorax in beagles after prolonged residence at 3100 m. *J. Appl. Physiol.* 59:1773–1782, 1985.

82. Johnson B.D., W. G. Reddan, K.G. Deow, and J.A. Dempsey. Mechanical constraints on exercise hyperpnea in a fit aging population. *Am. Rev. Respir. Dis.* 143:968–977, 1991.

83. Johnson B.D., M.A. Babcock, O.E. Suman, and J.A. Dempsey. Exercise-induced diaphragmatic fatigue in healthy humans. *J. Physiol. (Lond.)* 460:385–405, 1993.

84. Johnson B.D., M.S. Badr, and J.A. Dempsey. Impact of the aging pulmonary system on the response to exercise. *Clin. Chest Med.* 15:229–246, 1994.

85. Jones N.L., D.G. Robertson, and J.W. Kane. Difference between end-tidal and arterial PCO$_2$ in exercise. *J. Appl. Physiol.* 47:954–960, 1979.

86. Jones N.L. Our debt to Peter Stewart. *Can. J. Appl. Physiol.* 20:327–332, 1995.

87. Jones P.W., A. Huszczuk, and K. Wasserman. Cardiac output as a controller of ventilation through changes in right ventricular load. *J. Appl. Physiol.* 53:218–244, 1982.

88. Kao F.F. An experimental study of the pathway involved in exercise hyperpnea employing cross-circulation technique. In: *The Regulation of Human Respiration,* edited by D.J.C. Cunningham and B.B. Lloyd. Blackwell: Oxford, 1963, pp. 461–502.

89. Karas R.J., C.R. Taylor, H.J. Jones, S.L. Lindstedt, R.B. Reeves, and E.R. Weibel. Adaptive variation in the mammalian respiratory system in relation to energetic demand. *Respir. Physiol.* 69:101–115, 1987.

90. Kaufman M.P. and H.V. Forster. Reflexes controlling circulatory, ventilatory and airway responses to exercise. In: *Handbook of Physiology. Section 12, Exercise: Regulation and Integration of Multiple Systems,* edited by L.B. Rowell and J.T. Shepherd. New York: Oxford, 1996, pp. 381–447.

91. Kayser B., M. Narici, T. Binzoni, B. Grassi, and P. Cerretelli. Fatigue and exhaustion in chronic hypobaric hypoxia: influence of exercising muscle mass. *J. Appl. Physiol.* 76:634–640, 1994.

92. Kenyon C.M., S.J. Cala, S. Yan, A. Aliverti, G. Scano, R. Doratni, A. Pedotti, and P.T. Macklem. Rib cage mechanics during quiet breathing and exercise in humans. *J. Appl. Physiol.* 83:1242–1255, 1997.

93. Killian K.J., P. Leblanc, D.H. Martin, E. Summers, N.L. Jones, and E.J.M. Campbell. Exercise capacity and ventilatory, circulatory, and symptom limitation in patients with chronic airflow limitation. *Am. Rev. Respir. Dis.* 146:935–940, 1992.

94. Kim T.S., H. Rahn, and L.E. Farhi. Estimation of true venous and arterial PCO_2 by gas analysis of a single breath. *J. Appl. Physiol.* 21:1338–1344, 1966.

95. Koulouris N.G., I. Dimopoulov, P. Valta, R. Finkelstein, R. Cosio, and J. Milic-Emili. Detection of expiratory flow limitation during exercise in COPD patients. *J. Appl. Physiol.* 82:723–731, 1997.

96. Kowalchuk J.M. and B.W. Scheuermann. Acid-base regulation: a comparison of quantitative methods. *Can. J. Physiol. Pharmacol.* 72:818–826, 1994.

97. Krogh A. and J. Lindhard. The regulation of respiration and circulation during the initial stages of muscular work. *J. Physiol. (Lond.)* 47:112–136, 1913.

98. Lavoisier A.L. Alterations qu'eprouve l'air respire. In: *Oeuvres de Lavoisier.* Paris: Imprimaire Imperiale, 1862–1893, pp. 676–687.

99. Lewis S.M. and E.P. Hill. Effect of plasma carbonic anhydrase (C.A.) on ventilation in exercising dogs. *Fed. Proc.* 38:1034, 1970.

100. Lilienthal J.R., R.L. Riley, D.D. Proemmel, and R.E. Franke. An experimental analysis in man of the oxygen pressure gradient from alveolar air to arterial blood during rest and exercise at sea level and at altitude. *Am. J. Physiol.* 147:199–216, 1946.

101. Linnerholm G. Pulmonary carbonic anhydrase in the human, monkey, and rat. *J. Appl. Physiol.* 52:352–356, 1982.

102. Longworth K.E., J.H. Jones, J.E.P.W. Bicudo, C.R. Taylor, and E.R. Weibel. High rate of O_2 consumption in exercising foxes: large PO_2 difference drives diffusion across the lung. *Respir. Physiol.* 77:263–276, 1989.

103. Lugliani R., B.J. Whipp, C. Seard, and K. Wasserman. The effect of bilateral carotid body resection on ventilatory control at rest and during exercise in man. *N. Engl. J. Med.* 285: 1105–1111, 1971.

104. Manohar M. Blood flow to the respiratory and limb muscles and to abdominal organs during maximal exertion in ponies. *J. Physiol.* 377:25–35, 1986.

105. Margaria R., J.M. Milic-Emili, J.M. Petit, and G. Cavangna. Mechanical work of breathing during muscular exercise. *J. Appl. Physiol.* 15:354–358, 1960.

106. McCloskey D.I. and J.H. Mitchell. Reflex cardiovascular and respiratory responses originating in exercising muscle. *J. Physiol.* 224:173–186, 1972.

107. Mead J. Mechanical properties of the lungs. *Physiol. Rev.* 41:281–329, 1961.

108. Mitchell J.H., D.R. Reeves, H.B. Rogers, and N.H. Secher. Epidural anesthesia and cardiovascular responses to static exercise in man. *J. Physiol. (Lond.)* 417:13–24, 1989.

109. Mitchell J. and B. Saltin. Oxygen transport and the concept of maximal oxygen uptake. Chapter 6, this volume.

110. Neilsen M. Untersuchungen ueber die Atemregulation beim Menschen, besonders mit Hinblick auf die Art des chemischen Reizes. *Skand. Arch. Physiol.* 74(Suppl 10): 83–208, 1936.

111. Noakes T.D. Maximal oxygen uptake: "classical" versus "contemporary" viewpoints: a rebuttal. *Med. Sci. Sports Exer.* 30:1381–1398, 1998.

112. O'Donnell D.E., J.C. Bertley, L.K.L. Chow, and K.A. Webb. Qualitative aspects of exertional breathlessness in chronic airflow limitation:pathophysiologic mechanisms. *Am. J. Resp. Crit. Care Med.* 155:109–115, 1997.

113. Olafsson S. and R.E. Hyatt. Ventilatory mechanics and expiratory flow limitation during exercise in normal subjects. *J. Clin. Invest.* 48:564–573, 1969.

114. Orem J. and A. Netick. Behavioral control of breathing in the cat. *Brain Res.* 366: 238–253, 1986.

115. Otis A. Work of breathing. *Physiol. Rev.* 34:449–472, 1954.

116. Paintal A.S. Vagal sensory receptors and their reflex effects. *Physiol. Rev.* 53:159–227, 1973.

117. Pan L.G., H.V. Forster, A.G. Brice, T.F. Lowry, C.L. Murphy, and D. Wurster. Effect of multiple denervations on the exercise hyperpnea in awake ponies. *J. Appl. Physiol.* 79:302–311, 1995.

118. Patterson, D. Potassium and ventilation during exercise. *J. Appl. Physiol.* 72:811–820, 1992.

119. Pickar J.G., J.M. Hill, and M.P. Kaufman. Stimulation of vagal afferents inhibits locomotion in mesencephalic cats. *J. Appl. Physiol.* 74:103–110, 1993.

120. Piiper J. and P. Scheid. Model for capillary-alveolar equilibration with special reference to O_2 uptake in hypoxia. *Respir. Physiol.* 46:193–208, 1981.

121. Powers S., J. Lawler, D. Criswell, K.L. Fu, and D. Martin. Aging and respiratory muscle metabolic plasticity: effects of endurance training. *J. Appl. Physiol.* 72:1068–1073, 1992.

122. Rahn H. A concept of mean alveolar air and the ventilation-blood flow relationships. *Am. J. Physiol.* 158:21–30, 1949.

123. Rahn H. and W.O. Fenn. *A Graphical Analysis of the Respiratory Gas Exchange.* Washington, DC: American Physiological Society, 1955.

124. Rausch S. M., B.J. Whipp, K. Wasserman, and A. Huszczuk. Role of the carotid bodies in the respiratory compensation for the metabolic acidosis of exercise in humans. *J. Physiol. (Lond.)* 444:567–578, 1991.

125. Reinhard U., P.H. Mueller, and R.M. Schmuelling. Determination of anaerobic threshold by the ventilation equivalent in normal individuals. *Respiration* 38:36–42, 1979.

126. Riley R.L. and A. Cournand. 'Ideal' alveolar air and the analysis of ventilation-perfusion relationships in the lungs. *J. Appl. Physiol.* 1:825–847, 1949.

127. Riley R. Pulmonary function in relation to exercise. In: *Science & Medicine of Exercise & Sports,* edited by W.R. Johnson. New York: Harper & Brothers, 1960, pp. 162–177.

128. Riley R.L. The hyperpnea of exercise. In: *The Regulation of Human Respiration,* edited by D.J.C. Cunningham and B.B. Lloyd. Oxford: Blackwell, 1963, pp. 525–534.

129. Roussos C.S. and P.T. Macklem. Diaphragmatic fatigue in man. *J. Appl. Physiol.* 43:189–197, 1977.

130. Rowell L.B., H.L. Taylor, Y. Wang, and W.S. Carlson. Saturation of arterial blood with oxygen during maximal exercise. *J. Appl. Physiol.* 19:284–286, 1964.

131. Ryan U.S., P.L. Whitney, and J.W. Ryan. Localization of carbonic anhydrase on pulmonary artery endothelial cells in culture. *J. Appl. Physiol.* 53:914–919, 1982.

132. Ryffel J.M. Experiments on lactic acid formation in man. *J. Physiol.* 39:29, 1909.

133. Shepherd R.H. Effect of pulmonary diffusing capacity on exercise tolerance. *J. Appl. Physiol.* 12:487–488, 1958.

134. Smith C.A., C.A. Harms, K.S. Henderson, and J.A. Dempsey. Ventilatory effects of specific carotid body hypoxia and hypocapnia in awake dogs. *J. Appl. Physiol.* 82:791–798, 1997.

135. Stewart P.A. Independent and dependent variables of acid-base control. *Respir. Physiol.* 33:9–26, 1978.

136. Sutton J.R., J.T. Reeves, P.D. Wagner, et al. Operation Everest II: oxygen transport during exercise at extreme simulated altitude. *J. Appl. Physiol.* 64:1309–1321, 1988.

137. Sylvester J.T., B.J. Whipp, K. Wasserman. Ventilatory control during brief infusions of CO_2-laden blood in the awake dog. *J. Appl. Physiol.* 35:178–186, 1973.

138. Szydlo Z. *Water Which Does Not Wet Hands.* Warsaw: Polish Academy of Sciences, 1994, pp. 82–85.

139. Taylor R. and N.L. Jones. The reduction by training of CO_2 output during exercise. *Eur. J. Cardiol.* 9:53–62, 1979.

140. Tenney S.M., and J.E. Remmers. Comparative quantitative morphology of the mammalian lung: diffusing area. *Nature* 197:54–66, 1963.

141. Thurlbeck W.M. Lung growth and development. In: *Pathology of the Lung* (2nd ed.), edited by W.M. Thurlbeck and A.M. Chung. New York: Thiene, 1995.

142. Victor R.G., L.A. Bertocci, S.L. Pryor, and R.L. Nunally. Sympathetic nerve discharge is coupled to muscle cell pH during exercise in humans. *J. Clin. Invest.* 82:1301–1305, 1988.

143. Wagner P.D., H.A. Saltzman, and J.B. West. Measurement of continuous distributions of ventilation-perfusion ratios: theory. *J. Appl. Physiol.* 36:588–599, 1974.

144. Waldrop T.G., D.C. Mullins, and D. Millhorn. Control of respiration by the hypothalamus and by feedback from contracting muscles in cats. *Respir. Physiol.* 64:317–328, 1986.

145. Waldrop T.G. and R.W. Stremel. Muscular contraction stimulates posterior hypothalamic neurons. *Am. J. Physiol.* 256:R348–R356, 1989.

146. Ward S.A. Assessment of peripheral chemoreflex contributions to ventilation during exercise. *Med. Sci. Sports Exerc.* 26:303–310, 1994.

147. Wasserman K. and M.B. McIlroy. Detecting the threshold of anaerobic metabolism in cardiac patients during exercise. *Am. J. Cardiol.* 14:844–852, 1964.

148. Wasserman K., B.J. Whipp, and J. Castagna. Cardiodynamic hyperpnea: hyperpnea secondary to cardiac output increase. *J. Appl. Physiol.* 36:457–464, 1974.

149. Wasserman K., B.J. Whipp, R. Casaburi, W.L. Beaver, and H.V. Brown. CO_2 flow to the lungs and ventilatory control. In: *Muscular Exercise and the Lung*, edited by J.A. Dempsey and C.E. Reed. Madison: University of Wisconsin Press, 1977, pp. 105–135.

150. Wasserman K., R.A. Mitchell, A.J. Burger, R. Casaburi, and J.A. Davis. Mechanism of the isoproterenol hyperpnea in the cat. *Respir. Physiol.* 38:359–376, 1979.

151. Weibel E.R. *Morphometry of the Human Lung.* Berlin and New York: Spring and Academic, 1963.

152. Weibel E.R., L.B. Marques, M. Copnstantinopol, F. Doffey, P. Gehr, and C.R. Taylor. Adaptive variation in the mammalian respiratory system in relation to energetic demand. VI: The pulmonary gas exchanger. *Respir. Physiol.* 69:81–100, 1987.

153. Welch H. Effects of hypoxia and hyperoxia on human performance. *Exerc. Sport Sci. Rev.* 15:191–221, 1987.

154. West J.B. and O. Mathieu-Costello. Stress failure of pulmonary capillaries: role in lung and heart disease. *Lancet* 340:762–767, 1992.

155. West J.B. (ed). *Respiratory Physiology: People and Ideas.* New York: Oxford University Press, 1996.

156. Whipp B.J., N. Lamarra, S.A. Ward, J.A. Davis, and K. Wasserman. Estimating arterial PCO_2 from flow-weighted and time-averaged alveolar PCO_2 during exercise. In: *Respiratory Control: Modelling Perspective*, edited by G.D. Swanson and F.S. Grodins. New York: Plenum, 1990, pp. 91–99.

157. Zu F. and D.T. Frazier. Respiratory-related neurons of the fastigial nucleus in response to chemical and mechanical challenges. *J. Appl. Physiol.* 82:1177–1184, 1997.

158. Yamamoto W.S. and M.W. Edwards, Jr. Homeostasis of carbon dioxide during intravenous infusion of carbon dioxide. *J. Appl. Physiol.* 15:807–818, 1960.

159. Yamamoto W.S. Looking at the regulation of ventilation as a signalling process. In: *Muscular Exercise and the Lung*, edited by J.A. Dempsey and C.E. Reed. Madison, WI: University of Wisconsin, 1977, pp. 137–148.

chapter 5

THE AUTONOMIC NERVOUS SYSTEM

Charles M. Tipton

I T is a daunting task to present in a cohesive chapter a historical perspective on the functions of the autonomic nervous system during exercise by normal and physically trained subjects. The multiplicity of systems regulated by the autonomic nervous system is summarized as follows by Hamill (76, p. 12):

> The autonomic nervous system (ANS) is structurally and functionally positioned to interface between the internal and external milieu, coordinating bodily functions to insure normal homeostasis (cardiovascular control, thermal regulation, gastrointestinal motility, urinary and bowel excretory functions, reproduction, and normal metabolic and endocrine physiology), and adaptive responses to stress (fight or flight response).

Coupled with the limitations on space, this restricts the scope of the chapter to the roles of the sympathetic and parasympathetic nervous systems in characterizing the cardiorespiratory responses to exercise. The approach here will be to focus on selected ideas and the people contributing to their development, with a primary emphasis on dynamic exercise results from healthy populations. In general, animal findings will be incorporated when the experimental approach would be inappropriate for human investigations. Like other authors, I regret not being able to include many of the studies conducted by my contemporaries.

HISTORICAL OVERVIEW OF THE AUTONOMIC NERVOUS SYSTEM

The historical record of the structure and function of the autonomic nervous system began with the description by Galen of Pergamon (ca. 129–200) of the vagosympathetic trunk, the appearance of the rami communicans, and the existence of three separate ganglionic "swellings." Galen believed that "nerves" were hollow tubes and conducted animal spirits that went from one organ to another in a coordinated manner which was called "sympathy" (112, 160). In 1545, Charles Estienne of Paris (1504–1569) described the separate existence of the vagus and sympathetic nerves (112), and 119 years later Thomas Willis in England (1621–1675) observed that the vagus nerve innervated the heart, speculated that it was responsible for a change in the pulse, and introduced the concepts of involuntary and voluntary functions of the nervous system (160).

The Danish anatomist Jacque Benigne Winslow (1669–1760) reported the existence of *"les grands nerfs sympathiques"* along the spinal cord (the paravertebral ganglia) which he felt functioned as independent nerve centers. He is credited with being the first to use the term "sympathetic nerves" to describe his findings (160, 232). At the beginning of the nineteenth century, the French physician Marie Francois Xavier Bichat (1771–1802) believed that sympathetic ganglia functioned independently of the central nervous system and she expounded the idea that life consists of an animal component (*la vie animale*) and an organic component (*la vie organique*)—a notion which continues today in the distinction between visceral and somatic functions of the autonomic nervous system (112, 160). In 1854, Ernst Heinrich Weber (1795–1878) of Leipzig, Germany, and his younger brother Eduard F. Weber (1806–1871) effectively demonstrated the inhibitory effects of vagal stimulation on heart rate (112). Interestingly, Willis's assistant and collaborator at Oxford University, Richard Lower (1631–1691), had postulated this vagal inhibition in 1669 (112). At the time of the Webers' experiments, Claude Bernard (1813–1878) had initiated lesion and stimulation studies involving sympathetic nerves and reported vasomotor functions that included vasodilation and vasoconstriction (160, 174). Bernard also advanced the idea that all sympathetic reflexes are mediated through the spinal cord and that centers exist within the medulla oblongata that when stimulated initiate impulses carried by the sympathetic nerves (160, 174). Between 1851 and 1857, Bernard formulated the concept of the constancy of the internal environment (*le mileu interier*), and years later it was incorporated by Cannon as the foundation for his teaching on homeostasis (21, 160).

In 1866 M. de Cyon and Elie de Cyon (1843–1912), working in Dubois-Reymond's laboratory in Berlin, reported that stimulation of the sympathetic nerves to the heart would increase its rate (160, 174). Walter Holbrook Gaskell (1847–1914) of the Cambridge Medical School in England is remembered for his extensive anatomical and physiological studies on the autonomic nervous system at the turn of the century. He thought that nearly every tissue in the body is innervated by nerve fibers (autonomic) which have opposite characteristics and that these differences could be explained by chemical changes (66, 160). Gaskell is also remembered for the characterization of the cranial (bulbar), thoracolumbar, and sacral myelinated efferent outflows from the central nervous system and for his unsuccessful efforts to have the term "involuntary nervous system" describe the functions of the sympathetic and parasympathetic nervous systems (112, 160). John Newport Langley

(1852–1925), a contemporary of Gaskell at Trinity College in Cambridge, studied the effects of nicotine on pre- and postganglionic neurons. As a result of his studies, in 1898 he stated (115, p. 270): "I propose the term 'autonomic nervous system' for the sympathetic nervous system and the allied nervous system of the cranial and sacral nerves, and for the local nervous system of the gut." Through the findings of Gaskell and his contemporaries, it was known by the end of the nineteenth century that the thoracolumbar sympathetic activity went to virtually all regions whereas the neural outflow from the cranial and sacral structures was distributed to certain regions of the body. Moreover, it was understood that they had different influences on the tissues they innervated (160). As a result of his pharmacological studies with cholinergic agents, which produced findings similar to those recorded after stimulating cranial and sacral regions, Langley used the term "para-sympathetic" to characterize the efferent responses from these regions. In 1905 he stated (116, p. 403): "I use the word para-sympathetic for the cranial and sacral autonomic system." Although extensive anatomical, physiological, and pharmacological information has been gathered since the era of Gaskell and Langley, the terminology and classification system remain in use. A notable refinement has been the classification of the autonomic nervous system into central (brain stem, diencephalon, telencephalon) and peripheral (sympathetic, parasympathetic, enteric) components because integrative and regulatory functions occur inside and outside the central nervous system (76). The enteric component will not be discussed in this chapter. Readers interested in more detailed information on the history of the autonomic nervous system are encouraged to consult the texts of Kuntz (112), Pick (160), and Rothschuh (174).

THE CANNON INFLUENCE

It is noteworthy that authors of exercise physiology texts up to 1948 (19) confined their discussions of the functions of the autonomic nervous system during exercise to the physiological effects of epinephrine (adrenaline, adrenin). This reflected the influence of the studies conducted by Walter Bradford Cannon at Harvard University (Fig. 5.1) that were responsible for the idea of fight-or-flight responses to danger.

Adrenal Contributions

In early studies involving the gastrointestinal system, Cannon observed that the emotions experienced by experimental animals had a profound effect on gastric motility (11). He focused on the adrenal gland because in 1892 Jacobj from Strassburg, Germany, had demonstrated that electrical stimulation of splanchnic nerves supplying the adrenal gland resulted in the release of a "substance" that reduced the amplitude of contractile tissue (94). He was also interested in the adrenal gland because in 1895 George Oliver and E. A. Schafer from University College, London, had discovered an adrenal extract that produced physiological effects similar to the ones found by stimulation of the splanchnic nerves (153). In 1911, Cannon and de la Paz

Fig. 5.1. Walter B. Cannon, M.D. (1871–1945) was a renowned investigator of the functions of the autonomic nervous system who was affiliated with Harvard University. His contributions to this chapter can be found in references 15 and 21–27. The photograph was taken in 1928, shown on the cover of the *Physiologist* [24:(5), 1981], includes experimental tracings on a kymograph, and is provided through the courtesy of the American Physiological Society.

noted that barking dogs excited cats by stimulating the sympathetic nervous system to release epinephrine (25). They also found that "excited blood" had elevations in glucose, which Cannon felt would be used by skeletal muscles to increase muscle power in periods of stress and danger. He also felt that epinephrine would be important for the redistribution of blood flow that would be necessary for running or fighting for one's life (11).

Simulated-Exercise Studies

Cannon, with the aid of Leonard Nice, an instructor in physiology at Harvard University, used a cat nerve-muscle (tibialis anterior) preparation with an intact blood supply to study muscle responses to splanchnic nerve stimulation (26). They found that fatigued tibialis anterior muscles would improve their force production and suggested that this change would enhance muscle activity during emergency conditions (26). Moreover, Cannon thought that the presence of glucose and epinephrine during periods of "major excitements" which caused fight-or-flight responses would help to delay the onset of fatigue and improve the ability to continue exhaustive efforts (11). Included within the Cannon and Nice study (26) was the idea that muscular fatigue would be delayed by sympathetic stimulation. This possibility was discussed well in advance of the Orbeli-Ginetzinski experiments with frogs, addressing the same problem that began to be studied in Russia in the 1920s and led to the Orbeli phenomenón or effect (154).

The Emergency Function of the Adrenal Gland

In 1914 Cannon discussed the "efficiency" of the adrenal gland and the importance of glucose and epinephrine in situations involving fear, asphyxia, rage, and pain. He felt fear would initiate a flight response whereas rage and anger would elicit fighting reactions. Cannon concluded by stating (23, p. 372):

> The organism which with the aid of increased adrenal secretion can best muster its energies, can best call forth sugar to supply the laboring muscles, can best lessen fatigue, and can best send blood to the parts essential in the run or fight for life, is most likely to survive. Such according to the view here propounded, is the function of the adrenal medulla at times of great emergency.

Although Campos, Cannon, and associates had noted in 1929 that the sympathetic–adrenal apparatus was not a necessity if emergencies did not exist in the "outer world" and that it could be removed without consequences (20), the physiological effects of epinephrine under fight-or-flight conditions was used to explain the role of the sympathetic nervous system during exercise for several more decades.

AFTER CANNON: THE CHARACTERIZATION OF AN EXERCISE RESPONSE

Epinephrine

Until 1922, it had been assumed that epinephrine is released during exercise. Although Frank Hartman, a former teaching fellow at Harvard, and his associates at the University of Buffalo have been credited as being the first to measure epinephrine levels in exercising animals (cats). However, their conclusion on the release of epinephrine into the blood evolved from comparisons of pupillary responses between exercised cats, cats with denervated pupils, and cats that had received injections of epinephrine (82). A decade later, Masao Wada and coworkers at Tohoku Imperial University in Japan secured blood from the exteriorized suprarenal veins of exercising dogs and found venous concentrations of "epinephrine like substances" were elevated only when exercise was "fatiguing" (223). Wilhelm Raab from the University of Vermont in 1943 reported exercise would increase an "E like" substance by 49%, although he failed to specify the intensity of exercise being performed (167).

In 1960, Aado Vendsalu from the University of Lund in Sweden progressively exercised subjects and reported epinephrine values were not elevated until the exercise was "heavy" (215). More than a decade later, Kotchen and Harvard colleagues noted that exercise intensities had to exceed 40% $\dot{V}O_2$max and approach maximal conditions before significant changes in epinephrine would occur (109). However, it was the Galbo, Host, and Christensen investigation in 1975 at the University of Copenhagen that effectively demonstrated that an intensity level approaching 75% $\dot{V}O_2$max was the threshold for eliciting significant increase in the release of epineprhine from the adrenal medulla (64; Figs. 5.2 and 5.3).

Fig. 5.2. Heinrich Galbo, M.D., Dr. Med. Sc. (1946–), is a distinguished investigator and author currently located at the PANUM Institute of the University of Copenhagen in Copenhagen, Denmark. His contributions to this chapter can be found in references 58 and 63–65. The photograph was taken in 1978 and provided to the author upon request.

Fig. 5.3. Niels J. Christensen, M.D., Dr. Med. Sc. (1939–), is a distinguished investigator currently located at the Herlev Hospital of the University of Copenhagen in Copenhagen, Denmark. His contributions to this chapter can be found in references 64, 65, 188, and 169. The photograph was taken in 1993 and provided to the author upon request.

Norepinephrine

The primary reason why authors of exercise physiology texts before 1950 focused on the adrenal gland was the difficulty in differentiating the responses of the sympathetic nervous system from those associated with the release of epinephrine from the adrenal medulla. This was most apparent when attempting to explain the excitatory and inhibitory responses, and in 1930, Cannon and Zenon Bacq suggested the name "sympathin" be given to a substance present in tissues that responded to sympathetic nerve stimulation (24). Later, with Arturo Rosenblueth, they proposed that stimulation of the sympathetic nervous system releases epinephrine, which interacts with two reactive substances within the cell to form sympathin E (excitatory) or sympathin I (inhibitory [27]). No reactive substances were found and no exercise studies concerned with sympathin E and I were undertaken.

It was not until the 1940s that research conducted by the 1970 Nobel laureate Ulf S. von Euler (1905–1983) of Stockholm found a solution. He conclusively demonstrated that norepinephrine is the primary transmitter released from sympathetic nerve endings (52). It is of interest that in the 1930s, Bacq had suggested that sympathin I was epinephrine and that sympathin E could be norepinephrine (9, 233).

Von Euler in 1952, with the assistance of Hellner, demonstrated that urinary concentration of catecholamines were elevated in humans after exercise (53). This study confirmed the 1947 trend reported by Peter Holtz and colleagues from the University of Rostock in Germany, who had shown "urinary urosympathin" was elevated with exercise and suggested the changes noted would be related to exercise intensity (92). Ten years later, Irving Gray and William Beetham at Natick, Massachusetts, were the first to record plasma norepinephrine concentrations after prolonged running (29 km) and reported values 133% higher than pre-run conditions (69). In the 1960s, Vendsalu observed that venous norepinephrine concentrations rose in proportion to exercise intensity (215). A decade later, Haggendahl and colleagues at Gotenborg, Sweden, measured arterial norepinephrine concentrations of subjects and noted a linear increase up to 75% $\dot{V}O_2$max after which a curvilinear profile was observed (74). Similar to their findings with plasma epinephrine concentration, Galbo, Holst, and Christensen demonstrated in 1975 that only minimal changes in plasma norepinephrine concentrations would occur at exercise intensities between 40% and 50% $\dot{V}O_2$max whereas the threshold for the occurrence of significant elevations was between 60% and 70% of $\dot{V}O_2$max (63, 64).

Norepinephrine Spillover and Exercise

During late 1970s and early 1980s, Murray Esler (Fig. 5.4) and his colleagues, which included Garry Jennings at the Baker Medical Research Institute in Melbourne, Australia, developed the idea and perfected the techniques necessary to measure norepinephrine spillover (neuronal leakage into the plasma) from discrete regions and organs of the body (heart, kidneys, arms, legs, lungs, heptomesenteric [50]). The

Fig. 5.4. Murray Esler, M.B.B.S., Ph.D. (1943–), is a distinguished investigator currently located at the Baker Medical Research Institute in Melbourne, Australia. His contributions to this chapter can be found in references 50, 83, 95, 134, and 142. The photograph was taken in 1999 and provided to the author upon request.

necessary techniques enabled investigators to quantify sympathetic responses at rest and during exercise and to better differentiate the sympathetic response with normal and diseased populations. (50, 180). At rest, it is estimated that muscles account for 25%, kidneys 25%, lungs 33%, splanchnic region 10%, skin 5%, and the heart 3% of the total spillover value (50). In 1988 Hasking, including Esler and Jennings, had subjects exercise in the supine position at ~50% maximum voluntary working capacity and noted the total norepinephrine spillover value to plasma was elevated by 324% with the heart and kidneys representing 10- and fourfold increases, respectively (83). Although not recorded with this experiment, Esler and colleagues estimated that approximately 60% of the norepinephrine entering the plasma is derived from skeletal muscle sympathetic nerves (50). Between 1987 and 1989, Gabrielle Savard and associates from the University of Copenhagen measured norepinephrine spillover values from subjects performing dynamic exercise (188, 189). They demonstrated that the total norepinephrine spillover value was markedly elevated with changes in exercise intensity and that increasing the amount of muscle mass performing dynamic exercise, from one leg to two arms and two legs, caused a linear increase in norepinephrine (NE) total spillover value that was approximately sixfold-higher than the initial resting mean (188). An interesting finding was that inactive muscles released norepinephrine as the exercise intensity increased. They suggested that the "additional" NE's purpose was not to maintain vasoconstriction or to improve vascular conductance; rather it was for purposes of maintaining vascular smooth muscle tone or enhancing cellular metabolism in muscles (188). Leuenberger and colleagues from Pennsylvania State University confirmed that exercise intensity increased the norepinephrine spillover value in their exercising subjects and showed that the norepinephrine clearance value was unchanged until the intensity level became "heavy," which caused a decrease in clearance and a rise in plasma concentrations (120). Also associated with this finding was an interaction between exercise intensity and its duration.

Mazzeo from the University of Colorado with collaborators from the Baker Institute in Australia measured norepinephrine spillover values from the hepato-mesenteric region of subjects who exercised at 50% $\dot{V}O_2$max (134). They found the norepinephrine spillover percentage remained near the pre-exercise value of 5.5% for the younger subjects but was variable with the older individuals, presumably because of a reduction in clearance capacity.

One idea emerging from these innovative studies was that cardiac sympathetic outflow is selectively activated during progressive exercise. While skeletal muscle is responsible for the majority of the total norepinephrine spillover value, its percentage increase is similar to that reported for the renal bed (50).

Training and Catecholamine Changes with Acute Exercise

Several decades elapsed from the time when catecholamines were characterized to the time when effects of chronic exercise (training) on human responses were assessed. One of the first studies was conducted in 1972 by Hartley and associates at the U.S. Army Research Institute of Environmental Medicine in Massachusetts (80). Their training program produced a significant increase in $\dot{V}O_2$max (14%) without meaningful changes in norepinephrine and epinephrine concentrations. However, they did find significant reductions in norepinephrine concentrations in one submaximal exercise test (73% $\dot{V}O_2$max).

In an extensively cited 1978 study (228), Winder and coworkers at Washington University in St. Louis had subjects conduct a 7 week training program that significantly increased $\dot{V}O_2$max (22%) and power output (30%). After training, testing at the same exercise intensities revealed significant reductions in both norepinephrine (46%) and epinephrine concentrations (73%). Moreover, when the subjects were tested at the higher post-training $\dot{V}O_2$max value, norepinephrine and epinephrine values were notably elevated, suggesting the responsiveness of the sympathetic nervous system had increased with the higher intensities (228). In 1985, Peronnet and his Canadian colleagues extended the investigations of Winder and associates by having subjects perform a 20 week training program established at 80% of their maximum heart rate (158). They found significant increases in $\dot{V}O_2max$ (27%) and marked reductions in norepinephrine concentrations (51%) at the same pretraining power outputs. However, these differences did not prevail when comparisons were made at the same percentage of their maximum oxygen uptake (158).

Gary Jennings and his Australian colleagues obtained norepinephrine and epinephrine concentrations plus norepinephrine spillover results in a training study designed to examine risk factors (95). The training program lasted 4 weeks and consisted of subjects exercising at 60%–70% of their maximum work capacity. The authors chose to evaluate their results by responders (N = 8) or nonresponders (N = 2) and reported that the responders exhibited significant reductions in plasma norepinephrine concentrations (52%) and in norepinephrine spillover rate (65%) with epinephrine concentrations exhibiting no meaningful trends (95). Reasons for the existence of nonresponders were not pursued by the authors. Although trained populations have a different renal norepinephrine spillover rate at rest than nontrained subjects (142), the effects of training for this and related organ systems during exercise remain to be investigated. Hence, there is scientific support for the idea that endurance training will reduce the activity of the sympathetic nervous system during exercise as demonstrated by decreases in catecholamine concentrations and in the norepinephrine spillover rate. However, it is uncertain which organ system(s) (muscle, heart, kidney, liver, etc.) or cellular mechanisms are responsible for this effect.

ACTIVATION OF THE AUTONOMIC NERVOUS SYSTEM WITH EXERCISE

Early Studies

While authors of exercise physiology textbooks did not address this subject until after the middle of the twentieth century, investigators had ideas well in advance of this time. The possibility that the autonomic nervous system was "involved" with exercise is credited to the 1886 investigations of Nathan Zuntz (1847–1920) and J. Geppert in Berlin, Germany, which pertained to the hyperpnea of exercise (235). In 1893, Jons Erik Johansson in Stockholm, Sweden, simulated exercise in rabbits with passive leg movements and concluded that activation of the skeletal muscles was associated with the excitation of neural centers in the brain that would also increase heart rate (96). After the turn of the century, August Krogh (1862–1949, and Nobel laureate in 1920) and Johannes Lindhard (1870–1947) in Copenhagen, Denmark, were investigating the circulatory and respiratory changes that occurred with light or heavy exercise (111). They found some responses were present in "less than a second" that likely occurred because of a "neural mechanism" and supported Johansson's idea that the change in heart rate was due to the activity of neural centers in the brain. They also felt the respiratory changes favored an "irradiation of impulses from the motor cortex rather than a reflex from the muscles" which activated the respiratory centers in the brain (111, pp. 122–123).

Central Command and Activation of the Autonomic Nervous System

Krogh and Lindhard's proposal of cortical irradiation eliciting cardiorespiratory responses led to the 1972 idea that a central command existed during exercise (68). As defined by Waldorf and colleagues, central command refers to a "parallel, simultaneous excitation of neural circuits controlling the locomotor and cardiovascular systems thus serving a feed forward control mechanism" (224, p. 339). Central command includes neural output to the respiratory muscles and neural signals decending from higher (hypothalamus, mesencephalon, amyglada) and spinal centers which influence neurons in the medulla oblongata to influence sympathetic and parasympathetic efferent nerve activity to the heart and blood vessels (224). As discussed in a subsequent section, it was Rowell's idea in 1980 that central command would "set the basic patterns of effector activity, which are in turn modulated by baroreceptors, muscle mechanorecptors, and muscle chemoreceptors [actually chemosensitive afferent fibers] as error signals may develop" (176, p. 314).

Vagal Withdrawal with the Onset of Exercise

Heart-rate changes with exercise have intrigued physiologists for more than a century and in 1895, Henrich Ewald Herring (1886–1948) of Prague, Czechoslovakia, exercised rabbits and explained the increase in heart rate, in part, by the increased activity of the accelerator nerve and by the decreased activity of the vagus nerve, which had been influenced by the elevations in vagus activity (85). In 1904, Wilbur Bowen from the University of Michigan examined the latency periods after the first cycle of exercising subjects and concluded decreased vagal activity was responsible (14). Gasser and Meek during 1914 used atropine (muscarinic blocker), vagal sectioning, and adrenalectomy in exercising dogs and concluded that inhibition of vagal activity was the most "economical means" by which acceleration of heart rate could occur (67). Fifty-two years later (1966), Brian Robinson and associates at NIH in Maryland reexamined this issue by administering atropine and propranolol (β-adrenoceptor blocker) to subjects exercising on a cycle ergometer. They showed that "withdrawal of parasympathetic restraint on the sinoatrial node" occurred early in exercise (170, pp. 408–409) whereas sympathetic activity increased later and became more pronounced as the exercise intensity became greater (170), a point inferred by Cannon decades earlier.

In 1970 Freyschuss from the Karolinska Institute evaluated the single and combined effects of atropine and phentolamine (α-adrenoceptor blocker) in subjects performing isometric exercises. She found vagal withdrawal was "instantaneous," occurring at the same time as muscular contractions and responsible for the elevation in heart rate, whereas the increase in blood pressure was the result of sympathetic influences on vascular resistance (61). Later, Maciel and colleagues used an experimental approach similar to one followed by Freyschus and confirmed that the initial elevations in heart rate were of a "parasympathetic nature" (125). Victor and others in 1989 used the neuromusclular blocker tubocurarine to prevent force development when an attempted maximum handgrip was performed (central command [217]). Under these conditions, muscle sympathetic activity increased slightly, as did mean blood pressure, whereas the heart rate increase was similar to the value recorded when a 30% voluntary maximum contraction was performed. Collectively, these findings indicate central command will inhibit cardiac parasympathetic activity with the onset and continuation of exercise while having a minimal influence on increasing sympathetic nerve activity.

Sympathetic Nervous System Activation and Nerve Recordings

As elaborated earlier, characterization of sympathetic nervous system activation and responses before the 1980s was mostly done by analysis of catecholamine concentrations in blood and urine. However, this situation began to change after 1967 when Vallbo and Habarth from Uppsala, Sweden, percutanesously inserted tungsten microelectrodes in nerve fascicles of humans and demonstrated the ability to record nerve impulses from peripheral nerves (212). Subsequently, this innovative technique was used to make the first recordings of sympathetic nerve activity from postganglionic fibers (212). Investigations led by Wallin soon demonstrated that it was possible to identify and to record sympathetic activity in the trigeminal nerve to the face and in the peripheral nerves to limb muscles (median, tibial, peroneal) or muscle sympathetic nerve activity (227). Although the procedure provides direct evidence for sympathetic efferent activity, it can not distinguish whether the fibers are sudomotor, vasodilator, or constrictor in nature (177).

   In view of the previous discussion on the use of plasma catecholamines values to depict sympathetic activation during exercise, Seals and coworkers confirmed that increases in plasma norepinephrine concentrations can accurately depict increases in measured sympathetic nerve activity to muscle (193).

Sympathetic Nerve Activity and Dynamic Exercise

In 1987 when Ronald Victor, Douglas Seals, and Allyn Mark were at the University of Iowa (Figs. 5.5, 5.6, and 5.7), they reported that muscle sympathetic nerve activity (MSNA) with arm and leg exercise did not significantly increase with mild or moderate exercise even though heart rates were elevated (219). These findings were regarded as controversial when Saito and associates in Japan reported decreased MSNA activity at 20% $\dot{V}O_2$max that returned to resting baseline values at 40% $\dot{V}O_2$max before increasing at 60% $\dot{V}O_2$max (186). However, the uncertainty was removed by Rowell, who reminded the scientific community that restoration of central venous pressure and the cardiac output by the muscle pump would lower MSNA. Seals, Victor, and Mark did report the threshold necessary to significantly increase MSNA to be 40 W or 33% of the maximum power output of their subjects (193). It is noteworthy that Victor and Seals had subjects performing arm cycling at 30% of their maximum loads for a 6 minute period and MSNA increased signifi-

Fig. 5.5. Ronald G. Victor, M.D. (1952–) is a distinguished investigator currently located at the University of Texas Southwestern Medical Center in Dallas, Texas. His contributions to this chapter can be found in references 78, 79, 129, 166, 193, 216–219, and 221. The photograph was taken in the early 1990s and provided to the author upon request.

Fig. 5.6. Douglas R. Seals, Ph.D. (1954–) is a distinguished investigator currently located at the University of Colorado, Boulder, Colorado. His contribution to this chapter can be found in references 191–193, 218, and 219. The photograph was taken in the mid 1990s and provided to the author upon request.

Fig. 5.7. Allyn L. Mark, M.D. (1936–) is a distinguished investigator currently located at the University of Iowa at Iowa City, Iowa. His contributions to this chapter can be found in references 129, 193, and 219. The photograph was taken in 1995 and provided to the author upon request.

cantly within 2 minutes but remained at that level until exercise was terminated (218). Although not feasible for humans, direct measurements of sympathetic nerve activity from the renal nerves of rabbits performing light-to-moderate exercise have shown significant increases (150). In contrast to the human muscle sympathetic nerve recordings, renal sympathetic nerve activity in rabbits was markedly elevated within 10 seconds and exhibited a profile that was responsive to exercise intensity. After reviewing data from a variety of exercise related experiments including their own, Rowell and O'Leary introduced the idea that in humans vagal withdrawal is near completion when heart rates are approximately 100 beats/min while sympathetic activation (characterized by norepinephrine concentrations, plasma renin activity, and muscle sympathetic nerve activity) begins at approximately 90 beats/min and increases in a linear manner as the exercise intensity increases (179). Intrinsic to this idea is that a mismatch occurs between cardiac output and vascular conductance and the pressure "error" initiates the rise in sympathetic nerve activity.

Sympathetic System Nerve Activity and Isometric Muscle Contractions (Static Exercise)

Static exercise has been associated with changes in muscular strength and endurance since the time of Milo of Croton (~600 BC [104, p. 35]) although the scientific interest in its relationship to responses of the autonomic nervous system is a recent development. In 1920, Johannes Linhard (1870–1947) of Copenhagen University and a colleague of August Krogh recorded select circulatory and respiratory effects from subjects hanging with bent elbows and attributed the changes to the occlusion of muscle blood flow (122). Although the physiological effects of static exercise were extensively investigated for the next five decades by renowned investigators such as Asmussen and Hansen, Horvath and Tuttle, and Lind and associates, none had direct evidence for autonomic nervous system involvement or for sympathetic activation. As noted earlier when discussing vagal withdrawal, this situation changed in 1970 with the study conducted by Freychuss in Sweden, who attributed the increases in blood pressure to sympathetic activity on vascular resistance and the elevated heart rate to vagal withdrawal (61).

Using microneurographic techniques to monitor sympathetic activity to limb muscles, Allyn Mark and his colleagues in 1985 were the first to demonstrate that MSNA did not increase with the onset of isometric exercise that was established at 30% of a maximum voluntary contraction (129). Although heart rate and blood pressure were increasing during this time, MSNA was not elevated until ~120 seconds later. Subsequently, Saito and associates from Nagoya University showed that significant increases in MSNA could occur at a lower threshold (between 10% and 15% of a maximum voluntary contraction), that the increase in MSNA was higher as the percentage of a maximum voluntary contraction increased, proportional to the tension developed, and that the response time for MSNA was between 30 and 60 seconds (184). Microneurographic studies conducted with static exercise in the 1990s by Saito and colleagues (185) and by Vissing and associates from Dallas (221) demonstrated that sympathetic nerve traffic to the skin occurs abruptly at 10% of a maximum voluntary contraction. Although the response did not reveal the delay characteristic noted for MSNA recordings, they exhibited elevations in skin sympathetic nerve activity with increased percentages of a maximum contraction. Renal nerve sympathetic activity was measured by Matsukawa and others from Dallas in cats trained to perform static exercise to receive a food reward (132). They observed that renal nerve sympathetic activity increased initially at, or immediately before, force development and persisted for 10 seconds before a second activity level occurred which gradually increased during the experiment. Heart rates increased when renal nerve sympathetic activity appeared while the elevation in arterial blood pressure was delayed by ~10 seconds before exhibiting a gradual increase while force was maintained (132). Several ideas emerged from these experiments: the selective activation of the sympathetic nervous system in various structures of the body as a result of isometric exercise; the distinctive nature of isometric exercise and sympathetic activation; and the possibility that physical occlusion of blood vessels alters sympathetic control of blood flow.

Sympathetic Activity and Muscle Mass

While Lind and McClosky had examined the relationship between muscle mass and hemodynamic responses with exercise, especially isometric exercise, the role of the sympathetic nervous system in the process received little attention. In 1974, Davies and coworkers at the University of London measured catecholamine concentrations in subjects using varying amounts of muscle mass performing dynamic exercise and noted a differential response associated with muscle mass at a relative $\dot{V}O_2$ value but failed to specify the norepinephrine and epinephrine changes (35). When submaximal and maximal exercises were performed using one-arm, one-leg, and two-leg cycling exercises, Blomqvist and associates from Southwestern Medical Center in Dallas found plasma norepineprhine increased in a curvilinear fashion as muscle mass and exercise intensity were increased (13). Seals measured MSNA activity in subjects using one and then two forearms to perform isometric contractions (30% maximum voluntary concentration) and found a 70% increase when both arms became involved (191). Seals also measured MSNA activity from the first interosseous hand muscle and from the muscles of the forearm during isometric contraction and reported the MSNA activity was 450% higher in the larger muscles (192). While these studies were unable to differentiate hemodynamic responses between dynamic and static exercises, they did demonstrate that the amount of contracting muscle mass was an important factor in the activation of the sympathetic nervous system in both dynamic and static exercise.

Revisit: The Cannon Idea for Activation of the Sympathetic Nervous System with Exercise

To Cannon, the sympathetic nervous system was essential for the maintenance of homeostasis, and he proposed that the sympathetic nervous system and the adrenal gland respond as a single unit and uniformly impact all tissues of the body at virtually the same time (21). In 1935 he stated (22, p. 2): "It is a point of considerable importance the sympatho-adrenal system when strongly excited, appears to act as a unit. No matter what the stimulus, the effect is diffuse and extends throughout the body." To many teachers of exercise physiology after 1935, this doctrine was interpreted to mean that the sympathetic nervous system and adrenal responses with exercise were instantaneous, simultaneous, and uniformly distributed to all tissues of the body. However, because of the findings from the norepinephrine spillover studies and the microneurographic investigations and because sympathetic traffic to the brain and spinal cord does not occur, the Cannon doctrine must be modified to indicate that the response can be discrete and not always global, uniform, or simultaneous in nature.

THE SYMPATHETIC NERVOUS SYSTEM AS THE EFFERENT ARM OF REFLEXES THAT MODULATE PHYSIOLOGICAL RESPONSES

Overview and Reflexes from Active Muscles

It is apparent from the previous sections that the integrative functions of the autonomic nervous system when exercise begins and continues involves more than a sudden withdrawal of parasympathetic activity and delayed responses of the sympathetic nervous system. In fact, the possibility that a reflex initiated during muscular activity would elicit and modulate physiological responses was proposed by Nathan Zuntz and J. Geppert in Germany almost a century ago (235).

Chemoreflexes

In 1937, Alam and Smirk from the faculty of medicine of the Egyptian University in Cairo, Egypt, provided the first experimental evidence that muscle activity could reflexly initiate cardiovascular responses (3). They demonstrated that occlusion of the blood flow to the thigh or to the forearm during exercise would subsequently increase blood pressure to high values that remained elevated and even rose further after the cessation of exercise with continued occlusion. A year later, they showed that heart rate was elevated as well during occlusion; but, unlike blood pressure, it fell during recovery even though occlusion was continued. The pressor changes were attributed to an accumulation of metabolites and to an insufficient blood flow being distributed to the limbs (105). Approximately four decades later, Rowell and colleagues began a series of innovative experiments that confirmed the findings of Alam and Smirk and gave credence to the idea of Zuntz and Geppert that a reflex occurred when a "mismatch" existed between blood flow and oxygen demand (177, 178). They indicated that the reflex originates in muscles, involves chemosensitive muscle afferent fibers, and has little or no involvement with respiratory control. In addition, the reflex was dependent upon the muscle mass being activated (59) and related to the intensity of exercise being performed. More importantly, they felt it should be considered to be a chemoreflex (177, 178, 180). As these results were unfolding, creative experiments that have been summarized in the American Physiological Society *Handbook* on exercise by Kaufman and Forester effectively demonstrated that group III and IV muscle afferents were responding to chemical, metabolic, or mechanical stimuli created by static contractions that restricted blood flow or by dynamic exercise conditions that restricted blood flow below normal (105). Not only was there a rise in blood pressure and heart rate by activation of this reflex; there was also increased myocardial contractility, elevated cardiac output, and a redistribution of blood flow. Thus the finding that limb occlusion during exercise would reflexly elevate blood pressure (the "blood pressure raising reflex" of Alam and Smirk [3]) has been associated with the actions of receptors affiliated with groups III and IV afferents and re-named as the chemoreflex, pressor reflex, ischemic muscle pressor reflex, muscle metaboreflex, and combinations thereof.

Sympathetic Activation and the Exercise Threshold

In 1989, Victor and Seals had subjects perform rhythmic hand-gripping exercise at various percentages of their maximum load with and without venous occlusion (218). No-load arm cycling without occlusion caused significant increases in heart rate and mean arterial pressure but no significant changes in MSNA. In fact, when they repeated the experiment, only total occlusion (250 mmHg) yielded significant values for MSNA. With light exercise (20% maximal voluntary contraction) followed by postexercise occlusion, heart rate and mean blood pressure were significantly changed whereas MSNA was not. However, at moderate exercise (60% maximal voluntary contraction), heart rate and mean blood pressure were significantly increased and augmented while MSNA was markedly elevated by 160%. Although they did not couple their blood-flow results with MSNA, they felt moderate exercise was the threshold and that an "interplay" existed between muscle perfusion and exercise intensity to elicit the reflex (218). Rowell and associates measured catecholamine concentrations when they had subjects perform light-to-moderately-supine cycling exercise while experiencing graded levels of increased lower-body positive pressure to reduce muscle blood flow (181). They found norepinephrine concentrations followed the pattern exhibited by mean arterial blood pressure, which meant norepinephrine had its highest concentration with heavy exercise and when muscle blood flow was the lowest. Although they were unable to characterize a threshold for the chemoreflex, they did demonstrate that the reflex was operative with metabolically active muscles during mild dynamic exercise as vasoconstriction occurred when muscle perfusion was reduced (181).

Mark and associates in 1985 showed that hand-grip isometric exercise at 30% maximal voluntary contraction with occlusion would significantly increase leg MSNA by 25% when compared to static exercise without occlusion (129). Hanson from Victor's laboratory in Dallas had subjects perform isometric contractions at 20% maximal voluntary contraction of the large toe with and without occlusion while recording MSNA from active and inactive muscles. No increase in MSNA occurred without ischemia. However, with occlusion, MSNA was significantly increased in active and inactive legs. Hanson and colleagues thought the increase in MSNA would help to offset the metabolically induced vasodilation, maintain blood pressure, and limit the reflex in augmenting blood flow to active muscle (79). After analyzing results from a myriad of occlusion experiments including their own, Loring Rowell (Fig. 5.8) and Donal O'Leary proposed in 1990 that the delayed rise in MSNA activity during isometric exercise was due to the presence of the chemoreflex and partially responsible for the increase in blood pressure (179). However, the delayed rise was not regarded to be an important contributor to the subsequent elevation in heart rate.

Peridural Blocking, Exercise, and the Chemoreflex

In 1979, Freund and others from Rowell's laboratory were the first to investigate in humans the influence of peridural or epidural anesthetic blockage that was sufficient

Fig. 5.8. Loring (Larry) B. Rowell, Ph.D. (1930–), is a distinguished investigator and author currently located at the University of Washington at Seattle, Washington. His contributions to this chapter can be found in references 59, 60, 156, 177–181, and 195. The photograph was taken in the mid 1990s and provided to the author upon request.

to inhibit chemoreflex effects on cardiovascular responses without restricting muscle blood flow (60). Although they had no measures of sympathetic activity, they did find two subjects capable of performing mild exercise. Afferent nerve blockage did prevent the rise in mean blood pressure that occurs with occlusion but it had a minimal influence on heart rate and pulmonary ventilation. Since the responses to exercise with unrestricted blood flow were normal, despite the blockade, they felt the reflex was not normally tonically active (60, 177). More than a decade later, Fernandes and colleagues at Copenhagen, which included Heinrich Galbo and Jere Mitchell (Fig. 5.9), used epidural anesthesia in a design that included submaximal and maximal dynamic exercise and occlusion (58). At rest, afferent nerve blockage was associated with a reduction of approximately 30% in resting leg strength. During submaximal exercise and maximal exercise, sensory blockage was associated with a 17% and 47% reduction in mean blood pressure, respectively. Heart rate and pulmonary ventilation exhibited minimal changes with these conditions. Sympathetic efferent activity

Fig. 5.9. Jere H. Mitchell, M.D. (1928–), is a distinguished investigator and author currently located at the University of Texas Southwestern Medical Center in Dallas, Texas. His contributions to this chapter can be found in references 58, 59, 132, 169, 217, and 224. The photograph was taken in the mid 1990s and provided to the author upon request.

was considered to be unchanged as assessed by a cold pressor test and a Valsalva maneuver. Although afferent blockage was partial in completeness when the different levels of exercise were performed, and considerd by Rowell to be inadequate for the purposes of the experiment (177), the results did indicate that attenuation of afferent traffic within exercising muscle would modify the blood pressure response of the chemoreflex with little influence on heart rate or ventilation (58).

The Nature of the Stimulus for the Chemoreflex

The possibility that a "mismatch" between blood flow and oxygen demands would create a signal to elicit a muscle reflex has intrigued physiologists for more than a century (235). In 1971 Coote and coworkers at the University of Birmingham in England conducted muscle stimulation studies with cats and demonstrated that the stimulus for the pressor response was chemical rather than mechanical and that metabolic receptors for "this exercise reflex are the free endings of group III and IV sensory nerve fibres located around the blood vessels" (30). Hypoxia was an attractive candidate as were chemicals and various metabolites. Potassium was the choice of many until it was demonstrated that muscle afferents respond transiently to its presence even though the concentration remains elevated for several minutes (105).

A better understanding of the stimulus emerged from the 1987 chemoreflex investigation of Sheriff and coworkers at the University of Washington. They used dogs in an experiment that combined exercise, an occlusion cuff to gradually reduce the dimension of the terminal aorta, and inhalation of carbon monoxide to reduce arterial oxygen content (195). They found when oxygen delivery reached a critically low level, a marked rise in H^+ and lactate occurred. However, the study included no measurements of sympathetic activity (195). A year later, Rotto and Kaufman from the University of Texas Health Science Center measured the impulse activity of group III and IV afferents after arterial injections of various metabolic products of muscle contraction and found that lactic acid and some cyclooxygenase products such as prostaglandins and thromboxanes "are the most likely to be responsible for any metabolic stimulation of group III and IV afferents during muscular contraction" (175). Also in 1988, Victor and associates used phosphorous 31 nuclear magnetic resonance spectroscopy to estimate intramuscular pH, inorganic phosphate, adensosine diphosphate, and phosphocreatine while subjects performed rhymthmic handgripping exercises (50% maximal voluntary contraction) while recording MSNA (216). Muscle-nerve sympathetic nerve activity was closely associated with the decrease in pH and not with alterations concerning inorganic phosphate, adenosine diphosphate, or phosphocreatine concentrations.

Because the activation of the chemoreflex was associated with the acidosis produced by occlusion and static exercise (197), Sinoway and associates at Pennsylvania State University postulated that the release of lactic acid during exercise could be acting indirectly to acidify a metabolic product which in turn initiated the chemoreflex (198). Therefore, they combined static exercise, occlusion, MSNA, and nuclear magnetic resonance measurements of forearm H^+ and diprotonated phosphate ($H_2PO_4^-$). MSNA was markedly increased with ischemic exercise and was significantly correlated to $H_2PO_4^-$ levels in seven of the eight subjects while it had little relationship with pH changes (198). They concluded that dipronated phosphate "is a major stimulant of muscle afferents and in large part determines the degree of sympathoexcitation during exercise" (198, p. H770).

Using subjects with an inborn metabolic defect that impairs muscle glycogen degradation to lactic acid (McArdle's disease), Pyror and associates from the University of Texas Southwestern Medical Center in Dallas found that 2 minutes of static exercise (30% maximal voluntary contraction) was unable to elicit significant increases in MSNA (166). Since normal subjects exhibited a significant elevation of 130% with the same exercise, the data suggested that the hydrogen ions released with the formation of lactic acid were contributing to the sympathetic response. Vissing and colleagues from Copenhagen, however, recently conducted an experiment that included subjects with McArdle's disease and concluded that muscle acidification, as well as changes in interstitial ammonia concentrations, were not mediators of sympathetic activation during exercise (220). Thus as the twenty-first century begins, it brings uncertainty as to whether H^+ and $H_2PO_4^-$ should be designated as the primary initiators of the chemoreflex during exercise.

Activation of Muscle Reflexes and Their Opposition by Chemical–Metabolic Influences

In 1841, the possibility was raised by Alfred Wilhelm Volkmann (1800–1877) of Halle, Germany, and a student of Ernst H. Weber, that the products of muscle contraction could alter vascular responses because of a central effect (222). Almost a century later (1930), Rein and associates at the University of Freiburg in Germany evaluated the vasoconstricting effects of femoral arteries as a result of the electrical stimulation of gastrocnemius muscles of anesthetized dogs and noted an attenuation of the vasoconstricting response when comparisons were made to resting conditions (168). In 1962, Remensnyder, Jere Mitchell, and Sarnoff at the National Heart Institute conducted a simulated exercise study with anesthetized dogs and introduced the idea of a "functional sympatholysis" to the scientific community (169). In their study, they demonstrated that direct (lumbar chain) or reflex stimulation of the sympathetic nervous system (stimulation of the vagus nerve and of the carotid sinus) during electrical stimulation of the hindlimbs was associated with markedly attenuated vasoconstriction. The same effect was present when norepinephrine was infused during simulated exercise (169). One interpretation of these results was that the metabolic vasodilation of exercise had altered the capacity for vasoconstriction.

Proponents cite the 1965 investigation of Kjellmer from the University of Goteborg in Sweden who conducted simulated exercise experiments with cats (107). They electrically stimulated the lumbar sympathetic chain when they activated the muscles of the calf by stimulation of the sciatic nerve. Their blood-flow measurements and vascular resistance calculations supported the idea of sympatholysis and indicated that resistance vessels were sensitive to vasodilator metabolites whereas the reverse was true for the capacitance vessels (107). Savard and colleagues had subjects perform arm and leg exercises that elicited $\dot{V}O_2$ responses that ranged from 24% to 71% of their whole-body capacities (188). They reported an increase in norepinephrine spillover over value and in plasma catecholamine concentrations as the muscle mass was increased, but noted that they were not associated with "any significant neurogenic vasoconstriction and reduction in flow through the muscle vascular bed" (188). Motivated by their simulated exercise results with rats pertaining to sympatholysis, Hansen and coworkers from Dallas conducted muscle oxygenation studies and measured mitochondrial cytochrome a and a3 states in human forearm muscles that were contracting rhythmically at either 10% or 45% maximal voluntary contraction level (78). MSNA was recorded from the tibial nerve and the existence of sympatholysis was assessed by a decrease in "sympathetically mediated muscle oxygenation." (Unfortunately, flow and peripheral resistance were not measured.) They did find that sympathetic vasoconstrictor drive to skeletal muscle did abolish muscle oxygenation when the exercise intensity was higher than 10% of the voluntary maximal contraction value. Moreover, they suggested that muscle hypoxia per se could be the primary determinant of functional sympatholysis (78).

Opposition to the idea that metabolic vasodilator responses can inhibit or override sympathetic vasoconstriction during exercise began in 1967 with the study of Strandell and Shepherd at the Mayo Clinic, who showed a reduction in forearm blood flow (~16%) with supine leg exercise during lower-body negative pressure activation of the sympathetic vasomotor fibers (203). Three years later at the Mayo Clinic, Donald and associates had instrumented dogs perform staged treadmill exercise on a treadmill (0%–28% grade) and found that electrical stimulation of the implanted electrodes on the lumbar sympathetic chain effectively reduced limb blood flow at every stage of the exercise (42). In 1983 Thompson and Mohrman at the University of Minnesota at Duluth conducted a nerve-muscle simulated exercise experiment with anesthetized dogs and found that electrical stimulation of motor nerves and sympathetic nerves showed that vasoconstriction effectively reduced the oxygen consumption of active muscle (209). O'Leary from Wayne State University in Detroit evaluated the blood-flow results of Kjellmer (107) with regard to vascular conductance and found no evidence to indicate that sympatholysis had occurred (152).

Cognizant that sympatholysis was a controversial subject, Ruble and associates from the Medical College of Wisconsin and the Veterans Administration Medical Center in Milwaukee exercised instrumented dogs that had received atropine and a nicotinic postganglionic stimulant that would result in vasoconstriction of the muscle vasculature (182). Iliac blood flow and iliac vascular conductance results showed significant reductions throughout the various exercise stages and demonstrated dynamic exercise attenuated the vasoconstrictor response to sympathoexcitation. They concluded that their data supported the existence of "exercise sympatholysis." However, this study will not be the "last word" on the subject, and it is likely that sympatholysis will continue to be a subject of much interest as the twenty-first century unfolds.

Reflexes from Baroreceptors

Overview

It was known at the turn of the century that blood pressure would increase with exercise (135); however, it was several decades later when studies were initiated concerning its regulation. In 1923 Heinrich E. Herring (1886–1948) at the University of Prague in Czechoslovakia demonstrated that mechanical stimulation of the carotid sinus resulted in a reflex response leading to bradycardia and a fall in blood pressure (84). Since that time, resting studies in animals have shown that receptors sensitive to mechanical changes (baroreceptors) are located in the adentitia with afferent input to the nucleus tractus solitari that will elicit sympathetic and parasympathetic activity leading to a cascade of changes that alter resting heart rate, blood pressure, atroventricular conduction, myocardial contractility, and peripheral resistance on a beat-to-beat basis (44). However, it was not until the 1960s that attempts were made to explain why heart rate and blood pressure during exercise did not change in accordance with Marey' law (Etienne J. Marey [1830–1904] of Paris, France [128]).

Select Views on the Importance of Baroreceptors during Exercise and Its Onset

In 1966, Sture Bevegard and John Shepherd (Fig. 5.10) from the Mayo Clinic examined the influence of activating the carotid sinus baroreceptors during mild supine exercise. The baroreceptors were activated by raising their transmural pressure by neck suction (12). The reflex functioned during exercise as it did during supine rest; namely, reductions in heart rate and blood pressure occurred while forearm vascular resistance and cardiac output exhibited small but significant decreases. In their summary the researchers noted (12, p. 142): "Thus the carotid sinus mechanism continues to oppose, through negative feedback, the rise in heart rate and blood pressure

Fig. 5.10. John T. Shepherd (1919–) is a distinguished investigator and author currently affiliated with the Mayo Clinic and Foundation at Rochester, Minnesota. His contributions to this chapter can be found in references 12, 34, 41, 43, 203, and 226. The photograph was taken in the mid-1990s and provided to the author upon request.

during exercise, but is overcome by the exercise stimulus so that the net result is an increase in both heart rate and blood pressure."

During the same year, Vanhoutte and colleagues in Belgium conducted an extensive study on cardiovascular responses with intact and isolated carotid and aortic sinuses of unanesthetized and anesthetized dogs that exercised on a treadmill or whose muscles were electrically stimulated to mimic the treadmill results. The cardiovascular responses to exercise could occur very well when "these important arterial pressoreceptors are not actively inserted into the circulation." In essence, they were not involved in the regulation of blood pressure during exercise (213). Vatner and colleagues at San Diego in 1970 electrically stimulated the carotid sinuses of dogs that were instrumented to measure hemodynamic, cardiac output, and peripheral resistance change at rest and during mild exercise (214). The increase in pressure sensed by the carotid sinuses caused marked decreases in blood pressure and peripheral vascular resistance (iliac artery, renal and mesenteric beds) while causing "profound vasodilation" in the circulation of the hindlimbs. Since this was the first study to show contracting skeletal muscle received tonically active sympathetic vasoconstrictor outflow, they promoted the idea that a release of sympathetic constrictor tone had occurred with electrical stimulation even though the muscular bed was "already dilated on a metabolic basis" (214; see previous section entitled "Activation of Muscle Reflexes and Their Opposition by Chemical–Metabolic Influences").

Krasney and colleagues at Albany Medical College in New York in 1974 exercised dogs before and after sinoaortic denervation and reported that the hemodynamic responses were qualitatively similar to those recorded for intact animals except that the apparent hypotension at the onset of exercise was enhanced (110). Two years elapsed before McRitchie and coworkers in Boston, who included Vatner and Braunwald, measured the hemodynamic responses of dogs running on the treadmill before and after barorecptor denervation (139). Their focus was on the responses with heavy exercise (10–12 mph). From their results they concluded that the reflex was "turned off" during severe exercise and did not significantly modify the cardiovascular responses to exercise (139). Later (1982), Faris, Jamieson, and Ludbook at the University of Adelaide in Australia had rabbits perform heavy exercise on a treadmill for brief periods. The aortic baroreceptors of the animals had been denervated and they were instrumented with variable-pressure neck chambers so the carotid sinus could be stimulated over a wide range of sinusoidial transmural pressures. They noted that carotid reflex control for blood pressure was impaired, but not absent, with treadmill exercise (55). Ludbrook and Graham had rabbits run on a treadmill for brief periods at an intensity estimated to 90% of their maximum heart rate capacity (124). When exercised after denervation of the carotid and aortic sinuses, mean arterial pressure fell precipitously before returning to baseline values; the vascular resistance index decreased markedly where it remained while measurements of cardiac index, stroke volume, and heart rate were increased (124). Hales and Ludbrook also used rabbits whose arterial baroreceptors had been denervated to study cardiovascular changes with the onset of heavy exercise. They found when using radioactive microspheres that approximately 40% of the increased blood flow to the heart and to the skeletal muscles that had been diverted from splanchnic organs, kidneys, skin, and fat was lost by the denervation procedure. One of their conclusions was that vasoconstriction in select regions may be caused by "transient suppression of the reflex effects of arterial baroreceptor input at the onset of exercise" (75).

Using innovative procedures that will be described later, Anders Melcher and David Donald (Fig. 5.11) exercised dogs that had been selectively denervated so the contributions of the various baroreceptors could be evaluated (141). They first showed decreases in mean arterial blood pressure that occurred during the initial stages of either light or heavy exercise (~20%) when either the carotid or aortic or cardiopulmonary baroreceptors were removed. However, a precipitous (~36%) fall occurred when all baroreceptor input was removed that was sustained (light exercise) or somewhat sustained (heavy exercise) as the animals continued to run (141).

One idea emerging from these investigations was that the elevation in arterial pressure at the onset and to some degree during exercise required the presence of baroreceptors. Another notion worthy of attention is that baroreceptors will initiate sympathetic vasoconstriction of the vascular beds to achieve an elevation in pressure.

Fig. 5.11. David E. Donald (1921–) is a renowned investigator who retired after a distinguished career at the Mayo Clinic in Rochester, Minnesota. His contributions to this chapter can be found in references 41–43, 70, 141, 201, and 225. The photograph was taken in 1950 and graciously provided to the author by the Mayo Foundation.

The Role of Baroreceptors during Exercise

Concerned that existing baroreceptor denervation procedures used with exercising animals were providing transient information and were incomplete in securing carotid sinus pressures or stimulus–response curves, Stephenson and Donald at the Mayo Clinic developed an ingenious method to reversibly isolate and study the functions of the carotid sinuses in resting and exercise conditions (201). The technique enabled investigators to establish intersinus pressure at a level below the threshold, thereby suppressing baroreceptor discharge. The technique also allowed baroreceptors to be activated over their full range and allowed baroreflex stimulus–response curves to be assessed. In addition, the technique could be used to systematically evaluate arterial and cardiopulmonary baroreceptors (201).

As mentioned, Melcher and Donald used this approach when they reported reductions in mean arterial blood pressure with the onset of light and heavy exercise (141). When only the carotid baroreflex was operational, steady state light exercise was associated with a mean arterial pressure that had returned to pre-exercise level and that remained plateaued until the end of the exercise. A similar profile occurred when only the aortic cardiopulmonary baroreceptors were intact. Removal of all baroreflex input not only caused a precipitous fall in mean arterial blood pressure (~36%), it remained at that level until the cessation of exercise. With heavy exercise and only intact carotid baroreceptors, the fall in mean arterial pressure gradually returned to the pre-exercise level. On the other hand, with only intact aortic cardiopulmonary baroreceptors, the response profile with heavy exercise was similar to the one recorded for light exercise. When input from all baroreceptors were removed before heavy exercise, mean arterial fell markedly before gradually returning to a level that was below the pre-exercise values (~14%). A similar trend was reported by Krasney and others for their exercising sinoaortic denervated dogs (110). Melcher and Donald concluded that the carotid, aortic, and cardiopulmonary baroreceptors were involved in the regulation of arterial blood pressure during exercise, but that other mechanisms were also involved (141). To Rowell, the chemoreflex was a mechanism to consider for the return of arterial blood pressure after removal of baroreceptor input (177).

Stimulus Response Curves and Baroreceptor Resetting

Important roles for the baroreceptor are activating the sympathetic nervous system and elevating arterial blood pressure at the onset of exercise. Their function is best understood by analysess of stimulus–response curves which reveal threshold, operating point, gain or sensitivity, and especially the shifts in thresholds and operating points or resetting (108). However, to describe these functions in humans is difficult because numerous studies have been published in past decades using techniques (e.g., infusion of vasodilators or pressor drugs) or analysis procedures (e.g., electrographic R-R intervals) that are considered to be controversial.

Although Robinson and colleagues mentioned in 1966 that baroreceptors could be reset during exercise (170, p. 409), it has been the experimental animal studies that have provided the support for this idea. Wagenbach and Donald employed the reversible vascular isolation of carotid sinus technique to study baroreceptor functions in exercising dogs (225). Analysis of mean arterial pressure, heart rate, and blood flow results from intact and aortic denervated dogs with vascularly isolated carotid sinuses and sinus pressure maintained at baseline levels strongly suggested that the baroreflex had been reset upward to higher pressures in a stepwise manner during exercise (225). Of interest is that Rowell suggested these results occurred because the "resting" carotid sinus pressure was interpreted centrally as a hypotensive signal that became more pronounced as the intensity of exercise increased (177). As discussed earlier, Ludbrook and associates also conducted experiments with animals pertaining to baroreceptor functioning during the onset of exercise. In a 1985 investigation, they surgically denervated the baroreceptors while isolating the cardiac afferent (interpericardial procaine) before exercising rabbits on a treadmill. When only the arterial baroreceptors were studied, they found a sudden fall in mean arterial pressure that was rapidly returned to pre-exercise values within a 60 second exercise bout. With regard to the onset of exercise they state (124, p. 464):

> Yet the very opposite occurred at the onset of exercise, for within 10 s the tonic depressor effects of baroreceptor input was suppressed, rather than enhanced (Figs. 5.3 and 5.5). It is as though the properties of the arterial reflex were altered so that it was operating below its threshold, rather near its saturation point. In other words, during the first 20 s of exercise the set point for the reflex rose faster than did arterial pressure.

In 1993, DiCarlo from Northeastern Ohio University and Bishop from the University of Texas Health Science Center in San Antonio conducted an exercise study with rabbits to test the hypothesis that the elevation in sympathetic nerve activity and mean arterial pressure with the onset of exercise was dependent upon an upward shift of the operating point of the arterial baroreflex (37). Intravenous infusion of nitroglycerin was used to attenuate the pressure response while activity from the renal nerve was measured to denote sympathetic involvement. Drug infusion with exercise reduced the rise in mean arterial pressure (13%), elevated heart rate (16%), and markedly increased renal nerve sympathetic activity (105%) when compared to control exercise conditions. Their data indicated the onset of exercise shifted the operating point to higher pressures. In addition, they measured the responses of sinoaortic denervated rabbits with the onset of exercise and found notable decreases in mean arterial pressure (46%), heart rate (26%), and in renal nerve sympathetic activity (13%), which reinforced the viewpoint that an intact baroreflex is required to elevate blood pressure and to activate the sympathetic nervous system. Like others before them, they speculated that central command had changed the operating point of the baroreflex (37). When reflex gains with exercise were reported, Faris and colleagues found a decrease in their rabbit studies (55) while Melcher and Donald found no changes with their dogs (141).

Human studies conducted in the 1990s extended these animal investigations. Papelier and associates in France studied the effects of cycling exercise on the carotid baroreflex using a neck suction chamber that delivered pulsatile positive and negative pressures to the carotid sinus (156). Four stages of exercise were performed with the highest level at ~75% $\dot{V}O_2$max. Analysis of the baroreflex responses indicated heart rate and mean arterial pressure were higher at any given carotid sinus pressure or that resetting had occurred. Potts, Shi, and Raven in Fort Worth had subjects perform light and moderate exercise on a cycle ergometer at intensity levels between 25% and 50% to study the responsiveness of the carotid baroreflex (164). A neck collar placed over the carotid sinus regions was used to deliver positive pressures or to create negative pressures in a graded fashion. Heart rate, mean arterial pressure, and central venous pressure were measured throughout the experiments. At 50% $\dot{V}O_2$max, they found significant increases for the centering point, threshold, and saturation pressures, which supported their conclusion that resetting had occurred (164). Both studies reported baroreflex sensitivity or gain was unaffected by exercise (156, 164).

Exercise Training and the Baroreflex

The idea that exercise training could alter baroreflex functions originated from resting measurements obtained from a cross-sectional study of fit and nonfit subjects (202). In 1988, DiCarlo and Bishop demonstrated that training would attenuate baroreflex responses; however, the results were obtained from resting animals (36). Thus it remains for the future to determine whether training will have a significant influence on baroreceptor responses during acute exercise.

REFLEXES FROM CARDIOPULMONARY AND CARDIAC RECEPTORS

Overview

It is known that receptors (mechanoreceptors, baroreceptors) exist in the superior and inferior vena cavae, atria, ventricles, coronary arteries, and pulmonary arteries and veins that will respond to low pressure at their respective sites. They also respond to variations in central blood volume, central venous pressure, myocardial contractility, and stroke volume. Afferent fibers from these receptors course with sympathetic fibers to the spinal cord and with vagus nerves to the medulla oblongata where they converge on neuron pools in the nucleus tractus solitarii (44). Afferent input is associated with an inhibition of sympathetic nerve activity to vascular beds with the purpose of modulating regional vascular resistance. In resting conditions, changes in central venous pressure appear to be the most effective stimuli for activation of the cardiopulmonary baroreflex (180). Because of the uncertainty in securing direct measures of the reflex, discussion of their role during exercise will be confined to animals.

The Mayo Summary of the Importance of the Reflex in Animals During Exercise

In 1984 Walgenbach and Shepherd summarized animal exercise studies conducted with intact, sino-aortic denervated, and vagotomized dogs and/or dogs with isolated carotid sinus preparations that performed light to heavy exercise and concluded that cardiopulmonary baroreflexes were not important, at least for the control of blood pressure during exercise (226). During the same year, Daskalopoulds with Shepherd and Walgenbach studied chronic sino-aortic denervated dogs to determine the hemodynamic changes after altering the cardiopulmonary input by bilateral cervical vagotomy (34). Heart rate and mean arterial blood pressure data recorded during systematic graded exercise before and after vagotomy exhibited minimal changes, which led them to conclude that the "cardiopulmonary mechanoreflexes contributed little to the regulation of arterial pressure during exercise." However, this was not the case after the cessation of exercise.

Role of Cardiac Afferents

In their baroreceptor studies concerned with the onset of exercise, Ludbrook and Graham in 1985 surgically eliminated the influences of baroreceptors while procaine injections into the pericardial sac were given to remove input from the cardiac afferents (124). Removal of cardiac afferents alone had no significant influence on mean arterial pressure, but their removal significantly lowered heart rate (10%) and cardiac index (19%) while elevating systemic vascular resistance (26%). When baroreceptors and cardiac afferents were eliminated, systemic vascular resistance was the only measurement that did not fall below the value recorded from "intact" exercising animals. From these results the researchers promoted the idea that the inhibitory input from cardiac receptors on systemic vascular resistance was abolished with the onset of exercise (124). DiCarlo and Bishop in 1990 pursued the role of cardiac afferents on vascular resistance changes with exercise using instrumented rabbits capable of performing moderate-to-heavy exercise for sustained periods (38). They varied combinations of pharmacological agents administered in sequence to inhibit cardiac afferent (procainamide) and efferent (scopolamine methyl bromide and propranolol) activity. Their results indicated cardiac afferents during exercise were not significantly contributing to total systemic vascular resistance or to terminal aortic vascular resistance. However, they did show that cardiac afferents were important in controlling the vascular resistance of the mesenteric and renal vascular beds through their influence on vasoconstriction (38). In 1994, O'Hagan and others at Milwaukee used the exercising rabbit model to study the effects of pericardial infusions of saline and procaine to eliminate input from the cardiac afferents or of atenolol and methscopolamine to inhibit cardiac efferent activity (151). Compared to the responses obtained with saline infusions, heavy exercise was associated with significantly increased renal nerve sympathetic activity, indicating that cardiac sensory receptors had an inhibitory effect on the sympathetic drive to the kidney.

Training and Cardiac Afferents

In 1990, DiCarlo and Bishop studied rabbits that had been regularly exercised on a treadmill during an 8 week period (38). Significant reductions in submaximal heart rate values indicated they were trained. The rabbits were instrumented to measure arterial and venous pressures and blood flow in select vessels and beds while procedures described previously were used to selectively block cardiac afferent and efferent activity. Before training, submaximal exercise was associated with increased resistance values in the renal and mesenteric arteries. After training efferent blockage had no meaningful effect on resistance measures whereas elimination of the cardiac afferents significantly elevated both renal and mesenteric resistance values. Later, DiCarlo, Stahl, and Bishop extended their observations by measuring sympathetic activity in the renal nerve in rabbits also exercised for 8 weeks to produce a trained state (40). Evaluation of the influences of cardiac afferent and efferent activity was performed as described previously. Compared to control conditions, inhibition of the cardiac afferent activity had no meaningful influence on renal sympathetic nerve activity results from the nontrained animals whereas the renal sympathetic activity values were significantly increased by almost twofold (40). While these findings provided support for the idea that training will attenuate cardiac afferent influences during exercise, the responsible mechanism remains unknown.

Ablation of Autonomic Structures and Their Influence on Physiological Responses During Exercise

Background Information

The idea of removing anatomical structures in order to better understand their function is an ancient one, but its application to the study of exercise responses came much later. According to Monro, Francois Pourfour du Petit (1664–1741) was the first to surgically investigate the functions of the sympathetic nervous system when in 1727 he sectioned the cervical sympathetic change of a dog between the third and fourth cervical vertebrae and reported pupillary constriction and protraction of the nicitating membrane (143). He also observed that the vagosympathetic trunk was not connected to the brain, and like others who followed, observed that vagal sectioning altered the breathing rates of animals (112, 160). Claude Bernard repeated and extended the Pourfois du Petit studies with rabbits, dogs, and horses during 1851 and 1852 and observed that sectioning caused vasodilation, increased circulation, and warmth to the face of the animal. He also confirmed the pupillary findings of du Petit but failed to acknowledge him for the observation (143). During 1852, the Frenchman Charles Edouard Brown-Sequard (1817–1894) reported vasomotor paralysis with cervical sympathectomy and recognized the significance of his findings even though seven decades would pass before sympathectomies were used for clinical purposes. Therefore, in 1925, Addison and Brown began the era of clinical sympathectomy when they sectioned the sympathetic rami in the lumbar region in order to improve limb blood flow of patients with Raynaud's disease (143).

At the turn of the century and for several subsequent decades the interest in surgical removal of autonomic structures appeared to be focused on control of heart rate. Later, the emphasis was on the maintenance of homeostasis as defined by Bernard and Cannon (21).

Ablation, Heart Rate, and Work Performance

The interest in determining the effects of select ablations of sympathetic and parasympathetic nerves of exercising animals began with the 1895 study of Herring and others (85) and included Friedenthal in 1902 (62), Gasser and Meek in 1914 (67), and Campos and coworkers (20), who found the animals capable of exercising—but, with altered heart rates. The 1929 study by Campos and associates at Harvard University is notable because they used the methods of Zuntz (236) and of Slowtzoff (199) to quantify the energy expenditure of the dogs running on the treadmill (20). When the sympathetic nerves to the limbs of dogs were sectioned, it had no apparent effect on the ability to exercise. However, sectioning the sympathetic nerves to the liver decreased the work that could be performed. Campos and colleagues felt the surgical procedures used by Friedenthal to denervate the heart were incomplete, so they denervated the heart and eliminated adrenal influences (demedullation and removal) in their dogs. When exercised, both heart rate and work performed were reduced (nerves to the liver had been sectioned). Injections of epinephrine (they knew a denervated heart was sensitive to epinephrine) increased the amount of work performed. Consequently, they concluded that more work could be accomplished "when the pulse is accelerated" (20).

In 1935, Adi Samaan from the University of Ghent in Belgium extended the 1914 studies of Gasser and Meek (67) by incorporating grade running on the treadmill to quantify their work on responses of dogs (187). When the dogs were exercised after bilateral vagotomy, the amount of work performed was greatly reduced and the animals became rapidly "fatigued" early in the run. In addition, Samaan reported elevated exercise heart rates and postexercise respiratory rates. Samaan also investigated the single and combined influences of stellate ganglionectomy, splanchectomy, cardiac denervation, and atropine administration that subsequently led to the conclusion that "The maximum capacity for the performance of work is markedly augmented subsequent to the extirpation of the upper halves of thoracic sympathetic chains" (187, p. 329).

Samaan's procedures and conclusions, especially concerning the augmentation of work performance with sympathectomy, were questioned by Lucien Brouha and collaborators at Harvard University who included Walter Cannon and Bruce Dill. This motivated them to conduct studies on the matter between 1936 and 1939 (15, 16). Brouha and others felt that Samaan had neglected to consider the influences of the cardio-accelerator fibers from the mesencephalon which leave via the vagus nerve, enter the vagosympathetic trunk, and were not affected by sympathectomy but were blocked by atropine. In addition they believed that Samaan's use "of atropine to suppress vagal impulses in dogs deprived of sympathetic cardio-accelerators obviously did not complete the denervation of the heart as he supposed" (15, p. 357).

Therefore, the Harvard groups used the technique of Bacq, Brouha, and Heymans from the Institute Leon Fredericq in Liege, Belgium, which removed the sympathetic chains from the stellate ganglion to the second ganglion of dogs in a three-step procedure (10). It should be noted that one of the investigators responsible for this technique was Corneille Heymans (1892–1968), who was awarded the Nobel Prize in physiology and medicine for 1938 (174). After the dogs had recovered, Brouha and associates found that the dogs could perform intense exercise but that their ability for prolonged work was reduced, and there was absolutely no evidence for an increased capacity as suggested by Samaan. Their heart rates were approximately 70% of the presurgical value during moderate-to-heavy exercise. One year later, they noted that the cardiac acceleration with exercise approached 90%–95% of the presurgical mean and that these changes were not influenced by elevations in body temperature, metabolites, or epinephrine (or by sympathin; see section entitled "After Cannon: the Characterization of an Exercise Response" [16]). They found in sympathectomized dogs that when the effects of bilateral vagotomy were evaluated, there were marked reductions in exercise capacity, altered breathing patterns, and reduced panting. Bilateral vagotomy was also associated with a "striking reduction in the cardiac acceleration in response to muscular work" (16). Earlier they had concluded that cardiac acceleration "seems to be due to a reduction in the tonicity of the cardio-inhibitors of the vagi and to an increase in the tonicity of vagal cardio-accelerators of the dog" (15, p. 359). Hence an important result of these denervation studies was a better definition of the relationship between the sympathetic and parasympathetic systems in the control of the heart rate response to exercise. An additional result that was marginally emphasized in the 1936 study for sympathectomized animals was the importance of the respiratory system to explain the drastic reduction (100 minutes to as low as to 2 minutes) in the duration of work performed animals after unilateral and bilateral vagotomy (15).

Ablation, Myocardial Responses, and Exercise Performance

Hiram Essex and associates at the Mayo Clinic in Rochester, Minnesota, were stimulated by the studies of Samaan and in 1945 studied dogs during moderate exercise before and after staged vagotomy and sympathectomy. Their main finding was that increase in coronary blood flow during exercise was dependent more on changes in blood pressure than on the presence of an intact vagi or an innervated heart (51). In the 1960s, a national interest emerged in the use of cardiac transplants to replace functionally impaired hearts. Associated with this interest was an awareness that led to more research on myocardial functioning and on improved surgical techniques that would not interrupt the sympathetic nerve fibers to the lungs, eliminate the cardiopulmonary afferent fibers, or deplete the myocardium of its catecholamine concentrations (29). Therefore, a new surgical approach was developed and perfected by Theodore Cooper at the National Heart Institute in Bethesda, Maryland, to resolve these problems. It was called "extrinsic cardiac denervation" (29). Later, Donald and Shepherd at the Mayo Clinic used a modification of the Cooper procedure and tested the responses of cardiac denervated dogs (43). When compared to presurgical exercise values, they noted a much attenuated rise in heart rate before steady-state conditions prevailed, which was attributable to the increase in stroke volume. The stroke-volume change was related to Starling's law of the heart (note that the pericardium had been removed) and to the inotrophic effect of the circulating catecholamines acting on the myocardium. In addition, they found cardiac output to be increased as the exercise became more strenuous and subsequently evaluated the ability and willingness of cardiac denervated greyhounds to run 5/16ths of a mile. They found their speed of 33.6 mph was 10% slower than the track record and 3%–5% slower than the presurgical run time (41). Hence, they concluded in 1964 that the ability of the denervated heart to respond to brief maximal exercise is "but slightly reduced."

Gregg and coworkers at the Walter Reed Army Institute of Research in Washington, DC, in 1972 studied myocardial functioning in dogs whose hearts had been denervated by the Cooper technique before being subjected to moderate treadmill exercise (70). In contrast to what was expected from the Donald and Shepherd studies, they found coronary blood flow with the experimental animals was approximately 50% lower than values obtained from intact animals. In addition they noted a decrease in myocardial oxygen utilization.

Patrica Gwirtz and colleagues from the Texas College of Osteopathic Medicine in Fort Worth did not ablate the autonomic nerves to the heart, but they did surgically remove the pericardium from dogs and selectively denervate the sympathetic fibers to the ventricles 2 months before testing for changes in myocardial contractility, blood flow, and ventricular oxygen consumption in exercising dogs (71). Measurements obtained during moderate-to-heavy exercise revealed a significant reduction in left ventricular systolic blood pressure (~20%), contractility (dp/dt max, ~26%), and mean blood flow in the circumflex artery (~47%) when compared to "normal" dogs. Intact ventricles were associated with a marked increase in oxygen consumption during exercise whereas the denervated structures had elevations that were one-third of "normal." The decreases in coronary flow and in myocardial oxygen consumption with surgical removal of the sympathetic nerves to the ventricles did follow the trends reported four decades earlier by Gregg and coworkers for cardiac denervated animals (70).

Ablation and Changes in Pressure and Flow

In the 1960s, Edmundo Ashkar from the Experimental Institute of Biology and Medicine in Buenos Aires, Argentina, led a renewed interest in the effects of surgical removal of autonomic structures on the cardiovascular responses to exercise. In his first study with William Hamilton (1893–1964) (7) he compared his exercise results from sympathectomized, adrenal dernervated, and vagotomized dogs with results from the literature for normal dogs and commented on their lower blood pressures and the attenuated cardiac output values. In 1966 he reported findings from vagotomized dogs that had either complete or partial removal of the sympathetic nervous system (5). Specifically, the left stellate ganglion was present in some and the thoracic innervation from T8 to T12 to the splanchnic region was intact in others. Blood pressure during moderate exercise was noticeably elevated in intact animals and in the sympathectomized animals with intact splanchnic innervation whereas it was noticeably decreased in the vagotomized and totally sympathectomized animals and in the animals with intact stellate ganglions. He attributed this difference in blood pressure, in part, to a "lack of homeostatic control of the splanchnic area, presumably by the sympathoadrenaline mechanism" (5). Later with Stevens and Bernado Houssay (1887–1971, Nobel Prize winner in physiology and medicine for 1947), Ashkar (6) conducted cross-sectional investigations with vagotomized and partially sympathectomized dogs (T8 to T12 were intact) which included either adrenal inactivated (demedullation and adrenalectomized) or denervated cranial mesenteric and common hepatic arteries (eliminated innervation to the majority of splanchnic organs and to the spleen). Intact and experimental dogs had similar energy expenditures when exercised on the treadmill but the percent changes for cardiac index from rest to exercise for the heptomesenteric denervated dogs were twofold higher than in the intact dogs. This result demonstrated the importance of heptomesenteric vasoconstriction in maintaining a sufficient blood volume for the other regions of the body. The percentage change for the cardiac index for the medulloadrenalectomized dogs was midway between the two groups (6).

In 1970 Donald and coworkers electronically monitored iliac blood-flow changes in exercising dogs while measuring oxygen saturation values in iliac veins (42). The dogs had selective unilateral sympathectomy of the hindlimb. They found that limb blood flow in the sympathectomized leg was noticeably lower during the initial stages of exercise, after which the flow patterns and changes in both legs became similar as the exercise intensity approached maximal values. Hence, they promoted the idea that limb blood flow was determined more by the intensity of exercise than by the activation of the sympathetic nervous system (42). Donald was aware that changes in limb blood flow did not provide insights as to its distribution. Using radiolabeled microspheres, Fred Peterson, Robert Armstrong, and Harold Laughlin in 1988 at Oral Roberts University in Oklahoma City studied regionally sympathectomized rats (bilateral sectioning of the sympathetic innervation between L2 and L7). The animals exercised moderately for brief (2 minutes) and longer (15 minutes) periods of time (159) after which the researchers analyzed blood-flow distributions to 25 hindlimb muscles. During the initial stages large increases in blood flow were observed in many muscles, which was interpreted as demonstrating that the sympathetic nervous system had a minimal influence on muscle blood-flow distribution regardless of fiber type. Moreover, when steady-state conditions were present, total blood-flow values were significantly elevated in the sympathectomized animals, which suggested that the sympathetic nervous system was not limiting peripheral blood flow (159). Later, Laughlin and colleagues included these results in the American Physiological Society *Handbook* series on exercise and noted that sympathetic activation would decrease vascular conductance during sustained exercise (119). A generalization emerging from the Peterson and coworkers study (159) was that slow red oxidative fibers (type I) are more responsive to the influences of sympathectomy than the fast glycolytic fibers (type II).

In 1979 Roger Hohimer and Orville Smith at the University of Washington in Seattle studied baboons who performed dynamic leg exercise when provided a food reward (91). Renal blood flow was measured before and after the renal nerve of the contralateral kidney had its sympathetic innervation removed. Mild exercise resulted in a noticeable (15%) decrease in renal blood flow to the intact kidney whereas there was little change (1%) in the denervated kidney. An idea that emerged from these denervation and exercise studies was that vascular beds are selectively different in their responsiveness to sympathethic stimuli.

Chemical Ablation and $\dot{V}O_2$ max

In 1982, Ernst and coworkers from the University of California at Irvine chemically sympathectomized their rats with repeated injections of guanethidine monosulfate (depletes catecholamines and depresses function of postganglionic adrenergic nerves) and performed adrenal demedullations before comparing their maximum exercise results with those in intact animals (49). They found that their experimental animals had reduced $\dot{V}O_2$max and heart rate values (both 15%). With submaximal tests, the two groups had similar $\dot{V}O_2$ results but the sympathectomized-demedullated rats had markedly lower heart rates, which suggested that an augmented stroke volume was responsible for the similar $\dot{V}O_2$ results (49).

Chemical and Surgical Ablation Studies with Trained Animals

Although it was known in 1965 that immunologically sympathectomized rats could be endurance trained and would exhibit resting adaptations, no results were obtained during exercise (210). More than a decade later, Sigvardsson and his Swedish collaborators trained chemically sympatectomized rats (6-hydroxydopamine) on a treadmill and compared their heart rate changes with those of intact control animals (196). During the highest stage of the power test, the normal trained rats exhibited a marked reduction in exercise heart rate (–7%) and the trained sympathectomized animals (–4%) exhibited the same trend when compared to nontrained groups. Ordway and colleagues at the University of Kentucky in 1982 exercised dogs that had been cardiac denervated by the extrinsic method of Cooper (29) and compared their findings with those for sham-operated controls (155). Training had no significant effect on the heart rate values recorded during the submaximal test. MacIntosh and associates at the University of California at Irvine studied rats that were chemically sympathectomized (guanethidine) after birth and subsequently progressively trained on a motor-driven treadmill but using a prescribed exercise program less intense than the "normal" rats being trained (126). When tested for $\dot{V}O_2$max, both trained groups exhibited significant increases over their nontrained controls. Comparisons of absolute values between the normal and sympathectomized trained groups had no statistical significance. Measurements recorded during submaximal exercise indicated that the trained sympathectomized rats had markedly lower heart rates and left ventricle systolic blood pressures than the other groups. While more confirmation is needed and more extensive investigations are required, the collective results provide some support for the idea that endurance training by "whole-body" sympathectomized rats will result in select cardiac adaptations similar to those found with "normal" trained animals.

INHIBITION OF AUTONOMIC RECEPTORS THAT ALTER CARDIORESPORATORY RESPONSES DURING EXERCISE AND AFFECT PERFORMANCE CAPACITY

Overview and Sympathetic Inhibition

Autonomic drugs are substances that mimic, modify, or interfere with peripheral ganglionic or motor effector transmission (2). Identification of sympathetic agents began in 1895 with the study of adrenal extracts by Oliver and Schafer to elicit a pressor response (153) and by Sir Henry Dale (1875–1968, Nobel Prize in physiology and medicine for 1936 with Otto Loewi [174]) who in 1906 at Cambridge, England, used extracts of ergot (a plant alkaloid) to reverse the effects of epinephrine or sympathetic nerve stimulation (32). It is important to note that Dale suggested that these results could involve the actions of more than one type of receptor.

The field was advanced in the 1940s when it became known from the findings of von Euler that the transmitter released from the sympathetic nerve endings is norepinephrine (52) and with the epic publication by Raymond Ahlquist from the University of Georgia on the identification, classification, and characterization of sympathetic receptors (1). He proposed the existence of α- and β-adrenotropic receptors (adrenoceptors) after using sympathomimetics on a variety of species and preparations. In addition, he determined their order of potency in various physiological systems. Because the two-receptor system of Ahlquist had inconsistencies in explaining the responses from a single β-receptor, Lands and others from the Steerling-Winthrop Research Institute in Rensselaer, New York, promoted the idea that there are two β-receptors that should be classified as β_1 and β_2 (113). The foundation for this idea was related to whether epinephrine and norepinephrine were equipotent in receptor activation; if so, the receptor was considered to be β_1, whereas if the receptor had a greater affinity for epinephrine than norepinephrine, it was classified as β_2. For the cardiorespiratory systems, it was determined that β_1-adrenoceptors are in the atrium and ventricles of the heart, coronary arteries, kidneys, and bronchioles whereas β_2-adrenoceptors are postjunctional and located within the atrium, the resistance vessels, bronchi, and at the terminals of sympathetic nerves. Although norepinephrine predominates in stimulating β_1-adrenoceptors, circulating epinephrine will also have a limited influence. However, epinephrine will preferentially activate the β_2-adrenoceptor. It has been shown that a β_3-adrenoceptor is expressed in adipose tissue and the gastrointestinal tract, but its presence and role in the cardiorespiratory systems are uncertain (194). Therefore, the ensuing sections will pertain to the responses from the β_1- and β_2-adrenoceptors. While Ahlquist had characterized the response of an α-adrenoceptor, the 1974 results of Langer from the Instituto de Inestigaciones Farmacologicas in Buenos Aires, Argentina, led to the idea that α-adrenoceptors should be classified as either α_1 or an α_2 depending whether their location is presynaptic (α_2) or postsynaptic (α_1) in nature (114). In contrast to the α_1-adrenoceptor, the presynaptic α_2-adrenoceptor inhibits peripheral neuronal release of the transmitter (90, 114). The term "adrenergic blocking agent" refers to a compound that will selectively inhibit various responses to adrenergic nerve activity and to a sympathomimetic amine whose locus of action is the effector cell (146).

Given the availability of adrenergic blocking agents, countless exercise studies have been conducted, most concerned with diseased populations. Thus the focus here is on select studies that introduce new ideas or amplify or confirm ideas from the denervation investigations.

α-Adrenoceptor Inhibition and Hemodynamic Responses

Studies that administered pharmacological agents to inhibit the actions of α-receptors in healthy humans during dynamic exercise began in the 1980s and focused more on metabolic than on cardiorespiratory changes. Among those who recorded hemodynamic changes in humans with selective inhibition of α_1-adrenoceptors (prazosin) were McLeod and colleagues from Duke University. In 1984 they measured heart rate and arterial blood pressure changes during submaximal exercise performed on a cycle ergometer (136). Compared to placebo conditions, heart rate was significantly elevated at each stage and systolic blood pressure was unchanged, whereas the inhibition of vasoconstriction caused a significant reduction in diastolic blood pressure. The results also included plasma norepinephrine concentrations during peak exercise, which were more than twofold higher after blockade than recorded during placebo conditions, which indicated that augmented responses can be expected from structures whose β-adrenoceptors are stimulated by norepinephrine.

α-Adrenoceptor Inhibition and Myocardial Responses

The possibility that a sympathetic mediated vasoconstrictor tone existed in the coronary arteries during exercise was proposed in 1979 by Murray and Vatner from Harvard University, who measured coronary vascular responses from exercising dogs with paced heart rates while administering various blocking agents that included a nonselective α-adrenoceptor inhibitor (phentolamine) (145). Gwirtz and Stone at the University of Oklahoma pursued this possibility by having instrumented dogs perform staged submaximal treadmill exercise before and after the infusion of phentolamine. Compared to control exercise conditions, significant increases were found for coronary blood flow (42%), myocardial contractility (dP/dt max, 38%), and myocardial oxygen consumption (38%) before and after infusion of the blocking agent (73). In a subsequent study Heyndrickx and associates from the University of Gent in Belgium studied the selective effects of the inhibition of α-adrenoceptors (prazosin, yohimbine, and phentolamine) (89) in instrumented dogs that performed moderately heavy treadmill exercise. Blocking the actions of the postjunctional α_1-adrenoceptors by prazosin had little effect on coronary blood flow but significant elevations occurred when the nonselective α-adrenoceptor inhibitor phentolamine and the prejunctional α_2-adrenoceptors were infused (~95% for phentolamine and ~36% for yohimbine). In addition, they found "uncontrolled norepinephrine release" and marked elevations in contractility and heart rate when the α_2-adrenoceptors were inhibited which were attributed, in part, to the actions of norepinephrine on β-adrenoceptors. In 1996, Kim and associates from Fort Worth administered the specific α_1-adrenoceptor inhibitor prazosin into the coronary arteries of instrumented dogs that performed staged submaximal treadmill exercise (106). Compared to control exercise conditions with the blocking agent, significant increases were noted for coronary blood flow (14%) and myocardial contractility (16%). Additionally, the researchers found significant elevations in cardiac output (16%) and myocardial oxygen consumption (7%). These composite studies give credence to the idea that a sympathetic vasoconstrictor tone is present within the coronary arteries during exercise. In addition, they indicate that α-adrenoceptors are mediating this tone, and that when inhibited, myocardial functions will be altered directly by the myocardial actions of the elevated concentrations of norepinephrine and indirectly by its influence on myocardial β-adrenoceptors.

Characterization of α-Adrenoceptors

To date, there is no historical "message" concerning the characterization of α-adrenoceptors mediating cardiorespiratory responses in healthy humans during exercise because the available information has been obtained from circulating blood from measurements made on different types of blood cells obtained during or after exercise. Because these results are at best an approximation of vascular and tissue receptor responses, readers will have to wait until suitable methodology is available.

Training and α-Adrenoceptors

Animal studies have provided limited information on the effects of training. Gwirtz and Stone infused the nonselective α-adrenoceptor blocking agent phentolamine into exercising dogs that had been progressively exercised for 4–5 weeks. Before training began, coronary blood flow had significantly increased by 10% whereas after being "trained" a significant increase of 21% was noted, suggesting that training had diminished the existing vasoconstrictor tone (72). Because of the paucity of information, researchers of the future will have to determine whether endurance training will alter α-adrenoceptor-mediated vasoconstrictor tone during exercise.

β-Adrenoceptor Inhibition and Hemodynamic Responses

In 1958 Powell and Slater from Eli Lilly and Company in Indianapolis, Indiana, introduced the compound dichloroisoproterenol as a blocking agent of β-adrenoceptors (165). When it was discovered that the drug contained agonist and antagonistic properties, its use was discontinued. In the early 1960s, pronethalol and propranolol were introduced for experimental and clinical use as nonselective β_1- and β_2-adrenoceptor antagonists with propranolol emerging as the preferred drug of choice because pronethalol had an intrinsic influence, unpleasant side effects, and caused tumors in mice (146). Because the "invention" of propranolol had "revolutionized" the medical management of angina pectoris, Sir James Black was awarded the 1988 Nobel Prize in medicine and physiology (200). Cognizant that previous studies had used pronethalol and quanethidine (to deplete norepinephrine concentrations at nerve terminals), Epstein and colleagues at the National Heart Institute in 1965 administered propranolol to seven normal subjects performing submaximal and maximal exercise (propranolol is a nonselective blocking agent because it inhibits β_1- and β_2-adrenoceptors). From their results they concluded that the "acute induction of β-adrenergic blockagde impairs the circulatory response to exercise" (48, p. 1750).

In the study of Epstein and others, maximal exercise was associated with significant reductions in heart rate (19%). With the advent of selective β-adrenoceptor blocking agents, Anderson and associates from Sydney, Australia, administered a placebo, a β_1-adrenoceptor inhibitor (metoprolol), and propranolol to subjects who performed progressive power outputs on a cycle ergometer until they were unable to continue (4). Compared to the placebo results, maximal heart rates were significantly reduced by both agents (20% and 24%, respectively). Wilmore and others from the University of Arizona assigned normal subjects to a placebo, to a β_1-adrenoceptor inhibitor (atenolol), and to a nonselective β-adrenoceptor inhibitor group (propranolol [231]). When they performed submaximal exercise (60% $\dot{V}O_2$max) and compared the blocking results to those from the placebo subjects, the heart rates of the two groups were significantly reduced (22% and 25%, respectively). In 1986, Hespel and coworkers from the University of Leuven administered a placebo, a β_1-adrenorecptor inhibitor (atenolol), or a blocker of β_2-adrenoceptors (ICI 118551) to normal subjects who performed graded exercise until exhaustion (87). The exercise heart rates for the placebo and the β_2-adrenoceptor inhibitor groups were similar whereas blocking the responses of the β_1-adrenoreceptors was associated with a significantly lowered exercise heart rate (28 beats/min). These selective findings support the concept that β_1-adrenoceptors are significant mediators of heart rate changes with exercise. The idea's importance is demonstrated in the recent study of Rohrer and coworkers of Stanford University who exercised mice with β_1-adrenoceptors that were genetically eliminated (knockout) and found that their heart rates with maximal exercise were ~200 beats/min or ~30% lower than recorded for their wild-type controls (172).

Blood pressures are usually measured when β-adrenoceptor blocking agents are administered to exercising subjects. In the 1965 study of Epstein and others, propranolol was associated with a significant reduction (15%) in mean arterial blood pressure with maximum exertion (48). Submaximal exercise by the subjects receiving either propranolol or an inhibitor of β_1-adrenoceptors (atenolol) in the Arizona studies was associated with significant reductions in systolic blood pressure (~16% for both groups) and no meaningful changes in diastolic blood pressure when compared to placebo conditions (231). Heavy exercise (90% $\dot{V}O_2$max) by Arizona subjects participating in a related experiment indicated systolic blood pressure was reduced with both inhibitors (14% and 12%, respectively) while diastolic pressures were not significantly changed (100). In the experiments of Hespel and coworkers, inhibition of β_1-adrenoceptors (atenolol) had a marked effect on lowering systolic blood pressure (21 mmHg) but little influence on diastolic blood pressure. Blocking the activity of the β_2-adrenoceptors (ICI 118551) had no significant effect on systolic blood pressure at any stage of the exercise protocol. However, diastolic blood pressure was significantly elevated (8 mmHg), which was attributed to a decrease in β_2-adrenoceptor-mediated vasodilation and to an α_2-adrenoceptor-initiated vasoconstriction response (87). These composite results endorse the idea that β_1-adrenoceptors will mediate systolic blood pressure changes during exercise while having minimal influences on the changes associated with diastolic pressure. However, this statement receives little support from the results with the β_1-adrenoceptor knockout mice because they exhibited minimal changes in mean blood pressure as the exercise intensity increased (172).

β-Adrenoceptor Inhibition and Myocardial Function

It was evident from the sympathetic ablation studies conducted with exercising animals that myocardial function was altered. Insights on the receptors initiating the changes began with the 1965 study conducted by Epstein and others (48), who reported that maximal exercise was associated with a significant reduction in left ventricle minute work (34%) after the administration of propranolol. Several years later, Ekstom-Jodal and coworkers at the University of Goteborg in Sweden blocked both β-adrenoceptor types (sotalol) in subjects performing mild-to-moderate supine exercise (47). They found significant reductions in coronary blood flow (31%), myocardial oxygen consumption (25%), and cardiac output (24%) with significant increases in coronary resistance (31%) and oxygen extraction (5%). Jorgenson and colleagues at the University of Minnesota in 1973 studied the influence of β-adrenoceptor blockage (propranolol) on subjects performing light-to-moderate power outputs on a cycle ergometer (97). They also noted significant reductions in myocardial oxygen consumption (34%) and for myocardial blood flow (36%) when compared to control exercise conditions.

Studies with instrumented animals have helped to differentiate the contributions of the β-adrenoceptors. In 1980, Heyndrickx and coworkers from the University of Ghent in Belgium found that in dogs doing moderate treadmill exercise after receiving propranolol, there were significant reductions in cardiac contractility (56%), mean coronary flow (34%), myocardial oxygen consumption (24%), and in the ratio of oxygen delivery to oxygen consumption (27%) when compared to control conditions (88). Mass and Gwirtz from the Texas College of Osteopathic Medicine in Fort Worth (131) exercised dogs on a treadmill who had received on different occasions either a placebo or inhibitors of β_1-adrenoceptors (atenolol), β_2-adrenoceptors (ICI 118551), or these two types (propranolol). With heavy exercise, ventricular contractility and shortening were significantly reduced by atenolol (33% and 12%) and by propranolol (17% and 15%), which indicated the β_1-adrenoceptors were primarily responsible for such changes. As for coronary blood flow velocity with heavy exercise, it was significantly decreased by each of the β-adrenoceptor inhibitors (11% to 22%) (131). In 1988, DiCarlo and colleagues from San Antonio, Texas, measured coronary blood flow velocity and coronary resistance after the administration of a β_2-adrenoceptor inhibitor (ICI 118551) to exercising dogs (39). They reported marked reductions in coronary blood flow (14%) and increases (35%) for late diastolic coronary resistance measures with moderate-to-heavy exercise. Their study also included α-adrenoceptor inhibition (phentolamine) and the results indicated that β_2- and α-adrenoceptors were mediating responses which significantly influenced coronary blood flow and coronary vasoconstrictor tone (see previous section entitled "α-Adrenoceptor Inhibition and Myocardial Responses"). While these studies supported the idea that β-adrenoceptors were mediating myocardial responses during exercise and that β_1-adrenoceptors were initiating the changes in contractility, they did not eliminate a vasodilator function that could occur because of metabolic influences.

The importance of β-adrenoceptors for myocardial collateral blood flow was illustrated by the 1995 exercise study with dogs conducted by Traverse and others at the University of Minnesota (211). They used microspheres to measure blood flows in select regions of the heart that included sites with established collateral blood vessels. Inhibition of β-adrenoceptoers (propranolol) was responsible for significant reductions in flow to noncollateral regions (29%) and to those areas in which collaterals were present (22%). Increases in transcollateral vascular resistance and in small-vessel resistance within the collateral region were responsible for the changes noted. However, no attempt was made to determine the contributions of the respective receptor types.

β-Adrenoceptor Inhibition and Peripheral Blood Flow

The promotion of the 1962 idea concerning sympatholysis (170) and the 1970 negative blood flow results from exercising dogs whose hindlimbs had been selectively sympathectomized (42) raised the issue of the role and importance of the sympathetic nervous system in regulating blood flow during exercise.

In 1978, Astrom and Huhlin-Dammfelt from Huddinge, Sweden, had subjects perform cycle ergometry at approximately 50% $\dot{V}O_2$max and reported that nonselective inhibition of β-adrenoceptors (propranolol) resulted in virtually unchanged leg muscle blood flow and vascular resistance values (8). During the same year, McSorley and Warren from the University of Southhampton in England (140) conducted exercise investigations used xenon to assess leg muscle blood flow while blocking β_1-adrenoceptors (metoprolol) or β_2-adrenoceptors (propranolol) and found significant reductions in both instances (74% and 42%, respectively). When the same methodology was employed in a cross-sectional study to measure forearm blood flow before and after blocking β_1- (atenolol) or both β-adrenoceptors (propranolol) with moderate hand-grip exercise (50 contractions per minute), there were no significant differences between the groups with respect to forearm blood flow or to vascular resistance (81). In 1992 when discussions on sympatholyis were becoming intensified, Pawelczyk and associates from the University of Copenhagen and the University of Texas Southwestern Medical Center in Dallas had subjects perform repeated bouts of cycling exercise at 40%–84% of $\dot{V}O_2$max (157). Once the β_1-adrenoceptor blocking agent was administered (metoprolol), leg blood flow at the highest exercise intensity was significantly reduced (11%) as was leg vascular conductance (11%) and cardiac output (19%). Under these conditions, mean blood pressure did not differ from control values. Since they demonstrated that sympathetic vasoconstriction was occurring at the higher exercise intenstity via the β_1-adrenoceptors when cardiac output was notably compromised, they felt that functional sympatholysis was present but "incomplete" in skeletal muscles (157).

Laughlin and Armstrong from the Universities of Missouri and Georgia, respectively, measured blood flow in 32 hindlimb muscles of the exercising rat using radiolabeled microspheres after the administration of propranolol (118). During low-to-moderate treadmill exercise, muscle blood flow initially declined (17%), after which it returned to levels associated with placebo exercise conditions, whereas with heavy exercise, muscle blood flow did not significantly increase. In fact, the authors comment that "muscle blood flow during high-intensity exercise is not significantly influenced by β-receptor blockage" (118, p. 1471). Recently (1997), Buckwalter and associates from the Veterans Affairs Medical Center and the Medical College of Wisconsin in Milwaukee measured iliac blood flow in instrumented dogs performing moderate treadmill exercise. The administration of propranolol to these dogs had no significant effect on iliac blood flow and one of the researcher's conclusions was that β-adrenergic receptors were not involved in the control of blood flow to skeletal muscles in moderate exercise by dogs (17).

Disappointingly, these β-adrenoceptors blocking studies do not resolve issues created by the sympatholysis idea nor do they effectively delineate their selective roles during the spectrum of exercise.

β-Adrenoceptor Inhibition and Respiratory Responses

One of the first to determine whether blocking β-adrenoceptors would alter respiratory responses of normal subjects during exercise was Pirnay and his Belgium coworkers in 1970 (161). They reported that pulmonary ventilation was significantly reduced (12%) with maximum exercise when propranolol was administered. Anderson and others 9 years later (4) studied subjects performing graded cycling until they were unable or unwilling to continue after administration of selective β_1-adrenoceptor inhibition (metoprolol) or nonselective β-adrenoceptor blockade (propranolol). Their results indicated that there were no significant reductions in pulmonary ventilation, tidal volume, or frequency with any of the blocking agents. In the early 1980s, Tesh and Kaiser in Sweden reported no significant decreases in pulmonary ventilation with maximal exercise after their subjects had received propranolol (208). McLeod and coworkers from Duke University inhibited β_1-adrenoceptors (atenolol) of normal subjects during light-to-moderate steady-state exercise and found notable declines in peak respiratory flow rates (13%) and tidal volume (15%) with significant increases in frequency (137). Significant but less pronounced changes were observed with propranolol. In the Arizona β-adrenoceptor blockade experiments led by Joyer and associates (101), submaximal exercise (60% $\dot{V}O_2max$) by nontrained subjects had limited influences on pulmonary ventilation or tidal volume measures when inhibitors of β-adrenoceptors were administered (atenolol, propranolol). However at 100% $\dot{V}O_2max$, both pulmonary ventilation and tidal volume were noticeably reduced (9% and 8%, respectively) with blocking of β_1-adrenoceptors, and by a higher percentage (~14%) when β_2-adrenoceptors were studied. Since the inhibitory effort on tidal volume was significantly less with the β_1-adrenoceptors, the researchers felt the difference was the contributions of the airway β_2-adrenoceptor-mediated bronchodilation. When frequency was plotted against changes in $\dot{V}CO_2$ production, both inhibitors caused an increased rate for a given level of CO_2 production at the higher exercise intensities. These studies from normal subjects have yielded inconsistent and conflicting results, and uncertainty remains as to their importance in an exercise response.

β-Adrenoceptor Inhibition and Changes in $\dot{V}O_2$ and Exercise Performance

One of the first to test the possibility in healthy humans that blocking the actions of β-adrenoceptors would alter oxygen consumption and exercise performance was Epstein and colleagues (48). In 1965 they found that propranolol would significantly reduce $\dot{V}O_2max$ (6%) and endurance run time (~40%). Pirnay and coworkers at the University of Liege in Belgium in 1970 administered propranolol to healthy subjects and observed that oxygen consumption was lower at each power output during graded treadmill exercise that reached −15% at maximal conditions (161). In the Sydney experiments (4), Anderson and associates noted that β_1-adrenoceptor inhibition (metoprolol) resulted in significant reductions in $\dot{V}O_2max$ (13%), power output (12%), and in total work (20%) when compared to the placebo findings. Related decreases occurred when nonselective β-adrenoceptor inhibition occurred (propranolol) (4). In 1981, Tesch and Kaiser (208) extended the observations of their Stockholm colleagues and noted that subjects receiving β-adrenoceptor blockers (propranolol) had significant reductions in $\dot{V}O_2max$ (14%), power output (8%), and performance time (48%). Joyner and his colleagues at Arizona noted several years later that nontrained subjects had significantly lower reduced $\dot{V}O_2max$ values (6%) and shorter treadmill run times (10%) after inhibition of β_1-adrenoceptors (100).

Significant decreases were also noted for both parameters (4% and 7%, respectively) when the effects of the administration of propranolol were contrasted with the placebo changes. Pawelczyk and colleagues in their 1992 study demonstrated that inhibition of β_1-adrenoceptors resulted in significant reductions in cardiac output (19%), leg blood flow (11%), and oxygen consumption (~10%) at ~85% $\dot{V}O_2$ max (157). These investigations clearly support the idea that blocking the activity of β-adrenoceptors will impair the capacity of the oxygen transport system and its role in exercise performance; however, as emphasized by Galbo et al. (65) and by Kaiser and others (103), decreases in exercise performance with blocking of β-adrenoceptors is more than a cardiorespiratory occurrence as changes in metabolic, hormonal, and muscle receptor density factors must be considered. Although extensively mentioned, but poorly explained, subjects receiving β-adrenoceptor inhibitors, and especially propranolol, frequently "blame" fatigued muscles for their poor performance. Hence, clarification of the uncertainties on this subject must remain for the research of the future.

Endurance Training: Inhibition of β-Adrenoceptors and Changes in Heart Rate, Cardiac Output, Oxygen Consumption, Pulmonary Ventilation, and Exercise Performance

The development, refinement, and utilization of pharmacological compounds to alter β-adrenoceptor-mediated cardiorespiratory responses was done for clinical and rehabilitation purposes with little concern as to whether their presence would affect the adaptations associated with endurance training by healthy subjects. Although chemical and surgical sympathectomy procedures with animals had demonstrated attenuated exercise responses, the results were unable to define the roles of the β-adrenoceptors in the process. In fact, they raised the question of whether subjects could be trained when the β-adrenoceptors were inhibited. One of the first to investigate this matter in humans was Michael Maksud and associates from Milwaukee, Wisconsin, who in 1972 evaluated the responses of normal subjects before and after training (127). They had subjects perform a progressive cycle ergometry test that led to exhaustion before and after administration of a β-adrenoceptor blocking agent (propranolol). When compared to the changes noted with placebo conditions, neither $\dot{V}O_2$ nor pulmonary ventilation nor endurance run time exhibited significant changes that indicated the exercise capacity of trained subjects was either "impaired or enhanced" when these receptors were inhibited (127). Nearly a decade passed before training studies were conducted that administered (β-adrenoceptor antagonists to their subjects. Sable and associates from the University of Colorado assigned subjects to two aerobic training groups that lasted 5 weeks (183). Compared to the group that received a placebo, and after removal of the drug, subjects receiving chronic administration of propranolol exhibited no significant differences with regard to heart rate, $\dot{V}O_2$max, or pulmonary ventilation, but they did demonstrate a significant increase in treadmill run time (8%). Not unexpectedly, the researchers concluded that β-adrenoceptor stimulation was essential for exercise conditioning to occur (183).

McLeod and coworkers from Duke University incorporated β_1-adrenoceptor inhibition (atenolol) into an 8 week exercise training design that included groups receiving a placebo or a nonselective blocker of adrenoceptors (propranolol) (138). Results from a treadmill work capacity test at the end of the study indicated all groups exhibited significant increases with the highest gain (18%) occurring with the subjects receiving atenolol and the lowest from those on propranolol (8%). Since the percentages increased when subjects were tested after removal of the inhibitors, the researchers attributed, in part, the improvement to the elimination of the "muscle fatiguing effects" associated with β-adrenoceptor inhibitors.

Evy and collaborators at the University of Arizona (54) evaluated the maximal effects of a 13 week training program with two experimental groups, one of which received a nonselective inhibitor of β-adrenoceptors (sotalol). Maximum heart rate was significantly reduced (20%) at the end of the experimental period that increased beyond the starting period when the drug was removed. Values for $\dot{V}O_2$max and pulmonary ventilation were significantly elevated (7% and 5%, respectively) only after removal of the blocking agent. However, treadmill run time was significantly longer after training regardless of the presence of the inhibitor (54). Later, these Arizona investigators had normal subjects assigned to groups receiving a placebo, a β_1-adrenoceptor inhibitor (atenolol), or propranolol while undertaking a 15 week aerobic training program (231). All groups had significant decreases in maximum heart rate (4% for the placebo group) and the two experimental groups demonstrated reductions of approximately 25%. The groups also showed significant increases in maximum values for oxygen consumption, pulmonary ventilation, and treadmill run time. When the drugs were removed and the groups were retested, the subjects that previously received the inhibitors continued to exhibit evidence for being trained (231). In Sweden at the Karolinska Institute, Svedenhag and associates conducted an 8 week experiment using a cycle ergometer to train two groups while administering either propranolol or a placebo throughout the study (205). Heart rate values showed no training effect but both groups experienced significant increases in $\dot{V}O_2$max (8% and 9%, respectively). Savin and associates at Stanford University in 1985 studied the effects of 6 weeks of endurance training using normal subjects in a design that contained administration of a placebo, a β_1-adrenoceptor inhibitor (atenolol), and propranolol (190). At the completion of the investigation, the two experimental groups showed reductions in peak heart rate (~9% and 14%, respectively) while all groups exhibited significant training effects for peak oxygen consumption (~17% with each group), peak pulmonary ventilation (~26%, 34%, and 10%, respectively), and work time (~28%, 23%, and 15%, respectively). After removal of the inhibitors, the heart rate means were similar to their pretraining value whereas the increases in the group means in oxygen consumption and pulmonary ventilation remained or became elevated (190).

Several years later, in 1989, Sweeney and collaborators from Atlanta, Georgia, conducted a 9 week training study with subjects receiving either a placebo, a β_1-adrenoceptor inhibitor (atenolol), or the nonselective propranolol (206). As expected, exercise heart rates were significantly reduced for all groups. Significant increases were reported for maximal oxygen consumption for the group receiving a placebo

(11%) but not for the other two groups. Pulmonary ventilation means were significantly elevated for all groups tested (~7% to 20%) when compared to baseline values. The oxygen consumption measurements recorded a week after the inhibitors had been eliminated (washout) indicated that the groups receiving inhibitors exhibited a training effect with significant increases ranging from 9% to 11%. The same trend occurred with the pulmonary ventilation results (206).

Submaximal exercise was performed in the Maksud and associates study and oxygen consumption and pulmonary ventilation results did not show evidence for a training effect (127). In the 1985 study of Wilmore and coworkers, they found at 60% $\dot{V}O_2$max their three groups (placebo, atenolol, propranolol) exhibited no effect of training with regard to oxygen consumption. Chronic exercise did significantly lower heart rate and pulmonary ventilation values that were more pronounced in the groups receiving either atenolol or propranolol and training while receiving β-adrenoceptor blocking agents did result in significantly lower cardiac output values (7% and 9%, respectively) when compared to their pretraining means (231). Submaximal exercise (120 W) by subjects in the Svedenhag study concerning oxygen consumption and cardiac output after receiving propranolol revealed that no significant changes occurred (205). However, an effect was observed with heart rate (−8%) at this exercise intensity. These composite findings provide credence to the concept that subjects receiving β-adrenoceptor inhibitors can be endurance trained and will exhibit significant cardiorespiratory changes. Inherent in the acceptance of this idea is that the training effects will either be present with maximal exercise or "masked" until the inhibitors are removed. It is not certain whether similar changes will be demonstrated with submaximal exercise because of the variable nature of the studies to date.

Endurance Training: Myocardial β-Adrenoceptor Characterization in Animals

The search for a cellular explanation for the bradycardia of training led Williams of Duke University to measure β-adrenoceptors from rats that trained for 8 weeks by swimming. Compared to the results obtained from nonswimmers, there were no significant group differences in β-adrenoceptor number or in their affinity values (229). Moore and colleagues from Washington State University in 1982 sought to explain changes in myocardial contractility by measuring β-adrenoceptor characteristics in ventricles obtained from rats trained by treadmill running for 15–26 weeks (144). Using methodology followed by Williams, they could find no evidence to indicate training had significantly changed receptor number or binding. While Moore and coworkers were reporting negative results, Slyvestre-Gervais and associates from the University of Laval in Canada found that ventricle membranes from treadmill trained rats (10 weeks) had significantly fewer β-adrenoceptors (21%) but no meaningful changes in binding capacity (207). Fell and others from the University of Louisville in Kentucky exercised rats for 10 weeks and found no statistical evidence that β-adrenoceptor number or dissociation constant had changed (57).

In 1987, Hammond from the University of California at San Diego exercised Yucatan mini swine on a treadmill for 10–19 weeks (77). In their experiment they measured receptor characteristics in the same tissue before and after the completion of the study. Training was associated with a significant down-regulation of β-adrenoceptor number in membranes from the right atrium (44%) with no significant change in their binding capacity. In contrast, chronic exercise had no important effect on left ventricle β-adrenoceptor number. (Affinity values were not reported.) Plouride and coworkers from Laval University in Quebec studied rats that were exercised on a treadmill for 10 weeks (162). They found ventricle tissue from the trained group had a β-adrenoceptor density value that was significantly lower (13%) than their nontrained counterparts. No group differences existed when dissociation constants were compared. They included measurement of receptors in high- or low-affinity states and reported that training was associated with a significant decrease in the number of receptors (17%) in the high-affinity state, which indicated the coupling between the β-adrenoceptor and the guanine stimulating binding protein decreased (162).

Nietro and coworkers from Madrid, Spain, in 1995 also reported that endurance training would significantly lower β-adrenoceptor density (30%) in ventricles obtained from rats (147). They showed that β-adrenoceptor adenylyl cyclase activity was depressed with training. In fact, they characterized training as having caused "diminished sensitivity" of the cardiac β-adrenoceptor adenylyl cyclase system, a conclusion that was the opposite of the one reached almost two decades earlier by Wyatt and associates based on swimming rats (234). Nietro and coworkers also reported affinity values and there were no results that indicated training had made a difference. Roth and coworkers from the University of Colorado at Bolder recently (1997) conducted an extensive training and aging study with rats whose age range was 7–25 months. There was no significant evidence with any age group that exercise training had altered β-adrenergic receptor density or binding affinity values obtained from their left ventricles. Adenylyl cyclase activity results indicated that training had no significant effect on the older rats but was associated with an increase in the G-protein-dependent cyclic AMP production in the myocardiums of the youngest animals (173). Thus, the animal investigations do not provide adequate support for the attractive idea that exercise training would result in the down-regulation of β-adrenoceptors. However, it is apparent that training has no meaningful effect on β-adrenoceptor binding.

Endurance Training: β-Adrenoceptor Characterization and Skeletal Muscles

Whether acute exercise would alter the characteristics of β-adrenoceptors in skeletal tissues was addressed by Buckenmeyer and others in 1990 at the University of Maryland in College Park. They found that running on a treadmill by rats did not significantly alter β-adrenoceptor number, affinity values, or adenylate cyclase activityl (17). Whether the same effect would occur with humans is unclear.

Martin and associates from Washington University in St. Louis investigated whether 12 weeks of chronic exercise would alter β-adrenoceptor characteristics in humans. They obtained biopsy samples from gastrocnemius and soleus muscles and reported that type I fibers had a higher density (300%) than type II fibers. After training, there were no significant changes in β-adrenoceptor density to indicate up-regulation had occurred even when adjustments were made for the pretraining fiber type composition (130).

Animal studies on this issue began well in advance of the Martin and cowork-ers study. In 1984, Williams and his Duke University colleagues swam or ran rats for training purposes (230). Only the runners exhibited significant increases in β-adrenoceptor density values that were measured in soleus (~30%) and gastrocene-mius muscles (~14%). Affinity results were similar for the nontrained and trained animals. The next year Fell and others from the University of Louisville in Kentucky trained rats on a treadmill and secured β-adrenoceptor density and affinity results from various muscles (soleus, plantaris, vastus lateralis, and gastrocenemius) of dif-ferent fiber types (57). They found no significant increases that could be attributed to chronic exercise. In a 1990 study by Buckenmyer and coworkers that was men-tioned previously, the 18 week training study resulted in significant increases in β-adrenoceptor density in type I (19%) and type IIA (25%) but not in type IIB fibers. No meaningful changes were found for the affinity measures (17). Several years later, Plourde and associates from Laval University in Quebec also reported that en-durance training would be associated with significant increases in β-adrenoceptors in type I (48%) and type IIA fibers, with limited changes noted for the type IIB fibers (163). Mazzeo and others from the University of Colorado conducted an aging and training study with rats and measured β-adrenoceptor density and affinity values in animals representing a spectrum of ages. There were no results from soleus muscles (type I) that exhibited an effect from training (133). However, Farrar and coworkers from the University of Texas at Austin reported that a 6 month training program by rats considered to be young at the start of the program significantly increased skele-tal muscle β-adrenoceptor density values. This trend did not occur with the older an-imals (56).

The idea that training can up- or down-regulate β-adrenoceptors in different anatomical sites depending upon the functional and metabolic demands of the tissue is an interesting and important one. However, from the skeletal muscle data studied over the years, the idea has limited support, at least with humans.

SELECT CARDIORESPIRATORY EFFECTS DURING EXERCISE FROM INHIBITION OF THE PARASYMPATHETIC NERVOUS SYSTEM

Overview of the Parasympathetic Nervous System

The cholinergic era began in 1899 when Langley and Dickinson in Cambridge, En-gland, demonstrated during sympathetic stimulation that the application of nicotine to the superior cervical ganglion of a cat would block preganglionic but not postganglionic transmission (117). Dixon showed at the turn of the century that the alkoloid muscarine, found in mushrooms, had physiological actions similar to stimulation of the vagus nerve. Several years later, Sir Henry Dale at Cambridge began studies on ergot (an alkoid) and choline esters and reported acetylcholine would initiate physiological changes that resembled those associated with stimulation of the vagus nerve (32, 33). Subsequently, Dale began extensive investigations on the physiological effects of nicotine. In 1921, Otto Loewi (1873–1961) at the University of Graz in Germany conducted the research (123) in which he named the substance released from the stimulation of the vagus nerve to be *Vagusstoff* (later identified as acetycholine) and which enabled him to share the 1936 Nobel Prize with Sir Henry Dale (174). Emerging from these and related investigations was the cholinergic concept, which indicated muscarinic and nicotinic receptors existed within the parasympathetic nervous system with nicotinic receptors being located at preganglionic synapses while muscarinic receptors are found in postganglionic neurons and are present in cardiac muscle, endocrine and exocrine glands, smooth muscle cells of the gastrointestinal system, and on bronchopulmonary structures. They can also be found in sympathetic ganglia cells, sweat glands, and within the brain (76).

While it is known that subtypes of muscarinic and nicotinic receptors have been cloned and biochemical actions modulated, in part, by G-proteins, the focus here is on muscarinic receptors. Matters pertaining to the muscarinic subtypes or to nicotinic receptors are left for future authors.

Inhibition of Muscarinic Receptors and Cardiac Responses

The alkoids of the belladonna plants are recognized as naturally occurring antimuscarinic drugs and have been used since ancient times by physicians for a myriad of disorders (93). One example of a belladonna drug, atropine, was purified in 1831 and used as early as 1867 to block the cardiac effects of vagal stimulation by Bezold and Bloebaum (93) and, as mentioned previously, was used by Gasser and Meek in 1914 with exercising dogs to study the role of the parasympathetic nervous system in cardiac acceleration (67). Craig from the Army Medical Center in Maryland during 1952 was among the first to have subjects exercise after receiving atropine (31). They performed light exercise on a treadmill and recorded heart rates that were 38% higher than without the inhibitor. A year later, Sid Robinson (1903–1981) and colleagues from Indiana University in Bloomington administered atropine sulfate (intramuscular) to subjects who performed moderate, heavy, and exhaustive treadmill exercise and reported notable increases in heart rate with the elevated intensity levels (17% and 9%, respectively), but not with exhaustive exercise when compared to control conditions (171).

In the 1960s, Kahler and others at the National Heart Institute in Maryland studied subjects who had received intravenous injections of atropine when they performed light supine exercise on a cycle ergometer (102). Significant increases were noted for heart rate (14%) that were associated with reductions in stroke volume

(16%) and left ventricle stroke work (20%). No significant changes were noted for cardiac index, mean arterial blood pressure, and left ventricle minute work (102). This same laboratory several years later, with Brian Robinson leading the investigation, reported results from four subjects who had received infusions of atropine while exercising on a treadmill, and similar to Kahler and others (102), they found heart rate to increase (19%), which, not unexpectedly, was attributed to vagal withdrawal (170). The withdrawal effect was best demonstrated by Ekblom and associates in Stockholm when they reported heart rates after intravenous administration of atropine and noted the differential was 28 beats/min at 25% $\dot{V}O_2$max, 18 beats/min at 50% $\dot{V}O_2$max, 12 beats/min at 75% $\dot{V}O_2$max, and essentially no difference with maximal exercise when compared to control conditions (46).

Atropine was administered to exercising β_1-adrenoceptor gene knockout mice and the percentage of heart rate increase (~30%) was similar to the result found with the wild-type controls although the absolute mean value was markedly lower. These findings indicated that in "both genotypes, vagal withdrawal accounts for ~50% of the tachycardic response to exercise" (172, p. H1189).

Muscarinic Receptor Inhibition and Changes in Oxygen Consumption and Pulmonary Ventilation

One of the purposes of the Robinson and associates 1953 experiment was to determine whether exercise $\dot{V}O_2$ would be altered after injections of atropine. They found $\dot{V}O_2$ measures did not exhibit significant reductions (10%) until exhaustive strenuous exercise was performed (171). Negative results were noted reported by Kahler and associates when their subjects performed light supine exercise after receiving atropine (102). In contrast, $\dot{V}O_2$ was reduced (9%) after the administration of atropine when moderate exercise was performed on a treadmill by a small number of subjects in the Brian Robinsion study (170). Ekblom and Stockholm associates concluded, from their systematic study of submaximal and maximal exercise after providing various pharmacological agents to subjects which included atropine, that muscarinic receptor inhibition had no significant influence on oxygen consumption, pulmonary ventilation, or work time (46). Their conclusion is likely the appropriate "take home" idea on this subject.

Muscarinic Receptors and Training

In 1921, Herxheimer from Berlin, Germany, showed with resting subjects that the elevation in heart rate that occurs after the administration of atropine was attenuated when trained "sportsmen" were tested (86). This finding raised the possibility that endurance training had altered the characteristics of muscarinic receptors. Decades later, Williams from Duke University measured the number of binding sites and the binding characteristics of receptors secured from cardiac membranes obtained from nontrained and trained rats (229). They reported there were "no alterations in the properties of cardiac muscarinic–cholinergic receptors induced by physical conditioning." Muscarinic receptor number in tissue obtained from the right atrium of nontrained and trained pigs by Hammond and others showed no significant differences between the groups (77). Edwards and associates at the University of Arizona exercised rats deprived of anterior pituitary hormones for 8 weeks and measured muscarinic receptor number and binding affinity in atrial and ventricular tissue (45). Training was associated with a significant decrease in the number of receptors (33%) in the right atrium with no meaningful effect in the left atrium or both ventricles or affinity values. Recently (1998), Chen and Liao from Taiwan trained rats for 10 weeks before removing the thoracic aorta to assess muscarinic receptors located within the endothelium. They found the muscarinic receptor subtype that enhanced acetylcholine-induced vasodilation did not exhibit significant changes in receptor number or in high- or low-affinity values (28). Consequently, upregulation of muscarinic receptors was discarded as a viable mechanism. Although the information is sparse, there is minimal support for the idea that endurance training will induce significant changes in the characteristics of muscarinic receptors.

CARDIORESPIRATORY RESPONSES AND EXERCISE PERFORMANCE WITH THE SIMULTANEOUS INHIBITION OF β-ADRENOCEPTOR AND MUSCARINIC RECEPTORS

Overview and the Intrinsic Heart Rate

The availability of adrenergic blocking agents in the 1960s and notably the β-adrenoceptor inhibitors enabled investigators to include atropine to conduct human experiments that simulated the surgical denervation studies undertaken earlier with animals. In a 1962 study that preceded the use of propranolol as a nonselective blocker of β-adrenoceptors, Kahler and others at the National Heart Institute combined the injection of atropine with the chronic administration of quanethidine to evaluate select physiological responses of subjects performing light exercise on a cycle ergometer in the supine position (102). Compared to control conditions, significant reductions were noted for cardiac output (13%), stroke volume (21%), mean arterial blood pressure (17%), left ventricle stroke work (34%), and left ventricle minute work (28%) whereas significant elevations in heart rate occurred (9%). Since these changes were similar to those recorded with quanethidine alone, they felt their results had demonstrated the importance of the sympathetic nervous system for submaximal exercise response (102). Later, but in the same laboratory, Brian Robinson and coworkers conducted an experiment with four subjects performing moderate exercise after receiving atropine and propranolol and reported that heart rate was slightly reduced (4%) while mean arterial blood pressure was lowered by 17% (170). They combined their laboratory data with those published by Sid Robinson and associates (171) and by Epstein and associates (48) to present a graphic concept of cardiac acceleration during exercise which indicated that parasympathetic nervous system withdrawal progressively decreased while sympathetic nervous system activation progressively increased (170).

At the time when Brian Robinson was presenting the exercise profile of cardiac acceleration and vagal withdrawal, Anthony Jose from Sydney, Australia, introduced to the scientific community the concept of an intrinsic heart rate that evolved from cardiac studies with resting patients receiving inhibitors of muscarinic and β-adrenoceptors (98). It was proposed to be a valuable index for assessing "intrinsic cardiac function" especially for disease states. The mean value from 311 normal individuals aged 20–70 years was 104 beats/min. The following year, John Sutton and colleagues from Sydney reported that exercising healthy students on a treadmill (intensity not specified) caused a mean rise of 20 beats/min (204). Later, Jose and coworkers had healthy subjects perform supine light–moderate exercise on a bicycle ergometer and compared their double blockage results to control conditions. Heart rate was significantly lower (7%) (99) and no significant changes were recorded for oxygen consumption, cardiac output, or mean arterial pressure. However, stroke volume was significantly increased (7%). Changes in blood temperature could not explain the increase in the intrinsic heart rate from resting and its use during exercise to assess cardiac function in normal individuals had no apparent value. In the early 1970s, Nordenfelt had subjects perform moderate and moderately heavy exercise on a cycle ergometer (148). With moderate power output and doubleblockade (atropine and propranolol) he found significant decreases in heart rate (8%), cardiac output (20%), stroke volume (11%), mean arterial pressure (10%), and in mean systolic ejection rate (19%) with an increase in total peripheral resistance (18%). Nordenfelt extended these studies by measuring their effects on heart rate and muscle blood flow (xenon-clearance method) and reported significant reductions in heart rate (17%) and in blood flow (14%). Because his previous study observed a reduction in mean arterial blood pressure, he felt a decrease in perfusion pressure rather than increased vasodilation was responsible for the change (149).

In Stockholm, Ekblom and coworkers had subjects perform submaximal and maximal exercise after double blockade (atropine and propranolol) and found at maximal conditions heart rate was significantly reduced (~18%), as were $\dot{V}O_2$max (6%) and work time (24%) (46). In a small sample of subjects, cardiac output was also significantly lower (15%) when β-adrenoceptors and muscarinic receptors were inhibited. Using a cross-sectional design to examine the difference between nontrained and trained subjects with regard to the intrinsic heart rate and select cardiorespiratory responses, Lewis and Scandinavian coworkers recorded submaximal and maximal responses of subjects on a bicycle ergometer without or with the administration of atropine and propranolol, or with atropine and an inhibitor of β_1-adrenoceptors (metoprolol) (121). With the nontrained subjects performing submaximal exercise (50% $\dot{V}O_2$max), double blockage with atropine and propranolol was associated with significant reductions in heart rate (6%), $\dot{V}O_2$ (7%), systolic blood pressure (~17%), and pulmonary ventilation (13%). Similar results were noted when only the β_1-adrenoceptors were blocked. With maximal exercise after atropine and propranolol were injected, heart rate was significantly reduced (24%), as were $\dot{V}O_2$ (13%) systolic blood pressure (19%), and pulmonary ventilation (22%). Again, similar changes were noted when atropine was injected with metoprolol. Negative iliac blood-flow results from a recent animal study (1997) which used atropine and propranolol to block muscarinic and β-adrenoceptors, respectively, led Buckwalter and associates to conclude that these receptors were not involved in the control of blood flow during moderate exercise (in dogs) (18).

When Lewis and coworkers focused on the intrinsic heart rate change, the mean value increased notably with each change in intensity level, presumably because of unspecified nonautonomic factors (121). Although no attempt was made to use the intrinsic heart rate value to assess cardiac function, there was a low but significant negative correlation coefficient (–0.43) between the resting and the maximal intrinsic heart rate value. This study, when combined with previous ones, has reinforced the emerging belief that the intrinsic heart rate idea has limited value in explaining cardiorespiratory functions for exercising healthy subjects. In addition, numerous double blockade studies conducted since the time of Brian Robinson and others (170) have confirmed their heart-rate concepts for progressive exercise and indicate that attenuated maximal cardiorespiratory responses can be obtained by inhibition of β-adrenoceptors alone.

Lewis and associates suggested the intrinsic heart rate of their elite subjects exhibited nonautonomic components, but they failed to specify what they were. To be meaningful, further testing is need using an experimental design that includes nontrained subjects participating in a longitudinal study.

CLOSING COMMENTS

Scientific writers on the history of the autonomic nervous system usually start with the observations of Galen, but this is not the situation when the effects of exercise are considered because the findings are essentially contemporary in nature. Few, if any, historians of physiology would consider Walter Cannon to be an exercise physiologist. However, his investigations during the early decades of the twentieth century pertaining to the functions of the sympathetic nervous system and the responses of the adrenal gland during emergency conditions were the foundations followed by early authors to describe and to explain autonomic responses during exercise. The Cannon era prevailed well beyond the middle of the twentieth century, and its effects continue to be present.

To understand the contributions of the sympathetic nervous system in an exercise response, it was necessary to identify the primary transmitter being released from sympathetic nerve endings and to perfect the measurement of epinephrine being released from the adrenal gland. After von Euler had demonstrated that norepinephrine was the elusive transmitter being sought by Cannon and associates, this finding provided the impetus for an era of characterization that allowed investigators to verify sympathetic nervous system involvement and to indicate potential mechanisms. The peak activity of this era was during the 1970s and 1980s.

Insights into the selectivity of the sympathetic nervous system during an exercise response emerged during the characterization era with the introduction of the norepinephrine spillover technique and concept by Elser and associates. They effec-tively demonstrated in exercising healthy subjects that the sympathetic nervous sys-tem outflow being directed to different organs was variable in nature. This result, when combined with microneurographic recordings of sympathetic nervous activity to skeletal muscles and skin, has indicated that the concept of instantaneous, simul-taneous, and uniformly distributed sympathetic outflow to all tissues—the concept of Cannon—must be modified.

Surgical ablation of anatomical structures is a time-honored procedure for ex-perimental physiologists and it is no surprise that denervation and animal exercise experiments were conducted before 1900 to test select ideas of various investigators —for example, the functions of sympathetic and parasympathetic nerves. Although denervation and exercise studies were conducted during the Cannon era, the zenith for such studies was during the 1970s and 1980s—when Donald and Mayo col-leagues were extremely productive. It was evident from surgical and chemical abla-tion experiments that sympathectomized animals can exercise (an original concern of Cannon) and are able to exhibit select adaptations associated with endurance training. However, ablation studies have been unable to resolve the continuing con-troversy concerning the existence and importance of exercise sympatholysis.

Exactly how the autonomic nervous system integrates a central command and multiple afferent inputs to elicit efferent responses continues to intrigue physiolo-gists. It is evident that vagal withdrawal occurs early in the process with a delay in the time course for sympathetic activation. The degree and magnitude of the activa-tion events will be influenced by the nature of the sensory input, with those arising from skeletal muscles having a profound effect. Of the many reflex findings over the years, the resetting of the baroreflex to a higher level during exercise is likely the most important one.

Studying physiological responses after inhibiting (blocking) the actions of re-ceptors is also a time-honored procedure used by experimental physiologists. The re-finement of receptor classification systems has enabled pharmacologists to develop selective compounds that effectively inhibit muscarinic and adrenergic adrenocep-tors. In general, history has shown that cardiovascular responses in humans are at-tenuated during moderate-to-heavy exercise with inhibition of β-adrenoceptors. A similar generalization cannot be made for the exercise responses of the respiratory system exercise as consistent findings were not present in the studies mentioned. History has also shown the chronic administration of β-adrenoceptor inhibitors will not prevent training effects from occurring. Other than changes in heart rate, it is difficult to generalize on the other exercise results that would be obtained when muscarinic receptors are inhibited during exercise. However, this situation could change when exercise studies are initiated with selective inhibitors for the various subtypes that have been identified. It appears that future exercise studies with healthy subjects on the physiological responses after blockade of muscarinic and β-adrenoceptors (double blockade) are not needed as the results will be similar if only β-adrenoceptors are inhibited. Despite the extensive number of animal studies over the years on whether endurance training will result in the up- or downregulation of muscarinic and β-adrenergic receptors, there is no unequivocal historical "take home" message for the reader. However, it is evident that binding affinities for these receptors are not influenced by training.

Lastly, the chapter provides minimal information on the ideas associated with the exciting and promising era of autonomic genomics and leaves this responsibility for future authors.

REFERENCES

1. Ahlquist R. P. A study of the adrenotropic receptors. *Am. J. Physiol.* 154:586–600, 1948.
2. Ahlquist R. P. Effects of the autonomic drugs on the circulatory system. In: *Handbook of Physiology, Section 2: Circulation, Vol. III*, edited by W. F. Hamilton and P. Dows. Washington, DC: American Physiological Society, 1965, pp. 2457–2475.
3. Alam M. and F. H. Smirk. Observations in man upon a blood pressor raising reflex arising from the voluntary muscles. *J. Physiol. (Lond.)* 89:372–383, 1937.
4. Anderson S. D., P. T. P. Bye, C. P. Perry, G. P. Hamor, G. Theobald, and G. Nyberg. Limitation of work performance in normal adult males in the presence of beta-adrenergic blocakade. *Aust. N. Z. J. Med.* 9:515–520, 1979.
5. Ashkar E. Heart rate and blood pressure during exercise in dogs with autonomic denervation. *Am. J. Physiol.* 210:950–952, 1966.
6. Ashkar E., J. J. Stevens, and B. A. Houssay. Role of the sympathicoadrenal system in the hemodynamic response to exercise in dogs. *Am. J. Physiol.* 214:22–27, 1968.
7. Ashkar E. and W. F. Hamilton. Cardiovascular response to graded exercise in the sympathectomized-vagotomized dog. *Am. J. Physiol.* 204:291–296, 1963.
8. Astrom H. and A. Juhlin-Dannfelt. Effect of beta-blockade on leg blood flow and lactate release in exercising man. *Acta Med. Scand.* 625 (Suppl. 44):44–48, 1978.
9. Bacq Z. M. The metabolism of adrenaline. *Pharmacol. Rev.* 1:1–26, 1949.
10. Bacq Z. M., L. Brouha, and C. Heymans. Recherches sur la physiologie et la pharmacologie du systeme nerveux autonome. *Arch. Int. Pharmacodyn.* 48:429–456, 1934.
11. Benison S., A. C. Barger, and E. L. Wolfe. *Walter B. Cannon: The Life and Times of a Young Scientist.* Cambridge, MA: Harvard University Press, 1987.
12. Bevegard B. S. and J. T. Shepherd. Circulatory effects of stimulating the carotid arterial stretch receptors in man at rest and during exercise. *J. Clin. Invest.* 45:132–142, 1966.
13. Blomqvist C. G., S. F. Lewis, W. F. Taylor, and R. M. Graham. Similarity of the hemodynamic response to static and dynamic exercise of small muscle groups. *Cir. Res.* 48(Suppl. Part II):I87–I92, 1981.
14. Bowen W. P. Changes in heart-rate, blood pressure, and duration of systole resulting from bicycling. *Am. J. Physiol.* 11:59–77, 1904.
15. Brouha L., W. R. Cannon, and D. B. Dill. The heart rate of the sympathectomized dog in rest and exercise. *J. Physiol. (Lond.)* 87:345–359, 1936.
16. Brouha L., S. J. G. Nowak, and D. B. Dill. The role of the vagus in the cardio-accelerator action of muscular exercise and emotion in sympathectomized dogs. *J. Physiol. (Lond.)* 95:454–463, 1939.
17. Buckenmeyer P. J., A. H. Goldfarb, J. S. Partilla, J. S. Pineyro, and E. M. Dax. Endurance training, not acute exercise, differently alters β-receptors and cyclase in skeletal fiber types. *Am. J. Physiol.* (Endocrinol. Metab. 21) 258:E71–E77, 1990.
18. Buckwalter J. B., P. J. Mueller, and P. S. Clifford. Autonomic control of skeletal muscle vasodilation during exercise. *J. Appl. Physiol.* 83:2037–2042, 1997.

19. Buskirk E. R. and C. M. Tipton. Exercise physiology. In: *The History of Exercise and Sport Science,* edited by J. D. Massengale and R. A. Swanson. Champaign, IL: Human Kinetics, 1997, 357–438.

20. Campos F. A. de M., W. B. Cannon., H. Lundin, and T. T. Walker. Some conditions affecting the capacity for prolonged muscular work. *Am. J. Physiol.* 87:680–701, 1929.

21. Cannon W. B. Organization for physiological homeostasis. *Physiol. Revs.* 9:399–431, 1929.

22. Cannon W. B. Stresses and strains of homeostasis. *Am. J. Med. Sci.* 189:1–14, 1935.

23. Cannon W. B. The emergency function of the adrenal medulla in pain and the major emotions. *Am. J. Physiol.* 33:356–372, 1914.

24. Cannon W. B. and Z. M. Bacq. Studies on the conditions of activity in endocrine organs. XXVI. A hormone produced by sympathetic action on smooth muscle. *Am. J. Physiol.* 96:392–412, 1931.

25. Cannon W. B. and D. de la Paz. Emotional stimulation of adrenal secretion. *Am. J. Physiol.* 28:64–70, 1911.

26. Cannon W. B. and L. B. Nice. The effect of adrenal secretion on muscular fatigue. *Am. J. Physiol.* 32:44–60, 1913.

27. Cannon W. B. and A. Rosenblueth. Studies on conditions of activity in endocrine organs. XXIX. Sympathin E and sympathin I. *Am. J. Physiol.* 104:557–574, 1933.

28. Chen H. and Y.-L. Liao. Effects of chronic exercise on muscarinic receptor-mediated vasodilation in rats. *Chinese J. Physiol.* 41:161–166, 1998.

29. Cooper T., J. W. Gilbert, Jr., R. D. Bloodwell, and J. R. Crout. Chronic extrinsic cardiac denervation by regional neural ablation. *Cir. Res.* 9:275–281, 1961.

30. Coote J. H., S. M. Hilton, and J. F. Perez-Gonzalez. The reflex nature of the pressor response to muscular exercise. *J. Physiol. (Lond.)* 215:789–804, 1971.

31. Craig F. N. Effects of atropine, work and heat on heart rate and sweat production in man. *J. Appl. Physiol.* 4:826–833, 1952.

32. Dale H. H. On some physiological actions of ergot. *J. Physiol. (Lond.)* 34:163–206, 1906.

33. Dale H. H. The action of certain esters and ethers of choline, and their relationship to muscarine. *J. Pharmacol. Exp. Ther.* 6:174–190, 1914.

34. Daskalopoulos D. A., J. T. Shepherd, and S. C. Wagenbach. Cardiopulmonary reflexes and blood pressure in exercising sinoaortic-denervated dogs. *J. Appl. Physiol.* 57:1417–1421, 1984.

35. Davies C. T. M., J. Few, K. G. Foster, and A. J. Sargeant. Plasma catecholamine concentration during dynamic exercise involving different muscle groups. *Eur. J. Appl. Physiol.* 32:195–206, 1974.

36. DiCarlo S. E. and V. S. Bishop. Exercise training attenuates baroreflex regulation of nerve activity in rabbits. *Am. J. Physiol.* 255 (Heart Circ. Physiol. 24):H974–H979, 1988.

37. Dicarlo S. E. and V. S. Bishop. Onset of exercise shifts operating point of arterial baroreflex to higher pressures. *Am. J. Physiol.* 262 (Heart Circ. Physiol. 31):H303–H307, 1992.

38. DiCarlo S. E. and V. S. Bishop. Regional vascular resistance during exercise: role of cardiac afferents and exercise training. *Am. J. Physiol.* 258 (Heart Circ. Physiol. 27):H842–H847, 1990.

39. DiCarlo S. E., R. W. Blair, V. S. Bishop, and H. L. Stone. Role of β_2-adrenergic receptors on coronary resistance during exercise. *J. Appl. Physiol.* 64:2287–2293, 1988.

40. DiCarlo S. E., L. K. Stahl, and V. S. Bishop. Daily exercise attenuates the sympathetic nerve response to exercise by enhancing cardiac afferents. *Am. J. Physiol.* 273 (Heart Circ. Physiol. 42):H1606–H1610, 1997.

41. Donald D. E., S. E. Milburn, and J. T. Shepherd. Effect of cardiac denervation on the maximum capacity for exercise in the racing greyhound. *J. Appl. Physiol.* 19:849–852, 1964.

42. Donald D. E., D. J. Rowlands, and D. A. Ferguson. Similarity of blood flow in the normal and the sympathectomized dog hind limb during graded exercise. *Cir. Res.* 26:185–199, 1970.

43. Donald D. E. and J. T. Shepherd. Response to exercise in dogs with cardiac denervation. *Am. J. Physiol.* 205:393–400, 1963.

44. Eckberg D. L. High- and low-pressure baroreflexes. In: *Primer on the Autonomic Nervous System*, edited by D. Robertson, P. A. Low, and R. J. Polinsky. San Diego, CA: Academic Press, 1996, pp. 59–65.

45. Edwards J. G., D. D. Lund, T. G. Bedford, C. M. Tipton, R. D. Matthes, and P. G. Schmid. Metabolic and cardiovascular adaptations in trained hypophysectomized rats. *J. Appl. Physiol.* 53:448–454, 1982.

46. Ekblom B., A. N. Goldbarg, A. Kilbom, A. and P.-O. Astrand. Effect of atropine and propranolol on the oxygen transport system during exercise in man. *Scand. J. Clin. Lab. Invest.* 30:35–42, 1972.

47. Ekstrom-Jodal B., E. Haggendal, R. Malmberg, and N. Svedmyr. The effect of adrenergic-receptor blockade on coronary circulation in man during work. *Acta Med. Scand.* 10:245–248, 1972.

48. Epstein S. E., B. H. Robinson, R. L. Kahler, and E. Braunwald. Effect of beta-adrenergic blockade on the cardiac response to maximal and submaximal exercise in man. *J. Clin. Invest.* 44:1745–1753, 1965.

49. Ernst S. B., W. J. Mullin, R. E. Herrick, and K.M. Baldwin. Exercise and cardiac performance capacity in rats with partial sympathectomy. *J. Appl. Physiol.* 53:242–246, 1982.

50. Esler M., G. Jennings, G. Lambert, I. Meredith, M. Horne, and G. Eisenhofer. Overflow of catecholamine neurotransmitters to the circulation: Source, fate and functions. *Physiol. Rev.* 70:963–985, 1990.

51. Essex H. E., J. F. Herrick, E. J. Baldes, and F. C. Mann. Effects of exercise on the coronary blood flow, heart rate and blood pressure of trained dogs with denervated and partially denervated hearts. *Am. J. Physiol.* 138:687–697, 1943.

52. Euler U. S. von. *Noradrenaline: Chemistry, Physiology, Pharmacology, and Clinical Aspects.* Springfield, IL: Charles C. Thomas, 1956.

53. Euler U.S. von and S. Hellner. Excretion of noradrenaline and adrenaline in muscular work. *Acta Physiol. Scand.* 26:183–191, 1952.

54. Ewy G. A., J. H. Wilmore, A. R. Morton, P. R. Stanforth, S. H. Constable, M. J. Buono, K. A. Conrad, H. Miller, and C. F. Gatewood. The effect of beta-adrenergic blockade on obtaining a trained exercise state. *J. Cardiac Rehab.* 3:25–29, 1983.

55. Faris I. B., G. G. Jamieson, and J. Ludbrook. Effect of exercise on gain of the carotid-sinus reflex in rabbits. *Clin. Sci.* 63:115–119, 1982.

56. Farrar R. P., K. A. Monnin, D. E. Frodyce, and T. J. Walters. Uncoupling of changes in skeletal muscle β-adrenergic receptor density and aerobic capacity during the aging process. *Aging* 9:153–158, 1997.

57. Fell R. D., F. H. Lizzo, P. Chervoni, and D. L. Crandall. Effect of contractile activity on rat skeletal muscle β-adrenoceptor properties. *Proc. Soc. Expl. Biol. Med.* 180:527–532, 1985.

58. Fernandes A., H. Galbo, M. Kjaer, J. H. Mitchell, N. H. Secher, and S. N. Thomas. Cardiovascular and ventilatory responses to dynamic exercise during epidural anaesthesia in man. *J. Physiol. (Lond.)* 420:281–293, 1990.

59. Freund P. R., S. F. Hobbs, and L. B. Rowell. Cardiovascular responses to muscle ischemia in man-dependency on muscle mass. *J. Appl. Physiol.* 45:762–767, 1978.

60. Freund P. R., L. B. Rowell, T. M. Murphy, S. F. Hobbs, and S. B. Butler. Blockade of the response to muscle ischemia by sensory nerve block in man. *Am. J. Physiol.* 237 (Heart Circ. Physiol. 6):H433–H439, 1979.

61. Freyschuss U. Elicitation of heart rate and blood pressure increase on muscle contraction. *J. Appl. Physiol.* 28:758–761, 1970.

62. Friendenthal H. Ueber die Entfornung der extracardialen Herznerven bei Saugethieren. *Archiv fur Physiologie* 1:135–145, 1906.

63. Galbo H. *Hormonal and Metabolic Adaptation to Exercise.* Stuttgart: Georg Thieme Verlag, Stuttgart, 1983.

64. Galbo H, J. J. Host, and N. J. Christensen. Glucagon and plasma catecholamine responses to graded and prolonged exercise in man. *J. Appl. Physiol.* 38:70–76, 1975.

65. Galbo H., J. J. Host, N. J. Christensen, and J. Hilsted. Glucagon and plasma catecholamines during beta-receptor blockade in exercising man. *J. Appl. Physiol.* 40: 855–863, 1976.

66. Gaskell W. H. On the structure, distribution and function of the nerves which innervate the visceral and vascular systems. *J. Physiol. (Lond.)* 1–80, 1886.

67. Gasser H. S. and W. J. Meek. A study of the mechanisms by which muscular exercise produces acceleration of the heart. *Am. J. Physiol.* 34:48–71, 1914.

68. Goodwin G. M., D. I. McCloskey, and J.H. Mitchell. Cardiovascular and respiratory responses to changes in central command during isometric exercise at constant muscle tension. *J. Physiol. (Lond.)* 236:173–190, 1972.

69. Gray I. and W. P. Beetham, Jr. Changes in plasma concentration of epinephrine and norepinephrine with muscular work. *Proc. Soc. Exp. Biol. Med.* 96:636–638, 1957.

70. Gregg D. E., E. M. Khouri, D. E. Donald, H. S. Lowensohn, and S. Pasyk. Coronary circulation in the conscious dog with cardiac neural ablation. *Cir. Res.* 31:129–144, 1972.

71. Gwirtz P. A., H. J. Mass, J. R. Strader, and C. E. Jones. Coronary and cardiac responses to exercise after chronic ventricular sympathectomy. *Med. Sci. Sports Exerc.* 20:126–135, 1988.

72. Gwirtz P. A. and H. L. Stone. Coronary vascular response to adrenergic stimulation in exercise-conditioned dogs. *J. Appl. Physiol.* 57:315–320, 1984.

73. Gwirtz P. A. and H. L. Stone. Coronary blood flow and myocardial oxygen consumption after alpha adrenergic blockade during submaximal exercise. *J. Pharmacol. Exp. Ther.* 217:92–98, 1981.

74. Haggendal J., H. Hartley, and B. Saltin. Arterial noradrenaline concentration during exercise in relation to the relative work levels. *Scand. J. Clin. Lab. Invest.* 26:337–342, 1970.

75. Hales J. R. S. and J. Ludbrook. Baroreflex participation in redistribution of cardiac output at onset of exercise. *J. Appl. Physiol.* 64:627–634, 1988.

76. Hamill R. W. Peripheral autonomic nervous system. In: *Primer on the Autonomic Nervous System,* edited by D. Robertson, P. A. Low, and R. J. Polinsky. San Diego: CA: Academic Press, 1996, pp. 12–25.

77. Hammond H. K., F. C. White, L. L. Brunton, and J. C. Longhurst. Association of decreased myocardial β-receptors and chronotropic response to isoproterenol and exercise in pigs following chronic dynamic exercise. *Cir. Res.* 60:720–726, 1987.

78. Hansen J., G. D. Thomas, S. A. Harris, W. J. Parsons, and R. G. Victor. Differential sympathetic neural control of oxygenation in resting and exercising human skeletal muscle. *J. Clin. Invest.* 98:584–596, 1996.

79. Hansen J., G. D. Thomas, T. N. Jacobsen, and R. G. Victor. Muscle metaboreflex triggers parallel sympathetic activation in exercising and resting human skeletal muscle. *Am. J. Physiol.* 266 (Heart Circ. Physiol. 35):H2508–H2514, 1994.

80. Hartley L. H., J. W. Mason, R. P. Hogan, L. G. Jones, T. A. Kotchen, E. H. Mougey, F. E. Wherry, L. L. Pennington, and P. T. Ricketts. Multiple hormonal responses to graded exercise in relation to physical training. *J. Appl. Physiol.* 33:602–606, 1972.

81. Hartling O. J., I. Noer, T. L. Svendsen, J. P. Clausen, and J. Trap-Jensen. Selective and nonselective β-adrenorecptor blockade in the human foreman. *Clin. Sci.* 58:279–286, 1980.

82. Hartman F. A., R. H. Waite, and H. A. McCordock. The liberation of epinephrin during muscular exercise. *Am. J. Physiol.* 62:225–241, 1922.

83. Hasking G. J., M. D. Esler, G. L. Jennings, E. Dewar, and G. Lambert. Norepinephrine spillover to plasma during steady-state supine bicycle exercise. *Circulation* 78: 516–521, 1988.

84. Hering H. E. Der Karotisdruckversuch. *Munch. Med. Wochschr.* 70:1287–1290, 1923.

85. Hering H. E. Ueber die Beziehung der extracardialen Herznerven zur Steigerung der Herzchlagzahl dei Muskelthatigkeit. *Archiv fur die Gesammte Physiologie* 40:429–492, 1895.

86. Herxheimer H. Zur Bradykardie der Sportsleute. *Muench. Med. Wochschr.* 68:1515–1518, 1921.

87. Hespel P., P. Lijnen, L. Vanhees, R. Fagard, and A. Amery. β-Adrenoceptors and the regulation of blood pressure and plasma renin during exercise. *J. Appl. Physiol.* 60:108–113, 1986.

88. Heyndrickx G. R., J.-L. Pannier, P. Muylaert, C. Mabilde, and I. Leusen. Alteration in myocardial oxygen balance during exercise after β-adrenergic blockade in dogs. *J. Appl. Physiol.* 49:28–33, 1980.

89. Heyndrickx G. R., J-P. Vilaine, E. J. Moerman, and I. Leusen. Role of prejunctional α_2-adrenergic receptors in the regulation of myocardial performance during exercise in conscious dogs. *Circ. Res.* 54:683–693, 1984.

90. Hoffman B. B. and R. J. Lefkowitz. Alpha-adrenergic receptor subtypes. *N. Engl. J. Med.* 302:1390–1396, 1980.

91. Hohimer A. R. and O. A. Smith. Decreased renal blood flow in the baboon during mild dynamic leg exercise. *Am. J. Physiol.* (Heart Circ. Physiol. 5) 236:H141–H150, 1979.

92. Holtz P., K. Credner, and G. Kroneberg. Uber das sympathicomimetische pressorische Prinzip des Harns ("urosympathin"). *Naunyn-Schmiedebergs Arch.* 204:228–243, 1947.

93. Innes I. R. and M. Nickerson. Drugs inhibiting the action of acetylcholine on structures innervated by post ganglionic nerves (antimuscarinic or atropine drugs). In: *The Pharmacological Basis of Therapeutics* (4th ed.), edited by L. S. Goodman and A. Gilman. New York: The Macmillan Co., 1970, pp. 524–548.

94. Jacobj C. Beitrage zur physiologischen und pharmakologischen Kenntniss der Darmbewegungen mit besonderer Berucksichtigung der Beziehung der Nebenniere zu denselben. *Arch. Exp. Pathol. Pharmak.* 29:171–211, 1892.

95. Jennings G., L. Nelson, P. Nestel, M. Esler, P. Korner, D. Burton, and J. Bazelmans. The effects of changes in physical activity on major cardiovascular risk factors, hemodynamics, sympathetic function, and glucose utilization in man: a controlled study of four levels of activity. *Circulation* 73:30–40, 1986.

96. Johansson J. E. Ueber die Einwirkung der Muskelthatigkeit auf die Athmung und die Herzthatigkeit. *Skan. Arch. Physiol.* 5:20–66, 1893.

97. Jorgensen C. R., K. Wang, Y. Wang, F. L. Gobel, R. R. Nelson, and H. Taylor. Effect of propranolol on myocardial oxygen consumption and its hemodynamic correlates during upright exercise. *Circulation* 48:1173–1182, 1973.

98. Jose A. D. Effect of combined sympathetic and parasympathetic blockade on heart rate and cardiac function in man. *Am. J. Cardiol.* 18:476–478, 1966.

99. Jose A. D., F. Stitt, and D. Collison. The effects of exercise and changes in body temperature on the intrinsic heart rate in man. *Am. Heart J.* 79:488–498, 1970.

100. Joyner M. J., B. F. Freund, S. M. Jilka, G. A. Hetrick, E. Martinez, G. A. Ewy, and J. H. Wilmoe. Effects of β-blockade on exercise capacity of trained and untrained men: a hemodynamic comparison. *J. Appl. Physiol.* 60:1429–1434, 1986.

101. Joyner M. J., S. M. Jilka, J. A. Taylor, J. K. Kalis, J. Nittolo, R. W. Hicks, T. G. Lohman, and J. H. Wilmore. β-Blockade reduces tidal volume during heavy exercise in trained and untrained men. *J. Appl. Physiol.* 62:1819–1825, 1987.

102. Kahler R. L., T. E. Gaffney, and E. Braunwald. The effects of autonomic nervous system inhibition on the circulatory response to muscular exercise. *J. Clin. Invest.* 41:1981–1987, 1962.

103. Kaiser P., S. Rossner, and J. Karlsson. Effects of β-adrenergic blockade on endurance and short-time performance in respect to individual muscle fiber composition. *Int. J. Sports Med.* 2:37–42, 1981.

104. Karpovich P. V. *Physiology of Muscular Exercise* (5th ed.). Philadelphia: W. B. Saunders, 1959.

105. Kaufman M. P. and H. B. Forster. Reflexes controlling circulatory, ventilatory and airway responses to exercise. In: *Handbook of Physiology, Section 12: Exercise: Regulation and Integration of Multiple Systems*, edited by L. B. Rowell and J. T. Shepherd. New York: Oxford University Press, 1996, pp. 381–447.

106. Kim S.-J., G. Kline, and P. A. Gwirtz. Limitation of cardiac output by a coronary α_1-constrictor tone during exercise in dogs. *Am. J. Physiol.* 271 (Heart Circ. Physiol. 40): H1125–H1131, 1996.

107. Kjellmer I. On the competition between metabolic vasodilation and neurogenic vasoconstriction in skeletal muscle. *Acta Physiol. Scand.* 63:450–459, 1965.

108. Korner P. I. Central nervous control of autonomic function. In: *Handbook of Physiology, The Cardiovascular System. The Heart*, edited by R. M. Berne and N. Sperelakis. Bethesda, MD: American Physiological Society, 1979, pp. 691–739.

109. Kotchen T. A., L. H. Hartley, T. W. Rice, E. H. Mougey, L. G. Jones, and J. W. Mason. Renin, norepinephrine, and epinephrine responses to graded exercise. *J. Appl. Physiol.* 31:178–184, 1971.

110. Krasney J. A., M. G. Levitzky, and R. C. Koehler. Sinoaortic contribution to the adjustment of systemic resistance in exercising dogs. *J. Appl. Physiol.* 36:679–685, 1974.

111. Krogh A. and J. Lindhard. The regulation of respiration and circulation during the initial stages of muscular work. *J. Physiol. (Lond.)* 31:112–133, 1904.

112. Kuntz A. *The Autonomic Nervous System* (4th ed.). Philadelphia: Lea and Febiger, 1953, pp. 15–20.

113. Lands A. M., A. Arnold, J. P. McAuliff, F. P. Luduena, and T. G. Brown, Jr. Differentiation of receptor systems activated by sympathomimetic amines. *Nature* 214:597–598, 1967.

114. Langer S. Z. Presynaptic regulation of catecholamine release. *Biochem. Pharmacol.* 23: 1793–1800, 1974.

115. Langley J. N. On the union of cranial autonomic (visceral) fibres with the nerve cells of the susperior cervial ganglia. *J. Physiol. (Lond.)* 23:240–270, 1898.

116. Langley J. N. On the reaction of cells and of nerve-endings to certain poisons, chiefly as regards the reaction of striated muscle to nicotine and to curari. *J. Physiol. (Lond.)* 33:374–413, 1905.

117. Langley J. N. and W.L. Dickinson. On the local paralysis of peripheral ganglia, and on the connection of different classes of nerve fibers with them. *Proc. R. Soc. B.* 46:423–431, 1889.

118. Laughlin M. H. and R. B. Armstrong. Adrenoceptor effects on rat muscle blood flow during treadmill exercise. *J. Appl. Physiol.* 62:1465–1472, 1987.

119. Laughlin M. H., R. J. Korthius, D. K. Duncker, and R. J. Bache. Control of blood flow to cardiac and skeletal muscle during exercise. In: *Handbook of Physiology, Section 12: Exercise: Regulation and Integration of Multiple Systems*, edited by L. B. Rowell and J. T. Shepherd. New York: Oxford University Press, 1996, pp. 705–765.

120. Leuenberger U., L. Sinoway, S. Gubin, L. Gaul, D. Davis, and R. Zelis. Effects of exercise intensity and duration on norepinephrine spillover and clearance in humans. *J. Appl. Physiol.* 75:668–674, 1993.

121. Lewis S. F., E. Nylander, P. Gad, and N.-H. Areskog. Non-autonomic component in bradycardia of endurance trained men at rest and during exercise. *Acta Physiol. Scand.* 109:297–305, 1980.

122. Lindhard J. Untersuchungen uber statische Muskelarbeit. *Skand. Arch. Physiol.* 40:145–194, 1920.

123 Loewi O. Uber humorale Ubertragbartkeit der Herzenvenwirkung. *Pflugers Arch.* 189:239–242, 1921.

124. Ludbrook J. and W. F. Graham. Circulatory responses to onset of exercise: role of arterial and cardiac baroreflexes. *Am. J. Physiol.* 248 (Heart Circ. Physiol 17):H457–H467, 1985.
125. Maciel B. C., L. Gallo, Jr., J. A. Marvin Neto, and L. E. B. Martins. Autonomic nervous control of the heart rate during isometric exercise in normal man. *Pflugers Arch.* 408:173–177, 1987.
126. MacIntosh A. M., W. M. Mullin, D. P. Fitzsimons, R. E. Herrick, and K.M. Baldwin. Cardiac biochemical and functional adaptations to exercise in sympathectomized neonatal rats. *J. Appl. Physiol.* 60:991–996, 1986.
127. Maksud M. G., K. D. Coutts, F. E. Tristani, J. R. Dorchak, J. J. Barboriak, and L. H. Hamilton. The effects of physical conditioning and propranolol on physical work capacity. *Med. Sci. Sports* 4:225–229, 1972.
128. Marey J. Recherches sur le pouls au Moyen d'un Nouvel Appareil Enregistreur le Sphygmographe. *Mem. Soc. Biol. (Paris).* Ser. 3, 1:281–286, 1859.
129. Mark A. L., R. G. Victor, C. Nerhed, and B. G. Wallin. Microneurographic studies of the mechanisms of sympathetic nerve responses to static exercise in humans. *Circ. Res.* 57:461–469, 1985.
130. Martin III W. H., A. R. Coggan, R. J. Spina, and J. E. Saffitz. Effects of fiber type and training on β-adrenoceptor density in human skeletal muscle. *Am. J. Physiol.* 257 (Endocrinol. Metab. 20):E736–E742, 1989.
131. Mass H. and P. A. Gwirtz. Myocardial flow and function after regional β-blockade in exercising dogs. *Med. Sci. Sports Exerc.* 19:443–450, 1987.
132. Matsukawa K., J. H. Mitchell, P. T. Wall, and L. B. Wilson. The effect of static exercise on renal sympathetic nerve activity in conscious cats. *J. Physiol. (Lond.)* 434:453–467, 1991.
133. Mazzeo R. S., D. A. Podolin, and V. Henry. Effects of age and endurance training on β-adrenergic receptor characteristics in Fischer 344 rats. *Mech. Ageing Dev.* 84:157–169, 1995.
134. Mazzeo R. S., C. Rajkumar, G. Jennings, and M. Esler. Norepinephrine spillover at rest and during submaximal exercise in young and old subjects. *J. Appl. Physiol.* 82:1869–1874, 1997.
135. McCurdy J.H. The effect of maximum muscular effort on blood pressure. *Am. J. Physiol.* 5:95–103, 1901.
136. McLeod A. A., J. E. Brown, B. B. Kitchell, F. A. Sedor, C. Kuhn, D. G. Shand, and R. S. Williams. Hemodynamic and metabolic responses to exercise after adrenoceptor blockade in humans. *J. Appl. Physiol.* 56:716–722, 1984.
137. McLeod A. A., K. D. Knopes, D. G. Shand, and R. S. Williams. β₁-selective and nonselective β-adrenoceptor blockade, anaerobic threshold and respiratory gas exchange during exercise. *Br. J. Clin. Pharmacol.* 19:13–20, 1985.
138. McLeod A. A., W. E. Kraus, and R. S. Williams. Effect of beta₁-selective and nonselective beta-adrenoceptor blockade during exercise conditioning in healthy adults. *Am. J. Cardiol.* 53:1656–1661, 1984.
139. McRitchie R. J., S. F. Vatner, D. Boettcher, G. R. Heyndrickx, T. A. Patrick, and E. Braunwald. Role of aterrial baroreceptors in mediating cardiovascular response to exercise. *Am. J. Physiol.* 230:85–89, 1976.
140. McSorley P. D. and D. J. Warren. Effects of propranolol and metoprolol on the peripheral circulation. *B.M.J.* 2:1598–1600, 1978.
141. Melcher A. and D. E. Donald. Maintained ability of carotid reflex to regulate arterial pressure during exercise. *Am. J. Physiol.* 241 (Heart Circ. Physiol. 10):H838–H849, 1981.
142. Meredith I. T., P. Friberg, G. L. Jennings, E. M. Dewar, V. A. Fazio, G.W. Lambert, and M. D. Esler. Exercise training lowers resting renal but not cardiac sympathetic activity in humans. *Hypertension* 18:575–582, 1991.

143. Monro P. A. G. *Sympathectomy*. London: Oxford University Press, 1959, pp. 1–15.
144. Moore R. L., M. Riedy, and P. D. Gollnick. Effect of training on β-adrenergic receptor number in rat heart. *J. Appl. Physiol.* 52:1133–1137, 1982.
145. Murray P. A., and S. F. Vatner. α-Adrenoceptor attenuation of the coronary vascular response to severe exercise in the conscious dog. *Circ. Res.* 45:654–660, 1979.
146. Nickerson M. Drugs inhibiting adrenergic nerves and structures innervated by them. In: *The Pharmacological Basis of Therapeutics* (4th ed.), edited by L. S. Goodman and A. Gilman. New York: The Macmillan Co., 1970, pp. 549–584.
147. Nietro J. L., I. D. Laviada, A. Guillen, and A. Haro. Adenylyl cyclase system is affected differently by endurance physical training in heart and adipose. *Biochem. Pharmacol.* 51:1321–1329, 1996.
148. Nordenfelt I. Haemodynamic response to exercise after combined sympathetic and parasympathetic blockade of the heart. *Cardiovas. Res.* 5:215–222, 1971.
149. Nordenfelt I. Blood flow of working muscles during autonomic blockade of the heart. *Cardiovasc. Res.* 8:263–267, 1974.
150. O' Hagan K. P., L. B. Bell, S. W. Mittelstadt, and P. S. Clifford. Effect of dynamic exercise on renal sympathetic nerve activity in conscious rabbits. *J. Appl. Physiol.* 74:2099–2104, 1993.
151. O'Hagan K. P., L. B. Bell, S. W. Mittelstadt, and P. S. Clifford. Cardiac receptors modulate the renal sympathetic response to dynamic exercise in rabbits. *J. Appl. Physiol.* 76: 507–515, 1994.
152. O'Leary D. S. Regional vascular resistance vs. conductance: which index for baroreflex responses? *Am. J. Physiol.* 260 (Heart Circ. Physiol. 29):H632–H637, 1991.
153. Oliver G. and E. A. Schafer. The physiological effects of extracts of the suprarenal capsule. *J. Physiol. (Lond.)* 18:230–276, 1895.
154. Orbeli L. A. and G. Ginetzinski. Effect of the sympathetic nervous system on the function of muscle. *Sechenov. Physiol. J. USSR* 6:139–145, 1923 (Russian).
155. Ordway G. A., J. B. Charles, D. C. Randall, G. E. Billman, and D. R. Wekstein. Heart rate adaptation to exercise training in cardiac-denervated dogs. *J. Appl. Physiol.* 52:1586–1590, 1982.
156. Papelier Y., P. Escourrou, J. P. Gauthier, and L. B. Rowell. Carotid baroreflex control of blood pressure and heart rate in men during dynamic exercise. *J. Appl. Physiol.* 77:502–506, 1994.
157. Pawelczyk J. A., B. Hanel, R. A. Pawelczyk, J. Warberg, and N. H. Secher. Leg vasoconstriction during dynamic exercise with reduced cardiac output. *J. Appl. Physiol.* 73: 1838–1846, 1992.
158. Peronnet F., J. Cleroux, H. Perrault, D. Cousineau, J. De Champlain, and R. Nadeau. Plasma norepinephreine response to exercise before and after training in humans. *J. Appl. Physiol.* 51:812–815, 1981.
159. Peterson D. F., R. B. Armstrong, and M. H. Laughlin. Sympathetic neural influences on muscle blood flow in rats during submaximal exercise. *J. Appl. Physiol.* 65:434–440, 1988.
160. Pick J. *The Autonomic Nervous System*. Phildelphia: J. B. Lippincott Co., 1970, pp. 3–21.
161. Pirnay F., J. M. Delvaux, R. Deroanne, P. Wittamer, and J. M. Petit. Effet d'un bloqueur beta-adrenergique sur la reponse cardiaque pendant l'exercice musculaire. *Int. Z. Angew. Physiol.* 29:88–93, 1970.
162. Plourde G., S. Rousseau-Migneron, and A. Nadeau. β-Adrenoceptor adenylate cyclase system adaptation to physical training in rat ventricular tissue. *J. Appl. Physiol.* 70: 1633–1638, 1991.
163. Plourde G., S. Rousseau-Migneron, and A. Nadeau. Effect of endurance training on β-adrenergic system in three different skeletal muscles. *J. Appl. Physiol.* 74:1641–1646, 1993.

164. Potts J. T., X. R. Shi, and P. B. Raven. Carotid baroreflex responsiveness during dynamic exercise in humans. *Am. J. Physiol.* 265 (Heart Circ. Physiol. 34):H1928–H1938, 1993.

165. Powell C. E. and I. H. Slater. Blocking of inhibitory adrenergic receptors by a dichloro analog of isoproterenol. *J. Pharmacol. Exp. Ther.* 122:480–486, 1958.

166. Pryor S. L., S. F. Lewis, R. G. Haller, L. A. Bertocci, and R. G. Victor. Impairment of sympathetic activation during static exercise in patients with muscle phosphorylase deficiency (McArdle's Disease). *J. Clin. Invest.* 85:1444–1449, 1990.

167. Raab W. Adrenaline and related substances in blood and tissues. *Biochem. J.* 37:470–473, 1943.

168. Rein H. Die Interferenz der vasomotorischen Regulationen. *Klin. Wochen.* 9:1485–1489, 1930.

169. Remensnyder J. P., J. H. Mitchell, and S. J. Sarnoff. Functional sympatholysis during muscular activity. *Circ. Res.* 11:370–380, 1962.

170. Robinson B. F., S. E. Epstein, G. D. Beiser, and E. Braunwald. Control of heart rate by the autonomic nervous system. *Circ. Res.* 19:400–411, 1966.

171. Robinson S., M. Pearcy, F. R. Brueckman, J. R. Nicholas, and D. I. Miller. Effects of atropine on heart rate and oxygen intake in working man. *J. Appl. Physiol.* 5:508–512, 1953.

172. Rohrer D. K., E. H. Schauble, K. H. Desai, B. K. Kobilka, and D. Bernstein. Alterations in dynamic heart rate control in the β_1-adrenergic receptor knockout mouse. *Am. J. Physiol.* 274 (Heart. Circ. Physiol. 43):H1184–H1193, 1998.

173. Roth D. A. C. D. White, D. A. Podolin, and R. S. Mazzeo. Alterations in myocardial signal transduction due to aging and chronic dynamic exercise. *J. Appl. Physiol.* 84: 177–184, 1998.

174. Rothschuh K. E. *History of Physiology*. Huntington, NY: Robert E. Krieger Publishing Co., 1973.

175. Rotto D. M. and M. P. Kaufman. Effect of metabolic products of muscular contraction on discharge of group III and IV afferents. *J. Appl. Physiol.* 64:2306–2313, 1988.

176. Rowell L. B. What signals govern the cardiovascular responses to exercise? *Med. Sci. Sports Exerc.* 12:307–315, 1980.

177. Rowell L. B. *Human Cardiovascular Control*. New York: Oxford University Press, 1993.

178. Rowell L. B., L. Hermansen, and J. R. Blackmon. Human cardiovascular and respiratory responses to graded muscle ischemia. *J. Appl. Physiol.* 41:693–701, 1976.

179. Rowell L. B. and D. S. O'Leary. Reflex control of the circulation during exercise: chemoreflexes and mechanoreflexes. *J. Appl. Physiol.* 69:407–418, 1990.

180. Rowell L. B., D. S. O'Leary, and D. L. Kellog, Jr. Integration of cardiovascular control systems in dynamic exercise. In: *Handbook of Physiology, Section 12: Exercise: Regulation and Integration of Multiple Systems*, edited by L. B. Rowell and J. T. Shepherd. New York: Oxford University Press, 1996, pp. 770–838.

181. Rowell L. B., M. V. Savage, J. Chambers, and J. R. Blackmon. Cardiovascular responses to graded reductions in leg perfusion in exercising humans. *Am. J. Physiol.* 261 (Heart Circ. Physiol. 30):H1545–H1553, 1991.

182. Ruble S. B., Z. Valic, J. B. Buckwalter, and P. S. Clifford. Dynamic exercise attenuates sympathetic responsiveness of canine vascular smooth muscle. *J. Appl. Physiol.* 89: 2294–2299, 2000.

183. Sable D. L., H. L. Brammell, M. W. Sheehan, A. S. Nies, J. Gerber, and L. D. Horwitz. Attenuation of exercise conditioning by beta-adrenergic blockade. *Circulation* 65:679–684, 1982.

184. Saito M., T. Mano, H. Abe, and S. Iwase. Responses in muscle sympathetic nerve activity to sustained hand-grips of different tensions in humans. *Eur. J. Appl. Physiol.* 55:493–498, 1986.

185. Saito M. M. Naito, and T. Mano. Different responses in skin and muscle sympathetic nerve activity to static muscle contraction. *J. Appl. Physiol.* 69:2085–2090, 1990.
186. Saito M. A. Tsukanaka, D. Yanagihara, and T. Mano. Muscle sympathetic nerve responses to graded leg cycling. *J. Appl. Physiol.* 75:663–667, 1993.
187. Samaan A. Muscular work in dogs submitted to different conditions of cardiac and splanchnic innervations. *J. Physiol. (Lond.)* 83:313–331, 1935.
188. Savard G. K., E. A. Richter, S. Strange, B. Kiens, N. J. Christensen, and B. Saltin. Norepinephrine spillover from skeletal muscle during exercise in humans: role of muscle mass. *Am. J. Physiol.* 257 (Heart Circ. Physiol. 26):H1812–H1818, 1989.
189. Savard G., S. Strange, B. Kiens, E. A. Richter, N. J. Christensen, and B. Saltin. Noradrenaline spillover during exercise in active versus resting skeletal muscle in man. *Acta Physiol. Scand.* 131:507–515, 1987.
190. Savin W. M., E. P. Gordon, S. M. Kaplan, B. F. Hewitt, D. C. Harrison, and W. L. Haskell. Exercise training during long-term beta-blockade treatment in healthy subjects. *Am. J. Cardiol.* 55:101D–109D, 1985.
191. Seals D. R. Influence of muscle mass on sympathetic neural activation during isometric exercise. *J. Appl. Physiol.* 67:1801–1806, 1989.
192. Seals D. R. Influence of active muscle size on sympathetic nerve discharge during isometric contractions in humans. *J. Appl. Physiol.* 75:1426–1431, 1993.
193. Seals D. R., R. G. Victor, and A. L. Mark. Plasma norepinephrine and muscle sympathetic discharge during rhythmic exercise in humans. *J. Appl. Physiol.* 65:940–944, 1988.
194. Shen Y.-T., P. Cervoni, T. Claus, and S. F. Vatner. Differences in β_3-adrenergic receptor cardiovascular regulation in conscious primates, rats, and dogs. *J. Pharmacol. Exp. Ther.* 278:1435–1443, 1996.
195. Sheriff D. D., C. R. Wyss, L. B. Rowell, and A. M. Scher. Does inadequate oxygen delivery trigger pressor response to muscle hypoperfusion during exercise? *Am. J. Physiol.* 253 (Heart Circ. Physiol. 22):H1199–H1207, 1987.
196. Sigvardsson K., E. Svanfeldt, and A. Kilbom. Role of the adrenergic nervous system in development of training-induced bradycardia. *Acta Physiol. Scand.* 101:481–488, 1977.
197. Sinoway L., S. Prophet, I. Gorman, T. Mosher, J. Shenberger, M. Dolecki, R. Briggs, and R. Zelis. Muscle acidosis during static exercise is associated with calf vasoconstriction. *J. Appl. Physiol.* 66:429–436, 1989.
198. Sinoway L. I., M. B. Smith, B. Enders, U. Leuenberger, T. Dzwonczyk, K. Gray, S. Whisler, and R. L. Moore. Role of diprotonated phosphate in evoking muscle reflex responses in cats and humans. *Am. J. Physiol.* 267 (Heart Circ. Physiol. 36):H770–H778, 1994.
199. Slowtzoff B. Ueber die Beziehungen zwischen Korpergrosse und Stoffverbrauch der Hunde bei Ruhe und Arbeit. *Pflugers Arch.* 95:158–191, 1903.
200. Stapleton M. P. Sir James Black and propranolol. The role of the basic sciences in the history of cardiovascular pharmacology. *Texas Heart Inst. J.* 24:336–342, 1997.
201. Stephenson R. B. and D. E. Donald. Reversible vascular isolation of carotid sinuses in conscious dogs. *Am. J. Physiol.* 238 (Heart Circ. Physiol. 7):H809–H814, 1980.
202. Stegemann J., A. Busert, and D. Brock. Influence of fitness on the blood pressure control system in man. *Aerospace Med.* 45:45–48, 1974.
203. Strandell T. and J. T. Shepherd. The effect in humans of increased sympathetic activity on the blood flow to active muscles. *Acta Med. Scand.* 472:146–167, 1967.
204. Sutton J. R., A. Cole, J. Gunning, J. B. Hickie, and W. A. Seldon. Control of heart-rate in healthy young men. *Lancet* 2:1398–1400, 1967.
205. Svedenhag J., J. Henriksson, A. Juhlin-Dannfelt, and K. Asano. Beta-adrenergic blockade and training in healthy men—effects on central circulation. *Acta Physiol. Scand.* 120:77–86, 1984.

206. Sweeney M. E., B. J. Fletcher, and G. F. Fletcher. Exercise testing and training with β-adrenergic blockade: Role of the drug washout period in "unmasking" a training effect. *Am. Heart J.* 118:941–946, 1989.

207. Sylvestre-Gervais L., A. Nadeau, M. H. Nguyen, G. Tancrede, and S. Rousseau-Migneron. Effects of physical training on β-adrenergic receptors in rat myocardial tissue. *Cardiovasc. Res.* 16:530–534, 1982.

208. Tesch P. A. and P. Kaiser. Effect of β-adrenergic blockade on maximal oxygen uptake in trained males. *Acta Physiol. Scand.* 112:351–352, 1981.

209. Thompson L. P. and D. E. Mohrman. Blood flow and oxygen consumption in skeletal muscle during sympathetic stimulation. *Am. J. Physiol.* 245 (Heart Circ. Physiol. 14): H66–H71, 1983.

210. Tipton C. M. Training and bradycardia in rats. *Am. J. Physiol.* 209:1089–1094, 1965.

211. Traverse J. H., J. D Altman, J. Kinn, D. J. Duncker, and R. J. Bache. Effect of β-adrenergic receptor blockade on blood flow to collateral-dependent myocardium during exercise. *Circulation* 91:1560–1567, 1995.

212. Vallbo A. B., K.-E. Hagbarth, H. E. Torebjork, and B. G. Wallin. Somatosensory, proprioceptive, and sympathetic activity in human peripheral nerves. *Physiol. Rev.* 59: 919–957, 1979.

213. Vanhoutte P., E. Lacroix, and I. Leussen. The cardiovascular adaptation of the dog to muscular exercise. Role of the arterial pressoreceptors. *Arch. Int. Physiol. Biochem.* 74:201–222, 1966.

214. Vatner S. F., D. Franklin, R. L. Van Citters, and E. Braunwald. Effects of carotid sinus nerve stimulation on blood-flow distribution in conscious dogs at rest and during exercise. *Circ. Res.* 27:495–503, 1970.

215. Vendsalu A. Studies on adrenaline and noradrenaline in human plasma. *Acta Physiol. Scand.* 49 (Suppl. 173):8–123, 1960.

216. Victor R. G., L. A. Bertocci, S. L. Pryor, and R. L. Nunnally. Sympathetic nerve discharge is coupled to muscle cell pH during exercise in humans. *J. Clin. Invest.* 82:1301–1305, 1988.

217. Victor R. G., S. L. Pryor, N. H. Secher, and J. H. Mitchell. Effects of partial neuromuscular blockade on sympathetic nerve responses to static exercise in humans. *Circ. Res.* 65:468–476, 1989.

218. Victor R. G. and D. R. Seals. Reflex stimulation of sympathetic outflow during rhythmic exercise in humans. *Am. J. Physiol.* 257 (Heart Circ. Physiol. 26):H2017–H2024, 1989.

219. Victor R. R., D. R. Seals, and A. L. Mark. Differential control of heart rate and sympathetic nerve activity during dynamic exercise. Insights from intraneural recordings in humans. *J. Clin. Invest.* 79:508–516, 1987.

220. Vissing J., D. A. MacLean, S. F. Vissing, M. Sander, B. Saltin, and R. G. Haller. The exercise metaboreflex is maintained in the absence of muscle acidosis: insights from muscle microdialysis in humans with McArdle's disease. *J. Physiol. (Lond.)* 537:641–649, 2001.

221. Vissing S. F., U. Scherrer, and R. G. Victor. Stimulation of skin sympathetic nerve discharge by central command. *Circ. Res.* 69:228–238, 1991.

222. Volkmann A. W. Ueber die Bewegungen des Athmens und Schluckens, mit besonderer Berucksichtigung neurologischer Streitfragen. *Arch. Anat. Physiol. Wissens. Med.* 1: 332–360, 1841.

223. Wada M., M. Seo, and K. Abe. Effect of muscular exercise upon the epinephrine secretion from the suprarenal gland. *Tohoku J. Exp. Med.* 27:65–86, 1935.

224. Waldrop T. G., F. L. Eldridge, G. A. Iwamoto, and J. H. Mitchell. Central neural control of respiration and circulation during exercise. In: *Handbook of Physiology. Section 12, Exercise: Regulation and Integration of Multiple Systems,* edited by L. B. Rowell and J. T. Shepherd. New York: Oxford University Press, 1996, pp. 333–380.

225. Walgenbach S. C. and D. E. Donald. Inhibition by carotid baroreflex of exercise-induced increase in arterial pressure. *Circ. Res.* 52:253–262, 1983.

226. Walgenbach S. C. and J. T. Shepherd. Role of arterial and cardiopulmonary mechano-receptors in the regulation of arterial pressure during rest and exercise in conscious dogs. *Mayo Clin. Proc.* 59:467–475, 1984.

227. Wallin B. G. Peripheral sympathetic neural activity in conscious humans. *Ann. Rev. Physiol.* 50:565–576, 1988.

228. Winder W. W., J. M. Hagberg, R. C. Hickson, A. A. Ehsani, and J. A. McLane. Time course of sympathoadrenal adaptation to endurance exercise training in man. *J. Appl. Physiol.* 45:370–374, 1978.

229. Williams R. S. Physical conditioning and membrane receptors for cardioregulatory hormones. *Cardiovasc. Res.* 14:177–182, 1980.

230. Williams R. S., M. G. Caron, and K. Daniel. Skeletal muscle β-adrenergic receptors: variations due to fiber type and training. *Am. J. Physiol.* 246 (Endocrinol. Metab. 9): E160–E167, 1984.

231. Wilmore J. H., G. A. Ewy, B. J. Freund, A. A. Hartzell, S. M. Jilka, M. J. Joyner, C. A. Todd, S. M. Kinzer, and E. B. Pepin. Cardiorespiratory alterations consequent to endurance exercise training during chronic beta-adrenergic blockade with atenolol and propranolol. *Am. J. Cardiol.* 55:142D–148D, 1985.

232. Winslow J. B. Exposition Anatomique du corp humanin. Paris: G. Desprez, 1732.

233. Wolfe E. L., A. C. Barger, and S. Benison. *Walter B. Cannon, Science and Society.* Cambridge, MA: Boston Medical Library, 2000, pp. 321–328.

234. Wyatt H. L., L. Chuck, B. Rabinowitz, J. V. Tyberg, and W. W. Parmley. Enhanced cardiac responses to catecholamines in physically trained cats. *Am. J. Physiol.* 234 (Heart Circ. Physiol. 3):H608–H613, 1978.

235. Zuntz N. and J. Geppert. Ueber die Natur der normalen Athemreize und den Ort ihrer Wirkung. *Arch. Ges. Physiol.* 38:337–338, 1886.

236. Zuntz N. Einfluss der Geschwindigkeit, der Korpertemperatur und der Uebung auf den Stoffverbrauch bei Ruhe und bei Muskelarbeit. *Arch. Ges. Physiol.* 95:192–208, 1903.

chapter 6

THE OXYGEN TRANSPORT SYSTEM AND MAXIMAL OXYGEN UPTAKE

Jere H. Mitchell and Bengt Saltin

> However much the speed be increased beyond this limit, no further increase in oxygen intake can occur: the heart, lungs, circulation, and the diffusion of oxygen to the active muscle-fibers have attained their maximal activity.
> —Hill and Lupton, 1923

MAXIMAL oxygen uptake as we know it at the turn of the twenty-first century has a history that dates back to Archibald Vivian Hill and his work with Lupton and Long. In the 1923 article from which the above quote is taken, Hill and Lupton summarized data in the literature and added their own measurements of oxygen uptake obtained when running at increasing speeds in the open field, including one bout leading to exhaustion (60). They coined the term "maximum oxygen intake," which for A.V. Hill at the time was 4055 ml min^{-1}. They did point out that the maximum value for $\dot{V}O_2$ varied among individuals and they identified the main links in the oxygen transport system, which incorporated the combined and closely integrated functions of the respiratory, circulatory, and skeletal muscle systems. The concept of maximal oxygen consumption was fundamental to the use of exercise as a perturbation to understand basic physiological regulations.

The ability to study the oxygen transport system in exercising humans depended on many fundamental discoveries. These began with the isolation of oxygen independently in ~1774 by Joseph Priestly (1733–1804) in England and Carl Wilhelm Scheele (1742–1786) in Sweden, the latter naming this fraction of the air "fireair" (134). The findings of Priestly and Scheele were communicated to Antoine L. Lavoisier (1743–1794) in Paris, and it was Lavoisier who provided the definitive discovery of the respiratory gases and combustion as discussed in detail in Chapter 4 of this volume. In 1777 Lavoisier wrote: "respiration is nothing but a slow combustion of carbon and hydrogen similar in all aspects to that of a lamp or a lighted candle and from this point of view animals which breath are really combustible substances burning and consuming themselves" (90).

Lavoisier could not have defined respiration more clearly. However, not until the work performed by Justus von Liebig (1803–1873) in Germany on leg muscle of the frog was this concept experimentally proven (91). Moreover, further studies about a century later by Eduard Pflüger (1829–1910), also from Germany, had to be performed before it was generally accepted that combustion occurred in the tissues and that blood flow transported O_2 and CO_2 (109).

It was Lavoisier too who made the first attempt to measure pulmonary gas exchange not only at rest but also during exercise (Fig. 6.1). The results were published in *Memoir de Science* (1789) by Seguin and Lavoisier, and thanks to very detailed illustrations by Lavoisier's wife, Madame Marie-Anne Lavoisier, the experimental setup used in these experiments is quite clear (137). A face mask was used to measure not only the CO_2 produced, but also the O_2 consumed. The study was performed by the foot lifting a pedal and the work was estimated from the number of lifts of the weight × height. The observed values for oxygen intake and carbon dioxide exhaled were not accurate, but as later pointed out by Benedict and Cathcart "the results obtained [by Lavoisier] are not in accord with those obtained by the use of mod-

Fig. 6.1. Measurement of gas exchange during exercise by Lavoisier. The drawing was made by Madame Lavoisier (in 1772) and illustrates how the exercise was performed as well as how the respiration was measured. (For more details see text; illustration from ref. 137.)

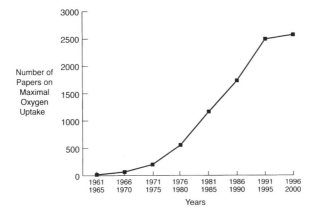

Fig. 6.2. Citation history from 1969 of papers on "maximal oxygen uptake" where this term or its equivalent is found in the title or as a key word in the abstract. From a search of Index Medicus, MEDLINE, and Current Contents® by Helen Mayo.

ern experimental techniques but it cannot dim in any way the brilliancy of the conception of the research" (17).

Since Lavoisier's research at the end of the eighteenth century, the understanding of the limits for humans in performing physical activity has intrigued investigators with both applied and basic science interests. But, before quantitative studies of gas exchange in humans could be performed, technical developments were crucial, such as the Tissot spirometer, Douglas's invention of a bag for air collection, and Haldane's absorption method for measuring CO_2 and O_2, all accomplished in the early 1900s (38, 52, 146).

Thus, some 140 years after Lavoisier, there was a basis for Hill to measure oxygen consumption and, more importantly, to provide an understanding of the meaning of such a measurement and its maximal value. After Hill, determinations of maximal oxygen uptake became the common method to characterize cardiorespiratory fitness in health and disease, and it is still the gold standard. In addition to this applied use of $\dot{V}O_2$max measurements, studies have been performed to define the functional role of each link in the oxygen transport chain and to determine their regulation and adaptability. The number of publications since 1966 that include the maximal oxygen uptake concept provides strong evidence of its steadily increasing importance (Fig. 6.2).

EARLY CONTRIBUTORS

More than 100 years after the observation of Lavoisier, Edward Smith (Fig. 6.3) began "the first systematical inquiry into the respiratory and metabolic response of the human subject to muscular exercise" (25). He published data on the quantity of air which is inspired (inspired volume), rate of respiration (respiratory rate), and

Fig. 6.3. Edward Smith (1818–1874). Received M.B. degree from Royal Birmingham Medical School and M.D. degree from University of London, Queens College, Birmingham (1843). After a trial in medical practice, he began a rather turbulent academic career in pulmonary physiology at the Charing Cross Hospital School of Medicine in London (25). Courtesy of National Library of Medicine.

pulsation (heart rate) during rest and various types of exercises including swimming, rowing, walking, and using the treadwheel (138). His findings clearly showed an increase in heart rate, respiratory rate, and inspiratory volume with increasing work intensity. The highest values he obtained were observed while walking at 3 miles per hour carrying a weight of 118 pounds. At this level of exercise, the inspiratory volume was 2141 cubic inches per minute (about 34 l min^{-1}), respiratory rate 24 breaths min^{-1} and heart rate 189 beats min^{-1} (139). He also reported his inspiratory volume data during exertion in relative terms compared to the inspiratory volume at rest. He stated that during exercise "it is possible that the quantity of air which we breathe may be increased at least seven-fold for a short period" (139). Expressing changes relative to values at rest being unity is now widely used in expressing the rate of metabolism during exercise (MET = O_2 consumption during rest). In later work, Smith also measured the production of carbonic acid during walking and treadwheel exercise (140). During walking at 2 miles/h, carbonic acid production was about 18 grains/min, at 3 miles/h it was 26 grains/min, and during treadwheel exercise it was up to 48 grains/min. The demonstration of a correlation between CO_2 production and intensity of physical activity was Smith's most important contribution.

In the last decades of the nineteenth century, Nathan Zuntz (Fig. 6.4) was active in Germany and made major contributions to exercise physiology (51). His main

Fig. 6.4. Nathan Zuntz (1847–1920). Received his doctoral degree in medicine from Bonn University in 1868. The title of the thesis was "Beiträge zur Physiologie des Blutes." In 1870 he became the assistant to professor Eduard Pflüger (1829–1910) in Berlin. Zuntz returned to Bonn in 1874 as professor in physiology and he also held a position in Cologne. In 1881 he became professor in physiology in the newly started Department of Animal Physiology at the Royal Agricultural University in Berlin, a position he held until his retirement in 1919 (picture from ref. 51).

area of interest was blood gases and he used altitude hypoxia as a main intervention in his investigations of the interaction between ventilation and the cardiovascular system for gas transport. However, exercise was another perturbation and with the construction of the first treadmill, he could perform his studies of humans at well-controlled and reproducible exercise intensities (154). Detailed measurements were performed of the energy costs when walking/marching (156). An interesting contribution was the first use of X-rays to study the size of the heart, not only at rest but also when walking on the treadmill (Fig. 6.5A; ref. 105). He also made energy consumption measurements on horses walking at different speeds and on dogs running on a treadmill (153, 155).

Francis G. Benedict (Fig. 6.6) made measurements of oxygen consumption during light, moderate, and severe exercise on a cycle ergometer (Fig. 6.5B; ref. 16). During his research training with Professor Wilbur O. Atwater (1844–1907) at Wesleyan College, Connecticut, Benedict changed his interest from biochemistry to

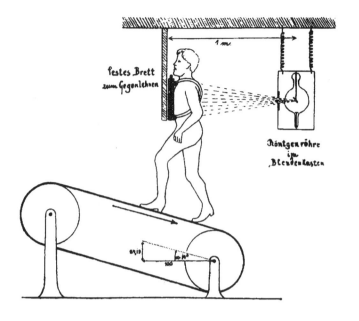

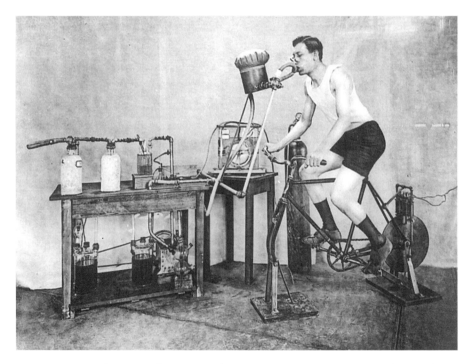

Fig. 6.5. *A*: Schematic illustration of Zuntz's treadmill with arrangements for measuring heart size with X-ray (105). *B*: Cycle ergometer and devices for measuring oxygen consumption, developed and used by Benedict and Cady (16). Courtesy of National Library of Medicine.

Fig. 6.6. Francis G. Benedict (1870–1957). Received his bachelor and master of art degrees from Harvard and Ph.D. degree from Heidelberg University (1895). First academic position in the Chemistry Department at Wesleyan University, Middletown, CN, with Prof. W.O. Atwater. In 1916 he became director of the Carneige Institute. Courtesy of National Library of Medicine.

physiology, a transformation that led to his outstanding contributions to the understanding of energy expenditure at rest and during work. Atwater had studied with the German chemists and physiologists Voigt and Rubner, and later with Zuntz, and he brought back the latest techniques for measuring respiration and metabolism. Benedict, as had Atwater, received training in Europe with Voigt. Benedict developed a highly sophisticated calorimeter that made it possible to very precisely measure heat production, energy turnover, and respiratory exchange ratios (15). He found that oxygen consumption progressively rose during exercise as work loads increased and that the highest value obtained in a 32-year-old former professional cyclist was 3265 ml/min (49.5 ml kg^{-1} min^{-1}, refs. 10, 15). However, he made no attempt to determine whether a maximal value was obtained in any of his studies. Benedict's other important contributions were the precise measurement of gross and net efficiency in bicycle exercise and the contributions of fat and carbohydrate as sources of energy (17).

Fig. 6.7. Göran Liljestrand (1886–1960). Received his doctoral degree from the Karolinska Institute and remained there for his whole research career. In 1910 he became the professor of pharmacology, a position he held until 1951. He was the secretary of the Nobel Prize Committee from 1928 to 1951. Courtesy of Karolinska Institute, Stockholm.

Göran Liljestrand (Fig. 6.7) studied at the Karolinska Institute with Erik Johan Johansson, who already in 1895 had suggested a coupling between central and peripheral neural mechanisms in regulating the heart rate in response to exercise (68). Liljestrand together with N. Stenström took these investigations to the field in an impressive series of studies on humans. They determined the oxygen uptake when walking, running, swimming, rowing, and cross-country skiing (95, 96). Also, measurements of blood pressure and cardiac output were made in some of these exercises—swimming and rowing (93, 94). They indirectly used the term "maximal oxygen intake" and demonstrated that slightly higher values were observed in cross-country skiing than in running (95). An unresolved question is whether these early measurements performed by Liljestrand and Stenström and others were true maximal values in the various exercise modes (see below).

August Krogh (Fig. 6.8), working in the laboratory of Christian Bohr (1855–1911), studied the mechanisms by which oxygen was transported from the alveoli to the blood in the pulmonary circulation. Bohr favored the notion, as did most physiologists at the time (Haldane, Douglas, Starling), that oxygen was secreted into the blood, whereas Krogh worked on the hypothesis that the O_2 transport was brought about by simple diffusion. His studies published in 1910 convincingly proved that O_2

Fig. 6.8. August Krogh (1874–1949). Received his doctor of science degree from the University of Copenhagen with Professor Christian Bohr as his mentor. In 1920 he was awarded the Nobel Prize in physiology and medicine for his studies on the regulation of capillary blood flow and transport mechanisms through the capillary wall (83). Courtesy of August Krogh Institute, Copenhagen.

even at large oxygen intakes could be transported by diffusion (84). August Krogh didn't get his Nobel Prize for this work. Instead it was given to him in 1920 for his equally outstanding work on the function of capillaries (83).

In the 1910s, Krogh's interest in using exercise as an intervention while studying cardiorespiratory regulation was initiated by Johannes Lindhard (1870–1947). Their cooperation resulted in several significant contributions over the next 15 years. A major one was the detailed elaboration of the role of "cortical irradiation" that instantaneously affects pulmonary ventilation and heart rate at the onset of exercise. The first study, published in 1913 (85), was followed by experiments on intense work that involved measurements at exhaustive exercise and in the recovery phase (86). They demonstrated that the oxygen uptake during recovery is far greater than the "lack" of it at the start of the exercise. Moreover, in a separate study they proved that not only carbohydrates, but also fat, could be directly combusted by contracting skeletal muscles (87).

DETERMINATION OF MAXIMAL OXYGEN CONSUMPTION

Methods of Producing Exercise

In the early studies of the eighteenth century, exercise was produced by lifting weights, walking/running, or bicycling—that is, the same three forms we use today. Lavoisier produced exercise by lifting weights but soon included walking at different speeds while collecting the expired air (90). Later, various treadmills were in use, as mentioned in Smith's work, and in 1899, Zuntz constructed his special treadmill for exercise studies (154). Various forms of braked cranks or turning wheels were in use in the latter part of the nineteenth century. Max-Josef Pettenkofer (1818–1901) in Munich placed in his respiratory chamber a flywheel over which a chain weighing 25 kg was placed, thereby increasing the friction and work of pedaling (108). Of some interest is the so-called "ergostat" that was originally developed by Speck and put into serial construction by Gaertner in Vienna in 1887 (46). This mechanically braked arm ergometer was primarily for the use of overweight people in their own homes to burn "calories."

The world's very first bicycle ergometer was developed by Bouny in Paris in 1896 (24). She put a mechanical brake directly on the rear wheel of an ordinary bicycle that was lifted up from the ground when used to determine exercise performance of bicyclists. As highlighted above, Atwater and Benedict, probably inspired by Pettenkofer and Voit, made the first stationary bicycle and later developed the electromagnetic brake (10). The Krogh bicycle, introduced a year later, was also based on the principle of braking the flywheel electromagnetically (82).

Thus, early in the 1900s there were well-functioning exercise machines in which one could precisely control speed and inclination (treadmills) or the brake on a flywheel either electronically or mechanically. Precision in setting this brake using a simple mechanical approach had to await the development made by von Döbeln in 1954 (39). He introduced the (sinus) pendulum with a known weight. By changing the angle of the axis from its vertical position via a ribbon, the pendulum precisely controlled the brake of the flywheel.

From early on, there was also an interest to measure oxygen uptake in several other forms of exercise. The expired air was collected in bags and analyzed for volume and O_2/CO_2 content. A similar procedure was also applied by Liljestrand and Stenström in their studies on swimming, skating, rowing, and cross-country skiing (95, 96). Together with Lindhard, Liljestrand also introduced tethered exercise: that is, the swimmer or the rowing boat was firmly attached to the bridge, thus making the subject exercise at one spot, which simplified the collection of the expired air (93,

94). Furthermore, the tethered exercise allowed for other measurements to be made, such as blood pressure and cardiac output, by using a foreign gas method.

Evaluation Criteria for $\dot{V}O_2max$

In 1923, *Archibald Vivian Hill* (Fig. 6.9) and Lupton reviewed (60) the work of Benedict and Catchart from 1913 (17), Liljestrand and Stenström's articles from 1920, and reports of others who had measured oxygen uptake in various "exercises" (17, 156). It was quite apparent that peak oxygen uptake was highest in cross-country skiing, slightly lower in running or bicycling, and even lower in swimming. In the running experiments by Liljestrand and Stenström (95), it appears likely that at least in some subjects $\dot{V}O_2max$ was reached, since with increasing speed no further elevation in oxygen uptake was observed. Indeed, a drop from the peak values was apparent at

Fig. 6.9. Archibald Vivian Hill (1886–1977). Finished the mathematical tripos at Trinity College, Cambridge, in 1907 and the natural sciences tripos in 1909. In 1910 as a fellow in physiology at Trinity he began his classical studies on the biophysics and biochemistry of skeletal muscle contraction, for which he was awarded the Nobel Prize in physiology and medicine in 1922. From 1920 to 1923 he was the chair of physiology at Manchester and during that time he began his first studies in humans performing severe muscular exercise. Courtesy of National Library of Medicine.

times. Hill and Lupton acknowledged this finding and highlighted the point that this leveling off in O_2 uptake at "super" high speeds for the subject does not mean an improved running efficiency as proposed by Liljestrand and Stenström (95)—but rather that aerobic metabolism cannot contribute more, and the anaerobic energy yield makes up for the difference in regard to energy demand at the actual speed (60). It is surprising that it took so long for the interplay between aerobic and anaerobic energy yields during heavy exercise to be defined since it was known in the previous century that lactic acid is formed during intense exercise (21). In contrast, ATP and phosphocreatine (PCr) and their role in providing the energy for the mechanical output of skeletal muscles had to await the work of Lundsgaard (99) and Lohmann (98) in the early 1930s (see also Barnard and Holloszy, 14).

It is noteworthy that although well-designed treadmills were available, Hill preferred walking and running in the field or on the track for his experiments. To determine the velocity of the runner, he developed a sophisticated electromagnetic system that provided split times for every 25 yards. In the experiments on Hill himself, a leveling off in $\dot{V}O_2$ was observed, not as a function of increasing speed of running, but with time at the highest velocity, which was 260 meters·min^{-1} (60).

Robert Herbst (Fig. 6.10) was the first to carefully apply the leveling-off criterion to establish $\dot{V}O_2$max in his studies in the late 1920s (55). His subjects ran in place with higher and higher step frequency. In 52 male subjects aged 19–24 years, he found $\dot{V}O_2$max values that ranged from 1815 to 4022 ml min^{-1}. He was also the

Fig. 6.10. Roger Herbst (1890–1962). He graduated from the medical school in Bonn in 1916 and continued his education and later research at the German "Sport Hochschule" in Cologne. In addition to his major contribution on maximal oxygen measurements, he had a general interest in how various exercise forms affect bodily functions, here illustrated (foreground) with his measurements on himself of respiratory variables during upper-body exercise. Courtesy of Deutsche Sporthochschule, Cologne.

first to express $\dot{V}O_2$max in ml kg^{-1} min^{-1}. The two principles to establish $\dot{V}O_2$max, a plateau in $\dot{V}O_2$ with time at an exercise intensity leading to exhaustion within 3–6 minutes or a plateau with increasing exercise loads, should give similar results. Although work by Taylor and associates in 1953 showed that using the latter principle was the better method (144), a study by Åstrand and Saltin (9) showed that both methods give similar maximal values provided that a sufficient warming-up period is allowed when using Hill's approach.

Noakes has challenged whether Hill actually demonstrated a plateau in $\dot{V}O_2$ and thus had measured a true $\dot{V}O_2$max (106). Hill appears to have accomplished this in the experiments conducted on himself; but more importantly, he was the one who conceived the physiological meaning of maximal $\dot{V}O_2$. Together with Lupton, he was the first to make a clear distinction between O_2 intake and demand. In a critical comment on the conclusion reached by Liljestrand and Stenström—that running economy improved at higher speeds—Hill and Lupton wrote (60):

> It was apparently more economical to run fast than slow! Now, the opposite is notoriously the case, and these observations of Liljestrand and Stenström (of which, on technical grounds, we have no criticism) obviously need an explanation. The explanation is simple: the subjects of their experiments were not in a genuine steady state at the higher speeds. In the case, e.g. of their subject N.S. [(95), p. 183] it is clear that the maximum oxygen intake of about 3.3 litres per min. was attained at a speed of about 186 metres per min. Hence, however fast N.S. ran above this speed he did not use more oxygen, not because he did not require it, but because he could not get it.

It is very fortunate that Hill decided in 1920 to apply his research talents to the study of exercising humans. His Nobel Prize–winning studies of isolated frog skeletal muscle made him uniquely qualified to study similar problems in exercising people. In this regard he wrote (58):

> As it proved, work on hard muscular exercise in man provided the same opportunity of getting accurate and reproducible results as is found in experiments on isolated muscle. Indeed, in some ways, man was the better experimental object … ; when trained, he can repeat the same performance again and again. And it remained a matter always of satisfaction, sometimes even of excitement, as the work evolved, to find how the experiments on man and those on isolated muscle confirmed and threw light on one another.

Skeletal Muscle Mass

Whether the muscle mass engaged in the exercise could play a role was not a major concern in the initial studies of oxygen transport. As mentioned earlier, Hill and others noted that the highest $\dot{V}O_2$ was commonly reported during cross-country skiing; however, the question of whether adding upper-body muscles was the cause of the observed higher $\dot{V}O_2$max was not discussed or answered.

Fig. 6.11. Henry Longstreet Taylor (1912–1983). Received his B.S. degree in 1935 from Harvard and began working with Dr. Ancel Keys in the Harvard Fatigue Laboratory. He followed Dr. Keys to the University of Minnesota where he earned his Ph.D. in physiology in 1941. He then joined the faculty of the Laboratory of Physiological Hygiene, where he remained a productive investigator for 46 years. Courtesy of Dr. Arthur Leon.

Henry Longstreet Taylor (Fig. 6.11) and his colleagues were the first to determine the role of adding arm to leg exercise in reaching $\dot{V}O_2$max. They established that arm cranking while running on a treadmill increased the value of maximal oxygen uptake. In later research—by Åstrand and Saltin (8), Hermansen (1933–1984) (56), Bergh and coworkers (18), and Secher (1946–) and associates (136), who more systematically evaluated the effect of adding arm exercise to leg exercise—a difference in $\dot{V}O_2$max in the two exercises was also found, but not as large as reported by Taylor and others (144). Arm-plus-leg exercise as compared to leg exercise alone most commonly produced a difference of only a few percent (0%–10%). It is noteworthy that when there was a difference, it was primarily explained by a wider a-vO_2 difference and not by an elevated cardiac output (142). The critical question that emerged was: what minimum amount of muscle mass must be engaged in the exercise to obtain a true measurement of $\dot{V}O_2$max? As shown by Andersen and Saltin (1), peak O_2 uptake by limb skeletal muscle is on the order of 0.2 (sedentary) to 0.4 (well trained) l min^{-1} kg^{-1}. Thus, a muscle mass of only 10–15 kg needs to be involved in order for the exercise to elicit $\dot{V}O_2$max. In healthy normal humans, this is accomplished when bicycling or running.

THE OXYGEN TRANSPORT SYSTEM

Determinants of Maximal Oxygen Uptake

In Hill's article with Long and Lupton in 1924 (59), the dominant discussion relates to the factors that determine maximal oxygen uptake (Fig. 6.12). They considered primarily the role of breathing in saturating the arterial blood with oxygen and the output of the heart, but they also touched upon the oxygen capacity of the blood and the degree of desaturation of the venous blood. In their experiments with O_2-enriched air, they saw consistently higher $\dot{V}O_2$max values than when breathing only air. Much later, Welch and associates (151) proposed that the variation and magnitude of the effect of hyperoxia suggested a methodological limitation; however, in the early work an elevation in performance was nevertheless observed. Based on this finding, Hill and his associates concluded that "It is necessary, however, to assume that a rich oxygen mixture works primarily by increasing the saturation of the blood with oxygen; there is no other way it may work" (59). Not having measured cardiac output, Hill and colleagues estimated what may be the flow rate. As they assumed a rather low oxygen carrying capacity of the blood (it appears as if they were unaware of hemoconcentration during severe exercise), a low arterial O_2 saturation (~85%; possibly correct as the respiratory valve and tubes added resistance to breathing) and a low O_2 extraction (~60%), the calculated demand on the heart was high. Stroke volume was estimated to be 170–220 ml, cardiac output to be 30–40 l min^{-1}, and coronary blood flow was thought to have had to be twice the volume of the heart. While these values are in the right range as we know them today for very well-

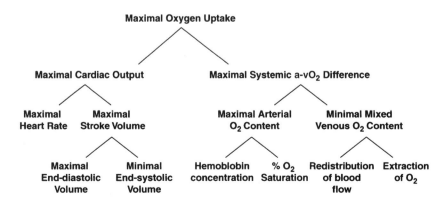

Fig. 6.12. Determinants of maximal oxygen uptake. Scheme indicating the key variables for oxygen transport and utilization by the human body.

trained people with high maximal oxygen uptakes (>5.0 l min^{-1}), they are too high for the subjects studied by Hill and others (59).

Herbst also discussed the reason for the variation in maximal oxygen uptake (55). He cites Hill and coworders (59) and points to the heart and its peak output, as well as citing a study by Himwich and Barr (62) suggesting that O_2 saturation may become reduced during exercise. Rowell and others (123) were the first to observe desaturation in intense exercise. Later, Dempsey and colleagues confirmed (35) that this occurs in some well-trained subjects and it appears from a wealth of more recent studies that it is a common phenomenon in many well-trained individuals (81). Moreover, those with the very highest $\dot{V}O_2$max values do not always desaturate. The phenomenon is most common in athletes (112) when the exercise engages a large fraction of the muscles of the body and breathing is coupled to the cadence of the movements of the extremities as in running and rowing.

Herbst also highlights another important point (55). Based on Hill's finding of an effect of O_2-enriched breathing on improved performance and $\dot{V}O_2$, Herbst concluded that skeletal muscle metabolism cannot be at its peak in normoxic exhaustive exercise, and thus cannot limit $\dot{V}O_2$max. This last conclusion is reiterated by Bainbridge, Boch, and Dill in a chapter (11) on the limits of muscular exertion in which they say that "this limit is imposed by the supply of oxygen to the muscles and brain rather than the functional capacity of skeletal muscles." They also base their conclusion on experiments where performance is enhanced when pure oxygen is given in connection with all-out exertion, in this case performed by Hill and Flack in 1910 (61).

A review article by Christensen, Krogh, and Lindhard in 1934 also addressed the question of what limits maximal oxygen uptake (29). The article essentially summarized all relevant studies in the field and included a section called "The Oxygen Supply in Work of Maximum Intensity." The article was a report to the predecessor of the World Health Organization in Geneva. They list the following factors as of importance for an individual's exercise capacity: (1) pulmonary ventilation, (2) the rate of diffusion of oxygen from alveolar air to the blood, (3) the minute volume of the heart, (4) the rate of blood flow through the muscles, and (5) the conditions for O_2 diffusion from the muscle capillaries to the tissue. On points 3–5, they reiterate what others had concluded before them, but some of the remarks under points 1 and 2 were novel and deserve to be mentioned. Liljestrand used estimates of cost of breathing based on experiments in which CO_2 rebreathing was used to elevate ventilation (92). From rest to a ventilation approaching 100 l min^{-1}, O_2 cost increased in proportion to ventilation with a factor of 4–5. Another argument for a possible limitation of the lungs in severe exercise came with the demonstration by August and Marie Krogh that the estimated maximum rate of diffusion of O_2 from alveoli to blood was 4.0–4.5, or possibly 5 l min^{-1} in exceptional cases (96). In a theoretical and experimental analysis, the value for peak O_2 diffusion in the lungs was later raised to 6.0 l min^{-1} (32). Thus, the points that were addressed in those early days in regard to the physiological meaning of and limits to $\dot{V}O_2$max were the same as today, and so were the arguments. The capacity of the lungs and of the heart was thought to be the most likely explanation for a limitation.

THE PHYSIOLOGICAL MEANING OF MAXIMAL OXYGEN UPTAKE

Cardiac Output and Systemic a-vO₂ Difference

As outlined in Figure 6.12, maximal oxygen uptake is a function of maximal cardiac output and systemic oxygen extraction. Foreign gas methods were used early in the twentieth century to estimate cardiac output at rest and during exercise up to exhaustive work (20, 97). Although the method became heavily criticized, especially when exercising, the early data have proven to be reasonably similar to what was later found using invasive techniques when the proper solubility constant for acetylene was applied (26, 125). The first to use dye dilution with arterial samples of blood during exercise were Asmussen (1907–1991) and Nielsen (1903–2000) in the early 1950s (2). They measured cardiac output during very intense bicycle exercise although maximal exertion may not have been achieved. Their subjects reached cardiac outputs of ~22 l min⁻¹ at an oxygen uptake of 3.2 l min⁻¹ giving an a-vO₂ difference of 150 ml l⁻¹. Similar values were provided by Freedman and colleagues some years later (45).

Mitchell, Sproule, and Chapman (102) were the first to provide data on the "missing determinants" at maximal oxygen uptake depicted in Figure 6.12. In their 1958 article entitled, "The Physiological Meaning of the Maximal Oxygen Intake Test" they determined the values for cardiac output (heart rate and stroke volume) and total body a-vO₂ difference at maximal oxygen uptake in untrained subjects. Peak cardiac output was 23.4 l min⁻¹ at a maximal oxygen uptake of 3.22 l min⁻¹. From the Fick principle, the estimated a-vO₂ difference was 143 ml l⁻¹, arterial oxygen content was 193 ml l⁻¹, and mixed-venous blood O₂ content was calculated to be 50 ml l⁻¹. The latter value was slightly above the oxygen content found in blood from the femoral vein during exercise, which was 35 ml l⁻¹. Thus, definite answers were given to questions which had been discussed for several decades. For a person with a moderate $\dot{V}O_2$max, oxygenation of blood was not impaired in going from rest to exhaustive exercise, as PaO₂ remained unchanged (87/88 mmHg). The small reduction in O₂ saturation from 97.1 to 94.7% was a function of a shift in acid–base balance, pH dropping to 7.19. A 9.5-fold elevation in oxygen uptake was brought about by a 2.3-fold widening of the a-vO₂ difference and a 4.3-fold increase in cardiac output. Heart rate and stroke volume contributed about equally to the elevation in the cardiac output when subjects were in an upright position both at rest and during exercise. Of particular note is that although O₂ extraction was markedly increased in these studies, blood draining an exercising limb still contained appreciable O₂, an indication of incomplete O₂ extraction by contracting muscle at $\dot{V}O_2$max. The large widening of the a-vO₂ difference in exercise is primarily a function of a larger fraction of the cardiac output being directed to the exercising limbs, which are able to more completely extract the O₂. The above findings have been confirmed in later studies with similar exercise paradigms and measurements: that is, the absolute values for maximal cardiac output and arterially transported O₂ vary among people and are the primary variables most closely related to $\dot{V}O_2$max and exercise capacity.

Mitchell and colleagues (102) also asked whether there was a fixed relation between $\dot{V}O_2$max and maximal cardiac output. In the study, cardiac output was also measured at an exercise intensity above that which elicited $\dot{V}O_2$max. At this intensity, the same maximal $\dot{V}O_2$ was achieved but with a 2.8 l min^{-1} lower cardiac output and a widening of the a-vO_2 from 14.3 to 15.7 ml l^{-1} (102). The explanation is that the more intense exercise elicits a pronounced increase in sympathetic activity, which reduces blood flow to skin and other nonactive tissues, thereby directing a larger fraction of the cardiac output to the contracting muscles. Additional support for this concept is obtained from measurements early (~3–4 minutes) and late (~7 minutes) during running (128, Table J). $\dot{V}O_2$max is the same while cardiac output is 2 l min^{-1} higher at the end of the run. Moreover, if the exercise is performed in a hot environment, cardiac output is also ~3 l min^{-1} higher at peak exercise, eliciting $\dot{V}O_2$max (125). Thus, there does not appear to be a fixed relationship between $\dot{V}O_2$max and maximal cardiac output in an individual. This should not detract from the early finding of a tight relationship between cardiac output and $\dot{V}O_2$, the ratio being 5 to 1: that is, for each liter of O_2 uptake above the resting value, there is an increase in cardiac output of about 5 l min^{-1}.

Cardiac Dimensions and Function

There is a very close relationship between the size of the heart and the cardiac output. This was suggested by very early studies on heart volume (44). Later, in 1899, Henschen in Sweden used the palpation technique to estimate heart size of cross-country skiers and found a positive relationship between this variable and success in skiing (54). This observation has since been confirmed with more direct measurements. Keys and Friedel (75) used X-ray in the posterior–anterior projection and observed larger diastolic and stroke areas in athletes compared to sedentary subjects, while systolic area was similar in the two groups. Further refinements were made by Jonsell (69) and Nylin (107) in Sweden in the 1930–1940s. They developed a method to simultaneously measure the size of the heart in diastole in two planes at a 90° angle. This method was later used in Sweden, but most extensively by Reindell (113) in Germany, more precisely linking heart size with the performance of the heart and exercise capacity than was previously possible. Stroke volume of the heart is larger in the best-trained subjects (148). While none of these groups measured $\dot{V}O_2$max, they all demonstrated very close relationships between heart size and their measures of exercise capacity. This relationship was first confirmed in the 1960s when $\dot{V}O_2$max was determined together with other circulatory variables (64). It has now been well established by echocardiography that eccentric hypertrophy is present in endurance-trained athletes (74). Also, a close correlation has been demonstrated between the magnitude of this adaptation of the left ventricle using magnetic resonance imaging and the level of maximal oxygen uptake in both males and females (101, 117).

Blood Volume and Hemoglobin

The role of blood volume and the total amount of hemoglobin in determining $\dot{V}O_2$max has been long debated. In 1965, Grande and Taylor (48) summarized matters in writing that, "at submaximal exercise, plasma volume affected the stroke volume and heart rate response, but it was more questionable whether this was the case also at peak exercise rates." They based this on the finding that plasma volume could be reduced acutely up to 20% without a decrease in cardiac output and $\dot{V}O_2$max (125). In contrast, previous studies in the 1940s, 1950s, and early 1960s found close relationships between both blood volume and total hemoglobin and maximal values for stroke volume, cardiac output, and maximal oxygen uptake (40, 64). Of note is that only a few training bouts may be needed to expand the plasma volume enough to enlarge peak stroke volume as demonstrated by Green and associates (49). Training also stimulates red cell production and thus increases the total amount of hemoglobin, a process which takes considerable time (months) as first convincingly shown in the late 1940s (76). However, there is little or no change in the hemoglobin concentration of the blood.

An elevation in hemoglobin concentration with only minor changes in blood volume also increases $\dot{V}O_2$max, as most clearly demonstrated in Ekblom (1937–) and coworkers' experiments in which they gave a red cell infusion to their subjects (43). Conversely, a lowering of the hemoglobin concentration—anemia—brings about a low exercise capacity and $\dot{V}O_2$max (13, 141). This is in line with systemic oxygen delivery at peak exercise being most closely related to $\dot{V}O_2$max (71).

FACTORS INFLUENCING THE LEVEL OF MAXIMAL OXYGEN UPTAKE

A question of great interest is: how high was the maximal oxygen uptake of our ancestors? A precise answer to this question cannot be given since such a measurement has been performed for less than 100 years. However, there are different approaches to providing a reasonable estimate.

One estimate was made during the International Biological Programme in the 1960s when various ethnic groups were to be physically characterized and the key measure was their maximal aerobic power (88). These studies demonstrated that $\dot{V}O_2$max varied markedly between ethnic groups of the world and appeared to be a function of the demand put upon them in their daily life. People with a sedentary lifestyle, as on Easter Island, achieved a $\dot{V}O_2$max of ~35 ml·kg^{-1}·min^{-1} (42) among the young adults. In contrast nomadic Lapps had, as long as they took care of their reindeer, a value of 50 ml·kg^{-1}·min^{-1} in $\dot{V}O_2$max, with a very narrow range around this mean value (89). This latter value is of interest as it may represent what was the "normal" $\dot{V}O_2$max in ancient time for active, healthy adults. Whipp's note on estimated aerobic capacity of Roman legionnaires indicated that it may have been in the range of 50 ml·kg^{-1}·min^{-1} (152). Whipp's estimation was based on reported values of body size of the soldiers, the distances they marched, and the speeds they maintained. Similar estimates of soldiers in the army of Alexander the Great, marching as far as India, indicate that $\dot{V}O_2$max was also likely in the range of 50–55 ml·kg^{-1}·min^{-1}.

Age and Gender

Sid Robinson (Fig. 6.13) in 1938 was the first to study the effect of age on $\dot{V}O_2$max (118). From the age of 7 years, boys increased their $\dot{V}O_2$max as a function of growth, reaching the highest level in young adulthood. Normalizing for body weight, the increase was small and thus reached just below 50 ml·kg^{-1}·min^{-1} as the peak $\dot{V}O_2$max. The oldest subjects were 75 years of age and they reached half the $\dot{V}O_2$max of the young adult, i.e., a reduction of ~0.5 ml·kg^{-1}·min^{-1} per year.

Fig. 6.13. Sid Robinson (1902–1981). Ran the 1500 meters in the 1928 Olympics and then coached track at the University of Indiana. He went to Harvard and in 1938 received his Ph.D. degree in physiology working with Dr. Bruce Dill in the Fatigue Laboratory. In 1939, he moved back to the University of Indiana and remained there for the rest of his academic career. Courtesy of Dr. Charles M. Tipton.

Per-Olof Åstrand (Fig. 6.14) was the first to include girls and young women in his investigations (in 1952), in addition to boys and young men (4). He found that up to puberty, girls had increases in $\dot{V}O_2$max similar to those of boys. Following puberty, however, he saw a graduated increase in the difference (~15%) between the sexes primarily due to a relatively low hemoglobin concentration and high percent of body fat in the girls. The first to perform studies on the effect of aging on $\dot{V}O_2$max in women was Irma Åstrand in Sweden, who confirmed that up to a 20% difference in $\dot{V}O_2$max existed between women and men (3). She also showed that the $\dot{V}O_2$max of females declined with age at a rate that was the same as that for men.

None of the above-cited studies used measurements from a random sample of subjects in a healthy population. During the last decades of the twentieth century, such measurements of $\dot{V}O_2$max have been performed in subsamples of large epidemiological studies. In these studies, mean values for $\dot{V}O_2$max are generally 5%–10% lower than those observed in previous studies of exercise capacity (22). This could be anticipated. Robinson (118), as well as the Åstrands (4), recruited healthy volunteers with a keen interest in being studied. In addition to this explana-

Fig. 6.14. Per-Olof Åstrand (1922–). He graduated from the Royal Gymnastics Institute in Stockholm in 1945. There he continued his exercise physiology studies with Erik Hohwü-Christensen (1904–1996) while he in parallel studied medicine at the Karolinska Institute. He defended his doctoral thesis in 1952. Together with his wife Irma Rhyming-Åstrand (1926–) they produced the nomogram for estimating $\dot{V}O_2$max from submaximal exercise testing (7). Courtesy of Dr. P.O. Åstrand.

tion for the difference, there could in more recent years be a general trend for a lower aerobic power in the population. Numerous follow-up studies performed throughout the world during the last three decades that have evaluated the $\dot{V}O_2$max status of various populations demonstrate mean values for young adult men more in the range of 40 to 48 ml·kg^{-1}·min^{-1}, and females are some 10% lower. More pronounced and also more important from a general health perspective, however, is the trend for a larger range in $\dot{V}O_2$max observed in these population studies, especially in the age groups from adolescence to the middle-aged (3). The focus on fitness in a small fraction of the population performing regular physical training as part of their lifestyle yields results reflecting the fact that they maintain quite high values up into old age (50, 53). However, the possibility of living an almost completely physically inactive life at work and in leisure time is also an option today in many countries, and it is the activity level preferred by most people. This adds a large group of people at the lower end of the range, which increases the risk for premature morbidity and mortality, as first shown by Morris and associates (104).

It is noteworthy that most studies mentioned above are cross-sectional—that is, at a given point in time, the various age groups were investigated. No population-based longitudinal studies exist on the effect of age on $\dot{V}O_2$max. Dill and others (37) were the first to report on 16 well-trained runners who were followed for 24 years. They found an average decline of 1.1 ml·kg^{-1}·min^{-1}·yr^{-1} (or 0.06 l·min^{-1}·yr^{-1}). This is a larger reduction with age than found later, where the decline is more on the range of 0.4–0.7 ml·kg^{-1}·min^{-1} both for sedentary subjects (53, 100) and physically active people, maintaining their activity level (5, 50, 73). Thus, the large drop in $\dot{V}O_2$max observed by Dill and others (37) is most likely explained by a reduced physical activity level adding to the effect of age.

Level of Habitual Activity

Inactivity

The first study to include accurate measurements of $\dot{V}O_2$max after bed rest in healthy volunteers was published in 1949 by Taylor and coworkers (145). $\dot{V}O_2$max was reduced by 17% or from 3.85 to 3.18 l min^{-1} after 20 days of bed rest. No detailed circulatory measurements were performed to determine the mechanism of this decline; however, such data are available in the Dallas bed rest and training study of 1968 (128). The $\dot{V}O_2$max of the five subjects in the Dallas study was reduced by 26% from 3.3 to 2.4 l min^{-1}. The slightly larger reduction than observed by Taylor and coworkers (145) was probably due to a more strict enforcement of inactivity. The reduction in $\dot{V}O_2$max was fully explained by a lowering of maximal cardiac output, which in turn was due entirely to a reduction in stroke volume (128). It is well known that bed rest causes people to become orthostatic, and it could be that this contributed to the lowering of stroke volume. In this regard, it is noteworthy that the reduction in stroke volume in the upright position was identical to that found during supine exercise. This finding suggests that an altered size and function of the heart may be the cause of the smaller stroke volume.

Physical Activity (Training)

Cross-sectional studies of $\dot{V}O_2$max in untrained and trained subjects and the effect of training were performed by Christensen along with an evaluation of the submaximal $\dot{V}O_2$, cardiac output, heart rate, and stroke volume responses (27). However, a true longitudinal study with $\dot{V}O_2$max measurements was published in 1941 when Robinson and Harmon (120) reported the changes in $\dot{V}O_2$max for nine men as a result of regular training for 26 weeks. The subjects were initially fit with an average $\dot{V}O_2$max of 53 ml·kg^{-1}·min^{-1}, which increased to just above 60 ml·kg^{-1}·min^{-1} or 17%. A parallel study performed at the Harvard Fatigue Laboratory by Knehr, Dill, and Neufeldt and published in 1942 demonstrated a rise in $\dot{V}O_2$max, but only by 4% (79). No circulatory variables were measured in these two studies.

In the 1960s, three investigations of exercise training were performed which included the measurement of circulatory variables. Those studies had similar designs, duration of training, and methods applied to evaluate, in detail, the cardiovascular response to exhaustive exercise before and after training. One was performed in Minneapolis by Rowell (122). Another study was performed in Dallas and included a period of inactivity (bed rest) before the training started (128). Ekblom and colleagues (41) performed the third study in Stockholm. Although the percentage increase in $\dot{V}O_2$max varied from 11% to 33% in the three studies, the mechanism by which the increase in $\dot{V}O_2$ was brought about was quite similar. In all three studies maximal cardiac output increased and maximal a-vO_2 difference widened. In the Dallas study, these two variables contributed equally, and this was also the case in the other two studies when considering the mean values of changes. Stroke volume alone contributed to the elevated maximal cardiac output as maximal heart rate was essentially unchanged in all three studies. A slightly more complete O_2 extraction by the exercising limb explained the enlarged systemic a-vO_2 difference. The larger heart volume contributed to the change in stroke volume.

The Endurance Athlete

Hill had an interest in the superb athlete and the highest $\dot{V}O_2$max value he observed was 65.7 ml·kg^{-1}·min^{-1} in a runner (59), a value similar to that of a Danish bicyclist (63.3 ml·kg^{-1}·min^{-1}) studied by Christensen (28). It was Robinson, Edwards, and Dill, however, who in 1937 made "human power" a special subject and reported that the best 2 mile runner at the time had a $\dot{V}O_2$max of 5.35 l min^{-1} or 81.5 ml·kg^{-1}·min^{-1} (119). Since that study, $\dot{V}O_2$max values of the same order of magnitude have often been reported, but recently, higher values have been found in some bicyclists and cross-country skiers. Ingjer has measured values greater than 90 ml·kg^{-1}·min^{-1} in the best Norwegian skiers (66). The first to report $\dot{V}O_2$max values for female athletes engaged in various sports were Saltin and Åstrand (127). They observed in the early 1960s values up to 70 ml·kg^{-1}·min^{-1}. Female elite athletes today are close to 80 ml·kg^{-1}·min^{-1} with the cross-country skiers having the highest values (66).

Genetics

In Grande and Taylor's chapter in the *Handbook of Physiology* (48), they comment upon a possible genetic explanation for the huge individual variation observed in $\dot{V}O_2$max at the time (30–80 ml·kg^{-1}·min^{-1}). Based on the available investigations of physical conditioning with 4%–17% (~9 ml·kg^{-1}·min^{-1}) increases in $\dot{V}O_2$max, they concluded that "the individual with a large $\dot{V}O_2$max per unit of body weight owes this characteristic primarily to the traits of the heart and circulatory system at the time when he reached his full growth."

Studies of twins further highlight the role of genetic factors influencing the $\dot{V}O_2$max and its response to training. To what extent genetics play a role, however, is still debated. In the early work of Klissouras, the heritage factor was low (77), but it was found to be strong in later studies by Bouchard (1939–) and associates, which indicate that up to 80% of the variation can be linked to heredity and only 20% to the level of physical activity (23). Some further support for a role for hereditary factors is found in the ongoing studies by Bouchard and colleagues (Family Heritage Study). Among many other physiological variables they are also investigating the response of $\dot{V}O_2$max to training (22). They found almost 5% of the sedentary people to be nonresponders to training in regard to $\dot{V}O_2$max—that is, in spite of the same training, no increase occurred. While there was a variation in the magnitude of the elevation in $\dot{V}O_2$max with a physical conditioning program in earlier studies, very seldom do the participants not improve at all (41, 120, 122, 128). A drawback with most longitudinal training studies is that they are not based on random samples of subjects but rather utilize people who have an interest in becoming fit. This may, at least in part, explain the difference between the Family Heritage Study and other training studies.

Another aspect of the response to training has also been highlighted. It has been found that the largest improvement in $\dot{V}O_2$max is observed in people with a low initial pretraining $\dot{V}O_2$max, a finding that applies to a wide range of age groups (124, 129). Those with a very high initial $\dot{V}O_2$max achieve a percentage-wise smaller improvement in $\dot{V}O_2$max than those with initial low levels of $\dot{V}O_2$max. The proposed explanation has been that the influence of training would be more dominant when the genetic potential is sufficient for a response to be expressed. It has not been defined where that limit in $\dot{V}O_2$max is and how it differs, most likely, among individuals. Based on a multitude of population studies, a $\dot{V}O_2$max of 45–50 ml·kg^{-1}·min^{-1} appears to be obtainable for healthy young individuals (124). The huge influence of physical activity below that level is underscored by the findings from the three sedentary subjects in the Dallas bed rest and training study (128). Following an inactivity period of 20 days, these three subjects lost 33% of their $\dot{V}O_2$max to reach a level of 24 ml·kg^{-1}·min^{-1}, a value which then doubled to 48 ml·kg^{-1}·min^{-1} with 2 months of subsequent training.

MAXIMAL OXYGEN UPTAKE AND ENDURANCE PERFORMANCE

In Hill's writings, the steady state concept was discussed, emphasizing "the equilibrium between processes building up versus breaking down" (57). If the latter was in excess of the former, there was no steady state and the exercise had to stop due to exhaustion. Hill also pointed out that the higher the $\dot{V}O_2$max, the higher the exercise intensity that could be performed under a steady-state condition. The first one to test a possible coupling between $\dot{V}O_2$max and performance was Herbst (55). His subjects performed a 3 mile run and Herbst noted an inverse relationship: the higher the subject's $\dot{V}O_2$max (expressed in $ml \cdot kg^{-1} \cdot min^{-1}$), the shorter was the running time. Herbst also plotted the running performance in relation to $\dot{V}O_2$max as a percent of the estimated energy demand for the running speed. Again, the higher the percent, the shorter the running time.

The percent of maximal oxygen uptake concept also appears in Robinson's study from 1938 of males of various ages (118), but he did not apply this concept to the capacity for endurance performance. However, in the parallel study performed at the Harvard Fatigue Laboratory, Knehr and coworkers (79) discussed the huge discrepancy that may exist in how much $\dot{V}O_2$max may change in comparison to changes in performance. They demonstrated an increase in $\dot{V}O_2$max of 4%, but Karpovich and Perstrecov (72) concomitantly demonstrated that time to exhaustion at a given exercise intensity not demanding $\dot{V}O_2$max increased tremendously after several weeks of training. The explanation is that time to exhaustion is *exponentially* related to the relative exercise intensity expressed in %$\dot{V}O_2$max (132). Thus, at a given workload, which may represent 90% $\dot{V}O_2$max before training, exhaustion may come after 20 minutes. With a 20% elevation in $\dot{V}O_2$max, the same exercise intensity now represents ~75% $\dot{V}O_2$max and time to exhaustion is elongated to 2–3 hours or at least four- to five-fold.

The related question is whether changes in activity level also affect the time one can perform at a given relative exercise intensity. P.-O. Åstrand and K. Rodahl have put forward this possibility (6), in part based on Irma Åstrand's studies published in 1960 (3). This is in contrast to Hill's explanation of the steady-state concept and accepts that expressing a given aerobic energy yield in percent of $\dot{V}O_2$max is a way to normalize people with different exercise capacities (57). Thus, performance should be quite similar at a given %$\dot{V}O_2$max regardless of training status. David Costill (1939–) and associates have worked extensively with these problems and found that elite marathon runners exercised at a similar %$\dot{V}O_2$max, in the range of 80%–90%, during a race as those finishing in the middle or at the end of the field (33).

The numerous studies highlighting the interplay between $\dot{V}O_2$max and relative exercise intensity, with endurance performance primarily being a function of the latter variable, should not overshadow the fact that among people with the same $\dot{V}O_2$max, endurance performance can be markedly different. Coyle and associates performed a study on this particular problem, with intriguing results (34). They argued that the explanation for a difference could be the metabolic capacity of the skeletal muscle engaged in the exercise. In bicyclists with a $\dot{V}O_2$max of 67 $ml \cdot kg^{-1} \cdot$

min^{-1}, he used the blood lactate response to submaximal exercise to assign well-trained bicyclists into two groups: low and high lactate responders. Endurance performance was markedly different in the two groups. The high lactate responders could only bicycle half the time at 88% $\dot{V}O_2$max compared to the low lactate responders. Leg muscles were studied for mitochondrial marker enzymes, fiber types and size, as well as capillaries. Most surprisingly, there was no difference between groups in regard to mitochondrial enzyme activity and capillaries per muscle fiber. Thus, neither the lactate response nor the performance could be related to these indices of muscle adaptation. Both variables are easily affected by changes in physical activity level and proposed to be the link to a more efficient metabolism of the muscle (47, 63, 103), although it may vary in spite of whether the exercise is performed at the same relative work load (12). Two other muscle indices differed. Those with a poor performance had less type I muscle fibers than the good performers (47 vs. 67%) and they had a 15% larger mean muscle fiber size. Muscle fiber types are not easily fully changed from type II to type I muscle fibers in humans with training although more IIa fibers will coexpress the myosin heavy chain I isoform (78). Nevertheless, it is difficult to relate the observed differences in lactate and performance responses to an effect of training. Rather, heredity may be critical as fiber-type composition is influenced by genetic endowment (80). It is difficult to envisage, however, how differences in myosin heavy-chain composition can explain the very different metabolic responses in the two groups as their mitochondrial enzyme profiles are so similar, taking into account the finding by Jansson and Kaijser that in well-trained endurance athletes, type II fibers are equally as oxidative as type I fibers (67). The larger fiber size in the group performing poorly probably plays a role as diffusion distances are large, but it can hardly be the whole explanation. We are then left with the conclusion that, generally speaking, expressing exercise intensity as a % $\dot{V}O_2$max normalizes the endurance performance; however, there is substantial individual variation. Very well-trained people are in the upper range of performance, whereas less trained or untrained are in the lower range of variation. The prevailing view to explain the variation that exists in time to exhaustion at a given % $\dot{V}O_2$max is then not only differences in capillaries and mitochondrial capacity of skeletal muscle, but also its fiber-type composition, which in turn has an effect on the metabolic response to exercise.

LIMITATIONS TO MAXIMAL OXYGEN UPTAKE

The view that Hill expressed in 1923–1924 on the meaning of the $\dot{V}O_2$max measurements became the unchallenged dogma (60). On the topic of what limits maximal oxygen uptake, no consensus has been reached, although almost a century of vigorous debate stimulated by the results obtained in many and occasionally brilliant experiments. A voice has been heard in these ongoing discussions arguing that every single link in the oxygen transport system is limiting. Others have argued that no single link can be pinpointed.

The strongest argument for skeletal muscle capacity being the limiting factor to $\dot{V}O_2$max comes from Weibel in Bern and Taylor at Harvard (143, 149). Their data are impressive and comprise studies relating $\dot{V}O_2$max and total volume of muscle mitochondria in various species of animals with weights ranging from 1 g to ~400–500 kg (horses). Based on a close relationship in a log-log plot of these two variables, they argue that the muscle mitocondrial content sets the ultimate limit for $\dot{V}O_2$max (65, 150). This may be the case in certain quadrupeds, but hardly in humans (121). Abrupt elevation in the maximal amount of oxygen delivered to the muscle results in elevated oxygen consumption. Several interventions have been used to achieve this goal. O_2-enriched gas has been mentioned above as well as red cell infusion, but reduction in blood volume or Hb concentration as well as changing heart rate in a subject with a pacemaker alters $\dot{V}O_2$max as a function of arterial oxygen delivery (131). Another approach to demonstrate that mitochondrial respiration has a marked overcapacity to utilize O_2 comes from experiments in humans who perform exercise with a small muscle mass. Muscle blood flow is several times higher than that provided when a large muscle mass is engaged in the exercise (bicycling, running). Muscle oxygen uptake follows the increase in blood flow and arterial oxygen delivery as does the power output. Peak muscle oxygen uptake per kilogram of muscle can be as high as 0.3–0.5 l min^{-1} (1, 116). Measurements of Vmax of aerobic enzymes such as pyruvate dehydrogenase and α-keto oxygluterate dehydrogenase, having the lowest activity of all mitochondria enzymes, also provide evidence for this conclusion (19).

Further evidence for skeletal muscle capacity being the limiting factor comes from the work of Jan Pretorius Clausen (1939–1976) in Copenhagen, which focused on the interplay between cardiac output and peripheral vasodilation during maximal exercise (30). These studies included both a cross-sectional examination of untrained and well-trained subjects and a longitudinal investigation of subjects before and after training. The main finding was that the lower the total peripheral resistance the higher the $\dot{V}O_2$max. Moreover, a low mixed-venous oxygen saturation was related to a low peripheral resistance. At the time of these studies, skeletal muscle adaptation to endurance training (increased capillarization and mitochondrial capacity; 63, 130) was paramount and influenced the interpretation of the finding by Clausen and his colleagues (30, 31). Thus, they concluded that the training-induced changes at the muscle level were a prerequisite for a larger muscle blood flow and an enhanced O_2 extraction by the muscle. A one-legged training study by Saltin and associates in 1976 strengthened this view (130). They found that the peak $\dot{V}O_2$ and endurance increased more in the trained leg than in the untrained leg. This was compatible with the observed muscle adaptation in regard to increased capillarization and mitochondrial capacity, which was highest in the endurance-trained leg. Thus, in the late 1970s, the prevailing view of what limits maximal oxygen uptake, especially in trained subjects, shifted from the output of the heart to the capacity of skeletal muscle to accommodate high blood flows and to extract O_2. This view has been extensively elaborated upon primarily by Wagner and associates (115, 147).

Further studies in Copenhagen by Niels H. Secher, however, swung the pendulum back toward oxygen delivery being the limiting factor. He demonstrated that adding arm exercise to ongoing heavy leg exercise induced vasoconstriction and reduced leg blood flow (135). The conclusion of this work was that the heart could not provide a sufficient cardiac output to maintain blood pressure and provide adequate perfusion of all capillary beds and thus an adequate oxygen delivery. Blood pressure at peak exercise was unaffected by adding active muscle mass, implying that the total peripheral resistance was reduced (135). Further support for a key role of increased cardiac output to explain the effect of endurance training on $\dot{V}O_2$max came from subjects who trained with either their arms or their legs (31). Arm training elevated peak arm $\dot{V}O_2$ but not leg $\dot{V}O_2$max, whereas leg training increased both peak arm and leg $\dot{V}O_2$max. The explanation proposed by Clausen and colleagues was that arm training was not adequate to increase cardiac output, but leg training was (31). Thus, when exercising with the untrained arms after leg training, a higher blood pressure was generated due to an elevation in peak cardiac output and a strong vasoconstriction of vessels in nonexercising limbs, forcing a larger perfusion through the arm muscle with an elevated peak arm $\dot{V}O_2$ as a result.

When the concept of a peripheral limitation to $\dot{V}O_2$max, in part overcome by training, was reexamined, it was proposed to consider the role of improved cardiac capacity from another perspective. A greater cardiac output would be a lesser threat to a drop in blood pressure and thus cause less of an effect of vasoconstrictor activity in the active muscle during the intense exercise, resulting in a lowering of the peripheral resistance. This would explain the findings of Clausen—that those with the highest $\dot{V}O_2$max have the lowest peripheral resistance (30). This interpretation should be viewed in the light of the concept of functional sympatholysis presented in 1958 by Remensnyder and associates (114), which is discussed in greater detail in Chapter 5 of this volume. During intense muscular work with a small muscle mass, locally produced metabolites in the muscle not only induce vasodilation by relaxing the smooth muscles but also inhibit the effect of an elevated sympathetic activity by partially or fully blocking the action of norepinephrine on the α (α_1)-receptor of vascular smooth muscle (70, 133). Such a regulatory control mechanism could play a major role in optimizing perfusion of skeletal muscle during exercise, and, at peak exercise, ensure that skeletal muscle conductance can be matched to the pumping capacity of the heart in order to maintain blood pressure. This effect may become attenuated in intense exercise involving a large muscle mass, especially in the untrained individual with a low cardiac capacity. Increasing the pump capacity of the heart with training would lessen the need for sympathetically mediated vasoconstriction, which could be brought about by lowering the sympathetic discharge to contracting skeletal muscle but also by a more potent functional sympatholysis. Moreover, in this perspective, the primary role of enlarging the capillary bed in skeletal muscle with training is to maintain or possibly increase mean transit time, which allows for a more complete O_2 extraction when O_2 delivery is elevated (126). The role for the enlarged mitochondrial volume and enhanced mitochondrial enzyme capacity in this process is still not fully clarified, but as it has been argued by several, it is of less importance for $\dot{V}O_2$max than it is for the choice of more lipid as substrate rather than carbohydrate in the metabolism of the skeletal muscle (149).

Continuing discussions of limits to maximal oxygen uptake commonly try to pinpoint one specific link to be *the* limitation. As advocated by di Prampero, such an approach is nonphysiological and overshadows the close interaction that exists between the links in securing a continuous oxygen supply to tissues, most especially to contracting muscle during exercise (111), without compromising blood pressure (136). He has analyzed the oxygen transport system by considering the various links in the chain as resistances in Ohm's law. By using cross-sectional as well as longitudinal (inactivity–activity) data, he has been able to assign a resistance value to each link. Instead of having one link being the limiting factor, he has estimated the relative role of each link. The largest roles (= lowest resistance) are assigned to cardiac output and muscle blood flow. Other links such as the lungs, [Hb], capillarization, and mitochondrial capacity also play a role, but their resistances are considerably higher. What di Prampero's estimations provide is information on the relative contribution and thereby the functional significance of a link rather than whether it "limits" the maximal oxygen uptake. From a physiological perspective, di Prampero's approach is a more constructive way to discuss the issue rather than to try to identify one single limiting factor.

With this perspective, we return to the relative role that some of the steps in the oxygen transport may constitute. In a subject who desaturates during exhaustive exercise, even if the reduction is very small and for which there could be several reasons, the lungs are to some extent limiting the maximal oxygen uptake. Thus, without this desaturation, maximal oxygen uptake would have been higher, as proven by Powers and coworkers (110). However, the magnitude of reduction in $\dot{V}O_2max$, due to incomplete oxygen saturation of the hemoglobin molecules may be 0.1–0.3 l min^{-1}, which may be up to 5% of the $\dot{V}O_2max$ in most humans. When examining the transport of O_2 that occurs from the red blood cells in the capillaries to the myoglobin and the mitochondria in the muscle cell, a similar point can be made. All the O_2 in the blood is not extracted in exercising muscle as 5–30 ml l^{-1} remains in the blood draining the muscle at $\dot{V}O_2max$. As in the lungs, there could be many explanations for the unextracted oxygen; however, if extraction were complete, $\dot{V}O_2$ would have been higher as elaborated upon by Wagner and colleagues (115, 147). They define this as a diffusion limitation in contracting skeletal muscles at $\dot{V}O_2max$ which may amount to ~0.1–0.4 l min^{-1} of O_2 that could have been extracted or less than 10% of $\dot{V}O_2max$. Thus, using di Prampero's approach, it is possible to evaluate to what extent differences in an individual's $\dot{V}O_2max$ can be explained by variation in incomplete oxygenation of the blood in the lungs, hemoglobin content, cardiac output, distribution of the systemic blood flow, and incomplete O_2 extraction by active muscle (111). Such an analysis demonstrates that the conclusions reached by the pioneers in the field were right, as peak cardiac output is the variable varying the most (two- to three-fold) in healthy subjects and thus can be said to have the largest impact on the achieved maximal oxygen uptake.

CONCLUSIONS

Studies to unravel the mechanisms involved in oxygen transport including its maximal capacity in humans during exercise have been performed for a little more than

200 years. During this time, major advances have been made by gifted scientists from all over the world. In the early days before rapid and widespread communication of scientific information, important discoveries were sometimes made in parallel, but more frequently they were made in series with an investigator being highly dependent on the previous discoveries of others.

In this context, it should be appreciated that most of the important intellectual concepts and hypotheses in our understanding of the oxygen transport system and its limitations were clearly proposed many years ago by giants in the field. A review of the historical record clearly distinguishes the major contributors to our knowledge in this area. Among the giants are A.L. Lavoisier, E. Smith, N. Zuntz, F.G. Benedict, A. Krogh, G. Liljestrand, A.V. Hill, R. Herbst, H.L. Taylor, S. Robinson, and P.-O. Åstrand. Subsequent discoveries have solidified these positions and provided better quantification of the important factors or links in the process. However, in order to reach a more fundamental understanding of the molecular and integrative aspects of the movement of oxygen from inspired air to energy-yielding mitochondria, major contributions still have to be made. They range from identifying genes of importance for $\dot{V}O_2$max and how they are activated to the very subtle and precise interplay between central nervous factors and reflexes to match and distribute the available cardiac output optimally to active muscle and other central organs at maximal exercise.

The authors gratefully acknowledge the helpful criticisms of Dr. George Ordway and the preparation of the text by Inge Holm.

REFERENCES

1. Andersen P. and B. Saltin. Maximal perfusion of skeletal muscle in man. *J. Physiol. (Lond.)* 366:233–249, 1985.
2. Asmussen E. and M. Nielsen. The cardiac output in rest and work determined simultaneously by the cetylene and the dye injection methods. *Acta Physiol. Scand.* 27:217–230, 1952.
3. Åstrand I. Aerobic work capacity in men and women with special reference to age. *Acta Physiol. Scand.* 49 (Suppl. 169):1–92, 1960.
4. Åstrand P.-O. *Experimental Studies of Physical Working Capacity in Relation to Sex and Age.* Copenhagen: Munksgaard, 1952.
5. Åstrand P.-O., U. Bergh, and A. Kilbom. A 33-year follow-up of peak oxygen uptake and related variables of former physical education students. *J. Appl. Physiol.* 82:1844–1852, 1997.
6. Åstrand P.-O. and K. Rodahl. *Textbook of Work Physiology. Physiological Bases of Exercise.* New York: McGraw-Hill, 1986, pp. 420–422.
7. Åstrand P.-O. and I. Ryhming. A nomogram for calculation of aerobic capacity (physical fitness) from pulse rate during submaximal work. *J. Appl. Physiol.* 7:218–221, 1954.
8. Åstrand P.-O. and B. Saltin. Maximal oxygen uptake and heart rate in various types of muscular activity. *J. Appl. Physiol.* 16:977–981, 1961.
9. Åstrand P.-O. and B. Saltin. Oxygen uptake during the first minutes of heavy muscular exercise. *J. Appl. Physiol.* 16:971–976, 1961.

10. Atwater W.O. and F.G. Benedict. Experiments on the metabolism of matter and energy in the human body. Bulletin 69, U.S. Dept. of Agriculture, Office of Experiment Stations. Washington: Government Printing Office, republished 1988.

11. Bainbridge F.A., A.V. Bock, and D.B. Dill. *The Physiology of Muscular Exercise* (1st ed. 1923, 2nd ed. 1931). London: Longmans, Green, pp. 171–176.

12. Baldwin J., R.J. Snow, M.F. Carey, and M.A. Febbraio. Muscle IMP accumulation during fatiguing submaximal exercise in endurance trained and untrained men. *Am. J. Physiol.* 277 (*Regul. Integr. Comp. Physiol.* 46):R295–R300, 1999.

13. Balke B., G.P. Grillo, E.B. Konecci, and U.C. Luft. Work capacity after blood donation. *J Appl. Physiol.* 7:231–238, 1954.

14. Barnard J. and J. Holloszy. Chapter 7 this volume.

15. Benedict F. An apparatus for studying the respiratory exchange. *Am. J. Physiol.* 24:345–374, 1909.

16. Benedict F.G. and W.G. Cady. *A Bicycle Ergometer with an Electric Brake*. Carnegie Institution of Washington, Publication 167. New York: Isaac H. Blanchard Company, 1912.

17. Benedict F.G and E.P. Cathcart. *Muscular Work*: A metabolic study with special reference to the efficiency of the human body as a machine. Carnegie Institution of Washington, publication 187, 1913.

18. Bergh U., I.-L. Kanstrup, and B. Ekblom. Maximal oxygen uptake during exercise with various combinations of arm and leg work. *J. Appl. Physiol.* 41:191–196, 1976.

19. Blomstrand E., G. Rådegran, and B. Saltin. Maximum rate of oxygen uptake by human skeletal muscle in relation to maximal activities of enzymes in the Krebs cycle. *J. Physiol.* (*Lond.*) 501:455–460, 1997.

20. Bock A.V., D.B. Dill, and J.H. Talbott. Studies in muscular activity. I. Determination of the rate of circulation of blood in man at work. *J. Physiol.* (*Lond.*) 66:121–132, 1928.

21. du Bois-Reymond E. Über angeblich saure Reaction des Muskelfleisches. *Gesammelte Abhandl. Zur allg. Muskel- u. Nervenphysik.* 2–36, 1877.

22. Bouchard C., P. An, T. Rice, J.S. Skinner, J.H. Wilmore, J. Gagnon, L. Perusse, A.S. Leon, and D.C. Rao. Familial aggregation of VO_{2max} response to exercise training: results from the Family HERITAGE Study. *J. Appl. Physiol.* 87:1003–1008, 1999.

23. Bouchard C., R. Lesage, G. Lortie, J.A. Simoneau, and P. Hamel. Aerobic performance in brothers dizygotic and monozygotic twins. *Med. Sci. Sports Exerc.* 18:639–646, 1986.

24. Bouny E. Mesure du travail dépénse dans l'emploi de la bicyclette. Paris Ac. Sci O.R. 122:1891–1895, 1896.

25. Chapman C.B. Edward Smith (?1818–1874). Physiologist, human ecologist, reformer. *J. Hist. Med. Allied Sci.* 22:1–26, 1967.

26. Chapman C.B., H.L. Taylor, C. Borden, R.V. Ebert, A. Keys, and W.S. Carlson. Simultaneous determinations of the resting arteriovenous oxygen difference by the acetylene and direct Fick methods. *J. Clin. Invest.* 20:651–659, 1950.

27. Christensen E.H. Beträge zur Physiologie schwerer körperlicher Arbeit. IV. Mitteilung: Die Pulsfrequenz während und unmittelbar nach schwerer körperlicher Arbeit. *Arbeitsphysiologie* 4:453–469, 1931.

28. Christensen E.H. Beträge zur Physiologie schwerer körperlicher Arbeit. V. Minutenvolumen und Schlagvolumen des Herzens während schwerer körperlicher Arbeit. *Arbeitsphysiologie* 4:470–502, 1932.

29. Christensen E.H., A. Krogh, and J. Lindhard. Recherches sur l'effort musculaire intense. Bulleting trimestriel de l'organisation d'hygiène. *Société des Nations* 3:407–439, 1934.

30. Clausen J.P. Circulatory adjustments to dynamic exercise and effects of physical training in normal subjects and in patients with coronary artery disease. *Prog. Cardiovasc. Dis.* 18:459–495, 1976.

31. Clausen J.P., K. Klausen, B. Rasmussen, and J. Trap-Jensen. Central and peripheral circulatory changes after training of the arms or legs. *Am. J. Physiol.* 225:675–682, 1973.

32. Cohn J.E., D.G. Carroll, B.W. Armstrong, R.H. Shepard, and R.L. Riley. Maximal diffusing capacity of lung in normal male subjects of different ages. *J. Appl. Physiol.* 6: 588–597, 1954.

33. Costill D.L., H. Thomason, and E. Roberts. Fractional utilization of the aerobic capacity during distance running. *Med. Sci. Sports* 5:248–252, 1973.

34. Coyle E.F., A. Coggan, M. Hopper, and T. Walters. Determinants of endurance in well-trained cyclists. *J. Appl. Physiol.* 64:2622–2630, 1988.

35. Dempsey J.A., P.G. Hamson, and K.S. Henderson. Exercise-induced arterial hypoxaemia in healthy human subjects at sea level. *J. Physiol. (Lond.)* 355:161–175, 1984.

36. Dempsey J.A. and B. Whipp. Chapter 4 this volume.

37. Dill D.B., S. Robinson, and J.C. Ross. A longitudinal study of 16 champion runners. *J. Sports Med. Phys. Fitness* 7:4–27, 1967.

38. Douglas C.G. and J.S. Haldane. The causes of absorption of oxygen by the lungs. *J. Physiol. (Lond.)* 44:305–354, 1912.

39. Döbeln W.v. A simple bicycle ergometer. *J. Appl. Physiol.* 7:222–224, 1954.

40. Döbeln W.v. Maximal oxygen intake, body size, and total hemoglobin in normal man. *Acta Physiol. Scand.* 38:193–199, 1956.

41. Ekblom B., P.-O. Åstrand, B. Saltin, J. Stenberg, and B. Wallström. Effect of training on circulatory response to exercise. *J. Appl. Physiol.* 24:518–528, 1968.

42. Ekblom B. and E. Gjessing. Maximal oxygen uptake of the Easter Island population. *J. Appl. Physiol.* 25:124–129, 1968.

43. Ekblom B., A.N. Goldbarg, and B. Gullring. Response to exercise after blood loss and reinfusion. *J. Appl. Physiol.* 33:175–180, 1972.

44. Fick A. Über die Messung des Blutquantums in den Herzventrikeln. *Physmed. Ges. Worzburg,* July 9, 1870.

45. Freedman M.E., G.L. Snider, P. Brostoff, S. Kimelblot, and L.N. Katz. Effects of training on response of cardiac output to muscular exercise in athletes. *J. Appl. Physiol.* 8:37–47, 1955.

46. Gaertner G. Concerning the therapeutical application of muscle work and a new device for its implementation [in German]. *Wiener medizinische Blätter.* 10:1554–82; Allgemeine Wiener meizinische Zeitung. 32:607–21, 1887.

47. Gollnick P.D., R.B. Armstrong, C.W. Saubert, K. Piehl, and B. Saltin. Enzyme activity and fiber composition in skeletal muscle of untrained and trained men. *J. Appl. Physiol.* 33:312–319, 1972.

48. Grande F. and H.L. Taylor. Adaptive changes in the heart, vessels and patterns of control under chronically high loads. In: *Handbook of Physiology,* edited by W.F. Hamilton and P. Dow. Washington, DC: American Physiological Society, Circulation, Sect. 2, Vol. III, 1965, pp. 2615–2678.

49. Green H.J., J.A. Thomson, M.E. Ball, R.L. Hughson, M.E. Houston, and M.T. Sharratt. Alterations in blood volume following short-term supramaximal exercise. *J. Appl. Physiol.* 56:145–149, 1984.

50. Grimby G. and B. Saltin. Physiological analysis of physically well-trained middle-aged and old athletes. *Acta Med. Scand.* 179:513–526, 1966.

51. Gunga H.-C. *Leben und Werk des Berliner Physiologen Nathan Zuntz* (1847–1920). Husum: Matthiesen Verlag: 1989.

52. Haldane J.S. Some improved methods of gasanalysis. *J. Physiol. (Lond.)* 22:465, 1897–1898.

53. Heath G.W., J.M. Hagberg, A.A. Ehsani, and J.O. Holloszy. A physiological comparison of young and older endurance athletes. *J. Appl. Physiol.* 51:634–640, 1981.

54. Henschen E.S. *Skiddlauf und Skidwettlauf. Eine medizinische Sportstudie. Mitt. Med. klin.* Upsala: Jena Fischer Verlag, 1899.

55. Herbst R. Der Gasstoffwechsel als Mass der körperlichen Leistungsfähigkeit. I. Mitteilung: Die Bestimmung des Sauerstoffaufnahmevermögens beim Gesunden. *Deut. Arch. Klin. Med.* 162:33–50, 1928.

56. Hermansen L. Respiratory and circulatory responses to different types of exercise. In: Oxygen transport during exercise in human subjects, chapter III. *Acta Physiol. Scand.* (Suppl.) 399:19–36, 1974.

57. Hill A.V. The role of oxidation in maintaining the dynamic equilibrium of the muscle cell. *Proc. R. Soc. Lond.* 103:138–162, 1928.

58. Hill A.V. *Trails and Trials in physiology.* Baltimore: Williams and Wilkins, 1966, p. 83.

59. Hill A.V., C.N.H. Long, and H. Lupton. Muscular exercise, lactic acid and the supply and utilization of oxygen, IV–VI and VII–VIII. *Proc. R. Soc. Lond.* 97:84–138 and 155–176, 1924.

60. Hill A.V. and H. Lupton. Muscular exercise, lactic acid, and the supply and utilization of oxygen. *Q.J.M.* 16:135–171, 1923.

61. Hill L. and M. Flack. On the influence of oxygen inhalations on muscular work. *J. Physiol.* (*Lond.*) 40:347–364, 1910.

62. Himwich H.E. and D.P. Barr. Studies in the physiology of muscular exercise. Relationships in the arterial blood. *J. Biol. Chem.* 57:363–378, 1923.

63. Holloszy J.O. Biochemical adaptation in muscle. Effects of exercise on mitochondrial oxygen uptake and respiratory enzyme activity in skeletal muscle. *J. Biol. Chem.* 242: 2278–2282, 1967.

64. Holmgren A. and P.-O. Åstrand. D_L and the dimensions and functional capacities of the O_2 transport system in humans. *J. Appl. Physiol.* 21:1463–1470, 1966.

65. Hoppeler H., S.R. Kayar, H. Claassen, E. Uhlmann, and R.H. Karas. Adaptive variation in the mammalian respiratory system in relation to energetic demand. III. Skeletal muscles: setting the demand for oxygen. *Respir. Physiol.* 69:27–46, 1987.

66. Ingjer F. Maximal oxygen uptake as a predictor of performance ability in women and men elite cross-country skiers. *Scand. J. Med. Sci. Sports* 1:25–30, 1991.

67. Jansson E. and L. Kaijser. Muscle adaptation to extreme endurance training in man. *Acta Physiol. Scand.* 100:315–324, 1977.

68. Johansson J.E. Über die Einwirkung der Muskel auf die Atmung und die Hertz. *Skandinaviesches Archiv für Physiologie* 5:20–66, 1895.

69. Jonsell S. A method for the determination of the heart size by teleroentgenographie. *Acta Radiol.* 20:325–340, 1939.

70. Joyner M.J. Sympathetic modulation of blood flow and O_2 uptake in rhythmically contracting human forearm muscles. *Am. J. Physiol.* 263 (*Heart Circ. Physiol.* 32): H1078–H1083, 1992.

71. Kanstrup I.-L. and B. Ekblom. Blood volume and hemoglobin concentration as determinants of maximal aerobic power. *Med. Sci. Sports Exerc.* 16:256–262, 1984.

72. Karpovich P.V. and K. Perstrecov. Effect of gelatin upon muscular work in man. *Am. J. Physiol.* 134:300–309, 1941.

73. Kasch F.W. and J.P. Wallace. Physiological variables during 10 years of endurance exercise. *Med. Sci. Sports* 8:5–8, 1976.

74. Keul J., H.-H. Dickhuth, G. Simon, and M. Lehmann. Effect of static and dynamic exercise on heart volume, contractility, and left ventricular dimensions. *Circ. Res.* 48: I162–I170, 1981.

75. Keys A. and H.L. Friedell. Size and stroke of the heart in young men in relation to athletic activity. *Science* 88:456–458, 1938.

76. Kjellberg S.R., U. Rudhe, and T. Sjöstrand. Increase of the amount of hemoglobin and blood volume in connection with physical training. *Acta Physiol. Scand.* 19:146–151, 1949.

77. Klissouras V. Heritability of adaptive variation. *J. Appl. Physiol.* 31:338–344, 1971.
78. Klitgaard H., O. Bergman, R. Betto, G. Salviati, S. Schiaffino, T. Clausen, and B. Saltin. Co-existence of myosin heavy chain I and IIa isoforms in human skeletal muscle fibers with endurance training. *Pflugers Arch.* 416:470–472, 1990.
79. Knehr C.A., D.B. Dill, and W. Neufeld. Training and its effects on man at rest and work. *Am. J. Physiol.* 136:148–156, 1942.
80. Komi P.V. and J. Karlsson. Physical performance, skeletal muscle enzyme activities, and fibre types in monozygous and dizygous twins of both sexes. *Acta Physiol. Scand.* (Suppl.) 462:1–28, 1979.
81. Koskolou M. and D.C. McKenzie. Arterial hypoxemia and performance during intense exercise. *Eur. J. Appl. Physiol.* 68:80–86, 1994.
82. Krogh A. A bicycle ergometer and respiration apparatus for the experimental study of muscle work. *Skand. Arch. Physiol.* 33:375–394, 1913.
83. Krogh A. The number and distribution of capillaries in muscles with calculations of the oxygen pressure head necessary for supplying the tissue. *J. Physiol. (Lond.)* 52:409–415, 1919.
84. Krogh A. and M. Krogh. On the tensions of gases in the arterial blood. *Skand. Arch. Physiol.* 23:179–192, 1910.
85. Krogh A. and J. Lindhard. The regulation of respiration and circulation during the initial stages of muscular work. *J. Physiol. (Lond.)* 47:112–136, 1913.
86. Krogh A. and J. Lindhard. The changes in respiration at the transition from work to rest. *J. Physiol. (Lond.)* 53:431–437, 1919/1920.
87. Krogh A. and J. Lindhard. The relative value of fat and carbohydrate as sources of muscular energy. *Biochem. J.* 14:290–363, 1920.
88. Lange-Andersen K. Ethnic group differences in fitness for sustained and strenuous muscular exercise. *Can. Med. Assoc. J.* 96:832–833, 1967.
89. Lange-Andersen L., R. Elsner, B. Saltin, and L. Hermansen. *Physical Fitness in Terms of Maximal Oxygen Uptake of Nomadic Lapps.* Technical documentary report AALTDR-61-53, Arctic Aeromedical Laboratory, Alaska, 1961–1962.
90. Lavoisier A.L. *Expériences sur la respiration des animaux, et sur les changements qui arrivent á l'air en passant par leur poumon,* 1777. Reprinted in Lavoisier, A.L.: *Oeuvres de Lavoisier.* Paris: Imprimerie Impériale, 1862–1893.
91. Liebig G. Über die Respiration der Muskeln. *Arch. Anat. Physiol.* 17:393–416, 1850.
92. Liljestrand G. Untersuchungen über die Atmungsarbeit. *Skand. Arch. Physiol.* 35:199–293, 1918.
93. Liljestrand G. and J. Lindhard. Über das Minutvolumen des Herzens beim Schwimmen. Studien über die Physiologie des Schwimmens. *Skand. Arch. Physiol.* 39:64–77, 1920.
94. Liljestrand G. and J. Lindhard. Zur Physiologie des Ruderns. *Skand. Arch. Physiol.* 39:215–235, 1920.
95. Liljestrand G. and N. Stenström. Respirationsversuche beim Gehen, Laufen, Ski- und Schlittschuhlaufen. *Skand. Arch. Physiol.* 39:167–206, 1920.
96. Liljestrand G. and N. Stenström. Studien über die Physiologie des Schwimmens. *Skand. Arch. Physiol.* 39:1–63, 1920.
97. Lindhard J. Über das Minutenvolumen des Herzens bei Ruhe und bei Muskelarbeit. *Pflugers Arch.* 161:233–383, 1915.
98. Lohmann K. Darstellung der Adenylphosphorsäure aus Muskulatur. *Biochem. Z.* 233:460–472, 1931.
99. Lundsgaard E. Untersuchungen über Muskelkontraktionen ohne Milchsäurebildung. *Biochem. Z.* 217:162–177, 1930.
100. McGuire D.K., B.D. Levine, J.W. Williamson, P.G. Snell, C.G. Blomqvist, B. Saltin, and J.H. Mitchell. A 30-year follow-up of the Dallas bed rest and training study. I. Effect of age on the cardiovascular response to exercise. *Circulation* 104:1350–1357, 2001.

101. Milliken M.C., J. Stray-Gundersen, R.M. Peshock, J. Katz, and J.H. Mitchell. Left ventricular mass as determined by magnetic resonance imaging in male endurance athletes. *Am. J. Cardiol.* 62:301–305, 1988.

102. Mitchell J.H., B.J. Sproule, and C.B. Chapman. The physiological meaning of the maximal oxygen intake test. *J. Clin. Invest.* 37:538–547, 1958.

103. Morgan T.E., L.A. Cobb, F.A. Short, R. Ross, and D.R. Gunn. Effects of long-term exercise on human muscle mitochondria. In: *Muscle Metabolism During Exercise*, edited by B. Pernow and B. Saltin. New York: Plenum Press, 1971, pp. 87–96.

104. Morris J.N., J. A. Heady, P.A.B. Raffle, C.G. Roberts, and J.W. Parks. Coronary heart disease and physical activity of work. *Lancet* 1:1053–1057, 1953.

105. Nicolai G.F. and N. Zuntz. Füllung und Entleerung des Herzens bei Ruhe and Arbeit. *Berl. Klein. Wschr.* 128:821–824, 1914.

106. Noakes T.D. Challenging beliefs: ex Africa semper aliquid novi. *Med. Sci. Sports Exerc.* 29:571–590, 1997.

107. Nylin G. On the amount of, and changes in, the residual blood of the heart. *Am. Heart J.* 25:598–608, 1943.

108. Pettenkofer M. von and C. Voit. Untersuchungen über den Stoffverbrauch des normalen Menschen. *Z. Biologie* II. Band. München, Oldenbourg, pp. 459–573, 1866.

109. Pflüger E. Über die Diffusion des Sauerstoffs, den Ort und die Gesetze der Oxydationsprocesse im thierischen Organismus. *Pflugers Arch.* 6:43–64, 1872.

110. Powers S.K., J.A. Dodd, J. Woodward, R.E. Beadle, and G. Church. Haemoglobin saturation during incremental arm and leg exercise. *Br. J. Sports Med.* 18:212–216, 1984.

111. di Prampero P.E. Metabolic and circulatory limitations to Vo_2max at the whole animal level. *J. Exp. Biol.* 115:319–331, 1985.

112. Rasmussen J., B. Hanel, B. Diamant, and N.H. Secher. Muscle mass effect on arterial desaturation after maximal exercise. *Med. Sci. Sports Exerc.* 23:1349–1352, 1991.

113. Reindell H. Über den Kreislauf der Trainierten. Über den Restblutmenge des Herzens und über die besondere Bedeutung röntgenologischer (kymographischer) hämodynamischen Beobachtungen in Ruhe und nach Belastung. *Arch. Kreislaufforsch.* 12:265, 1943.

114. Remensnyder J.P., J.H. Mitchell, and S.J. Sarnoff. Functional sympatholysis during muscular activity: observations on influence of carotid sinus on oxygen uptake. *Circ. Res.* 11:370–380, 1962.

115. Richardson R.S., E.A. Noyszewski, K.F. Kendrick, J.S. Leigh, and P.D. Wagner. Myoglobin O_2 desaturation during exercise. Evidence of limited O_2 transport. *J. Clin. Invest.* 96:1916–1926, 1995.

116. Richardson R., D.C. Poole, D.R. Knight, S.S. Kurdak, M.C. Hogan, B. Grassi, E.C. Johnson, K.F. Kendrick, B.K. Erickson, and P.D. Wagner. High muscle blood flow in man: is maximal O_2 extraction compromised? *J. Appl. Physiol.* 75:1911–1916, 1993.

117. Riley-Hagan M., Peshock, R.M., J. Stray-Gundersen, J. Katz, T.W. Ryschon, and J.H. Mitchell. Left ventricular dimensions and mass using magnetic resonance imaging in female endurance athletes. *Am. J. Cardiol.* 69:1067–1074, 1992.

118. Robinson S. Experimental studies of physical fitness in relation to age. *Arbeitsphysiologie.* 10:251–323, 1938.

119. Robinson S., H.T. Edwards, and D.B. Dill. New records in human power. *Science* 85: 409–410, 1937.

120. Robinson S. and P.M. Harmon. The effect of training and of gelatin upon certain factors which limit muscular work. *Am. J. Physiol.* 133:161–169, 1941.

121. Rose R.J., D. Cluer, and B. Saltin. Some comparative aspects of the camel as a racing animal. In: *The Racing Camel*. Physiology, metabolic functions and adaptations. *Acta Physiol. Scand.* 150 (Suppl 617):87–95, 1994.

122. Rowell L.B. Factors affecting the prediction of maximal oxygen intake from measurements made during submaximal work with observations related to factors which may limit maximal oxygen uptake. Ph.D. diss., University of Minnesota, 1962.

123. Rowell L.B., H.L. Taylor, Y. Wang, and W.S. Carlson. Saturation of arterial blood with oxygen during maximal exercise. *J. Appl. Physiol.* 19:284–286, 1964.

124. Saltin B. Cardiovascular and pulmonary adaptation to physical activity. With a note on the effect of ageing. In: *Proceedings of the International Conference on Exercise, Fitness and Health,* edited by C. Bouchard. Champaign, IL: Human Kinetics Publishers, 1990, pp. 187–204.

125. Saltin B. Circulatory response to submaximal and maximal exercise after thermal dehydration (with a comparison between the acetylene and the dye-dilution methods for determination of cardiac output). *J. Appl. Physiol.* 19:1125–1132, 1964.

126. Saltin B. Malleability of the system in overcoming limitations: functional elements. *J. Exp. Biol.* 115:345–354, 1985.

127. Saltin B. and P.-O. Åstrand. Maximal oxygen uptake in athletes. *J. Appl. Physiol.* 23: 353–358, 1967.

128. Saltin B., G. Blomqvist, J.H. Mitchell, R.L. Johnson, K. Wildenthal Jr., and C.B. Chapman. Response to exercise after bed rest and after training. a longitudinal study of adaptive changes in oxygen transport and body composition. *Circulation* 37/38 (Suppl. VII): 1–78, 1968.

129. Saltin B., L.H. Hartley, Å. Kilbom, and I. Åstrand. Physical training in sedentary middle-aged and older men. II. Oxygen uptake, heart rate, and blood lactate concentration at submaximal and maximal exercise. *Scand. J. Clin. Lab. Invest.* 24:323–334, 1969.

130. Saltin B., K. Nazar, D.L. Costill, E. Stein, E. Jansson, B. Essén, and P.D. Gollnick. The nature of the training response; peripheral and central adaptations to one-legged exercise. *Acta Physiol. Scand.* 96:289–305, 1976.

131. Saltin B. and S. Strange. Maximal oxygen uptake: "old" and "new" arguments for a cardiovascular limitation. *Med. Sci. Sports Exerc.* 24:30–37, 1992.

132. Savard G., B. Kiens, and B. Saltin. Central cardiovascular factors as limits to endurance: with a note on the distinction between maximal oxygen uptake and endurance fitness. In: *Exercise: Benefits, Limits and Adaptations,* edited by D. Macleod, R. Maughan, M. Nimmo, T. Reilly, and C. Williams. London: E. & F.N. Spon, 1987, pp. 162–180.

133. Savard G.K., E.A. Richter, S. Strange, B. Kiens, N.J. Christensen, and B. Saltin. Norepinephrine spillover from skeletal muscle during dynamic execise in man: role of muscle mass. *Am. J. Physiol.* 257 (*Heart Circ. Physiol.* 26):H1812–H1818, 1989.

134. Scheele C.W. *Chemische Abhandlung von der Luft und dem Feuer.* Uppsala: Leipzig, Verlegt von Magn. Swederus Buchhandler, zu finden bey S.L. Crusius, 1777, translation in *The Discovery of Oxygen,* pt. 2: Experiments by Carl Wilhelm Scheele (1777). Alembic Club Reprints, No. 8, Edinburgh: Clay, 1894, reprinted 1952.

135. Secher N.H., J.P. Clausen, K. Klausen, I. Noer, and J. Trap-Jensen. Central and regional circulatory effects of adding arm exercise to leg exercise. *Acta Physiol. Scand.* 100: 288–297, 1977.

136. Secher N.H., N. Ruberg-Larsen, R.A. Binkhorst, and F. Bonde-Petersen. Maximal oxygen uptake during arm cranking and combined arm plus leg exercise. *J. Appl. Physiol.* 36:515–518, 1974.

137. Seguin A. and A. Lavoisier. Premier mémoire sur la transpiration des animaux. In: *Oeuvres de Lavoisier,* Vol. II, Paris: Imprimerie Impériale, 1862–93, pp. 704–714.

138. Smith E. Inquiries into the quantity of air inspired throughout the day and night, and under the influence of exercise, food, medicine, temperature, etc. Proc. R. Soc. 8: 451–454, 1857.

139. Smith E. On the influence of exercise over respiration and pulsation; with comments. *Edinburgh Med. J.* 4:614–623, 1858–59.

140. Smith E. Experimental Inquiries into the chemical and other phenomena of respiration, and their modifications by various physical agencies. *Philos. Trans.* 149:681–714, 1859.

141. Sproule B.J., J.H. Mitchell, and W.F. Miller. Cardiopulmonary physiological responses to heavy exercise in patients with anemia. *J. Clin. Invest.* 39:378–388, 1960.

142. Stensberg J., P.-O. Åstrand, B. Ekblom, J. Royce, and B. Saltin. Hemodynamic response to work with different muscle groups sitting and supine. *J. Appl. Physiol.* 22:61–70, 1967.

143. Taylor C.R., R.H. Karas, E.R. Weibel, and H. Hoppeler. Adaptive variation in the mammalian respiratory system in relation to energetic demand. *Resp. Physiol.* 69:1–127, 1987.

144. Taylor H.L., E. Buskirk, and A. Henschel. Maximal oxygen intake as an objective measure of cardio-respiratory performance. *J. Appl. Physiol.* 8:73–80, 1955.

145. Taylor H.L., L. Erickson, A. Henschel, and A. Keys. The effect of bed rest on the blood volume of normal young men. *Am. J. Physiol.* 144:227–232, 1945.

146. Tissot J. A new method for measuring and registering the respiration of humans and animals (in French). *J. Physio. Path. Gen.* 6:688, 1904.

147. Wagner P.D. Gas exchange and peripheral diffusion limitation. *Med. Sci. Sports Exerc.* 24:54–58, 1992.

148. Wang Y., J.T. Shepherd, R.J. Marshall, L.B. Rowell, and H.L. Taylor. Cardiac response to exercise in unconditioned young men and in athletes. *Circulation* 24:1064, 1961 (Abstract).

149. Weibel E.R. *Symmorphosis. On Form and Function in Shaping Life*. Harvard University Press: Cambridge, Massachusetts, 2000.

150. Weibel E.R., C.R. Taylor, and H. Hoppeler. The concept of symmorphosis: a testable hypothesis of structure-function relationship. *Proc. Nat. Acad. Sci.* 88:10357–10361, 1991.

151. Welch H.G. Hyperoxia and human performance: a brief review. *Med. Sci. Sports Exerc.* 14:253–261, 1982.

152. Whipp B.J., S.A. Ward, and M. Hassall. Estimating the metabolic rate of marching Roman Legionaries. *J. Physiol. (Lond.)* 491:60P, 1996.

153. Zuntz N. Über den Stoffverbrauch des Hundes bei Muskelarbeit. *Pflugers Arch.* 68:191–211, 1897.

154. Zuntz N. Zwei Apparate zur Dosierung und Messung menschlicher Arbeit. *Arch. Anat. Physiol. Abt.* 372:1899.

155. Zuntz N. and C. Lehmann. Untersuchungen über den Stoffwechsel des Pferdes bei Ruhe und Arbeit. *Landw. Jb.* 18:1–156, 1889.

156. Zuntz N. and W. Schumburg. Studien zu einer Physiologie des Marsches. In: *Schjerning, O. Bibiliothek v. Coler. Sammlung von Werken aus dem Bereiche der medizinischen Wissenschaften.* 6. Bd, Berlin 1901. Besprochen in: *Zbl. Physiol.* 15:327–330, 1901.

chapter 7

THE METABOLIC SYSTEMS: AEROBIC METABOLISM AND SUBSTRATE UTILIZATION IN EXERCISING SKELETAL MUSCLE

R. James Barnard and John O. Holloszy

THIS section by J. O. H. deals with the early era of aerobic metabolism study, from the 1780s to the 1960s, and is presented with a historical perspective derived from the 1911 review of Nathan Zuntz (137) and the 1950 review of Alexander von Muralt (129).

THE EARLY ERA

The Discovery That Animals Consume Oxygen

Antoine Laurent Lavoisier (1743–1794) of Paris has been referred to as the founder of the science of chemistry. He can just as appropriately be considered the founder of the biological study of exercise. Building on the discovery by Joseph Priestly (1733–1804) and Karl Scheele (1742–1786) (132), Lavoisier, in an impressive tour de force, showed that oxygen is necessary for the survival of animals, measured the oxygen consumption in a human being during both rest and exercise, and recognized that respiration is oxidation (90). He also found with Pierre S. Laplance (1749–1827) that a guinea pig when placed in an ice calorimeter will produce heat and erroneously

292

concluded that the heat was generated by combustion in the lungs of a substance carried in the blood (90). This brilliant Frenchman's career ended abruptly when he was guillotined in 1794 by the radicals who were in power during and briefly after the French Revolution. Lavoisier's studies were done between ~1780 and 1790 and almost 100 years elapsed before it was shown conclusively by Eduard F. W. Plfuger (1829–1910) of Bonn, Germany, that oxygen is consumed by all the tissues, not just the lungs (109).

Substrate Utilization

By the latter part of the nineteenth century, methods had been developed to accurately measure oxygen consumption and carbon dioxide production. Nathan Zuntz (1847–1920) of Bonn, Germany (see Fig. 6.4), utilized this methodology to determine the respiratory exchange ratio (RER) of humans at rest and during exercise. In a series of experiments done in the 1890s and 1900, Zuntz and his coworkers found that the fuel used during exercise depends on the diet. They accurately deduced from the results of studies where subjects were fed fat or carbohydrate diets low in protein that working muscle can oxidize either fat or carbohydrate to provide energy and that the relative amounts of these substrates that are oxidized are determined by their availability (137).

In the century between the studies of Lavoisier and Zuntz, there was a considerable amount of research regarding substrate metabolism during exercise. Landmark discoveries during this period include the 1807 finding by Jons J. Berzelius (1779–1849) of Stockholm, Sweden, that muscles accumulate lactate when they are strenuously exercised (98) and the 1859 report by Emil H. du Bois-Reymond (1818–1896) of Bonn that muscles become acidic as a consequence of strenuous activity (98), a finding he related to the Berzelius report of an increase in lactic acid (129). In 1865 Adolph Fick (1829–1901) and Wislicenus of Zurich, Switzerland, reported that the small amount of nitrogen excreted in urine indicated that protein could account for only a negligible amount of the energy they utilized while climbing Mount Faulhorn (1956 m) in their country (15). This finding meant that the idea by Justus von Liebig (1803–1873) of Giessen, Germany, that protein is the source of energy for contracting muscles had to be discarded (129). Another important discovery during this period was that of Otto Weiss of Konigsberg, Germany, who found that muscle glycogen decreased during exercise (129). It had already been shown during the 1850s by Claude Bernard (1813–1878) of Paris that the liver contains glycogen, which is converted to blood glucose (15). This process was discussed in detail in 1911 by Zuntz, who knew that during very strenuous exercise more glucose is released into the blood by the liver than is used by the working muscles, resulting in an increase in blood glucose (137). As a result of his extensive laboratory research and a thorough understanding of what he referred to as the "neuere Biochemie," Zuntz's concepts of substrate metabolism during exercise were remarkably modern. Zuntz wrote that in the cells of living organisms chemical reactions such as the oxidation of fat and carbohydrate in contracting muscle are catalyzed by enzymes. He presented the concept that a fraction of the energy released from these substrates when they are broken down is used by muscles to perform work, while the rest is lost as heat.

Zuntz reviewed the study performed in 1909 by Ryffel (114), who found that a large amount of lactate could be obtained from a subject 30 minutes after he had sprinted for 2 minutes. By contrast, there was no lactate in the urine of subjects who had performed a 24 hour endurance march (137). He also discussed studies by Max Rubner (1854–1932) of Berlin and by Frey, conducted in 1885 in which they perfused mammalian muscles with oxygenated blood and found that, despite the presence of oxygen, the muscles produced lactate. Nevertheless, Zuntz was convinced that lactate was produced only under hypoxic conditions and suggested that portions of the muscle may have been hypoxic due to formation of arterial blood clots or narrowing. He was of the opinion that lactate is produced during exercise only when the work rate exceeds the capacity of the cardiovascular system to provide the working muscles with sufficient oxygen. This appears to be the only point on which Zuntz was mistaken, and his interpretation was very reasonable considering the complete lack of information regarding the regulation of glycogenolysis. What is remarkable is that some exercise physiologists continue to hold this view more than 30 years after the mechanisms involved in the activation of glycogenolysis and regulation of glycolysis and pyruvate oxidation in contracting muscle were elucidated, as discussed in Chapter 8 of this volume.

With regard to substrate metabolism during aerobic exercise, Zuntz stated that it was clear from studies by a number of investigators, including his own group, that the role of protein as an energy-providing substrate was negligible. The first study to clearly show that protein does not serve as an energy source during exercise was that of M. von Pettenhofer and Carl Voit (1831–1909) of Munich, Germany, in 1866 (130). Based on early studies, such as that of Sczelkow in 1862, it was, according to Zuntz, generally believed that carbohydrate was the only substrate used by working muscle. Sczelkow measured O_2 and CO_2 in the arterial blood entering, and the venous blood leaving, contracting muscles, and found that the respiratory quotient (RQ) increased in response to contractile activity. He followed up this study by measuring the RER in tracheotomized dogs at rest and during electrical stimulation of the hindlimb muscles that caused them to contract vigorously. As in his RQ measurements, he found a large increase in RER, which sometimes attained values above 1.0. Zuntz interpreted these and other earlier studies, along with his own extensive data obtained on exercising humans and animals, as indicating that resting muscles oxidize primarily fat and that glycogen and blood sugar become the preferred substrates during strenuous exercise. He explained the finding that the RER sometimes increased above 1.0 as being due to hyperventilation. To avoid this problem and obtain accurate information regarding the substrates that muscles oxidize during normal exercise, Zuntz said that it is necessary to use an exercise intensity that can be maintained continuously for a prolonged period without excessive fatigue. Zuntz and his colleagues and students used this approach in numerous studies on humans, as well as in an experiment on dogs and on a horse (137). They found that during prolonged exercise such as a 25 km march with a pack by a well-nourished individual, there was a progressive decline in RER. Similarly, RER decreased during more vigorous continuous or intermittent exercise performed for 90 minutes to 2 hours. They also found that when subjects performed 25 km marches carrying a military pack on 3 successive days on a constant food intake, RER at rest, before the march, as well as during the march, was lower on the second than on the first day, and lower on the third than on the second day. Zuntz interpreted these findings to indicate that as the body glycogen stores are progressively depleted, an increasing amount of energy is provided by oxidation of the body's fat stores.

Despite this incontrovertible evidence that fat oxidation can provide the energy required by working muscles, some physiologists still claimed that only carbohydrate can be used as a direct energy source for muscle contraction. The major advocate for this point of view when Zuntz wrote his review was the French physiologist J. B. Auguste Chauveau (1837–1917), who thought that fat must be converted to sugar before it could serve as the energy source for contracting muscle (23). He estimated that ~30% of the energy content of fat was lost during this conversion process. He felt that this concept was supported by experiments in which he measured the amount of weight lost by dogs during a bout of exercise. Zuntz pointed out that loss of body weight had no relevance to this question, and that the only way to test Chauveau's hypothesis was to measure $\dot{V}O_2$ and RER. To this end, Jacques Loeb (1857–1924) and Zuntz did experiments in which they compared $\dot{V}O_2$ of dogs fed starch and sugar before and during exercise with that of a dog that was fasted after induction of phlorizin diabetes to cause urinary glucose loss. The carbohydrate-depleted dog had an RER of 0.71 during exercise and utilized ~7% more O_2/kg body weight than the carbohydrate-fed dogs (137). Zuntz also reviewed studies by Heineman, Frentzel, and Reach and by Atwater, Benedict, and Carter, performed on humans, which also showed that $\dot{V}O_2$ was slightly higher when fat was the primary substrate. He concluded that the slightly lower efficiency of fat as a substrate for working muscle clearly was not compatible with the concept that fat must first be converted to carbohydrate. Another aspect of substrate metabolism during exercise addressed by Zuntz was the maintenance of blood glucose. He knew from the work of Seegen and others that sugar is stored in the liver as glycogen and that during exercise liver glycogen is converted to glucose, which is released into the blood (137).

Zuntz was a remarkably insightful scientist who was far ahead of his time. Shortly after Zuntz wrote his review, this area of research suffered serious setbacks and was kept in a state of confusion for many years by what the European physiologists referred to as the "British school of physiologists," which consisted of Fletcher, Hopkins, George Winfield, and most importantly, Archibald V. Hill (1886–1977) (see Fig. 6.9). In a paper published in 1913, Hill reported a series of experiments in which he measured heat production by frog muscles contracting in oxygen or in nitrogen (56). Although he made no biochemical measurements, Hill concluded from his heat measurements that the "lactic acid precursor is not glucose but something of noticeably more total energy." He proposed that the process of muscle contraction is due to liberation of lactic acid from some precursor, and that lactic acid increases the tension in some colloidal structure of the tissue, resulting in contraction. He hypothesized that in the presence of oxygen the lactate acid precursor is resynthesized.

By 1915, when they gave the Croonian lecture, Fletcher and Hopkins, who had also been of the opinion that glycogen (glucose) is not the precursor of lactate, had accepted as fact that lactate is formed from breakdown of the glycogen stores in contracting frog muscles (44). During the previous year, Parnas and Wagner had rediscovered the fact that glycogen breakdown accounts for the lactate production in (isolated frog) muscle (103). Parnas had worked with them at Cambridge before the war, and Fletcher and Hopkins accepted his findings. Like A. V. Hill, however, they were still convinced that the formation of lactic acid mediated muscle contraction. The following excerpt from their Croonian lecture provides an example of the flowery rhetoric that they used to argue their point:

> With a normal oxygen supply, however, the lactic acid is promptly removed after each contraction, and each successive stimulus, with its associated breakdown, is followed by a normal contraction. The removal of the lactic acid might be thought of as a direct oxidation, . . . and it might be supposed to be burnt, so to speak, to waste, when the energy liberated by its combustion would supply nothing to the mechanical energy of contraction. But this simplest view we are driven at once to forego, and perhaps not unwillingly, since it would be unwelcome to believe that a body of such high energy value as lactic acid can be only a waste product yielding nothing in its discharge except the indirect benefits of heat production unconnected with the muscle machinery. . . . If our picture of events is the true one, and if the machinery of contraction is of the kind we have suggested, then carbohydrate metabolism in muscle takes on an aspect of peculiar interest . . . in the muscles, which after all form the chief seat of metabolism, the acid intermediary product appears, if we are right, at such a stage and place as to have more than a purely chemical significance. It marks, on the one hand, an obligatory stage in a particular set of successive chemical reactions; but, on the other hand, it has here its special role to play in connection with the muscle machinery. In the evolution of muscle it would appear that advantage, so to speak, has been taken of this acid phase in carbohydrate degradation, and that by appropriate arrangement of the cell elements the lactic acid, before it leaves the tissue in its final combustion, is assigned the particular position in which it can induce those tension changes upon which all the wonders of animal movement depend.

Whether they were unaware of the (then available) evidence that prolonged exercise can be performed without lactate production and with fat oxidation providing most of energy or just chose to ignore it is not clear. They and A. V. Hill appear to have also ignored the facts that muscles relax between closely spaced contractions even while rapidly accumulating lactate and that anoxic, resting muscle can accumulate much lactate without contracting.

The increasing influence of the British school of physiologists' view that lactate mediates muscle contraction and that carbohydrate must, therefore, be the only energy source for working muscles, stimulated August Krogh (1874–1949) (see Fig. 6.7) and his colleagues in Copenhagen to reevaluate the role played by fat oxidation (77). In their introduction Krogh and Lindhard (77) stated that research on isolated muscles during the previous decade by the British school of physiologists had revived the old problem about the source of muscular energy. They pointed out that if lactic acid, which can only be formed from carbohydrate, mediates muscle contraction, then the view of Zuntz that a fraction of the energy liberated by oxidation of any substrate can be used to support muscle contractions must be wrong. It should be mentioned that the British group had supported their view with experiments by Winfield showing that excised frog muscles do not oxidize fats (136).

Like Chauveau (24), Krogh and Lindhard (77) believed that fat could be converted to glucose with a loss of 25%–30% energy. They therefore argued that if the British physiologists were correct, then the same exercise should require 25%–30% more oxygen when fat is being oxidized than when carbohydrate is the energy source. To evaluate this possibility they used the same approach as Zuntz and studied subjects repeatedly while they were maintained on either a carbohydrate or a fat diet. They found that more oxygen was used during the same exercise when fat was oxidized than when carbohydrate was the substrate. However, the differences were small, averaging about 11%, and Krogh and Lindhard concluded that their results, which confirmed the findings of Zuntz, provided no support for the concept that fat must be converted to glucose before it can provide the energy for working muscles. However, in contrast to Zuntz, who doubted that fat could be converted to carbohydrate, Krogh and Lindhard believed in this possibility and misinterpreted some of their results as indicating that at RERs below 0.90 some fat is converted to carbohydrate. When one calculates how much more oxygen is required to perform the same work when fat is oxidized than when carbohydrate is oxidized, using the ratio of moles of ATP produced per mole of oxygen utilized, the values obtained are in the range of 7% to 13%. The value varies depending on whether glucose or glycogen is used and on whether the glycerol phosphate shuttle or the malate–aspartate shuttle moves the reducing equivalents formed in the cytosol into the mitochondria. The value obtained by Krogh and coworkers was in the middle of this range, demonstrating the remarkable accuracy with which they were able to measure respiratory gas exchange using their Jacque type of respiration chamber (77).

The High-Energy Phosphate (~P) Compounds

In 1890, Ernest Salkowski (1844–1923) of Strasbourg, France, described the liberation of inorganic phosphate from an organic compound during muscle contractions (137). This finding was confirmed by MacLeod in 1899 (129). These findings did not make much of an impression at the time, and it was not until the 1920s that this finding was confirmed by Embden and Adler (37) and Meyerhof and Lohmann (91). The first of the organic phosphate compounds to be identified was creatine phosphate, which was discovered by Cyrus H. Fiske and Lapragada Subbarow of Harvard University (40, 41) and by Philip and Grace Palmer Eggleton of University College in London (35, 36), who also showed that creatine phosphate decreased during muscle contractions and was resynthesized during recovery. At the time, Hill's concept that muscle contraction was mediated by the formation of lactic acid was still the dominant dogma. In a paper published in 1928, in which he measured the recovery heat production in oxygen after a series of muscle twitches and determined an "oxidative quotient," Hill stated that, "The consistency of such results makes it difficult to imagine that lactic acid will be deposed from its key position in the theory of muscular contraction" (56).

The lactic acid era, which began with A.V. Hill's paper published in 1913 (55), ended abruptly in 1930 when Ejnar Lundsgaard of Copenhagen reported studies in which he poisoned glycolysis with iodoacetate (IAA) and showed that the poisoned muscles could perform a series of contractions in the absence of any lactic acid production (88). In a series of studies (86–90), he showed that the IAA-poisoned muscles contracted until their phosphocreatine (PC) store was exhausted and that PC disappearance correlated with the total amount of tension developed. He also found that following contractile activity, PC resynthesis in anoxic muscles is coupled with lactate production and that PC breakdown during contractile activity is less in normal muscles than in poisoned muscle. He concluded from these findings that PC resynthesis is normally coupled to the conversion of glycogen to lactate under anaerobic conditions. He further discovered that aerobic resynthesis of PC occurs in IAA-poisoned muscles. He reasoned on the basis of this finding that under normal conditions PC resynthesis can be supported by aerobic metabolism during exercise without the intermediate formation of lactate. It is evident from his papers that Lundsgaard clearly understood the general relationships between PC, glycolysis, and oxidative metabolism. Although the enzymatic steps in the glycolytic and oxidative pathway had not yet been worked out, and ATP had not been discovered, Lundsgaard's insightful interpretation of his findings regarding the roles played by glycolysis, aerobic metabolism, and ~P in muscle energetics is still generally accepted. Like A. V. Hill, Otto Meyerhoff (1884–1951) of Heidelberg, Germany, believed that formation of lactic acid supplied the energy for muscle contraction. Severo Ochoa, who was working as a postdoctoral fellow in Meyerhoff's laboratory at the time of Lundsgaard's discovery, recounts that Meyerhoff received Lundsgaard's findings with scepticism (100). Lundsgaard took the very sensible step of going to Heidelberg and repeating his experiment in Meyerhoff's own laboratory, convincing him that the lactic acid concept was wrong (100)

Soon after the identification of PC, Karl Lohmann (82, 84) and Fiske and Sublarow (39) discovered ATP. Subsequently, Lohmann discovered that cell-free muscle extracts could hydrolyze PC only when adenine nucleotides were added (83). He proposed, on the basis of his findings, that the reactions ATP $\rightarrow$ AMP + 2Pi, and AMP + 2PC $\rightarrow$ ATP + 2 creatine were involved. Lohmann suspected that myokinase activity was present in Lohmann's muscle extract and was able to show that the correct reactions are ATP $\rightarrow$ ADP + Pi and ADP + PC $\rightarrow$ ATP + creatine (80). Based on his findings with muscle extracts, Lohmann (82–84) hypothesized that ATP hydrolysis provides the energy for muscle contraction and that PC serves as an energy reserve that functions to resynthesize ATP via the creatine kinase reaction. This hypothesis was strongly supported by the discovery in 1939 by Engelhardt and Ljubimowa (38) that myosin has ATPase activity—that is, that one of the contractile proteins, myosin, is an enzyme that, when activated by actin, breaks down ATP. Based on the findings of Lohmann (83), H. Lehmann (80), and W. A. Engelhardt and M. N. Ljubimowa (38), biochemists generally accepted the concept that ATP provides the energy for muscle contraction. However, many physiologists, led by A. V. Hill, who continued to fight a vigorous rearguard action for many years, did not accept ATP as the energy source because of the inability to demonstrate a decrease in ATP concentration in response to a single muscle contraction. This state of affairs is summarized by comments that Lundsgaard made during a discussion on muscular contraction and relaxation published in the *Proceedings of the Royal Society* in 1949 (86).

> In his introduction, A. V. Hill mentioned that at present we do not even know if ATP under any conditions is broken down in the intact muscle. Undoubtedly A. V. Hill has deliberately exaggerated his statement on this point to some degree, probably to apply a kind of shock therapy to the biochemists interested in muscular contraction. I do not think it can be doubted that ATP constantly is broken down and rebuilt in the intact muscle. Experiments with labeled phosphate demonstrate this convincingly. It is true, that it has never been demonstrated that the rate of turnover of phosphate in ATP is increased by stimulation. This, however, is simply due to the fact that the rate of turnover is so rapid that it keeps pace with the transfer of labeled phosphate from the surrounding fluids to the interior of the muscle cells. As ATP is the universal phosphate-donor I have no doubt myself that the increased rate of metabolism which accompanies a muscle twitch must involve an increase in the turnover of ATP-phosphate.

As late as 1957, Britton Chance (22) concluded, from studies on perfused frog muscles in which ADP levels near mitochondria were measured with a double-beam spectrophotometer, that the increment in ADP following a muscle twitch was only 2%–3% of the expected value. On this basis he concluded that ATP "need not be broken down during the contraction phase of muscle activity." The demonstration that ATP is broken down during muscle contractions is generally credited to Cain and Davies (19), who used 2,4-dinitrofluorobenzene to poison creatine kinase, glycolysis, and oxidative phosphorylation. However, in a paper published 7 years earlier, Lange (79) had shown in frog rectus abdominous muscles poisoned with both IAA and nitrogen mustard that ATP decreased and ADP increased, while PC decreased minimally, in response to contractions induced either with acetylcholine or electrical stimulation. The reason for the inability to detect an increase in ADP in response to a single-twitch contraction, which puzzled physiologists for nearly 30 years, is explained by the kinetics of the creatine phosphokinase reaction, first described in detail by Henry Lardy's group (78). They calculated that the maximal velocity of the reaction would permit all of the ATP in a rabbit skeletal muscle to be resynthesized from ADP and creatine phosphate in 0.03 seconds.

Further Studies of Substrate Utilization during Exercise

Relatively little research was conducted on the biology of exercise during the period from ~1910 to 1960 as the rate-of-living theory proposed by Max Rubner

(1854–1932) of Berlin (113), and further popularized by Pearl (105), and by Hans Selye (1907–1982) of Montreal, Canada, who considered exercise a harmful stress (116), was rather generally accepted, and exercise was out of fashion. However, a few noteworthy studies on substrate metabolism during exercise were conducted during this period that confirmed and extended the information obtained by Zuntz and others prior to 1910 (137). Of these, the study by Krogh and Lindhard (77) has already been mentioned. Edwards, Margaria, and Dill (34) measured RER and blood glucose in a well-nourished, lean young man who ran intermittently for a period of 6 hours, at 11.3 km/h for 25 minutes and 9.3 km/h for 25 minures of each hour. They found that his RER was higher at 11.3 km/h than at 9.3 km/h, and fell progressively from 0.96 initially to 0.77 during the last 25 minutes at 9.3 km/h. During the 6 hours, he utilized ~415 g of glycogen and 175 g of fat. They concluded that glycogen stores are used preferentially, particularly during more strenuous exercise, and that fat provides progressively more of the energy as glycogen stores are depleted. Supporting the interpretation that glycogen stores were depleted was the finding that blood glucose level fell from ~100 mg/dl to ~58 mg/dl.

In 1939 Erik Hohwu Christensen (1904–1998) and Hansen in Copenhangen published a series of studies that serve as a link between the classical and the modern eras of investigation of substrate metabolism during exercise (24–26). Like earlier investigators, they found that the RER increased as exercise intensity increased and decreased during prolonged exercise. They confirmed that both fat and carbohydrate are utilized during exercise and that the relative amounts of carbohydrate and fat used during moderate exercise are largely determined by diet. When subjects were fed a high-fat diet their performance decreased while a high carbohydrate diet increased endurance. They also found that hypoglycemia sometimes occured during prolonged exercise in the carbohydrate-deprived state and could be the cause of exhaustion.

THE MODERN ERA

This section by R. J. B. emphasizes the point that the modern eras of exercise biochemistry and exercise metabolism began in the middle 1960s primarily as a result of activities attributed to several individuals in the United States and within Scandinavia. The American scientists were Philip D. Gollnick (1935–1991) (see Fig. 2.15) of Washington State University, who conducted research and initiated a graduate program in exercise biochemistry, Charles M. Tipton (1927–) (Fig. 7.1) of the University of Iowa, who established a research and graduate program with an emphasis in exercise physiology and biochemistry, and John O. Holloszy (1933–) (Fig. 7.2) of Washington University in St. Louis, who conducted biochemical research, established a biochemistry laboratory, and initiated a postdoctoral training program for individuals interested in exercise biochemistry and metabolism. In Scandinavia, Per-Olf Astrand (1922–) (see Fig. 6.14) of the Karolinska Institute of Stockholm was conducting research on exercise metabolism and mentoring Bengt Saltin (1933–) before he

Fig. 7.1. Charles M. Tipton prior to his retirement from teaching in 1998. He has, however, remained active professionally, including editing this and other books. Picture courtesy of Charles M. Tipton.

moved to Denmark. Saltin initiated and conducted extensive metabolic and biochemical investigations on exercising subjects and collaborated with Gollnick, who spent a sabbatical with Saltin. It is of interest that several of Tipton's students took their postdoctoral training with Holloszy before assuming research positions at major institutions. These individuals were Kenneth Baldwin at the University of Cal-

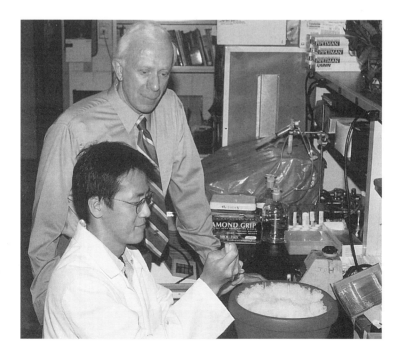

Fig. 7.2. John O. Holloszy (standing) and Kentaro Kawanaka (sitting) in the laboratory.

ifornia at Irvine, Frank Booth at the University of Texas Medical Center in Houston, Ronald Terjung at the University of Illinois at Champaign/Urbana, and Robert Conlee at Brigham Young University in Utah.

One of Tipton's Ph.D. students, R. James Barnard, took his postdoctoral training with James B. Peter, a rheumatologist, at the University of California, Los Angeles, School of Medicine, who was interested in muscle biochemistry. They, along with V. Reggie Edgerton from the Physical Education Department at UCLA, embarked on a series of studies that eventually led to the establishment of three fiber types in skeletal muscle (9, 106). The classification was based on histochemical, biochemical, and contractile function studies initially conducted on guinea pig and rabbit muscle. The three fiber types they described were slow twitch, oxidative (SO), fast twitch, glycolytic (FG), and fast twitch, oxidative glycolytic (FOG), a classification still in use today. These results showed that the old classification of muscle as red or white, based on visual inspection, was invalid. However, it is now recognized that there is more of a continuum as opposed to distinct fiber types. This is especially true for the metabolic profile between FG and FOG fibers. Similar histology and biochemical studies were conducted by Kenneth Baldwin and colleagues (6), postdoctoral fellows in Holloszy's laboratory, on rat muscle, focusing on aerobic capacity. They reported that the soleus and the deep, red portion of the quadriceps had higher oxidative capacity than did the superficial, white portion of the quadriceps. The higher oxidative capacity was demonstrated not only with pyruvate as a substrate but also with palmitate. A subdivision for the contractile chartacteristics in fast-twitch fibers was suggested in 1970 by Michael Brooke and Kenneth Kaiser (18) at the University of Colorado based on qualitative myosin ATPase histochemical techniques incubating at different pH. In 1977, Geraldine Gauthier at Wellesley College and Susan Lowey (47) at Brandeis University, using affinity-purified polyclonal antibodies, revealed differences in myosin composition in rat fast-twitch muscle fibers. Lowey and Laura Silberstein (117) then went on to study individual muscle fibers in chicken pectoralis, a fast-twitch muscle, and found that not only were there different myosin isoforms between fast-twitch fibers but that any given myofibril could have coexisting different myosin isoforms. These results were confirmed and extended by the use of antibodies against specific myosin heavy chains (MHCs). It is now recognized that adult rodent skeletal muscles used for postural activity and/or locomotion are made up of four different isoforms of MHC, one slow type and three fast types all with different ATPase activities and shortening velocities. Donald Thomason (124), a doctoral student in Kenneth Baldwin's laboratory at the University of California, Irvine, studied several different rat skeletal muscles and reported a wide variation in the distribution of MHC isoforms in different muscles. Slow-twitch muscles—that is, soleus and adductor longus—had almost exclusively ($\geq 97\%$) slow-type MHC (type I) while the three types of MHC in fast-twitch muscles (types IIa, IId/x, IIb) varied considerably. In humans there appear to be only two fast isoforms (types IIa, IIb) in addition to the slow type I MHC according to Geoffrey Goldspink and associates (115) at the Royal Free Hospital School of Medicine, London. A more detailed discussion of fiber types and motor units is given in Chapter 2.

Substrate Utilization

By the end of the 1930s it had been well established that both fat and carbohydrate could be used as fuel for aerobic metabolism during exercise. In more recent times it has also been demonstrated that lactate, generally associated with anaerobic metabolism, can also be used as a fuel for aerobic metabolism. Kenneth Baldwin and associates (5) at the University of California, Irvine, demonstrated in rat muscle homogenates that heart and FOG muscles had the highest capacity to utilize lactate for aerobic metabolism, followed by SO and FG muscles. They concluded that the ability of lactate to be used for aerobic metabolism correlated with the content of heart-type lactate dehydrogenase (LDH). J. B. Peter and others (108) had previously reported that both heart and soleus muscle had high levels of $LDH_{1,2}$ while FG and FOG muscles, with high glycogen content and anaerobic capacity, had the highest LDH activity and $LDH_{4,5}$ content. $LDH_{4,5}$ are considered to be the isoforms favoring lactate formation while $LDH_{1,2}$ isoforms favor pyruvate formation, which could be used for aerobic metabolism. This assumption is based on the observation that pyruvate has been shown to suppress LDH_1 but not LDH_5. However, G. Van Hall (127) from the Copenhagen Muscle Research Center recently pointed out that the concentration of pyruvate needed to suppress LDH_1 is far greater than the concentration ever achieved in muscle. He suggests that the compartmentalization of LDH in muscle cells, as opposed to the isoform of LDH, is really what determines the conversion of lactate to pyruvate for aerobic metabolism.

As early as 1923 David Barr and Harold Himwich (11) at Yale University reported that venous blood draining from an inactive arm had a lower concentration of lactate than arterial blood during leg exercise. Using muscle biopsies Jan Karlsson and coworkers from the Gymnastik-och Idrottshögskolan in Stockholm (72) demonstrated a rise in arm lactate concentration during leg exercise. More recently George Brooks and colleagues from the University of California, Berkley, as well as others, using tracer techniques have documented that lactate is indeed used as a substrate for aerobic metabolism in well-oxygenated muscle. These studies are detailed in Chapter 8 of this volume.

The heart relies on aerobic metabolism for its normal function and has the capacity to utilize different substrates. At rest the heart uses free fatty acids, glucose, and lactate. However, during exercise when myocardial oxygen demands greatly increase, A. Carlsten and associates (21) at the University of Göteborg and Joseph Keul and coworkers (73) at the University of Freiburg showed that lactate becomes the substrate accounting for almost two-thirds of the aerobic metabolism.

Although the early studies had documented that carbohydrate is a major substrate for aerobic metabolism during exercise and that diet plays an important role in determining endurance, there was renewed interest in this area with the introduction of the needle biopsy technique by Jonas Bergström and Eric Hultman at St. Erik's Hospital, Stockholm, and other Scandinavian investigators. The classic study of Bergström and others (14) clearly showed that the work time to exhaustion was highly correlated with muscle glycogen content and could be easily altered by con-

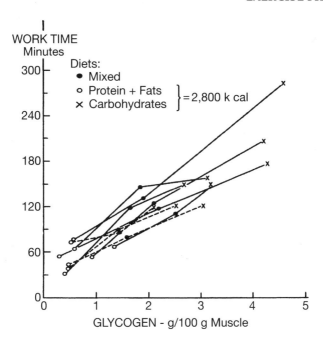

Fig. 7.3. Effect of diet on muscle glycogen and work time to exhaustion. From Bergström et al. (14).

suming various types of diets (Fig. 7.3). Using isocaloric diets they were able to re-duce muscle glycogen and endurance time by feeding a high-protein-and-fat diet. Conversely, when they fed a high-carbohydrate diet muscle, glycogen and exhaus-tion time increased greatly. These investigators also showed that at exhaustion there was almost complete depletion of muscle glycogen. In a review, Hultman (65) stated, "From these studies we conclude that the limiting factor in the performance of long term heavy muscular work is the preformed glycogen store in the working muscle." This conclusion has been universally accepted and is the basis for the concepts of "carbohydrate loading" and "glycogen supercompensation." Bergström and Hult-man (13) found that if they exercised to exhaustion and depleted muscle glycogen, 3 days of recovery on a high-carbohydrate diet resulted in more than a doubling of muscle glycogen content compared to the contralateral, sedentary leg—that is, glycogen supercompensation.

Although the early studies showed that fatigue occurred when muscle glycogen was severely reduced, glycogen levels never reached zero, indicating that there was still some substrate available. Robert Conlee (28) from Brigham Young University in his 1987 review concluded, but with some caution, that the fact that glycogen at ex-haustion is not zero could be related to fiber-type recruitment during exercise. The periodic acid–schiff (PAS) stain was used by many investigators to assess motor-unit recruitment as described in Chapter 2. Using this method, it was well documented that during moderate to heavy workloads many muscle fibers were completely devoid of glycogen, supporting the view that muscle fatigue could be caused by a lack of intramuscular glycogen substrate.

Although muscle glycogen seemed to be a limiting factor for exhaustive performance, earlier studies by David Bruce Dill (1891–1986) and colleagues at the Harvard Fatigue Laboratory (32) and by Christensen and Hansen (24) reported hypoglycemia at exhaustion, suggesting the use of blood glucose as a substrate. Recent tracer studies, as described in Chapter 9, have confirmed that blood glucose is utilized but that muscle glycogen is more important. In a study with frog muscle, John Holloszy in H. T. Narahara's laboratory at Washington University in St. Louis (59) reported that electrical stimulation led to increased permeability of glucose. John Ivy, a postdoctoral fellow with Holloszy (69), then confirmed in rat muscle that exercise led to increased glucose uptake. J. B. Peter and associates (7, 108) at UCLA reported that an acute bout of exercise increased hexokinase activity in red and white skeletal muscle, which may have been a factor in the increased glucose uptake. In 1985, in an attempt to understand how exercise increased muscle permeability and glucose uptake, Glen Grimditch, a doctoral student with Barnard at UCLA, developed an isolated sarcolemmal membrane preparation to study glucose transport (52). They, along with Eric Sternlicht, another doctoral student, confirmed that just like in fat cells, insulin increased muscle glucose transport by increasing the number of glucose transporters recruited to the sarcolemma membrane (119). The development of the sarcolemmal membrane preparation was an important advance that was adopted by other laboratories to study glucose transport (see Chapter 9), by David Roth, a doctoral student with George Brooks (112), to characterize the lactate transporter, and more recently by Arend Bonen and coworkers (16) at the University of Waterloo in Canada to characterize a fatty acid transporter in skeletal muscle.

Eric Sternlicht and Glen Grimditch (118) in Barnard's laboratory at UCLA then confirmed that an acute bout of exercise increased sarcolemma glucose transport to the same maximum level achieved with an injection of insulin. They concluded that exercise and insulin increased glucose transport by different mechanisms. While the conclusion was correct it was based on faulty data because they did not observe an increase in glucose transporters recruited to the membrane with exercise as they did with insulin. Their studies utilized cytochalasin B binding to quantitate glucose transporters as it was still not known at the time whether different isoforms of glucose transporters existed. With the discovery of the different isoform, and the production of specific antibodies for the transporters, several laboratories demonstrated that exercise does increase glucose transport by recruiting GLUT-4 from an intracellular pool to the plasma membrane as discussed in Chapter 9. Although exercise does increase glucose transport into muscle it still is not a major factor for aerobic metabolism during exercise of shorter duration. Blood glucose does become an important factor during prolonged, exhaustive exercise. In 1930 D. B. Dill and his colleagues (32) at the Harvard Fatigue Laboratory reported that a dog running on a treadmill could increase exhaustive run time by almost fourfold if the dog was given regular glucose supplements to prevent hypoglycemia. Many studies in humans then confirmed that prolonged (≥2 h) activity could be enhanced by carbohydrate ingestion during the activity and that this could slow the rate of muscle glycogen depletion as reviewed by Andrew Coggan and Edward Coyle (27). These observations led to the development of popular "sport drinks" which in the vast majority of cases are probably used by individuals where they would have little, if any, impact on performance.

In his 1987 review, Conlee (28) also emphasized that in spite of all the research on muscle glycogen since 1967, no one had determined why glycogen depletion causes fatigue. Why can't free fatty acids (FFA), blood glucose, or amino acids substitute for glycogen? Today we still cannot give a precise answer. Another interesting question that has eluded a precise answer is why FFA metabolism predominates at lower workloads while carbohydrate metabolism predominates at higher workloads? Based on respiratory quotient measurements, early studies by Zuntz (137) and later confirmed by E. H. Christensen and O. Hansen of Copenhagen (24) reported that as exercise intensity increased so did RER, indicating a shift in metabolism from a mixture of fat and carbohydrate to predominantly carbohydrate metabolism at $\dot{V}O_2$max, which confirmed earlier studies. Using tracer methods Johannes Romijn and colleagues (111) at the University of Texas reported that in subjects exercising for 30 minutes, the rate of fat oxidation was 27 μmol/kg/min at 25% $\dot{V}O_2$max, 43 μmol/kg/min at 65% $\dot{V}O_2$max, and dropped to 30 μmol/kg/min at 85% $\dot{V}O_2$max.

Factors involved in the regulation of fat metabolism are very complex and far from being completely understood. A number of sites could be rate-limiting, including the availability of fatty acids, the rate of transport into muscle cells, the rate of formation of fatty acyl-CoA, the rate of transport of fatty acyl-CoA into mitochondria by the carnitine palmitoyl-transferase system, the rate of β-oxidation, and finally the rate of oxidation of acetyl-CoA generated by β-oxidation. William "Will" Winder (135) from Brigham Young University also pointed out that the rate-limiting step may not always be the same under different workloads. Winder (135) as well as Blake Rasmussen and Robert Wolfe (110) indicate that the rate at which fatty acyl-CoA is transported into the mitochondria by the carnitine palmitoyl transferase system may be the major rate-limiting step under most circumstances.

As shown in Figure 7.4 from Winder's review (135), three stores of triglycerides could provide fatty acids for metabolism during exercise—adipocytes, plasma triglyceride lipoproteins, and intramuscular triglyceride stores. Adipose tissue is by far the biggest and probably the most important store. The fact that plasma triglycerides can be utilized by exercising muscle was documented by Norman Jones and Richard Havel (71) at the University of California Medical Center in San Francisco in rats injected with ^{14}C-chylomicrons. With exercise the uptake of ^{14}C into muscle was significantly increased. Plasma triglyceride may not be an important source of fuel for muscle in young, healthy subjects with low triglyceride levels but it may be more important in older individuals with hypertriglyceridemia who are at increased risk for coronary heart disease. Holloszy and colleagues (61) in Thomas Cureton's laboratory at the University of Illinois reported that 6 months of an endurance training program significantly reduced serum triglycerides. Based on a few individuals,

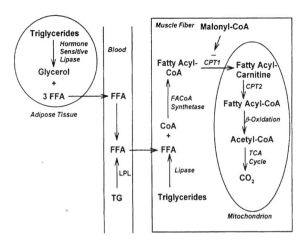

Fig. 7.4. Pathways for mobilization and metabolism of fatty acids. From Winder (135).

they suggested the response to exercise was observed within a few days and was transient with the effect being lost within a few days of exercise cessation. Larry Oscai and associates (101) at the University of Illinois, Chicago, then reported in six subjects with hypertryglyceridemia that 4 successive days of aerobic walk–jog exercise covering 3–4 miles resulted in a normalization of triglycerides and the lipoprotein electrophoretic pattern. Triglycerides dropped farther each day and within 4 days of exercise cessation had returned to preexercise levels. The ability of muscle to break down serum triglyceride and liberate free fatty acids for metabolism depends on the activity of lipoprotein lipase (LPL). In 1972 Eldon Askew and others (4) at the U.S. Army Medical Research and Nutrition Laboratory in Denver reported that a single, exhaustive bout of exercise increased LPL activity in both rat skeletal muscle and heart. The increase in muscle LPL during exercise was then confirmed in humans by Hans Lithell and coworkers (81) at the University of Uppsala, Sweden, using muscle biopsies. The increase in LPL activity would make more fatty acid available to the muscle cells.

Although the pool of intramusclular triglyceride is not large, compared to adipose tissue, it may be a source of fuel for metabolism during exercise. From studies on ^{14}C fatty acid turnover, Paule Paul and Bela Issekutz (67, 104) at Lankenau Hospital in Philadelphia concluded that only half of the fat oxidized during exercise came from plasma free fatty acids and suggested that intramuscular triglyceride stores must supply the remainder of fat that was oxidized. Sven Fröberg (45) at the University of Uppsala confirmed, in rats run to exhaustion, that muscle triglyceride concentration was decreased. The decrease was greater in the red compared to white portion of the gastrocnemius. Lars Carlson and others (20) from the University of Uppsala then confirmed in humans, by using muscle biopsies, that intramuscle triglyceride is also utilized during exercise. In fact, they reported that the amount of work performed to exhaustion on a bicycle at 70% max heart rate was correlated to both muscle glycogen and triglycerides. They concluded, "These two muscle sub-strates could be used to predict the work performance." With all of the more recent emphasis on muscle glycogen we seem to have lost sight of the importance of intra-muscular triglyceride stores as an possible predictor of exhaustive performance.

Using tracer methodology, Johannes Romijn and colleagues (111) at the University of Texas conducted a comprehensive study of substrate metabolism at various intensities and durations of exercise. Their results are summarized in Figure 7.5. Their data indicate that muscle triglyceride is important for metabolism, especially at 65% $\dot{V}O_2$max. However, this was a calculated value based on total fat metabolism and glycerol measurements. More recently Bryan Bergman, a graduate student of Brooks from the University of California at Berkeley studied fat metabolism in-depth using not only tracer methodology but also A–V difference and muscle biopsy measurements (12). They concluded that muscle triglyceride contributes very little to total fat metabolism and pointed out that there is a lot of disagreement in the re-ported literature. Thus, while it is well documented that fatty acid oxidation plays an important role in metabolism, especially at lower work rates, the source of the fatty acids (adipose, lipoprotein, or muscle) is still undetermined.

Winder (135) suggested that the rate-limiting step in fat metabolism may be different under different exercise conditions. One of the first studies to give some insight into rate-limiting steps for fat metabolism was the work of Paul Molé (1939–2001) and coworkers (92), postdoctoral fellows in Holloszy's laboratory. These in-vestigators showed that, with isolated mitochondria, increasing the substrate con-centration of palmitic acid from 0.1 to 1.0 mM increased mitochondrial aerobic me-tabolism, suggesting the availability of free fatty acid could be a limiting factor. These results were consistent with whole-animal studies previously reported by D. T. Armstrong and others (3) at the Brookhaven National Laboratory in New York and by Bela Issekutz and associates (68) at Lankenau, showing that as plasma free fatty acid concentration was increased so was fatty acid metabolism. Thus, these studies established that availability of free fatty acids may be a limiting factor in fat metabolism. Since saturation of fatty acid uptake was observed in these studies, the entry of fatty acid into the muscle cell may also be a limiting factor. For many years it was assumed that fatty acids enter cells by diffusion through the lipid bilayer membrane. However, Arend Bonen at the University of Waterloo (16) and Larraine Turcotte (125) at the University of Southern California have recently discovered a fatty acid binding protein/transporter in the sarcolemmal membrane that may also be a limiting factor.

Molé and colleagues reported that adding CoA or carnitine along with palmitic acid increased aerobic metabolism in both mitochondrial and whole homogenate preparations (92). This observation suggested that the formation of fatty acyl-CoA and the transport of fatty acid into mitochondria by the carnitine palmitoyl trans-ferase system (see Fig. 7.4) could be rate limiting. Winder and colleagues (135) con-ducted a series of experiments on factors that might be rate limiting for fatty acid metabolism during muscle contraction and proposed the following mechanism. The onset of exercise induces a rise in AMP which allosterically activates AMP kinase

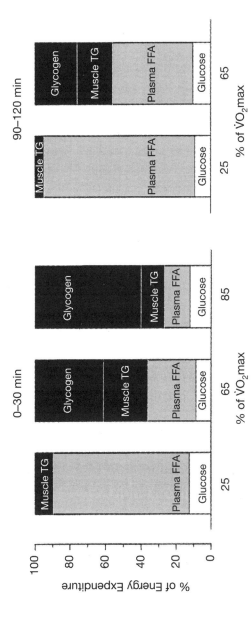

Fig. 7.5. Effect of intensity and duration of exercise on substrate utilization. Adapted from data by Romijn et al. (111).

(AMPK) and an AMPK kinase; the activated AMPK kinase phosphorylates and further activates AMPK, which phosphorylates and inactivates acetyl-CoA carboxylase (ACC) and the production of malonyl-CoA, an inhibitor of carnitine palmitoyl transferase, to increase the rate of fatty acyl-carnitine formation and transport of fatty acid into the mitochondria for β-oxidation.

EFFECTS OF TRAINING ON AEROBIC METABOLISM

As discussed in Chapter 6, "maximal oxygen uptake" ($\dot{V}O_2$max) was coined by A. V. Hill. At that time it was assumed that $\dot{V}O_2$max occurred because the heart, lungs, circulation, and the diffusion of oxygen to the active muscle fibers had attained maximal activity. Apparently, no consideration was given to the possibility that the ability of muscle to utilize the O_2 for aerobic metabolism might be an important factor. Over the years $\dot{V}O_2$max measurement became the gold standard to establish a training effect and over the years a debate has continued as to what exactly the limiting factor(s) is for $\dot{V}O_2$max.

Animal Studies

The cardiovascular improvements in response to training were the first to be identified. However, in the 1930s a Russian biochemist by the name of Alexander V. Palladin (1895–1972) became interested in the effects of physical activity on skeletal muscle. At the Institute of Biochemistry at Harkov, Palladin and his colleagues used electrical stimulation (two 15 minute periods per day for 15 days) to study biochemical adaptations in muscle. (He called it training.) They found that "training" increased the levels of phosphocreatine and glycogen in muscle and that with fatiguing work decreased the levels of lactic acid while increasing the work duration. These were subsequently used as markers for a training effect. Additional studies by Palladin's colleagues E. T. Sorenyi and O. P. Tchepinova found that training increased the flavin content of muscle and that Warburg studies showed an increase in respiration capacity. In 1945 Palladin was invited by the U.S. Academy of Sciences to address one of its meetings. The contents of his presentation were then published in Science (102). Unfortunately, most textbooks on exercise physiology have failed to recognize the early contribution of these Russian scientists.

In 1956 George R. Hearn, Philip Gollnick's mentor at Rutgers University, was aware of the Russian findings and along with W. W. Wainio (53) studied the effects of 5–8 weeks of swim training on muscle succinic dehydrogenase (SDH) activity. These investigators noted no difference between the trained and control muscles. Then in 1967 Holloszy (58) reported that rats subjected to a progressive program of treadmill training had a 60% increase in mitochondrial protein and almost a doubling in the capacity to metabolize pyruvate. These results were confirmed by Kraus and colleagues (76) in 1969 using rats and by Barnard, Edgerton, and Peter in 1970 (7) in a training study involving guinea pigs. The mitochondria in these studies showed a high level of respiratory control with tightly coupled oxidative phosphorylation indicating an enhanced capacity for oxidative phosphorylation. Realizing that the increase in mitochondrial protein reported in their initial studies might be due to better isolation from the trained muscle, both Holloszy and associates and Barnard and coworkers studied cytochrome c in whole homogenates and concluded that the increase in muscle mitochondria was real (57). Philip Gollnick and his student Douglas King (49) had also confirmed this by examining electron micrographs from trained and sedentary rats. Not only did the trained rats have more mitochondria; they were larger. Several studies were then undertaken by Holloszy's group to study individual enzymes involved in aerobic metabolism and they showed that not all mitochondrial enzymes increased their activity to the same extent. Some enzymes such as succinic dehydrogenase and isocitrate dehydrogenase showed a twofold increase while mitochondrial malate dehydrogenase only increased by 50% and glutamate dehydrogenase by 35% (57).

Although early studies by Hultman (65) reported that at a given submaximal workload trained individuals had a lower RER, suggesting a greater reliance on fat for metabolism, it was not until 1970 that Molé and Holloszy (93) showed in whole muscle homogenates that training resulted in a doubling in the capacity for muscle to metabolize fatty acids. They then confirmed the observation and also found the same increase in the ability of muscle to metabolize fatty acids when studied in isolated mitochondria (92). The increase in fatty acid metabolism with training was associated with increases in the activity of carnitine palmityl transferase, palmityl CoA dehydrogenase, and mitochondrial ATP-dependent palmityl CoA synthase. The results were confirmed by Baldwin and others (6) in Holloszy's laboratory, who showed that training increased both pyruvate and fatty acid oxidation in all fiber types.

In addition to an increased capacity of muscle to utilize fat as substrate following training, studies have also reported enhanced release of fatty acids from adipocytes. The release of free fatty acids from adipocytes is achieved primarily by sympathetic nervous system activation of hormone-sensitive lipase to degrade the stored triglycerides. Gollnick in 1967 (48) reported that in rats run on a treadmill for 1 hour, plasma free fatty acid levels were higher in trained animals. When fat tissue was removed from the rats immediately after exercise and placed in an incubation flask, free fatty acid release over 3 hours was significantly greater in the tissue from trained rats. These results were confirmed by Sven Fröberg and coworkers (46) and Eldon Askew and colleagues (4) in 1972 with adipose tissue removed from control and trained rats and then stimulated with norepinephrine or epinephrine. More recently a fatty acid binding protein has been identified in the sarcolemmal membrane and is thought to enhance fatty acid uptake by muscle as discussed earlier. Lorraine Turcotte and others (126) at the University of Southern California reported that training in rats increased the fatty acid binding protein content of the sarcolemmal membrane and enhanced free fatty acid uptake at rest and during contraction. Thus, all aspects of fatty acid metabolism from release to oxidation are increased with training.

In addition to examining the effects of exercise training on biochemical aspects of muscle metabolism, exercise physiologists began to study the effects of training on muscle fiber types as they are defined in part by metabolic characteristics determined by histochemical techniques. V. Reggie Edgerton (33) as a doctoral student in Rex Carrow's Laboratory at Michigan State University reported in 1969 that swim training in rats resulted in an increase in the number of plantaris fibers with high malate and succinate dehydrogenase activities with no change in the myosin ATPase staining pattern. They concluded that muscle fibers can be changed in a "nonpathologic condition." That same year K. Kowalski and others (75) from Michael Reese Hospital and Medical Center, Chicago, reported the effects of wheel running and climbing (weight lifting) on the histochemistry of quadriceps muscle using a subjective rating for stain intensity. They concluded that both running and climbing increased the percentage of fibers rated as "high activity" for phosphorylase, succinate dehydrogenase, or cytochrome oxidase. In 1970 Barnard and colleagues (7) reported that treadmill training in guinea pigs resulted in a significant increase in the percent of fibers classified as red and a decrease in the percent of fibers classified as white using NADH-diaphorase staining. Using myosin ATPase staining they concluded there was no change in the percentage of intermediate (slow-twitch) fibers. In a follow-up paper they measured contractile properties and concluded that there was no significant change in the twitch characteristics, which would agree with the myosin ATPase results (8). There was, however, a significant reduction in the rate of fatigue development during isometric contractions at a rate of 5/s, which could reflect the increased oxidative potential predicted by the increase in "red" fibers. In a more recent study using electrophoretic procedures to isolate different myosin isozymes Daniel Fitzsimons and colleagues (42) in Baldwin's laboratory reported that endurance training in rodents resulted in a shift in myosin isozymes to a slower-type ATPase in some but not all muscles and suggested that the shift was "designed to economize contractile activity without altering force production during repetitive limb movements."

Human Studies

Many of the adaptations to exercise training in skeletal muscle initially reported for rodents have been confirmed in humans. Thomas Morgan and associates (95) at the University of Washington reported an increase in mitochondrial protein in quadriceps muscle following 1 month of cycle ergometer training with one leg. Compared to the contralateral control, there was a 45% increase in aerobic capacity. These results were confirmed by E. Varnauskas and coworkers (128) from the University of Gothenburg, Sweden and by Phillip Gollnick and others (50).

Gollnick and others (50) also reported that histochemical analysis revealed that training resulted in a decrease in glycolytic fibers, an increase in aerobic fibers, and no change in the percentage of fast-twitch/slow-twitch fibers. In an earlier cross-sectional study of different types of athletes Gollnick and colleagues (51), reported that high-level endurance athletes (runners, cyclists, swimmers) not only had high levels of aerobic enzyme activities but also had a higher percentage of slow-twitch fibers compared to untrained individuals or high-level weight lifters. They concluded that the high percentage of slow-twitch fibers in the endurance athletes could be due to "natural endowment." In 1977 Per Andersen and Jan Henriksson (2) at the August Krogh Institute and E. Jansson and L. Kaiser (70) at the Karolinska Sjukhuset and Serafimer-lasarettet reported that endurance training in humans resulted in a decrease in fast-twitch IIB fibers with an increase in fast-twitch IIA fibers using the pH-sensitive myosin ATPase histochemical methods of Brooke and Kaiser (18). Gregory Adams (1) in conjunction with Kenneth Baldwin confirmed that the change in type II fiber type with training was the result of changes in the myosin heavy chain.

In addition to the increase in oxidative capacity and the number of oxidative fibers following training in humans, studies have also documented an increase in substrate availability. In 1967 Lars Hermansen and associates (54) from the Gymnastik-och Idrottshögskolan in Stockholm reported that trained individuals had a higher muscle glycogen content compared to untrained individuals. Then in 1972 Albert Taylor and coworkers (121) at the University of Alberta reported that 5 months of training increased glycogen synthase activity and glycogen content of skeletal muscle. The increase in muscle glycogen following training was also confirmed by Gollnick and others (50).

Thomas Morgan and colleagues (94) at the University of Washington reported an increase in muscle triglyceride following training while Ben Hurley and associates (66) in Holloszy's laboratory found no significant increase. Hurley and others did, however, find a greater triglyceride depletion with acute exercise at the same workload after training. They suggested a possible increase in hormone-sensitive lipase in the trained muscle. Hans Lithell and others (81) from the University of Uppsala reported that the best-trained cross-country skiers had the largest triglyceride stores before a 85 km race and had the greatest depletion at the end of the race.

In addition to enhancing the utilization of intramuscular fat stores, training also enhances the availability of fat from the circulation and adipose tissue. Bente Kiens and coworkers (74) at the Copenhagen Muscle Research Center recently (1997) reported that 3 weeks of intense, one-legged endurance training increased membrane-associated 40 kDa fatty acid binding protein by 49%. In 1978 Esko Nikkilä and associates (99) from the University of Helsinki reported that muscle lipoprotein lipase activity was significantly elevated in both men and women distance runners compared to sedentary controls. Jan Svedenhag and colleagues (120) from the Karolinska Institute in Stockholm reported an increase in muscle lipoprotein lipase activity following 8 weeks of training in normal, healthy men. Jean-Pierre Després and colleagues (31) at Laval University found that 20 weeks of training increased epinephrine-stimulated lipase activity in adipose tissue. This observation was confirmed by F. Crampes and others at Faculté de Medicine, Toulouse-Purpan for both normal (29) and obese (30) individuals.

Mechanisms of Adaptation

The first study to explore mechanisms involved in the increase in aerobic enzymes (mitochondria) resulting from training was conducted by Ronald Terjung and associates (123) as postdoctoral fellows in Holloszy's laboratory. Studying the turnover of ^{14}C-labeled cytochrome c they concluded that training resulted in a small increase in the synthesis rate but had a greater effect on decreasing the rate of degradation. Frank Booth, a postdoctoral fellow in Holloszy's laboratory (17), used a different method to study cytochrome c turnover as he felt that the method of ^{14}C labeling used by Terjung and coworkers may have resulted in reutilization of the ^{14}C, thus providing false data. Using the time course of decline following the cessation of exercise training they concluded that the higher levels of cytochrome c and other aerobic enzymes were due primarily to increased synthesis and not delayed degradation as proposed by Terjung and colleagues. Holloszy and his postdoctoral fellow William Winder (60) then showed that exercise training increased the activity of δ-aminolevulinic acid synthase, thought to be the rate-limiting enzyme for heme synthesis and incorporation into cytochromes.

Frank Booth realized that to adequately study mechanisms of adaptation to exercise training one would have to learn molecular biology techniques. After completing his postdoctoral training in St. Louis, he moved to the University of Texas at Houston and established the first molecular biology laboratory to study adaptations to physical activity. Booth and his students first focused on inactivity (immobilization) to study actin and cytochrome c mRNAs (96, 131). They then studied the effects of treadmill training and found that after 2 weeks of daily running, muscle cytochrome c mRNA increased by 17%–56%, which was associated with a 30%–41% increase in aerobic enzyme activity (cirtrate synthase). These studies confirmed that increased synthesis of cytochrome c was the major adaptation in the increase in mitochondrial enzymes resulting from exercise training. While most protein synthesis in the cell is determined by the nuclear genome, the mitochondria is unique in that it also contains a genome that regulates the synthesis of thirteen proteins essential for normal mitochondria function. R. Sanders Williams and colleagues (134, 135) at the University of Texas Southwestern Medical Center were the first to show that chronic contractile activity leads to increases in mRNA levels encoding both nuclear and mitochondrial gene products. They concluded that muscle contraction produces reciprocal changes in the expression of the two genomes at a transcriptional level but via different regulatory mechanisms.

The exact stimulus responsible for activation of protein synthesis is not known but may be associated with the development of local hypoxia. The fact that hypoxia might be the stimulus was suggested from studies conducted by J. Holm and others (63) at Sahlgrenska Sjukhusetin in Göteborg on patients with unilateral intermittent claudication, showing a higher mitochondrial oxidative capacity in the diseased, ischemic leg. Holm and coworkers (64) then reported that low-level walking in claudication patients induced an increase in succinic oxidase activity in the diseased legs with no change in normal legs. More recently N. Terrados and others (123) at the

Karolinska Hospital conducted a study with healthy individuals training under hypobaric conditions. Mitochondrial enzymes increased to a much greater extent with the same training under the hypobaric conditions, showing that hypoxia may be the major stimulus to activate protein synthesis and increase muscle aerobic enzymes. It is well recognized that exercise increases mitochondria only in active muscles, ruling out some humoral factor. However, David Hood and colleagues (64) at York University in Toronto suggest that in addition to a metabolic or ischemia stimulus, other factors, including electrical transmission via voltage-sensitive membrane proteins, ionic fluxes including calcium, or mechano transduction via the activation of integrin signaling might also be important. It is obvious that more research is needed to fully understand the mechanisms responsible for exercise-induced mitochondria biogenesis.

REFERENCES

1. Adams G. R., B. M. Hather, K. M. Baldwin, and G. A. Dudley. Skeletal muscle myosin heavy chain composition and resistance training. *J. Appl. Physiol.* 74:911–915, 1993.
2. Andersen P. and J. Henriksson. Training induced changes in the subgroups of human Type II scheletal muscle fibers. *Acta Physiol. Scand.* 99:123–125, 1977.
3. Armstrong D. T., R. Steele, N. Altszuler, A. Dunn, J. S. Bishop, and R.C. de Bodo. Regulation of plasma free fatty acid turnover. *Am. J. Physiol.* 201:9–14, 1961.
4. Askew E. W., G. L. Dohm, R. L. Huston, T. W. Sneed, and R. P. Dowdy. Response of rat tissue lipases to physical training and exercise. *Proc. Soc. Exp. Biol. Med.* 141:123–129, 1972.
5. Baldwin K. M., A. M. Hooker, and S. E. Herrick. Lactate oxidative capacity in different types of muscle. *Biochem. Biophys. Res. Com.* 83:151–157, 1978.
6. Baldwin K. M., G. H. Klinkerfuss, R. L. Terjung, P. A. Molé, and J. O. Holloszy. Respiratory capacity of white, red and intermediate muscle: adaptive response to exercise. *Am. J. Physiol.* 222:373–378, 1972.
7. Barnard R. J., V. R. Edgerton, and J. B. Peter. Effect of exercise on skeletal muscle. I. Biochemical and histochemical properties. *J. Appl. Physiol.* 28:762–766, 1970.
8. Barnard R. J., V. R. Edgerton, and J. B. Peter. Effect of exercise on skeletal muscle. II. Contractile properties. *J. Appl. Physiol.* 28:767–770, 1970.
9. Barnard R. J., V. R. Edgerton, T. Furukawa, and J. B. Peter. Histochemical, biochemical and contractile properties of red, white and intermediate fibers. *Am. J. Physiol.* 220: 410–414, 1971.
10. Barnard R. J. and J. B. Peter. Effect of training and exhaustion on hexokinase activity of skeletal muscle. *J. Appl. Physiol.* 27:691–694, 1969.
11. Barr D. P. and H. E. Himwitch. Studies in the physiology of muscular exercise. Comparison of arterial and venous blood following vigorous exercise. *J. Biol. Chem.* 55:525–537, 1923.
12. Bergman B. C., E. E. Wolfe, G. E. Butterfield, G. Lopaschuk, G. A. Casazza, M. A. Horning, and G. A. Brooks. Active muscle and whole body lactate kinetics after endurance training in man. *J. Appl. Physiol.* 87:1684–1696, 1999.
13. Bergström J. and E. Hultman. Muscle glycogen synthesis after exercise: an enhancing factor localized to the muscle cells in man. *Nature* 210:309–310, 1966.
14. Bergström J., L. Hermansen, E. Hultman, and B. Saltin. Diet, muscle glycogen and physical performance. *Acta Physiol. Scand.* 71:140–150, 1967.

15. Bernard C. Sur le mecanisme physiologique de la formation du sucre dans le foie. C.R. *Acad. Sci.* 44:578–586, 1325–1331, 1857.

16. Bonen A., J. J. Luiken, S. Liu, D. J. Dyck, B. Kiens, S. Kristiansen, L. P. Turcotte, G. J. Van Der Vusse, and J. F. Glatz. Palmitate transport and fatty acid transporters in red and white muscles. *Am. J. Physiol.* 275 (*Endocrinol. Metab.* 38):E471–E478, 1998.

17. Booth F. W. and J. O. Holloszy. Cytochrome c turnover in rat skeletal muscles. *J. Biol. Chem.* 252:416–419, 1977.

18. Brooke M. H. and K. K. Kaiser. Three "myosin ATPase" systems: the nature of their pH lability and sulfhydryl dependence. *J. Histochem. Cytochem.* 18:670–672, 1970.

19. Cain D. F. and R. E. Davies. Breakdown of adenosine triphosphate during a single contraction of working muscle. *Biochem. Biophys. Res. Commun.* 8:361–366, 1962.

20. Carlson L. A., L.-G. Ekelund, and S. O. Fröberg. Concentration of triglyceride, phospholipids glycogen in skeletal muscle and of free fatty acids and β-hydroxybutyric acid in blood in man in response to exercise. *Eur. J. Clin. Invest.* 1:248–254, 1971.

21. Carlsten A., B. Hallgren, R. Jagenburg, A. Svanborg, and L. Werkö. Myocardial metabolism of glucose, lactic acid amino acids, and fatty acids in healthy human individuals at rest and at different workloads. *Scand. J. Clin. Lab Invest.* 13:418–428, 1961.

22. Chance B. and C. M. Connelly. A method for the estimation of the increase in concentration of adenosine diphosphate in muscle sarcosomes following a contraction. *Nature* 4572:1235–1237, 1957.

23. Chauveau A. Source et nature du potentiel directment utilisé dans le travail musculairs d'après les exchanges respiratoires, chez l'hormme en état d'abstinence. *C.R. Acad. Sci. (Paris)* 122:1163–1221, 1896.

24. Christensen E. H. and O. Hansen. Arbeitsfahigkeit und Ernahrung. *Skand. Arch. Physiol.* 81:160–171, 1939.

25. Christensen E. H. and O. Hansen. Hypoglykame, arbeitsfahigkeit und ermudung. *Skand. Arch. Physiol.* 81:172–179, 1939.

26. Christensen E. H. and O. Hansen. Respiratorischen Quotient und O_2—Aufnahme. *Skand. Arch. Physiol.* 81:180–189, 1939.

27. Coggan A. R. and E. F. Coyle. Carbohydrate ingestion during prolonged exercise: effects on metabolism and performance. *Exerc. Sport. Sci. Rev.* 19:1–40, 1991.

28. Conlee R. K. Muscle glycogen and exercise endurance: a twenty-year perspective. *Exerc. Sport Sci. Rev.* 15:1–20, 1987.

29. Crampes F., M. Beauville, D. Rivière, and M. Garrigues. Effect of physical training in humans on the response of isolated fat cells to epinephrine. *J. Appl. Physiol.* 61:25–29, 1986.

30. De Glisezinski I., F. Crampes, I. Harant, M. Berlan, J. Hejnova, D. Laugin, D. Riviève, and V. Stich. Endurance training changes in lipolytic responsiveness of obese adipose tissue. *Am. J. Physiol.* 275 (*Endocrinol. Metab.* 38):E951–E956, 1998.

31. Després J. P., C. Bouchard, R. Savard, A. Tremblay, M. Marcotte, and G. Theriault. The effect of 20-week endurance training on adipose tissue morphology and lipolysis in men and women. *Metabolism* 33:235–239, 1984.

32. Dill B. D., H. T. Edwards, and J. H. Talbott. Studies in muscular activity. VII. Factors limiting the capacity for work. *J. Physiol. (Lond.)* 77:49–62, 1932.

33. Edgerton V. R., L. Gerchman, and R. Carrow. Histochemical changes in rat skeletal muscle after exercise. *Exp. Neurol.* 24:110–123, 1969.

34. Edwards H. T., R. Margaria, and D. B. Dill. Metabolic rate, blood sugar and the utilization of carbohydrate. *Am. J. Physiol.* 108:203–209, 1934.

35. Eggleton P. and G. P. Eggleton. The physiological significance of "phosphagen." *J. Physiol.* 63:155–161, 1927.

36. Eggleton P. and G. P. Eggleton. XXV. The inorganic phosphate and a labile form of organic phosphate in the gastrocnemius of the frog. *Biochem J.* 21:190–195, 1927.

37. Embden G. and E. Adler. Uber die physiologische Bedeutung des Wechsels des Permeabilitatszustandes von Muskelfasergrenzschichten. *Z. Physiol. Chem.* 118:1–11, 1922.
38. Engelhardt W. A. and M. N. Ljubimowa. Myosine and adenosinetriphosphatase. *Nature* 144:668–669, 1939.
39. Fiske C. H. and Y. Subbarow. Phosphocreatine. *J. Biol. Chem.* 81:629–679, 1929.
40. Fiske C. H. and Y. Subbarow. Phosphorus compounds of muscle and liver. *Science* 70: 381–382, 1929.
41. Fiske C. H. and Y. Subbarow. The nature of the "inorganic phosphate" in voluntary muscle. *Science* 65:401–403, 1927.
42. Fitzsimons D. P., G. M. Diffee, R. E. Herrick, and K. M. Baldwin. Effects of endurance exercise on isomyosin patterns in fast- and slow–twitch skeletal muscles. *J. Appl. Physiol.* 68:1950–1955, 1990.
43. Fletcher W. M. and F. G. Hopkins. Lactic acid in amphibian muscle. *J. Physiol.* 35:247–309, 1907.
44. Fletcher W. M. and F. G. Hopkins. The respiratory process in muscle and the nature of muscular motion. *Proc. Roy. Soc. B.* 89:444–467, 1917.
45. Fröberg S. O. Effect of acute exercise on tissue lipids in rats. *Metabolism* 20:714–720, 1971.
46. Fröberg S. O., I. Ostman, and N. O. Sjöstrand. Effect of training on esterified fatty acids and carnitine in muscle and on lipolysis in adipose tissue in vitro. *Acta Physiol. Scand.* 86:166–174, 1972.
47. Gauthier G. F. and S. Lowey. Distribution of myosin isozymes among skeletal muscle fiber types. *J. Cell Biol.* 81:10–25, 1979.
48. Gollnick P. D. Exercise, adrenergic blockade and free fatty acid mobilization. *Am. J. Physiol.* 213:734–738, 1967.
49. Gollnick P. D. and D. W. King. Effect of exercise and training on mitochondria of rat skeletal muscle. *Am. J. Physiol.* 216:1502–1509, 1969.
50. Gollnick P. D., R. B. Armstrong, B. Saltin, C. W. Saubert, W. L. Sembrowich, and R. E. Shepherd. Effect of training on enzyme activity and fiber composition of human skeletal muscle. *J. Appl. Physiol.* 34:107–111, 1973.
51. Gollnick P. D., R. B. Armstrong, C. W. Saubert, K. Piehl, and B. Saltin. Enzyme activity and fiber composition in skeletal muscle of untrained and trained men. *J. Appl. Physiol.* 33: 312–324, 1972.
52. Grimditch G. K., R. J. Barnard, and S. A. Kaplan. Insulin binding and glucose transport in rat skeletal muscle sarcolemma vesicles. *Am. J. Physiol.* 249 (*Endocrinol. Metab.* 12): E398–E408, 1985.
53. Hearn G. R. and W. W. Wainio. Succinic dehydrogenase activity of the heart and skeletal muscle of exercised rats. *Am. J. Physiol.* 185:348–350, 1956.
54. Hermansen L., E. Hultman, and B. Saltin. Glycogen during prolonged severe exercise. *Acta Physiol. Scand.* 71:129–139, 1967.
55. Hill A. V. The energy degraded in the recovery processes of stimulated muscles. *J. Physiol.* 46:28–80, 1913.
56. Hill A. V. The recovery heat-production in oxygen after a series of muscle twitches. *Proc. R. Soc. Lond.* 103:183–191, 1928.
57. Holloszy J. O. and F. W. Booth. Biochemical adaptations to endurance exercise in muscle. *Ann. Rev. Physiol.* 38:273–281, 1976.
58. Holloszy J. O. Biochemical adaptations in muscle. Effects of exercise on mitochondrial O_2 uptake and respiratory enzyme activity in skeletal muscle. *J. Biol. Chem.* 242:2278–2282, 1967.
59. Holloszy J. O. and H. T. Narahara. Studies of tissue permeability. X. Changes in permeability to 3-methylglucose associated with contraction of isolated frog muscles. *J. Biol. Chem.* 240:3492–3500, 1965.

60. Holloszy J. O. and W. W. Winder. Induction of δ-aminolevulinic acid synthase in muscle by exercise or thyroxine. *Am. J. Physiol.* 236 (*Regulatory Integrative Comp. Physiol.* 5):R180–R183, 1979.

61. Holloszy J. O., J. S. Skinner, G. Toro, and T. K. Cureton. Effects of a six month program of endurance exercise on the serum lipids of middle-aged men. *Am. J. Cardiol.* 14:753–760, 1964.

62. Holm J., A. G. Dahllöf, P. Björntorp, and T. Scherstén. Enzyme studies in muscle of patients with intermittent claudication. Effect of training. *Scand. J. Clin. Lab. Invest. Suppl.* 128:201–205, 1973.

63. Holm J., P. Björntorp, and T. Scherstén. Metabolic activity in human skeletal muscle. Effect of peripheral arterial insufficiency. *Europ. J. Clin. Invest.* 2:321–325, 1972.

64. Hood D. A., M. Takahashi, M. K. Connor, and D. Freyssenet. Assembly of the cellular powerhouse: current issues in muscle mitochondrial biogenesis. *Exerc. Sport Sci. Rev.* 28:68–73, 2000.

65. Hultman E. Physiological role of muscle glycogen in man, with special reference to exercise. *Circ. Res.* 20–21 (Suppl. I):I99–I114, 1967.

66. Hurley B. F., P. M. Nemeth, W. H. Martin III, J. M. Hagberg, G. P. Dalsky, and J. O. Holloszy. Muscle triglyceride utilization during exercise: effect of training. *J. Appl. Physiol.* 60:562–567, 1986.

67. Issekutz B. and P. Paul. Intramuscular energy sources in exercising normal and pancreatectomized dogs. *Am J. Physiol.* 215:197–204, 1968.

68. Issekutz B., H. I. Miller, P. Paul, and K. Rodahl. Source of fat oxidation in exercising dogs. *Am. J. Physiol.* 207:583–589, 1964.

69. Ivy J. L., and J. O. Holloszy. Persistent increase in glucose uptake by rat skeletal muscle following exercise. *Am. J. Physiol.* 241 (*Cell Physiol.* 10):C200–C203, 1981.

70. Jansson E. and L. Kaisser. Muscle adaptation to extreme endurance training in man. *Acta Physiol. Scand.* 100:315–324, 1977.

71. Jones N. L. and R. L. Havel. Metabolism of free fatty acids and chylomicron triglycerides during exercise in rats. *Am. J. Physiol.* 213:824–828, 1967.

72. Karlsson J., F. Bonde-Petersen, J. Henriksson, and H. G. Knuttgen. Effects of previous exercise with arms or legs on metabolism and performance in exhaustive exercise. *J. Appl. Physiol.* 38:763–767, 1975.

73. Keul J., E. Doll, H. Steim, U. Fleer, and H. Reindell. Uber den stoffwechsel des herzen bei hochleistungssportlern III. Der oxidative stoffwechsel des trainierten meuschlichen herzens unter verscheidenen arbeits bedingungen. *Ztschr. Kreislaufforsch.* 55:477–488, 1966.

74. Kiens B., S. Kristiansen, P. Jensen, E. A. Richter, and L. P. Turcotte. Membrane associated fatty acid binding protein ($FABP_{pm}$) in human skeletal muscle is increased by endurance training. *Biochem. Biophys. Res. Commun.* 231:463–465, 1997.

75. Kowalski K., E. E. Gordon, A. Martinez, and J. Adamek. Changes in enzyme activities of various muscle fiber types in rat induced by different exercises. *J. Histochem. Cytochem.* 17:601–607, 1969.

76. Kraus H., R. Kristen, and J. R. Wolff. Die Wirkung von Schwimm- und Lauftraining auf die Cellulare Funktion und Strukur des Muskels. *Arch. Ges. Physiol.* 308:57–79, 1969.

77. Krogh A. and J. Lindhard. The relative value of fat and carbohydrate as sources of muscular energy. *Biochem. J.* 14:290–363, 1920.

78. Kuby S. A., L. Noda, and H. A. Lardy. Adenosinetriphosphate-creatine transphosphorylase III. Kinetic studies. *J. Biol. Chem.* 210:65–82, 1954.

79. Lange G. Uber die dephosphorylierung von adenosintriphosphat zu adenosindiphosphat wahrend der kontraktionsphase von foschrectusmuskeln. *Biochem. Z.* 326:172–186, 1955.

80. Lehmann H. Über die umesterung des adenylsäuresystems mit phosphagenen. *Biochem. Z.* 286:336–343, 1936.

81. Lithell H., J. Orlander, R. Schéle, B. Sjödin, and J. Karlsson. Changes in lipoprotein-lipase activity and lipid stores in human skeletal muscle with prolonged heavy exercise. *Acta Physiol. Scand.* 107:257–261, 1979.

82. Lohmann K. Darstellung der adenylpyrophosphorsäure aus muskulatur. *Biochem. Z.* 233: 460–469, 1931.

83. Lohmann K. Über die enzymatische Aufspaltung der Kreatinphosphorsäue; zugleich ein Beitrag zum Chemismus der Muskelkontraktion. *Biochem. Z.* 271:264–277, 1934.

84. Lohmann K. Uber die pyrophosphatfraktion im muskel. *Die Naturwissenschaften* 17: 624–625, 1929.

85. Lundsgaard E. Phosphagen und Pyrophosphatumsatz in jodessigsäurevergifteten Muskeln. *Biochem. Z.* 269:308–328, 1934.

86. Lundsgaard E. The ATP content of resting and active muscle. In: A discussion on muscular contraction and relaxation, chaired by A. V. Hill. *Proc. R. Soc. B.* 137:73–79, 1949.

87. Lundsgaard E. Uber die Energetik der anaeroben Muskelkontraktion. *Biochem. Z.* 233: 322–343, 1931.

88. Lundsgaard E. Untersuchungen uber muskelkontraktionen ohne milchsäurebildung. *Biochem. Z.* 217:162–177, 1930.

89. Lundsgaard E. Weitere Untersuchungen über Muskelkontraktionen ohne Milchsaurebildung. *Biochem. Z.* 227:51–83, 1930.

90. McKie D. Antoine Lavoisier. New York: Henry Schuman, 1952, pp. 142–146.

91. Meyerhof O. and K. Lohmann. Uber die naturlichen guanidinophosphorsauren (phosphagene) in der quergestreiften muskulatur. *Biochem. Z.* 196:22–48, 1928.

92. Molé P. A., L. B. Oscai, and J. O. Holloszy. Adaptation of muscle to exercise. Increase in levels of palmityl CoA synthase, carnitine palmityl transferase and palmityl CoA dehydrogenase, and in the capacity to oxidize fatty acids. *J. Clin. Invest.* 50:2323–2330, 1971.

93. Molé P. A. and J. O. Holloszy. Exercise-induced increase in the capacity of skeletal muscle to oxidize palmitate. *Proc. Soc. Exp. Biol. Med.* 134:789–792, 1970.

94. Morgan T. E., F. A. Short, and L. A. Cobb. Effect of long-term exercise on skeletal muscle lipid composition. *Am. J. Physiol.* 216:82–86, 1969.

95. Morgan T. E., L. A. Cobb, F. Short, R. Ross, and D. R. Gonn. Effect of long-term exercise on human muscle mitochondria. In: *Muscle Metabolism During Exercise*, edited by B. Pernow and B. Saltin. New York: Plenum, 1971, pp. 87–85.

96. Morrison P. R., J. A. Montgomery, T. S. Wong, and F. W. Booth. Cytochrome C protein synthesis rate and mRNA levels during atrophy and regrowth of skeletal muscle. *Biochem. J.* 241:257–263, 1987.

97. Morrison P. R., R. B. Biggs, and F. W. Booth. Daily running of 2 wk and mRNAs for cytochrome c and α-actin in rat skeletal muscle. *Am. J. Physiol.* 257 (*Cell Physiol.* 26): C936–C939, 1989.

98. Needham D. M. Discovery of lactic acid and glycogen in muscle. In: *Machina Carnis*. London: Cambridge U. Press, 1971, pp. 41–42, 612, 679.

99. Nikkilä E. A., M.-R. Taskinen, S. Rehunen, and M. Härkönen. Lipoprotein lipase activity in adipose tissue and skeletal muscle of runners: relation to serum lipids. *Metabolism* 27:1661–1667, 1978.

100. Ochoa S. The pursuit of a hobby. *Ann. Rev. Biochem.* 49:1–30, 1980.

101. Oscai L. B., J. A. Patterson, D. L. Bogard, R. J. Beck, and B. L. Rothernel. Normalization of serum triglycerides and lipoprotein electrophoretic patterns by exercise. *Am. J. Cardiol.* 30:775–780, 1972.

102. Palladin A. V. The biochemistry of muscle training, *Science* 102:576–578, 1945.

103. Parnas J. K. and R. Wagner. Uber den Kolenhydratumsatz in isolierter Amphibien-muskeln und uber die Beziehungen Zwischen Kohlenhydratschwund und Milchsaure-bildung in Muskeln. *Biochem. Z.* 61:387–426, 1914.

104. Paul P. and B. Issekutz. Role of extramuscular energy sources in the metabolism of the exercising dog. *J. Appl. Physiol.* 22:615–622, 1967.

105. Pearl R. *The Rate of Living.* New York: Knopf, 1928.

106. Peter J. B., R. J. Barnard, V. R. Edgerton, C. A. Gillespie, and K. Stempel. Metabolic pro-files of three fiber types of skeletal muscle in guinea pigs and rabbits. *Biochemistry* 11:2627–2633, 1972.

107. Peter J. B., R. N. Jeffress, and D. R. Lamb. Exercise: effects on hexokinase activity in red and white skeletal muscle. *Science* 160:200–201, 1968.

108. Peter J. B., S. Sawaki, R. J. Barnard, V. R. Edgerton, and C. A. Gillespie. Lactate dehydro-genase isoenzymes: distribution in fast-twitch red, fast-twitch white and slow-twitch intermediate fibers of guinea pig skeletal muscle. *Arch. Biochem. Biophys.* 144:304–307, 1971.

109. Pfluger E. F. W. Uber die physiologisches Verbrennung in die lebendig organismes. *Pflugers Arch.* X:251–301, 1872.

110. Rasmussen B. B. and R. R. Wolfe. Regulation of fatty acid oxidation in skeletal muscle. *Annu. Rev. Nutr.* 19:463–484, 1999.

111. Romijn J. A., E. F. Coyle, L. S. Sidossis, A. Gastaldelli, J. F. Horowitz, E. Eudert, and R. R. Wolfe. Regulation of endogenous fat and carbohydrate metabolism in relation to exer-cise intensity and duration. *Am. J. Physiol.* 265 (*Endocrinol. Metab.* 28):E380–E391, 1993.

112. Roth D. A. and G. A. Brooks. Lactate transport is mediated by a membrane-borne carrier in rat skeletal muscle sarcolemmal vesicles. *Arch. Biochem. Biophys.* 279:377–385, 1990.

113. Rubner M. *Das Problem der Lebensdauer und seine Beziehungen zur Wachstum und Ernahrung.* Munich: Oldenbourg, 1908.

114. Ryffel J. H. Experiments on lactic acid formation in man. *J. Physiol. (Lond.)* 39:xxix–xxxii, 1909–1910.

115. Sant' Ana Pereira J. A., S. Ehnion, A. J. Sargeant, A. F. Moorman, and G. Goldspink. Com-parison of the molecular, antigenic and ATPase determinants of fast myosin heavy chains in rat and human: a single-fiber study. *Pflugers Arch.* 435:151–163, 1997.

116. Selye H. and G. Tuchweber. Stress in relation to aging and disease. In: *Hypothalamus, Pituitary and Aging*, edited by A. V. Everitt and J. A. Burgess. Springfield, IL: Thomas, 1976, pp. 553–569.

117. Silberstein L. and S. Lowey. Isolation and distribution of myosin isoenzymes in chicken pectoralis muscle. *J. Mol. Biol.* 148:153–189, 1981.

118. Sternlicht E., R. J. Barnard, and G. K. Grimditch. Exercise and insulin stimulate skeletal muscle glucose transport through different mechanisms. *Am. J. Physiol.* 256 (*Endo-crinol. Metab.* 19):E227–E230, 1989.

119. Sternlicht E., R. J. Barnard, and G. K. Grimditch. Mechanism of insulin action on glucose transport in rat skeletal muscle sarcolemmal vesicles. *Am J. Physiol.* 254 (*Endocrinol. Metab.* 17):E633–E638, 1988.

120. Svedenhag J., H. Lithell, A. Juhlin-Dannfelt, and J. Henriksson. Increase in skeletal mus-cle lipoprotein lipase following endurance training in man. *Atherosclerosis* 49:203–207, 1983.

121. Taylor A. W., R. Thayer, and S. Rao. Human skeletal muscle glycogen synthase activities with exercise and training. *Can. J. Physiol. Pharmacol.* 50:411–415, 1972.

122. Terjung R. L., W. W. Winder, K. M. Baldwin, and J. O. Holloszy. Effect of exercise on the turnover of cytochrome C in skeletal muscle. *J. Biol. Chem.* 248:7404–7406, 1973.

123. Terrados N., E. Jansson, C. Sylvén, and L. Kaijser. Is hypoxia a stimulus for synthesis of oxidative enzymes and myoglobin? *J. Appl. Physiol.* 68:2369–2372, 1990.

124. Thomason D. B., K. M. Baldwin, and R. E. Herrick. Myosin isozyme distribution in rodent hindlimb skeletal muscle. *J. Appl. Physiol.* 60:1923–1931, 1986.

125. Turcotte L. P., A. K. Srivastava, and J.-L. Chiasson. Fasting increases plasma membrane fatty acid-binding protein (FABP$_{pm}$) in red skeletal muscle. *Mol. Cell. Biochem.* 166: 153–158, 1997.

126. Turcotte L. P., J. R. Swenberger, M. Z. Tucker, and A. J. Yee. Training-induced elevation in FABP (PM) is associated with increased palmitate use in contracting muscle. *J. Appl. Physiol.* 87:285–293, 1999.

127. Van Hall G. Lactate as a fuel for mitochondria. *Acta Physiol. Scand.* 168:643–656, 2000.

128. Varnauskas E., P. Björntorp, M. Fahlén, I. Prerovsk, and J. Stenberg. Effects of physical training on exercise blood flow and enzymatic activity in skeletal muscle. *Cardiovasc. Res.* 4:418–422, 1970.

129. von Muralt A. The development of muscle-chemistry, a lesson in neurophysiology. *Biochim. Biphys. Acta* 4:126–129, 1950.

130. von Pettenkofer M. and C. Voit. Untersuchungen über den Stoffverbrauch des normalen Menschen. *Z. Biol.* 2:459–537, 1866.

131. Watson P. A., J. P. Stein, and F. W. Booth. Changes in actin synthesis and actin mRNA content in rat muscle during immobilization. *Am. J. Physiol.* 247 (*Cell Physiol.* 16): C39–C44, 1984.

132. West J. B. Pulmonary blood flow and gas exchange. In: *Respiratory Physiology: People and Ideas*, edited by J. B. West. New York: Oxford University Press, 1996, pp. 155–158.

133. Williams R. S., M. Garcia-Moll, J. Mellor, S. Salmons, and W. Harlan. Adaptation of skeletal muscle to increased contractile activity. Expression nuclear genes encoding mitochondrial proteins. *J. Biol. Chem.* 262:2764–2767, 1987.

134. Williams R. S., S. Salmons, E. A. Newsholme, R. E. Kaufman, and J. Mellor, Regulation of nuclear and mitochondrial gene expression by contractile activity in skeletal muscle. *J. Biol. Chem.* 261:378–380, 1986.

135. Winder W. W. Malonyl-CoA–regulator of fatty acid oxidation in muscle during exercise. *Exerc. Sport Sci. Rev.* 26:117–132, 1998.

136. Winfield G. The fate of fatty acids in the survival processes of muscle. *J. Physiol.* 49: 171–184, 1915.

137. Zuntz V. N. Umsatz der nahrstoffe. XI. Betrachtungen uber die beziehungen zwischen nahrstoffen und leistungen des korpers. In: *Handbuch der Biochemie des Menschen und der Tiere*, edited by K. Oppenheimer. Jena, FRG: Fischer, 1911, pp. 826–855.

chapter 8

THE METABOLIC SYSTEMS: ANAEROBIC METABOLISM (GLYCOLYTIC AND PHOSPHAGEN)

George A. Brooks and L. Bruce Gladden

I N this chapter, we address the history of the study of anaerobic metabolism during exercise. Our emphasis is on the early origins and progression of studies on phosphagen and glycolytic metabolism. Further, although our topic is "anaerobic" metabolism, we consider the interrelationship between oxygen and lactic acid production to be of primary importance. Other areas of this broad topic receive less attention; space limitations required us to focus on what we believe to be the more important events and their consequences. For information on studies and scientists before approximately 1900, we relied heavily on Fletcher and Hopkins (63), Keilin (99), Leicester (105), von Muralt (153), Rothschuh (141), Williams (164), and Zuntz (168).

THE PRELACTIC ACID ERA

Fermentation and the Pasteur Effect

Despite the fact that von Muralt (153) called the study of muscle metabolism before 1907 the "prelactic acid era," numerous important observations relating to glycolysis and lactic acid were made during this period. The notion that lactic acid is formed as the result of oxygen lack can be traced to alcohol fermentation technology of the

322

eighteenth century. David Keilin (99, pp. 65–72) and Henry M. Leicester (105, pp. 177–179) describe scintillating controversies in the 1850s and 1860s among Louis Pasteur (1822–1895), Justus von Liebig (1803–1873), Claude Bernard (1813–1878), and Marcellin Berthelot (1827–1907) about the nature of fermentation. The key observations of Pasteur were that some microorganisms can live and proliferate in the absence of air and cannot use oxygen. In fact, O_2 poisons those organisms. He also found that some facultative cells are capable of living in both the presence and absence of oxygen. Moreover, Pasteur found that these facultative cells respire normally in the presence of oxygen and cause very little fermentation, but in anoxia they show very active fermentation (99, p. 68). Not only did Pasteur recognize the existence of aerobic and anaerobic organisms (130), but he also recognized that different types of the organisms which he called yeast formed very different products (129). For instance, one kind of organism fermented sugar to lactic acid whereas another organism produced alcohol.

Lactic Acid

In 1808, Jöns Jacob Berzelius (1779–1848), "almost the final authority on chemical matters of his day" (105), found an elevated concentration of lactate in "the muscles of hunted stags" (124). In 1845, Hermann von Helmholtz (1821–1894) reported findings that were consistent with lactic acid formation at the expense of glycogen (153). Shortly thereafter (1859), Emil H. Du Bois-Reymond (1818–1896) (63, 153) noted that activity caused muscles to become acidic and actually related this finding to the increased lactic acid reported by Berzelius. Soon after (1864), Rudolph P.H. Heidenhain (1834–1897) (153) reported that the amount of lactic acid increased with the amount of work done. In the next year, Ranke (1865) (63) indicated that resting muscle is alkaline but becomes acid as it survives. As emphasized by R. James Barnard (1937–) and John O. Holloszy (1933–) in Chapter 6, Nathan Zuntz (1847–1920) (168) in 1911 reviewed other significant papers of the 1800s. For example, Zuntz noted that Minot's paper in 1876 reported that the hindlimbs of dogs perfused with anoxic serum produced lactate during electrical stimulation.

Zuntz was a disciple of Eduard Friedrich Wilhelm Pflüger (1829–1910). Through the invention of the portable Zuntz-Geppert breathing machine, "Zuntz became involved in those physiological events which were associated with sports, hiking and high altitude … He was a popular lecturer who could present in simple terms a series of rather complex biological phenomena, a gift which made him contribute to the dissemination of physiological knowledge to the persons interested in sports activities" (141).

Despite all of the evidence that seems obvious from our modern-day perspective, other studies including those of Astaschewsky (1880), Warren (1881), Blome (1891), Heffter (1893), and von Furth (1903) (63) found little or no increase in lactic acid over time in surviving muscle. The results of Astaschewsky and of Warren were probably explainable by the presence of an intact circulation that removed the lactic acid (63). Nevertheless, confusion persisted until the classic studies of Walter Morley Fletcher (1873–1910) and Frederick Gowland Hopkins (1861–1947) in 1907 (63). Fletcher was mentor to A.V. Hill and in 1929 Hopkins was awarded the Nobel Prize for "his discovery of growth-stimulating vitamins" (152).

These early studies of lactic acid should be viewed in the context of the theory of muscle contraction in the late 1800s. Ludimar Hermann (1838–1914) (63, 105) postulated that a hypothetical *inogen* molecule was an unstable precursor of both lactic acid and carbon dioxide. Presumably, oxygen was already combined into the inogen molecule and was ready to combine with carbon and hydrogen in an explosive breakdown of inogen. In recovery, fresh carbon bodies perhaps along with lactic acid would be combined again into a reoxygenated unstable inogen molecule. Pflüger supported this theory, referring to a giant molecule (inogen) which was unstable due to the fact that it contained what he called "intramolecular oxygen" (63).

Major advances were made by Fletcher and Hopkins in their classic studies of 1907 (63). Much of the uncertainty surrounding lactic acid formation in muscle at the beginning of the twentieth century was due to methodological problems mainly associated with muscle glycogenolysis prior to study. Fletcher and Hopkins (63) developed a method which prevented significant lactic acid formation in resting muscles before the extraction and analysis of the lactate. Accordingly, they were able to show that: (1) freshly excised resting muscle contains only a small amount of lactic acid, (2) lactic acid concentration increases in excised, resting, anaerobic muscles, (3) lactic acid accumulates to high levels during stimulation of muscles to fatigue, and (4) when fatigued muscles are placed in oxygen-rich environments, lactic acid disappears. Interestingly, Fletcher and Hopkins (63) were not certain that glycogen was the precursor of lactate at this time and they further assumed that muscles could form more lactate than could be provided by their glycogen stores. In 1910, Archibald Vivian (A.V.) Hill (1886–1977) (81) published data indicating that the immediate processes of muscle contraction did not require or involve the consumption of oxygen, the heat of contraction being the same in the presence or absence of O_2. However, extra heat was liberated in recovery, but only if O_2 was present.

Glycolysis

In 1912 Gustav Georg Embden (1874–1933) and associates (55) showed that yeast and working muscle produced the same intermediate which was believed to be a glyceraldehyde. Subsequently, in 1914 Embden and Laquer (56) found a phosphorus compound in muscle that caused production of lactic acid. They called this substance "lacticidogenen" and afterward identified it as a hexose monophosphate (57). As reviewed by Leicester (105, pp. 203–204) this "Embden Ester" turned out to be a mixture of hexose monophosphates that were subsequently identified: fructose-6-P (Neuberg Ester, 1918), glucose-6-P (Robison Ester, 1922), and eventually glucose 1-P (Cori Ester, 1936). In 1920, using frog muscle preparations (Fig. 8.1) Otto Fritz Meyerhof (1884–1951) (118–120) definitively identified glycogen as the precursor of lactic acid and also provided evidence strongly linking contraction to lactate formation and oxidative recovery to glycogen restoration.

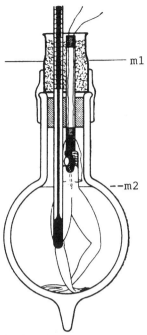

Abb. 2. Calorimeter mit Anord-
nung zur Indirekten elektrischen
Reizung eines Schenkelpaares.
Der horizontale Strich entspricht
dem Wasserstand des Thermo-
staten; der Strich im Gefäß
dem Spiegel der Ringerlösung.

Fig. 8.1. Meyerhof's frog muscle calorimeter, 1920. Translated: "Figure 2. Calorimeter with electrode for indirect electrical stimulation of a hemicorpus. The first
horizontal mark (m1, inserted) shows the level of the thermostatically controlled
water bath. The mark in the inside of the vessel (--m2, added) shows the level of
the Ringers solution." See (120).

Elaboration of the full Embden-Meyerhof (glycolytic) pathway took another
two decades (121). However, in the early twentieth century, the essence of the phenomena described by Pasteur was confirmed on lactic-acid-producing cells and tissues from vertebrates; when yeast cells were incubated or muscles were made to contract without oxygen, lactic acid accumulated. And, even though 1931 Nobel laureate
Otto Heinrich Warburg (1883–1970) (152) reported extensively on glycolysis leading to lactate accumulation in some types of well-oxygenated cells (e.g., cancer cells)
(155), the phenomenon describing glucose–oxygen–lactic acid interactions came to
be known as the "Pasteur effect" in textbooks of biochemistry (e.g., 104, p. 408). Ret-
rospectively, because lactate removal is primarily by oxidation we now understand
that in the absence of oxygen, cultured facultative cells and isolated frog muscles
could only produce, but not remove, lactate and other glycolytic or fermentation
products. *In this context the notion of a causative relationship between hypoxia and
lactate accumulation evolved.*

Current Knowledge

Today we know that some types of facultative cells can be cultured with lactate as a preferred fuel because their mitochondria possess means to consume and oxidize lactate directly without conversion to pyruvate in the cytosol. For instance mitochondria of yeast (*Saccharomyces cerevisiae*) contain flavocytochrome b_2, a lactate–cytochrome c oxidoreductase (45) that couples lactate dehydrogenation to reduction of cytochrome c (46). In fact, the association between cytochrome b_2 and LDH in yeast can be traced to the 1940s (99, p. 274). A similar phenomenon occurs in mammalian muscle (31, 151). Contemporary studies have reported a balance of lactic acid production and oxidation in exercising men at sea level and 4,300 m altitude (136); such data were simply not obtainable with the technology of the 1920s. Similarly, the nuclear magnetic resonance (NMR) technology that allows us to know that contractions, independently of O_2 availability, stimulate glycolysis in muscle (41), and that lactate is oxidized in working skeletal (20) and cardiac (38) muscle, is a recent development.

THE O_2 DEBT HYPOTHESIS

Following the 1907 work of Fletcher and Hopkins (63), in 1913 Hill (78) provided data indicating that the heat of recovery following muscle contraction was greater when O_2 was present. Hill concluded that the "processes of muscle contraction are due to the liberation of lactic acid from some precursor" and that "the lactic acid precursor is rebuilt after the contraction is over in the presence of, and by use of oxygen with the evolution of heat" (78, p. 79).

As already noted, using frog muscle preparations in 1920, Meyerhof (120) definitively identified glycogen as the precursor of lactic acid. Further, Meyerhof observed that after contractile activity, when lactate disappeared, glycogen reappeared in a corresponding amount, less a quantity which, calculated from heat production and O_2 consumption, approximated the enthalpy of a fraction (one-fourth to one-third) of the lactate that disappeared. Independently of Meyerhof, in 1914 Hill (79) found that the recovery heats of isolated frog muscles could account for combustion of about one-sixth to one-fifth of the lactate removed in recovery. The fraction of frog muscle lactate oxidized in recovery to restore the remainder to glycogen came to be known as the "combustion coefficient." While the results of Meyerhof (Fig. 8.2), and Hill (Fig. 8.3) contained quantitative differences, their results and conclusions were broadly the same: a minor fraction (roughly one-fifth) of lactate in muscle after contractions is oxidized to provide energy for the reconversion of the majority (about four-fifths) to its precursor, glycogen.

In 1922, Meyerhof and Hill shared the Nobel Prize in physiology and medicine. Meyerhof was recognized "for his discovery of the fixed relationship between the consumption of oxygen and the metabolism of lactic acid in the muscle," whereas Hill was recognized "for his discovery relating to the production of heat in the muscle" (152). The awards were richly deserved, but the studies were on ex vivo prepa-

Fig. 8.2. Portrait of Otto Meyerhof, 1922. Source, Nobel Prize Archives.

rations. Linking the experimental results to human physiology was a challenge that Hill took up in the early 1920s although, as noted later, he and others were either unaware of, or ignored, the efforts of others in this area.

In 1920, Danes, Shack August Steenberg Krogh (1874–1949) and Jens Peter Johannes Lindhard (1870–1947), were the first to report the exponential decline in O_2 consumption in men after exercise (102). August Krogh was a 1920 Nobel Prize winner "for his discovery of the capillary motor regulating mechanism" (152). Following the report of Krogh and Lindhard, Hill and associates turned their attention to studies of humans in an attempt to unite the new knowledge of muscle biochemistry and human metabolism. In 1923 Hill and Lupton (86) articulated the O_2 debt hypothesis, and the following year Hill, Long, and Lupton (83–85) published a series of noteworthy reports. The "oxygen debt" was defined (86, p. 142) as the "total amount of oxygen used, after cessation of exercise in recovery therefrom." Recognizing that during exercise onset and maximal exercise conditions there was a "deficit" in oxygen consumption, Hill and associates sought to measure the "excess post-exercise O_2 consumption" (O_2 debt) (64) to obtain an energy equivalent of the anaerobic lactate-

Fig. 8.3. Portrait of Sir Archibald Vivian (A.V.) Hill, 1922. Source, Nobel Prize Archives.

producing work done during exercise. Hill and associates attributed the first (fast) phase of decline in postexercise $\dot{V}O_2$ to oxidative removal of lactate in the muscles of formation and the second (slow) phase of decline in postexercise $\dot{V}O_2$ to oxidation of lactate that had escaped from muscles by diffusion.

In formulating their hypothesis, Hill and associates assumed that the energetics and metabolism in muscles of healthy humans were the same as in isolated frog muscles. With regard to the assumption of similar energetics, the supposition was appropriate as the laws of thermodynamics apply regardless of species differences. With regard to the phenomena of extra heat released from isolated muscles when they recovered in the presence, versus the absence, of O_2, and the extra O_2 consumed by persons recovering from exercise, it was logical for Hill and associates to link the two sets of phenomena together. That metabolic pathways, activities, and strategies differ significantly between mammals and lower vertebrates, particularly with regard to pathways of lactate disposal, is more recent knowledge (163).

As noted by Peter Harris in 1969 (73), given their experimental results and assumptions, it was natural for Hill and associates to conclude that in the human body, just as in isolated frog muscle, the "recovery volume of oxygen" (i.e., the O_2 debt) was caused by the oxidation of lactate. Further, because a "combustion coefficient" of approximately one-fifth could be assumed, a stoichiometry between the extra O_2 consumed in recovery and the lactate produced, but not removed, during exercise could be established. Thus, by their definitions, measurement of the O_2 debt provided a means to quantitate the energy provided by anaerobic mechanisms during human exercise. Though simple in concept, assessment of anaerobic metabolism during exercise from postexercise O_2 consumption measurements proved difficult because of problems associated with establishing the postexercise baseline O_2 uptake (see ref. 64 for explanation). With the benefit of hindsight, investigators should have been suspicious of extrapolating results obtained on nonperfused, highly glycolytic, frog white muscles that are poorly circulated and seldom recruited in nature, to mammalian systems containing extensive capillary beds, elaborate mitochondrial reticula, and high metabolic rates, cardiac outputs, and muscle and liver perfusion rates. Nonetheless, until recently, O_2 debt theory dominated scientific thinking; the theory still dominates some clinical fields.

PARTITIONING THE O_2 DEBT

As described in more detail in the section The Phosphagen Buffer System, in 1930 Einar Lundsgaard (112) discovered "phosphagen" (later shown to be ATP + PCr) and demonstrated the role of that energy system in sustaining muscle contraction. Additionally, Lundsgaard established the terms "lactacid" and "alactacid" in describing nonoxidative energy systems in muscle. Working in the Harvard Fatigue Laboratory in 1933, Rodolfo Margaria (1901–1983), Harold T. Edwards (1897–1937), and David Bruce (D.B.) Dill (1891–1986) (114) adopted Lundsgaard's terminology and used it to reinterpret the biphasic curve describing whole-body $\dot{V}O_2$ during recovery from exercise. Using human subjects and running protocols of 3–10 minutes duration, Margaria et al. (114) observed that immediately after hard exercise, blood [lactate] remained elevated while $\dot{V}O_2$ fell rapidly during the first, fast O_2 debt period. Subsequently, blood [lactate] declined during the second, slow O_2 debt period. Therefore, Margaria et al. (114) concluded that the rapid O_2 debt phase was a result of the restoration of phosphagen in recovering muscle. This rapid O_2 debt phase was termed "alactacid"—that is, not having to do with lactic acid removal. The investigators also termed the second, slow O_2 debt phase that coincided with the decline in blood [lactate] the "lactacid" O_2 debt. Though consistent with Lundsgaard's' terminology, Margaria et al. had no data on muscle lactate or PCr concentrations or flux rates. Yet, from their analyses of the changes in O_2 consumption and blood lactate in recovery they knew that the classic Hill–Meyerhof O_2 debt theory had to be revised.

There are several reasons why the 1933 paper of Margaria et al. became one of the most influential papers in the history of exercise physiology. Over the course of decades, this seminal work on oxygen debt withstood numerous challenges and is still fundamental to ideas expressed by proponents of the "anaerobic threshold." Of the several reasons responsible for longevity of the O_2 debt concept in a scientific enterprise in which hypotheses are discarded daily, the distinction of the investigators

Fig. 8.4. Portrait of Rodolfo Margaria, 1960, age 59. Courtesy P. Cerretelli.

must be recognized as primary. August Krogh, who first described recovery O_2 consumption in humans, was awarded a Nobel Prize in 1920, and as already mentioned, A.V. Hill and Otto Meyerhof shared a Nobel Prize in 1922. In the United States, D.B. Dill (of Margaria et al.) replaced Lawrence Joseph Henderson (1878–1942) as director of the Harvard Fatigue Laboratory; eventually Dill served as president of the American Physiological Society. When the Fatigue Laboratory closed after World War II, those trained there took positions of eminence in physiology departments and laboratories around the United States. Similarly, as professor of physiology in Milan, Rodolfo Margaria (Fig. 8.4) ascended to a position of preeminence in Europe. World War II intervened; work in the field halted in favor of more applied efforts; and the O_2 debt model was ensconced in textbooks of physiology and biochemistry.

It is appropriate to note here that despite the many contributions of the "British School" physiologists led by A.V. Hill, in some ways this group also hindered understanding of muscle metabolism during exercise. While their focus on reconciliation of their studies of isolated amphibian muscle with their studies of humans during exercise was laudable, it may also have led them astray. As Barnard and Holloszy emphasized in Chapter 7, Zuntz in 1911 was convinced that glycogen is the precursor of lactic acid and further that both fat and carbohydrate are substrates for energy during exercise in humans. Following the experiments of Fletcher and Hopkins and of A.V. Hill, as well as Hill's hypothesis that lactate mediates muscle contraction (thereby indicating that carbohydrate is the sole energy source for muscular contraction), in 1920 Krogh and Lindhard (103) reinvestigated Zuntz's findings. Their (103) results were largely confirmatory of Zuntz's view that fat could also serve as a fuel for exercise, findings that are now established (14, 15). Hill must have been aware of these studies since he cited other work of Krogh and Lindhard. Nevertheless, he and Lupton stated that "It would seem probable that carbohydrate alone provides the energy for the excess metabolism of exercise: this certainly appears to be the case in isolated muscle" (86). In 1939, Erik Hohwü Christensen (1904–1996) and Ove Hansen (1907–1990) (40) resurrected the idea that both fat and carbohydrate can be used as energy substrates for muscular exercise.

CHALLENGES TO O_2 DEBT THEORY

Ole Bang (1901–1988) was the first to question the fundamental assumptions and conclusions of O_2 debt theory (3, 4). His work benefited greatly from development of a micro-method for sampling and analysis of lactate content in arterialized capillary blood. This technical advance permitted repeated blood sampling during as well as after exercise. Through the study of exercises of varied intensities and durations, Bang showed that the results of Hill and Margaria and their colleagues were fortuitous consequences of the duration of their experiments. Using exercise bouts lasting only a few minutes, Bang found that the concentration of lactic acid in the blood reached a maximum after exercise had ended, and depending on the intensity of exercise, remained elevated long after the oxygen consumption had returned to preexercise levels. With prolonged exercise, Bang showed that the blood lactate level reaches a maximum after about 10 minutes of exercise and then declines whether exercise ceases or not. In some cases, basal lactate levels can be achieved during exercise itself. After exercise, however, there was always an "O_2 debt," with predictable kinetics to be "paid." The results could not be reconciled with the idea that lactic acid determines oxygen consumption after exercise (4).

In two letters to George A. Brooks (1944–) (September 10, 1972, and December 18, 1972) Bang offered opinions on why his work appeared to have been overlooked. Firstly, A.V. Hill did not acknowledge the work. Secondly, Bang's clinical commitments and the events of World War II served to deflect his efforts from exercise physiology. Likely important also was the fact that the German language journal selected for publication of results (*Skand. Arch. Physiol.*) ceased publication.

In addition to Bang's experiments on exercising humans, three notable studies on anesthetized dogs also cast doubt on the traditional O_2 debt concept. As early as 1927 Abramson et al. (1) tested the lactic acid/O_2 debt hypothesis by infusing lactate into anesthetized dogs. The change in $\dot{V}O_2$ was not different from that observed in their $NaHCO_3$ control. Those results were corroborated decades later (1954) by Norman R. Alpert (1922–) and Walter S. Root (2), who found no correlation between $\dot{V}O_2$ and the lactate dose given to anesthetized dogs. Alpert and Root induced O_2 deficits in dogs by means of cardiac tamponade and having them breathe O_2-deficient gases. Magnitude of the subsequent O_2 debt was poorly correlated with the preceding O_2 deficit. As the result of those experiments and subsequent data obtained in 1964 on lactate removal in anesthetized dogs after electrically stimulated exercise, Herbert L. Kayne and Alpert (98, p. 1004) stated that

> the basic assumptions of the 'oxygen debt' hypothesis are that lactate, unknown anaerobic metabolites, or 'excess lactate' act as security for the debt and force repayment of the debt during recovery from the stress. The experiments described above indicate that oxygen missed [i.e., O_2 deficit] during stress does not relate to the recovery oxygen and the removal of lactate or 'excess lactate' which are [sic] infused at rest or produced during exercise are [sic] not causally related to recovery oxygen consumption [i.e., O_2 debt].

While space prevents a complete discussion, it should be noted that Meyerhof's results from amphibian muscle in vitro were later shown to differ from the results obtained on lower vertebrates in vivo (13, 69). Further, we now know that lower vertebrates differ from mammals in the mechanisms of clearance of an exercise-induced lactate load (163). The important insight is that the isolated frog muscle systems studied by Hill and Meyerhof did not directly apply to humans, mammals in general, or even lower vertebrates in vivo.

O_2 AVAILABILITY, LACTATE PRODUCTION, AND O_2 DEBT

In 1923, David Barr and Harold Himwich (10) reported comparisons of arterial and venous blood drawn simultaneously from the arm 3 minutes following vigorous leg exercise. They (10) also took blood samples after vigorous exercise with the forearm. The conclusion was that less active tissues actually removed lactic acid from the blood. Very little was made of this remarkable finding for over 30 years. In 1955, Drury, Wick, and Morita (51) reported that a significant fraction of the respiratory CO_2 is derived from lactate in eviscerated and nephrectomized rabbits. Presumably, most of the metabolism in this preparation was due to skeletal muscle. In the very next year, Omachi and Lifson (126) studied $^{13}CO_2$ derivation from ^{13}C-labeled lactate in isolated, perfused dog gastrocnemius preparations at rest and during contractions. They found that lactate oxidation was greater when the muscles were contracting. While both of these studies in the 1950s suggested consumption of lactate by skeletal muscle, these results went largely unnoticed, perhaps because both preparations were far removed from normal physiological conditions.

It was left to Wendell N. Stainsby (1928–) (Fig. 8.5) to challenge the foundations of traditional theory regarding the relationship between oxygen and lactate production. Stainsby did his dissertation work on the effect of passive stretch on resting $\dot{V}O_2$ of the surgically isolated canine gastrocnemius (GP) in situ. Subsequently, over a period of 40 years (1957–1997), Stainsby and associates would use the canine GP preparation to study phenomena such as the "critical oxygen tension" in

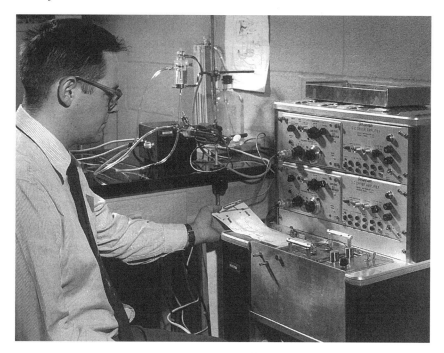

Fig. 8.5. Wendell Stainsby, University of Florida, Gainesville Laboratory, 1958. Courtesy of Wendell Stainsby.

muscle and the relationships among muscle O_2 delivery, lactate production and release, and O_2 debt.

In 1966, Stainsby and his doctoral student, Hugh G. Welch (1937–) reported the surprising result that net lactate output by the contracting GP was always transient. The net output peaked and declined within 15–20 minutes, even becoming net lactate uptake as the contractions continued at a constant stimulation rate (147). Subsequently, Welch's dissertation work (162) further established the transient nature of net lactate output by the contracting dog GP and additionally demonstrated that recovery O_2 uptake (O_2 debt) in the anesthetized dog GP was not related to lactate metabolism.

If oxygen deficiency was the cause of lactate production in contracting muscle, then why would net lactate output decline to near zero or even revert to net uptake with continued contractions? To answer this question, Stainsby collaborated with Frans Jöbsis (1929–) who with Britton Chance (1913–) had invented the surface fluorescence method for NAD/NADH detection. In 1968, Jöbsis and Stainsby (94) reported that NAD/NADH was highly reduced at rest and became highly oxidized quickly after the onset of contractions in the canine GP; NAD/NADH was becoming oxidized at a time when lactate production was known to be prevalent. Lactate production appeared *not* to be the result of muscle hypoxia. Although this study (94) has been criticized over the years, it became one of the key pieces of evidence against classic ideas of lactate production associated with O_2 insufficiency.

For years it has been known that the critical mitochondrial oxygen tension is very low, on the order of 1 mm Hg (e.g., ref. 142); see Gladden (67) for review. The critical oxygen tension is the partial pressure below which there is insufficient oxygen for mitochondria to achieve maximal rates of respiration. While the critical mitochondrial O_2 tension for isolated mitochondria has been known to be low, a major technical limitation has been the inability to measure the oxygen tension in working skeletal muscle as well as to understand the relationship between working muscle Po_2 and lactate formation. In 1984, Richard Connett (1943–), Thomas Gayeski (1946–), and Carl Honig (1925–1993) (42) observed that the intramuscular Po_2 remained above the critical O_2 tension in contracting canine muscles in situ. Their results suggested that lactate was produced in exercising muscle under fully aerobic conditions.

More recently (1998), Russell Richardson (1965–), Peter Wagner (1944–), and associates (137, 138) (Fig. 8.6) utilized progressive exercise protocols and the dual technologies of (1) NMR spectroscopy to measure myoglobin saturation and (2) classical (a−v) concentration measures to evaluate lactate balance in resting and exercising human quadriceps muscle. Their results for resting muscle indicated that the intracellular Po_2 is probably very close to that in the venous effluent from the muscle, approximately 40 torr, a value that is actually too high for accurate assessment by the NMR method. Nevertheless, these resting, well-perfused muscles released

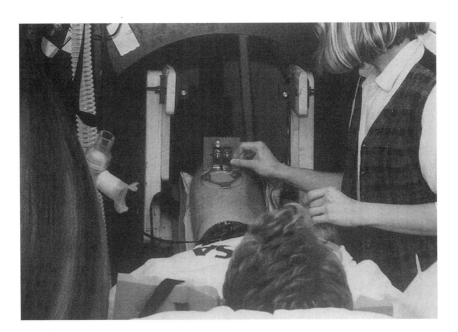

Fig. 8.6. Application of a surface coil to Russ Richardson (University of California, San Diego) by Liz Noyszewski for NMR studies of muscle oxygenation at the University of Pennsylvania. Note equipment for pulmonary gas-exchange measurements (left), ergometer support flanking coil, and magnet in background. Courtesy of R. Richardson and P.D. Wagner (see ref. 137).

lactate on a net basis (138). However, at exercise onset, Richardson et al. observed a dramatic decline in muscle P_{O_2} to approximately 4 torr, a value well within the range of detection but well above the critical mitochondrial P_{O_2} as determined in cellular and mitochondrial suspensions. Moreover, the intracellular P_{O_2} was maintained well above the critical mitochondrial P_{O_2} as power output increased to maximum.

In contrast to intramuscular P_{O_2} that fell rapidly at exercise onset and then remained constant as power output was progressively increased, Richardson et al. (138) reproduced the result that muscle lactate release changed little at low exercise intensities. However, at a power output corresponding to approximately 65% of $\dot{V}_{O_2}$max, coinciding with a rise in arterial epinephrine, muscle lactate release began a steep rise, again an observation made previously. When studies were performed on subjects breathing hypoxic gas mixtures (11% O_2), intramuscular P_{O_2} changed little, but lactate net release accelerated. Net lactate release was accentuated when epinephrine began to rise under both normoxic and hypoxic conditions. Like results obtained previously on men using isotope tracers and working limb lactate balance (36), results obtained using NMR spectroscopy are consistent with the interpretation that lactate production is the consequence of glycolysis, but not necessarily O_2 lack.

In 1969 and 1970, R. James Barnard, Kenneth M. Baldwin (1942–), and Merle Foss (1936–), working in the University of Iowa laboratory of Charles M. Tipton (1927–), approached the O_2 debt hypothesis with different types of experiments. Barnard and coworkers (7–9) studied recovery oxygen kinetics in intact dogs after treadmill running. Postexercise $\dot{V}_{O_2}$ was measured via a tracheotomy tube surgically implanted prior to studies. Further, they used pharmacological agents to block processes such as gluconeogenesis and the β-adrenergic functions. At the University of Michigan John A. Faulkner (1923–) and his graduate student George Brooks were impressed by several aspects of work in the Tipton laboratory, including the size of the postexercise O_2 consumption volumes. Brooks postulated that the tracheotomy procedure used in the dog studies of Barnard et al. affected the ability of the dogs to cool during exercise. Further, Brooks hypothesized that an elevated body temperature due to exercise might loosen the coupling of oxidative phosphorylation in mitochondria, thus increasing O_2 consumption during recovery and giving rise to at least part of the O_2 debt. In 1971 Brooks and associates (33) reported experimental results that largely confirmed his hypothesis; at high incubation temperatures similar to those found in working muscle the rate of mitochondrial O_2 consumption was raised and the efficiency of coupling was loosened. Further, in rats during recovery from exhausting exercise, the first, fast phase of whole-body O_2 consumption tracked muscle temperatures while the second, slow recovery O_2 phase tracked rectal temperature in recovery (34).

In combination, the work of several groups: (1) Stainsby and coworkers' dissociation of O_2 availability from muscle lactate production (94, 147, 162), (2) data on the effects of temperature on mitochondrial respiration (33), (3) the correlation between tissue temperatures and rate of O_2 consumption during recovery from exercise (34), and (4) the effects of β-blockade on recovery O_2 kinetics (8) led to the conclusion that exercise causes several perturbations in homeostasis (e.g., tissue temperatures, hormones, and ions) that individually and collectively affect postexercise $\dot{V}O_2$ recovery volume (64). Hence, data acquired in the 1960s and 1970s led to loss of confidence that the postexercise recovery O_2 uptake could be used as a surrogate marker of the anaerobic metabolism which occurred during exercise. Measurement of anaerobic ATP provision by any means, invasive or noninvasive, remains a major challenge to muscle and exercise physiologists.

After the work of Meyerhof it was long assumed that reconversion to glycogen was the primary fate of lactate during recovery from physical exercise. Since it was believed that mammalian muscles lacked the necessary enzymes to reverse glycolysis, it was presumed that most of the reconversion of lactate to glycogen or glucose would occur in the liver as a part of the Cori cycle (101). In 1973, Brooks et al. (28) at the University of Wisconsin, ran rats to exhaustion, injected them with [U-^{14}C]lactate, placed them in a metabolism chamber, and collected [14]CO_2 in recovery. The results were clear; most of the lactate was oxidized in recovery and injected lactate tracer yielded [14]CO_2 almost like an injection of tracer bicarbonate. Further, parallel experiments indicated that lactate was cleared with little muscle glycogen restoration in recovery (28). Subsequently at Berkeley in 1978, Brooks and Ph.D. student Glenn A. Gaesser (1950–) combined the techniques of whole-animal calorimetry, two-dimensional paper chromatography, autoradiography, and ^{14}C isotope infusion to trace the pathways of lactate removal during recovery from exhausting exercise (32). Again, lactate disposal in recovering rats was largely by means of oxidation; glycogen restoration depended on carbohydrate feeding.

In 1977, Lars Hermansen (1933–1984) and Odd Vaage (1939–) (76) revived the idea that lactate could be resynthesized into muscle glycogen during recovery from intense exercise in humans. Their estimate that ≈70% of the lactate remaining in muscle after short-term, intense exercise was converted back to glycogen was most likely inflated by errors in their assessment of blood flow. In 1991, also studying human subjects, Jens Bangsbo (1957–) and colleagues (6) attempted to quantify the fate of lactate during recovery from short-term, high-intensity exercise performed by a single muscle group, the quadriceps muscles. While they calculated that 13%–27% of the muscle lactate present at the end of the exercise was reconverted to glycogen, this still left a majority of the lactate accounted for by oxidation. Although the percentage of lactate converted to glycogen might be higher in whole-body exercise with higher blood lactate concentrations (128), the results of these human studies are in general agreement with the radioactive tracer studies on the rat; the primary fate of lactate in recovery from intense exercise is oxidation.

Oxidation of lactate is enhanced by moderate exercise during recovery from intense exercise. As early as 1928, Otto Jervell (1893–1973) (93) observed that blood lactate concentration declined more rapidly in recovery when moderate exercise was performed instead of a passive resting recovery. This observation was made in two experiments on one subject. In 1937, Dill's group at the Harvard Fatigue Laboratory (125) extended Jervell's work, finding that the rate of blood lactate decline during recovery increased with the intensity of recovery exercise up to a critical level of activity. They (125) speculated that the increased rate of lactate decline during an exercising recovery might be due to (1) an increased blood flow with more rapid transport of lactate to removal centers and (2) an increased utilization of lactate as a fuel for the exercise.

THE PHOSPHAGEN BUFFER SYSTEM

In the late 1920s, Embden's group (58, 59) discovered muscle adenylic acid, but its significance was not immediately obvious (123, p. 14). In 1928, Karl Lohmann (1898–1978) found a substance in trichloroacetic acid extracts of muscle that could be hydrolyzed into inorganic phosphate (107, 108). Lohmann thought the substance might be inorganic pyrophosphate, but his experiments did not definitively support that point of view (123, p. 14). Lohmann (109, 110) and independently, Cyrus H. Fiske and Yellapragada SubbaRow (1895–1948) (62) solved the puzzle in 1929 when both groups deduced the structure of ATP in muscle. Meanwhile, parallel experiments were uncovering important information about another muscle phosphate compound. In 1927, Philip Eggleton (1903–1954) and Grace Palmer Eggleton (1901–1970) (54) observed this second labile phosphate compound in trichloroacetic acid extracts of muscle and called it phosphagen. Shortly afterward, it was again Fiske and SubbaRow (61, 62) who observed the same behavior, and subsequently identified the compound as phosphocreatine (PCr).

In the 1930s Einar Lundsgaard (1899–1968) (111,113) demonstrated that muscles were able to contract even though lactic acid production was prevented by iodoacetate poisoning. Instead, there was a decrease in a phosphorylated compound (later shown to be PCr) and an increase in inorganic phosphate. Schwartz and Oschmann (145) had made similar observations in 1924, but their results were apparently overlooked. On the basis of Lundsgaard's work, it was hypothesized by 1932 that the energy for muscle contraction was derived from the breakdown of phosphocreatine and that this phosphocreatine was resynthesized in recovery via a mixture of anaerobic (lactic acid formation) and aerobic metabolism. These new ideas concerning muscle energetics in 1932 were so profoundly different from the state of knowledge in 1926 that Hill (82) referred to the experiments over that period of time as "the revolution in muscle physiology."

The discoveries of ATP and PCr had weakened the prevailing theories concerning the mechanism and energetics of muscle contraction (22, 123). There were two phosphagens in muscle, ATP and PCr, with PCr being the more quantitatively important (22, 123: p. 17). However, this information alone provided no insight into which of the compounds might be the direct energy donor for muscle contraction (123, p. 17). Once again, Lohmann came to the forefront. He found that PCr was only dephosphorylated in muscle extracts when AMP or ATP was present (109). In the late 1930s and early 1940s, investigations by Englehardt and colleagues (60) which were extended by reports from Banga and Szent-Gyorgi (5) and Straub (150), illustrated that actomyosin contracts only with the addition of ATP. This research established the idea that direct energy donation for muscle contraction is from ATP hydrolysis and that PCr serves as a buffer for ATP resynthesis (22; 123: p. 18).

According to Bessman and Geiger (22), Parnas in 1934 demonstrated that PCr and ADP could be synthesized from ATP and creatine; this meant that the Lohmann reaction was reversible. In 1941, Fritz Albert Lipmann (1899–1986) (99), who had worked in Meyerhof's laboratory from 1927 to 1931, surveyed the available information on the interactions of ATP, PCr, and energy metabolism in general and concluded that ATP stands as the central link between energy-using and energy-yielding reactions in cells. Lipmann (106) also coined the phrase "energy rich phosphate compounds" and originated the notation of "~P" to indicate readily convertible energy from such phosphate compounds. In 1953, Lipmann won a Nobel Prize "for his discovery of co-enzyme A and its importance for intermediary metabolism" (152). However, despite the abundance of circumstantial evidence, the role of ATP relative to PCr in muscle contraction was not yet finalized; A.V. Hill was not completely satisfied. Hill's 1950 "A Challenge to Biochemists" (80) stated the following:

> In the lactic acid era the evidence that the formation of lactic acid was the cause and provided the energy for contraction seemed pretty good. In the phosphagen era a similar attribution to phosphagen appeared even better justified. Now, in the adenosine triphosphate era lactic acid and phosphagen have been relegated to recovery and ATP takes their place. Those of us who have lived through two revolutions are wondering whether and when the third is coming.
>
> It may very well be the case, and none will be happier than I to be quit of revolutions, that the breakdown of ATP really is responsible for contraction or relaxation: but in fact there is no direct evidence that it is. Indeed, no change in the ATP has ever been found in living muscle except in extreme exhaustion, verging on rigor.

Ultimately, Hill's challenge was answered with finality by Dennis Francis Cain (1930–) and Robert E. Davies (1919–1993) (37) in 1962 when they provided the proof for ATP as the immediate energy donor for muscle contraction. They (37) inhibited creatine kinase in muscles with the poison 1,fluoro-2,4-dintobenzene (DFNB) and then immediately froze the muscles after a series of contractions. Under these conditions in which ATP resynthesis from PCr was prevented, a decline in ATP concentration was observed. And so it was that PCr came to be viewed as an energy buffer for the replenishment of ATP (22). To be complete, it should be noted that in 1943 Herman M. Kalckar (1908–1991) (97) discovered the enzyme adenylate kinase (called myokinase in muscle), which can resynthesize ATP from the combination of two ADP molecules; AMP is also a product. The myokinase reaction is thought to play a role in intense exercise, but that role has not yet been firmly established.

After a long period of work from the 1950s to the 1970s by numerous investigators, Samuel P. Bessman (1921–) formally proposed the existence of a phosphocreatine shuttle in 1972 (21). In this role, the PCr system serves as a spatial as well as temporal buffer of ATP. The original proposal of the shuttle suggested that as ATP from oxidative phosphorylation was transported out of the mitochondria, a mitochondrial-bound isoenzyme of creatine kinase (CK) immediately used that ATP to synthesize PCr. This PCr would diffuse to the myofibrils, where a second creatine kinase isoenzyme bound to the M band would replenish ATP from ADP + PCr, thus providing ATP to the contractile apparatus. The resulting creatine would diffuse back to the mitochondria to complete the shuttle. At present, two different models of the CK–PCr system are debated. One of these, an amplification of the original proposal, asserts that CK isoenzymes are highly compartmentalized at sites of energy production and utilization (154, 166). The alternative model of Martin Kushmerick (1937–) and colleagues proposes that specific localization of CK is not necessary for muscle performance; rather the CK–PCr system establishes facilitated diffusion that maintains CK-catalyzed fluxes near equilibrium in the cytosol (116, 117, 159, 166).

Changes in phosphagen concentration in human muscle during exercise were first studied by Eric Hultman (1925–) and colleagues (Fig. 8.7) (70, 90). Their results showed that PCr concentration declined linearly with increases in exercise intensity while changes in ATP concentration were comparatively minor. In 1968, Pietro di Prampero (1940–) and Margaria (49) and Piiper, di Prampero, and Paolo Cerretelli (1932–) (132) reported similar observations for the isolated dog gastrocnemius in situ. These findings have been borne out by numerous investigations to the present day (48, 146). Several studies have also shown that PCr degradation is instantaneous

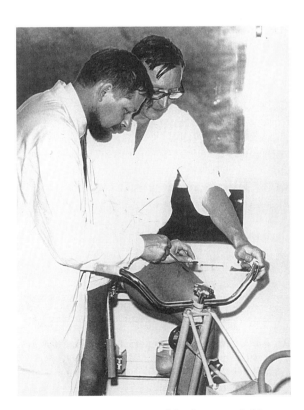

Fig. 8.7. Eric Hultman (left) conducts one of the first muscle biopsy studies on colleague Aasmund Roch Norlund, 1967. Courtesy of E. Hultman.

at the onset of exercise (91, 95). The winner of the 2002 Olympic Prize in Sport Sciences, Bengt Saltin (1935–), and colleagues (143) found elevations in human muscle lactate concentration after only 10 seconds of intense cycling exercise; this led them to suggest that glycolysis is activated at the onset of exercise also. Subsequent studies have supported this hypothesis (146) so now we realize that essentially all of the metabolic pathways of energy metabolism are turned on at the initiation of exercise.

GLYCOGEN

Almost 150 years ago, it was established that glycogen is the body's carbohydrate storage form. Perhaps the first scientist to report this finding was Claude Bernard in 1855 (19). As noted above, the relationship between a decrease in glycogen content and an increase in lactate concentration was then clarified by Meyerhof in the early 1920s. Also, as briefly reviewed in the earlier section, The Prelactic Acid Era, the steps in the glycolytic pathway were determined in the late 1930s largely by Embden and Meyerhof but with important contributions by Warburg in Germany and Carl Ferdinand Cori (1896–1984) and Gerty Theresa Cori (1896–1957) in the United States (104, pp. 317–318). A question that was being investigated in parallel was to what extent glycogen and the glycolytic pathway were employed during exercise in humans.

As early as the 1860s, studies of substrate metabolism during exercise were underway. Max von Pettenkofer (1818–1901) and Carl Voit (1831–1909) (131) measured urinary nitrogen output and proposed that protein was not an important source of fuel during exercise. Before the turn of the nineteenth century, Jean Baptiste Auguste Chaveau (1827–1917) in France (39) and Zuntz and Loeb in Germany (167) were measuring the respiratory exchange ratio (RER) in subjects during work. Zuntz studied mild work and found RER not to be very different from rest, supporting the idea that both fats and carbohydrates were being used. Chaveau, however, studied heavier work, obtained RER values near 1.0, and argued that carbohydrates are the sole source of energy for muscular exercise (39). The next important studies were those of Francis Gano Benedict (1870–1957) and Edward Provan Cathcart (1877–1954) in 1913 (12) at the Carnegie Institution of Washington. They found that RER increased with exercise intensity. Subsequently, in 1920 Krogh and Lindhard (102) published the results of very careful experiments on the use of fats and carbohydrates during exercise; their findings supported the notion that both substrates were used during rest and exercise. Recall that in 1920, the ruling theory of muscle contraction from the work of Hill and his contemporaries was that lactic acid formation, likely from a carbohydrate source, was the immediate source of energy. Accordingly, Krogh and Lindhard (102) assumed that if fat was used during exercise, it must be transformed into carbohydrate first. The next landmark study in the area of substrate metabolism and exercise was the report by Christensen and Hansen (40) in 1939. From measurements of respiratory gas exchange, they confirmed that both carbohydrates and fats are metabolized during exercise. Further, they showed that carbohydrate utilization decreases with increasing exercise duration in moderate exercise while carbohydrate usage increases with increasing exercise intensity. As noted above, these advances ran counter to the hypotheses of the "British School" of physiologists.

In a series of landmark studies that are summarized in a 1967 supplement to the *Scandinavian Journal of Clinical & Laboratory Investigation* (89, 90), Hultman, Jonas Bergström (1929–2001) and their colleagues reported the effects of various types of exercise and diets on muscle glycogen concentration. Since glycogen breakdown measurements in muscle biopsies provide no direct information on the actual fate of the glucosyl units, it was important that Hultman and colleagues were careful to determine carbohydrate oxidation by indirect calorimetry in order to verify the linkage between glycogen degradation and glucosyl oxidation. Though limited by the inability to distinguish between oxidation of glycogen and other carbohydrates or to identify the pathways of glycogen disposal, the studies of Hultman and colleagues confirmed the importance of muscle glycogen availability for exercise performance and established the theoretical basis for glycogen loading.

The pioneering studies of Hultman, Bergström, and coworkers, as well as numerous studies since, demonstrated clearly that glycogen is not only the most important energy source in contracting muscles but also that glycogen can be almost completely depleted during cycle exercise to fatigue at 60%–80% of $\dot{V}O_2$max. Performance and fatigue appear to be closely linked to muscle glycogen depletion in this range of exercise intensities (44).

Tracers and Lactate Flux during Rest and Exercise

In science, there is typically an interaction between the evolution of new ideas and technology; new ideas spawn technological advances that in turn yield information permitting additional ideas to be formulated. Such is the case with the application of isotope tracers to the study of metabolism. As reiterated by Hetenyi and associates (77) from Schoenheimer (144):

> the concept of the dynamic steady state has been one of the most far-reaching ideas in biomedical science in this century. According to this concept, all constituents of the body are continuously formed and utilized.... At the cellular level substances are taken up or formed in the cell as well as metabolized or released into extracellular fluid. Substances released into the extracellular fluid reach the bloodstream and are carried to all organs. At the same time they are excreted or taken up by other cells; therefore substances undergo turnover in the circulation.

Isotope tracers have application for the study of metabolic processes because the turnover of isotopically labeled metabolites can be measured. Most studies using tracers have measured glucose flux rates because of the obvious importance of maintaining glucose homeostasis in health and disease. The same basic methodology has also proven useful in measuring the effects of exercise and training on free fatty acid, glycerol, amino acid, and lactate flux rates (14–18, 29).

In dynamic steady states such as rest or sustained exercise, metabolites like lactic acid and glucose are continuously formed and enter the circulation. Because lactate is formed continuously, at greatly variable rates in anatomically distributed tissues (26), measurements of tissue lactate content provide only meager information about rates of production and removal. For this reason, when the technology became available, radioactive isotope tracers were applied to the study of lactate flux rates in animals. More recently, stable (nonradioactive) tracers have been used to study lactate metabolism in humans. All the results on humans and other mammals are consistent; lactate turnover is prominent at rest and scales exponentially to metabolic rate during exercise. Lactate is actively exchanged both within as well as between cells, organs, and tissues (25–27).

Pioneer work in the field of lactate kinetics in exercising mammals was that of Florent Depocas (1923–) and colleagues. In 1969 using continuous infusion of [U-^{14}C]lactate into dogs during rest and continuous steady-state exercise Depocas and coworkers (47) made several key, fundamental findings regarding lactate metabolism. These findings included: (1) there is active lactate turnover during the resting postabsorptive condition; (2) a large fraction, approximately half, of lactate formed during rest is removed through oxidation; (3) the turnover rate of lactate increases during exercise as compared to rest even if there is only a minor change in blood lactate concentration; (4) the fraction of lactate disposal through oxidation increases to approximately three-fourths during exercise; and (5) a minor fraction (one-tenth to one-fourth) of lactate removed is converted to glucose via the Cori cycle during exercise. Though the fractions are subject to species and experimental variations, the essential results have been reproduced in rats (50) (Fig. 8.8), dogs (92), horses (161), and humans (18, 115, 148).

A most relevant aspect of the issue of blood lactate accumulation during exercise concerns the role of net release from active muscle. Surprisingly, while net lactate release from resting muscle is common, net release from working muscle is usually transient if power output or stimulation rate is held constant. As shown first by Stainsby and Welch in 1966–1967 (147, 162) using dog muscle preparations contracting in situ, this "Stainsby Effect" (35) of transient muscle net lactate release at exercise onset followed by a switch to net uptake from the blood by working muscle has been confirmed in exercising humans (35, 36). Thus, it is certain that working skeletal muscle is not the sole source of blood lactate in humans during whole-body exercise. Epinephrine is more likely to signal glycolysis and lactate production in noncontracting tissues than working muscle; in working muscle epinephrine augments glycolysis leading to increased lactate accumulation (160). Additionally, recent studies show that epinephrine reduces net glucose and lactate uptake in working humans (160) and canine gastrocnemius (72) muscles in situ, respectively. Lactate uptake by exercising human skeletal muscle has been well documented, both with (18) and without tracers (35). L. Bruce Gladden's (1951–) studies on dog muscles contracting in situ (68) clearly show that lactate uptake is concentration (substrate), and not O_2 dependent, a finding that also appears to be the case in human muscle (18) (Fig. 8.9).

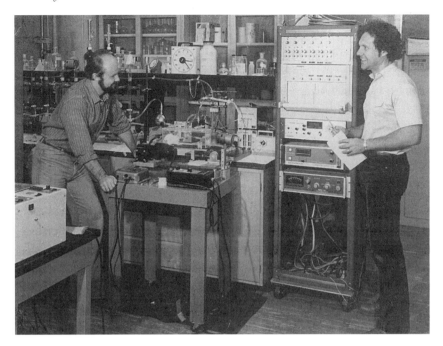

Fig. 8.8. Casey Donovan (1952–) (left) and George Brooks (right) and the Plexiglas metabolism chamber with mixing manifold suspended over hand-machined treadmill. At right are chart recorder and amplifiers for O_2 and CO_2 analyzers as well as vibrating reed electrometer in ionization chamber (for on-line detection of $^{14}CO_2$, left of metabolism chamber). Behind Donovan are the glass trapping columns to quantitate $^{14}CO_2$ excretion in expired air. The coiled wire above the equipment rack went to the Human Performance Laboratory where A-D interfaces to a digital computer served both animal and human experimentation (see refs. 30, 50).

Training

Space constraints do not permit adequate treatment of the effects of training on lactate turnover and its parameters (concentration, production, removal, clearance). In brief, it is well established that trained individuals maintain lower lactate levels during a given exercise task than do untrained, but highly trained athletes can achieve and sustain exercise at higher lactate concentrations than can the untrained. First measurements of lactate kinetics using radioactive tracers in rats (50) showed that training lowered circulating lactate levels by increasing clearance (Fig. 8.8). More recently, using a longitudinal training design, and a combination of stable isotope tracers, arterial–venous difference measurements, and muscle biopsies on humans, training has been found to have a small, but significant effect on blood lactate appearance, but a major effect on improving lactate clearance, especially during hard exercise (Fig. 8.9) (18). Training improves the capacity of gluconeogenesis in rats (30) and humans (17), and that effect helps exercising individuals maintain glucose homeostasis during prolonged exercise. However, the major effect of training is to improve lac-

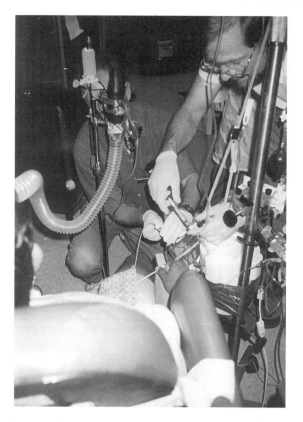

Fig. 8.9. Simultaneous femoral arterial and venous blood sampling during a resting phase of the 1996 lactate study in Gail Butterfield's Palo Alto laboratory. David Guido (left, obscured by the pneumotachometer) handles the venous side while Gene Wolfel takes arterial blood. Also visible on left are apparatus for thermal dilution measurement of limb blood flow and green dye measurement of cardiac output. A left arm vein was used for tracer glucose and lactate infusions into the seated subject (see refs. 15–18, 52).

tate oxidation within working muscles (17), an effect that may be partly attributable to the effects of training on increasing expression of lactate transporters in muscle sarcolemmal (52, 133) and mitochondrial (52) membranes. These adaptations facilitate operation of cell–cell and intracellular lactate shuttles (i.e., exchange of lactate between glycolytic and oxidative fibers, and the oxidative disposal of lactate within oxidative fibers, *vide infra*).

In summary, the use of isotopic tracers has either shown or confirmed that (1) lactate turnover is prominent at rest and scales exponentially to metabolic rate during exercise, (2) lactate is actively exchanged both within as well as between cells, organs and tissues, (3) lactate disposal through oxidation increases with exercise and is the dominant pathway, (4) only a minor fraction of lactate removed is converted to glucose via the Cori cycle during exercise, (5) net release from exercising muscle is usually transient at a constant power output, (6) contracting skeletal muscle is not the sole source of blood lactate, and (7) training lowers tissue lactate levels during exercise mainly by increasing clearance.

THE LACTATE SHUTTLE AND LACTATE TRANSPORT

In 1985, George Brooks took a unique approach to explaining phenomena related to lactate responses to exercise when he articulated the "lactate shuttle hypothesis" (26). Key elements of the hypothesis were described as follows: "the shuttling of lactate through the interstitium and vasculature provides a significant carbon source for oxidation and gluconeogenesis during rest and exercise." As such, the lactate shuttle hypothesis represented a model of how the formation and distribution of lactate represents a central means by which the coordination of intermediary metabolism in diverse tissues and different cells within tissues can be accomplished. The initial hypothesis was developed from results of original isotope tracer studies conducted on laboratory rats in Brooks's own laboratory along with numerous other studies, many of which were cited in the previous section. Thus, the working hypothesis was developed that much of the glycolytic flux during exercise passed through lactate.

According to the cell–cell lactate shuttle hypothesis, lactate is a metabolic intermediate rather than an end product (25, 26). Lactate is continuously formed in and released from diverse tissues such as skeletal muscle, skin, and erythrocytes, but lactate also serves as an energy source in highly oxidative tissues such as the heart and is a gluconeogenic precursor for the liver. Lactate exchanges among these tissues appear to occur under various conditions ranging from postprandial to sustained exercise (25).

If lactate does serve as a key metabolic intermediate that shuttles into and out of tissues at high rates, particularly during exercise, then transmembrane movement is critical. For many years, lactate was assumed to move across membranes by simple diffusion. It is now known that cell–cell lactate exchanges among all tissues are facilitated by membrane-bound monocarboxylate transporters (MCTs) (66, 135, 139). Lactate and pyruvate (monocarboxylate, MCT) transport were first described by Andrew P. Halestrap (1949–) in red blood cells in 1974 (71). It was not until 1990 that the characteristics of membrane transport of lactate in skeletal muscle were first described by David A. Roth (1953–) and Brooks (139, 140). The study of cell membrane lactate transport proteins took a major leap in 1994 when, looking for the mevalonate (Mev) transporter gene in Chinese hamster ovary (CHO) cells, Christine Kim Garcia (1968–) and associates Michael S. Brown (1941–) and Joseph L. Goldstein (1940–) (66) cloned and sequenced a monocarboxylate transporter which they termed MCT1. In 1985, Goldstein and Brown shared the Nobel Prize "for their discoveries concerning the regulation of cholesterol metabolism" (152). Garcia et al. found that MCT1 was abundant in erythrocytes, heart, and basolateral intestinal epithelium. MCT1 was detectable only in oxidative muscle fiber types, and not in liver. With an interest to describe a role for MCT isoforms in the Cori cycle, Garcia et al. (65) subsequently described isolation of a second isoform (MCT2) by screening of a Syrian hamster liver library; MCT2 was found in liver and testes.

Independently of Garcia et al. (65, 66), Halestrap and colleagues (71, 135) identified another MCT isoform (now known as MCT4) in 1998. In human skeletal muscle, two MCT isoforms (MCT1 and MCT4) with different kinetic properties have been described (135, 165). Recently (1999), the knowledge that lactate was formed and oxidized continuously within muscle and heart in vivo, led to extension of the cell–cell lactate shuttle to include an intracellular component, the "intracellular lactate shuttle," which is based on dual discoveries by Brooks and colleagues of the presence of lactate dehydrogenase (LDH) in liver, cardiac, and skeletal muscle mitochondria (31, 52), as well as a functional relationship between LDH and MCT1 in mitochondria (27), thus allowing mitochondria to directly oxidize lactate (31).

Understanding of the molecular biology and physiology of lactate transporters is in a state of rapid change, making it difficult to place the sequence of discovery in a clarifying perspective. However, this section would be incomplete without mentioning the contributions of Arend Bonen (1946–) and Carsten Jeul (1948–) to the field. Bonen is noted for his efforts to describe the effects of training (23) on expression of MCT isoforms. Additionally, Bonen and colleagues are pioneers in studying the role of transcription on expression of MCT isoforms (24). This work has relied heavily on use of laboratory animal models. Juel is distinguished for his development of the so called "giant sarcolemmal vesicles" technique, which permits lactate transport to be studied on small muscle samples, including human muscle biopsies (96). Further, Juel and associates have contributed to our understanding of the effects of training on MCT isoform expression in human skeletal muscle (134).

THE ANAEROBIC THRESHOLD

The concept of the "anaerobic threshold" is counter to the views of the authors of this chapter and to many of the views referenced here, but representative of an extensive body of scientific literature that has been adopted in clinical fields as diverse as cardiopulmonary assessment (156), treatment of septic shock (53), and field testing of athletes (88). Although the weight of evidence was provided by Karlman Wasserman (1927–) and associates (157, 158), very similar ideas were advanced independently by Wilfried Kindermann (1940–), Joseph Keul (1933–2000), and others (88, 100, 149) in Germany. Regardless of whether the anaerobic threshold concept is ultimately proven correct, partially correct, or incorrect, the originators of the concept have made notable contributions to the advancement of exercise physiology. Science is a self-correcting process that depends critically upon testable hypotheses. In this context, the anaerobic threshold has been one of the centerpieces of exercise physiology in the latter half of the twentieth century.

In the model of anaerobic threshold articulated by Wasserman and associates (156, 157), the assumption is that a series of metabolic and ventilatory events is precipitated by inadequate O_2 delivery. Wasserman and associates developed elaborate technological and experimental protocols to identify the anaerobic threshold noninvasively; a key element was determination of the time at which respiratory compensation for metabolic acidosis (lactic acidosis) occurred. It was argued that the subsequent respiratory exchange alterations established the work rate and metabolic rate at which O_2 supply became inadequate; this was accomplished without the necessity of taking a blood sample. In parallel experiments, instead of relying on pulmonary gas-exchange determinations, Kindermann and associates took the more direct approach of using micromeasurements of capillary blood to assess the onset of lactic acidosis (100).

Proponents of anaerobic threshold theory are in far better position to argue their hypotheses than is possible here. Therefore, readers are referred to the text of Wasserman's Cannon Lecture given to the American Physiological Society in 1994 (156). Regardless of one's opinion concerning tissue O_2 lack as the precipitator of lactic acid production, Wasserman's review stands as an important descriptor of how metabolism and cardiopulmonary function are linked. One reason for the widespread acceptance of the anaerobic threshold concept may be the model developed by Wasserman and colleagues to illustrate the coupling of ventilation to the circulation and of both of those to muscle metabolism (letter from Wasserman to Brooks, 2/21/2000). The central tenet of anaerobic threshold theory had wide appeal to physiologists, especially against the backdrop of the well-known and established Pasteur effect. A PubMed search on "anaerobic threshold" in January 2002 retrieved 1911 documents published since 1973.

The basic notion begins with the observation that during a progressive incremental exercise test, there is a range of mild work rates at which there is only a minimal or no increase in blood (and presumably muscle) lactate concentration. With further increases in work rate, a "threshold" is reached beyond which there is an abrupt or rapid increase in blood lactate concentration as work increases further. This finding dates back to W. Harding Owles (127), who made the following conclusion, "*A critical metabolic level* {italics are ours} was found below which there was no increase in blood lactate as a result of the exercise, although above this level such an increase did occur." Interestingly, there were two subjects in Owles' study: one was Owles himself while the other was apparently his mentor, the famous physiologist Claude Gordon Douglas (1882–1963), inventor of the "Douglas bag," used to collect respiratory gases.

Within the anaerobic threshold concept, the abrupt rise in blood lactate concentration is attributed to the development of muscle hypoxia. In the words of Wasserman and colleagues (158), "if the number of muscle units which must contract to generate the required power exceeds the oxygen delivery and exhausts the O_2 stores, the oxygen level will drop to critical levels in each muscle unit and prevent the ATP, which is needed for muscle contraction, from being generated at an adequate rate by the respiratory enzymes in the mitochondria." In other words, as exercise intensity increases, O_2 delivery to muscle mitochondria is unable to keep pace, allowing P_{O_2} in the neighborhood of at least some mitochondria at least some of the time to decline below the critical level required for adequate oxidative phosphorylation. Further elevations in work rate increase the number and duration of hypoxic sites. With insufficient O_2 to accept electrons at the end of electron transport chains within mitochondria, there would be a "back-up" of reducing equivalents. Ultimately, the tricarboxylic acid (TCA) cycle would be inhibited and increased cytosolic concentrations of both pyruvate and NADH would result; pyruvate would become the preferred H^+ acceptor in this scenario and increased lactate production would occur.

Disregarding any polemic over the meaning of the ventilatory or blood lactate inflection points during graded exercise (87, 88, 100, 149, 156, 158), it is clear that valuable information has been obtained on the parameters which determine the kinetics of O_2 transport and delivery and the control of breathing. The predicament remains, however, that the fundamental assumption of O_2 lack in working muscle at the "anaerobic threshold" has never been shown conclusively (*vide supra*). Further, the assumption is in conflict with all existing data on the state of muscle oxygenation when net lactate release occurs (36, 41, 42, 122, 137). Nonetheless, arguments in favor of the anaerobic threshold have been sufficiently compelling to find widespread acceptance. For a possible reconciliation of the ideas of hypoxia versus adequate oxygenation regarding lactate production, see Gladden's chapter in the *Handbook of Physiology* (67).

In a letter of 2/21/2000 to G.A. Brooks, Wasserman (Fig. 8.10) described aspects of his career that motivated his investigations into noninvasive assessment techniques for pulmonary medicine. In 1959 Wasserman was a postdoctoral fellow under Julius H. Comroe (1911–1984) in the University of California, San Francisco (UCSF) Cardiology Research Institute. Recognizing the "epidemic" in cardiovascular disease, in November of 1960 Comroe challenged Wasserman to develop procedures for early detection of heart disease. Wasserman's response was that evaluation

> would be best done during exercise when the heart was being stressed.... The first sign of heart failure would be reflected in the failure of the circulation to deliver adequate O_2 to the metabolizing tissues (exercising muscles). Since the muscle O_2 requirement would be markedly increased by exercise, the failure of the heart to transport O_2 adequately would result in lactic acidosis (Pasteur Effect). (letter of 2/21/2000)

Aware of the earlier reports of Harrison and Pilcher (74, 75) who showed increased CO_2 production and reduced O_2 uptake in heart-failure patients, Wasserman thought "it possible to investigate how to detect the $\dot{V}O_2$ at which lactic acidosis developed during exercise, using non-invasive gas exchange techniques." His experience at UCSF led to an appointment at Stanford, where Wasserman co-authored a paper with Malcolm B. McIlroy (1921–) that appeared in 1964 in which "anaerobic threshold" was defined as the "$\dot{V}O_2$ above which lactate systematically increased in response to increase in work rate" (157). It is intriguing to note that Wasserman discussed his research with D.B. Dill prior to coining of the term "anaerobic threshold" in Wasserman and McIlroy's 1964 paper. Wasserman showed Dill his data and the method of gas exchange he was using to "detect the work rate at which O_2 transport was inadequate to prevent an exercise lactic acidosis." Wasserman states, "As close[ly] as I can remember, his [Dill's] comment was that I was detecting the *thresh-*

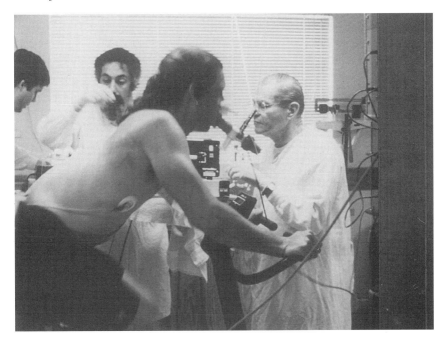

Fig. 8.10. Almost high noon during a study in Karl Wasserman's laboratory, Harbor General Hospital, 1992. Left to right: William Stringer, Richard Casaburi, subject, and Karl Wasserman. Investigators employed rapid arterial blood sampling to relate changes of pH, blood gases, and lactate to respiratory gas-exchange changes during exercise transitions. Courtesy K. Wasserman (see ref. 156)

old of anaerobic metabolism [*italics* are ours] during exercise by measuring exercise gas exchange. Thus, he put into words succinctly what I was trying to measure" (letter of 2/21/2000). In 1973, Wasserman, his colleague Brian J. Whipp (1937–), and other coworkers refined the concept in their classic paper, which generated tremendous interest in the topic (158).

For the subsequent span of over three decades Wasserman and colleagues vigorously pursued the development of technologies to identify the presence of lactic acidosis in graded clinical exercise protocols. For instance, early emphasis on using the RER to identify acidosis was abandoned in favor of using $V_E/\dot{V}_{O_2}$ because the RQ of glycogen oxidation is unity and the increased reliance on glycogen during progressive exercise forced the RER to increase even without lactic acidosis (158). Subsequently, because some normal subjects and patients lacked the chemosensitivity necessary to respond to the CO_2 released from buffering of lactic acid, the $V_E/\dot{V}_{O_2}$ method of threshold detection was reevaluated. Also, he found that obese patients and those with chronic obstructive pulmonary disease had limitations to pulmonary mechanics that limited their ability to develop ventilatory compensation to lactic acidosis. Therefore, Beaver, Wasserman, and Whipp developed the V-slope method to detect the "anaerobic threshold" (11). The V-slope method remains in common use today.

In a long and impressively productive career that included development of on-line computerized technologies for cardiopulmonary assessment and studies of ventilatory control, Wasserman remains best known for his work on using noninvasive means for detection of "anaerobic threshold." Clearly the work has led to methods for detection of lactic acidosis in pulmonary and cardiovascular medicine, but whether muscle hypoxia is identified remains an open question (43). As stated in his recent letter, Wasserman assumed that a Pasteur effect caused lactic acidosis during exercise both in patients with cardiovascular and cardiopulmonary limitations and in normal subjects. Again, justification for Wasserman's assumption of a Pasteur effect giving rise to lactic acidosis is found in his Cannon Lecture (156), a paper which is recommended reading for its description of physiological control in response to metabolic perturbations.

SUMMARY

By way of a summary of the history of anaerobic systems during exercise, we offer the following time line of key ideas, events, and eras:

1808–1907:   *The prelactic acid era.* Lactic acid is discovered and related to physical activity and hypoxia. The prevailing theory of muscle contraction is that a giant molecule, inogen, contains bound oxygen, which combines with carbon and hydrogen in an explosive breakdown. In the late 1800s and early 1900s, Nathan Zuntz measures the respiratory exchange ratio in exercising humans and provides evidence that both fats and carbohydrates are substrates for energy during exercise.

1907–1926:   *The lactic acid era.* This era was begun by the classical paper of Walter M. Fletcher and F. Gowland Hopkins detailing the formation of lactic acid in amphibian muscle. Archibald Vivian (A.V.) Hill and the "British School" of physiologists believed that the "processes of muscle contraction are due to the liberation of lactic acid from some precursor." Otto Meyerhof further detailed the relationship between glycogen and lactic acid formation. Hill developed the O_2 debt concept in human studies in the 1920s. August Krogh and Johannes Lindhard provided evidence supporting Zuntz's assertion that both fat and carbohydrate are substrates for energy during exercise in humans, but these studies were outside the prevailing theory of the "British School."

1926–1932:   *The revolution in muscle physiology.* Hill refers to experiments on the "phosphagens" between 1926 and 1932 as "the revolution in muscle physiology." Gustav Embden's group discovers muscle adenylic acid. Independently, Karl Lohmann and Cyrus H. Fiske and Yellapragada SubbaRow deduce the structure of ATP. The Eggletons discover phosphagen, which is subsequently identified as PCr by Fiske and

SubbaRow. Einar Lundsgaard demonstrates muscle contraction with a decrease in PCr while glycolysis is inhibited.

1933: Rodolfo Margaria, Harold T. Edwards, and David Bruce Dill of the Harvard Fatigue Laboratory apply the new knowledge of the phosphagens to O_2 debt theory in humans.

1941: Fritz Lipmann postulates that ATP stands as the central link between energy-using and energy-yielding reactions in cells.

1962: D.F. Cain and R.E. Davies provide proof that ATP is the immediate energy donor for muscle contraction. Also in this year, Jonas Bergström reintroduces the needle biopsy technique. Over the next 5 years, Bergström and Eric Hultman and colleagues confirm the importance of muscle glycogen availability for human exercise performance and establish the theoretical basis for glycogen loading. Hultman and colleagues also use the needle biopsy technique to discern the behavior of muscle PCr and ATP during exercise.

1964: Karlman Wasserman and Malcolm B. McIlroy coin the term "anaerobic threshold."

1966–1968: Wendell Stainsby demonstrates the transient nature of muscle lactic acid output in canine muscle in situ and presents evidence that argues for the O_2 independence of lactic acid formation during muscle contractions. Also using canine muscle in situ, Margaria, Piiper, di Prampero, and Cerretelli confirm the relationship between muscle PCr concentration and contraction intensity.

1969: Florent Depocas and colleagues use radioactive tracers to study lactate turnover and oxidation in resting and exercising dogs. Fundamental discoveries of Depocas et al. showing active lactate turnover in resting and exercising individuals have been replicated in numerous species, including humans.

1972: Samuel Bessman proposes the phosphocreatine shuttle.

1973: Brooks and associates give ^{14}C-lactate to rats after exhausting exercise and find, contrary to classic O_2 debt theory, little incorporation into glycogen but major disposal as $^{14}CO_2$.

1974: Andrew Halestrap describes monocarboxylate transport in red blood cells.

1977: Lars Hermansen and Odd Vaage revive the idea of lactate resynthesis into glycogen during recovery from intense exercise in humans.

1980: Glenn Gaesser and George Brooks use ^{14}C-tracers, indirect calorimetry and two-dimensional chromatography to trace the paths of lactate and glucose disposal during recovery from exhausting exercise. Again, they find little incorporation of lactate-derived carbon into glycogen but major disposal as $^{14}CO_2$. In mammals, oxidation, not reconversion to glycogen, is the major fate of lactate after exercise.

1984:   Richard Connett, Tom Gayeski and Carl Honig observe lactate production in canine muscle in situ when intramuscular P_{O_2} is apparently above the critical value for mitochondrial oxidative phosphorylation.

1985:   George Brooks proposes the lactate shuttle.

1990:   David Roth and George Brooks describe the characteristics of sarcolemmal lactic acid transport.

1991:   Using human subjects, Jens Bangsbo and colleagues attempt to quantify the fate of lactate during recovery from short-term, high-intensity exercise performed by a single muscle group. Their results support radioactive tracer studies on rats, indicating that the primary fate of lactate in recovery from intense exercise is oxidation.

1994:   Christine Kim Garcia, Michael S. Brown, Joseph L. Goldstein, and colleagues sequence and clone the gene encoding for a muscle cell membrane monocarboxylate transport protein (MCT). Subsequently, they identify a second isoform, MCT2, found mainly in liver.

1998:   George Brooks proposes the intracellular lactate shuttle.

1998:   Russ Richardson, Peter Wagner, and colleagues use magnetic resonance spectroscopy (MRS) to show lactate production and net release from fully aerobic, working human skeletal muscle.

1998:   Andrew Halestrap and colleagues clone and sequence four new MCT isoforms and describe tissue variability in MCT isoform expression.

1999:   Henriette Pilegaard (1962–), Andrew Halestrap, and Carsten Juel show MCT1 and MCT4 distribution in human skeletal muscle.

1999 & 2000: Brooks, Marcy Brown, Hervé Dubouchaud, and colleagues show LDH and MCT1 in muscle mitochondria of rats and humans.

We appreciate the assistance of Dr. Preben Pedersen in obtaining information on dates of birth and death, Dr. Ron Meyer for his insights on the phosphocreatine shuttle, and Kristeene Knopp for technical assistance.

REFERENCES

1. Abramson H.A., M.G. Eggleton, and P. Eggleton. The utilization of intravenous sodium r lactate. III. Glycogen synthesis by the liver. Blood sugar. Oxygen consumption. *J. Biol. Chem.* 75:763–778, 1927.

2. Alpert N.R. and W.S. Root. Relationship between excess respiratory metabolism and utilization of intravenously infused sodium racemic lactate and sodium L-(+)-lactate. *Am. J. Physiol.* 177:455–462, 1954.

3. Bang O. Blutmilchsaure und Sauerstoffaufnahme wahrend und nach Muskelarbeit beim Menschen. *Arbeitsphysiologie* 7:544–554, 1934.

4. Bang O. The lactate content of the blood during and after muscular exercise in man. *Skand. Arch. Physiol.* 74:46–82, 1936.

5. Banga I. and A. Szent-Gyorgyi. Preparation and properties of myosin A and B. *Stud. Inst. Med. Chem. Univ. Szeged* I:5–16, 1941–1942.

6. Bangsbo J., P.D. Gollnick, T.E. Graham, and B. Saltin. Substrates for muscle glycogen synthesis in recovery from intense exercise in man. *J. Physiol. (Lond.)* 434:423–440, 1991.

7. Barnard R.J. and K.M. Baldwin. The effect of training and various work loads on the lactacid-alactacid oxygen debt response of exercising dogs. *Int. Z. Angew. Physiol.* 28: 120–130, 1970.

8. Barnard R.J. and M.L. Foss. Oxygen debt: effect of beta-adrenergic blockade on the lactacid and alactacid components. *J. Appl. Physiol.* 27:813–816, 1969.

9. Barnard R.J., M.L. Foss, and C.M. Tipton. Oxygen debt: involvement of the Cori cycle. *Int. Z. Angew. Physiol.* 28:105–119, 1970.

10. Barr D.P. and H.E. Himwich. Studies in the physiology of muscular exercise. II. Comparison of arterial and venous blood following vigorous exercise. *J. Biol. Chem.* 55: 525–537, 1923.

11. Beaver W.L., K. Wasserman, and B.J. Whipp. A new method for detecting anaerobic threshold by gas exchange. *J. Appl. Physiol.* 60:2020–2027, 1986.

12. Benedict F.G. and E.P. Cathcart. *Muscular Work; A Metabolic Study with Special Reference to the Efficiency of the Human Body as a Machine.* Washington, DC: Carnegie Institution of Washington, 1913.

13. Bennett A.F. and P. Licht. Relative contributions of anaerobic and aerobic energy production during activity in amphibia. *J. Comp. Physiol.* 87:351–360, 1973.

14. Bergman B.C. and G.A. Brooks. Respiratory gas-exchange ratios during graded exercise in fed and fasted trained and untrained men. *J. Appl. Physiol.* 86:479–487, 1999.

15. Bergman B.C., G.E. Butterfield, E.E. Wolfel, G.A. Casazza, G.D. Lopaschuk, and G.A. Brooks. Evaluation of exercise and training on muscle lipid metabolism. *Am. J. Physiol.* 276 (*Endocrinol. Metab.* 39):E106–E117, 1999.

16. Bergman B.C., G.E. Butterfield, E.E. Wolfel, G.D. Lopaschuk, G.A. Casazza, M.A. Horning, and G.A. Brooks. Muscle net glucose uptake and glucose kinetics after endurance training in men. *Am. J. Physiol.* 277 (*Endocrinol. Metab.* 40):E81–E92, 1999.

17. Bergman B.C., M.A. Horning, G.A. Casazza, E.E. Wolfel, G.E. Butterfield, and G.A. Brooks. Endurance training increases gluconeogenesis during rest and exercise in men. *Am. J. Physiol.* 278 (*Endocrinol. Metab.* 2):E244–E251, 2000.

18. Bergman B.C. E.E. Wolfel, G.E. Butterfield, G.D. Lopaschuk, G.A. Casazza, M.A. Horning, and G.A. Brooks. Active muscle and whole body lactate kinetics after endurance training in men. *J. Appl. Physiol.* 87:1684–1696, 1999.

19. Bernard C.M. Sur le mechanisme de la formation du sucre dans le foie. *C.R. Acad. Sci. (Paris)* 41:461–469, 1855.

20. Bertocci L.A. and B.F. Lujan. Incorporation and utilization of [3-^{13}C]lactate and [1,2-^{13}C]acetate by rat skeletal muscle. *J. Appl. Physiol.* 86:2077–2089, 1999.

21. Bessman S.P. Hexokinase acceptor theory of insulin action. New evidence. *Isr. J. Med. Sci.* 8:344–352, 1972.

22. Bessman S.P. and P.J. Geiger. Transport of energy in muscle: the phosphorylcreatine shuttle. *Science* 211:448–452, 1981.

23. Bonen A., K.J. McCullagh, C.T. Putman, E. Hultman, N.L. Jones, and G.J. Heigenhauser. Short-term training increases human muscle MCT1 and femoral venous lactate in relation to muscle lactate. *Am. J. Physiol.* 274 (*Endocrinol. Metab.* 37):E102–E107, 1998.

24. Bonen A., M. Tonouchi, D. Miskovic, C. Heddle, J.J. Heikkila, and A.P. Halestrap. Isoform-specific regulation of the lactate transporters MCT1 and MCT4 by contractile activity. *Am. J. Physiol.* 279 (*Endocrinol. Metab.* 5):E1131–E1138, 2000.

25. Brooks G.A. Current concepts in lactate exchange. *Med. Sci. Sports. Exerc.* 23:895–906, 1991.

26. Brooks G.A. Lactate glycolytic end product and oxidative substrate during sustained exercise in mammals—the lactate shuttle. In: *Proceedings in Life Sciences. Circulation, Respiration, and Metabolism: Current Comparative Approaches; First International Congress of Comparative Physiology and Biochemistry, Liege, Belgium, Aug. 1984,* edited by R. Gilles. Secaucus, New York: Springer-Verlag, pp. 208–218.

27. Brooks G.A. Mammalian fuel utilization during sustained exercise. *Comp. Biochem. Physiol. B Biochem. Mol. Biol.* 120:89–107, 1998.

28. Brooks G.A., K.E. Brauner, and R.G. Cassens. Glycogen synthesis and metabolism of lactic acid after exercise. *Am. J. Physiol.* 224:1162–1166, 1973.

29. Brooks G.A., G.E. Butterfield, R.R. Wolfe, B.M. Groves, R.S. Mazzeo, J.R. Sutton, E.E. Wolfel, and J.T. Reeves. Increased dependence on blood glucose after acclimatization to 4,300 m. *J. Appl. Physiol.* 70:919–927, 1991.

30. Brooks G.A. and C.M. Donovan. Effect of endurance training on glucose kinetics during exercise. *Am. J. Physiol.* 244 (*Endocrinol. Metab. 7*):E505–E512, 1983.

31. Brooks G.A., H. Dubouchaud, M. Brown, J.P. Sicurello, and C.E. Butz. Role of mitochondrial lactate dehydrogenase and lactate oxidation in the intracellular lactate shuttle. *Proc. Natl. Acad. Sci. U.S.A.* 96:1129–1134, 1999.

32. Brooks G.A. and G.A. Gaesser. End points of lactate and glucose metabolism after exhausting exercise. *J. Appl. Physiol.* 49:1057–1069, 1980.

33. Brooks G.A., K.J. Hittelman, J.A. Faulkner, and R.E. Beyer. Temperature, skeletal muscle mitochondrial functions, and oxygen debt. *Am. J. Physiol.* 220:1053–1059, 1971.

34. Brooks G.A., K.J. Hittelman, J.A. Faulkner, and R.E. Beyer. Tissue temperatures and whole-animal oxygen consumption after exercise. *Am. J. Physiol.* 221:427–431, 1971.

35. Brooks G.A., E.E. Wolfel, G.E. Butterfield, A. Cymerman, A.C. Roberts, R.S. Mazzeo, and J.T. Reeves. Poor relationship between arterial [lactate] and leg net release during exercise at 4,300 m altitude. *Am. J. Physiol.* 275 (*Regulatory Integrative Comp. Physiol. 4*):R1192–R1201, 1998.

36. Brooks G.A., E.E. Wolfel, B.M. Groves, P.R. Bender, G.E. Butterfield, A. Cymerman, R.S. Mazzeo, J.R. Sutton, R.R. Wolfe, and J.T. Reeves. Muscle accounts for glucose disposal but not blood lactate appearance during exercise after acclimatizatioin to 4,300 m. *J. Appl. Physiol.* 72:2435–2445, 1992.

37. Cain D.F. and R.E. Davies. Breakdown of adenosine triphosphate during a single contraction of working muscle. *Biochem. Biophys. Res. Commun.* 8:361–366, 1962.

38. Chatham J.C., C. Des Rosiers, and J.R. Forder. Evidence of separate pathways for lactate uptake and release by the perfused rat heart. *Am. J. Physiol. (Endocrinol. Metab.)* 281: E794–E802, 2001.

39. Chauveau A. Source et nature du potentiel directement utilise dans le travail musculaire, d'apres les echanges respiratoires, chez l'homme en etat d'abstinence.. *C.R. Acad. Sci. (Paris)* 122:1163–1169, 1896.

40. Christensen E.H. and O. Hansen. I. Zur Mehodik der respiratorischen Quotient-Bestimmungen in Ruhe und Arbeit. II. Untersuchungen über die Verbrennungs-Vorgange bei langdauernder, schwere Muskelarbeit. III. Arbeitsfahigkeit und Ernahrung. *Skand. Arch. Physiol.* 81:137–171, 1939.

41. Conley K.E., M.J. Kushmerick, and S.A. Jubrias. Glycolysis is independent of oxygenation state in stimulated human skeletal muscle in vivo. *J. Physiol. (Lond.)* 511:935–945, 1998.

42. Connett R.J., T.E. Gayeski, and C.R. Honig. Lactate accumulation in fully aerobic, working, dog gracilis muscle. *Am. J. Physiol.* 246 (*Heart Circ. Physiol. 15*):H120–H128, 1984.

43. Connett R.J., C.R. Honig, T.E. Gayeski, and G.A. Brooks. Defining hypoxia: a systems view of VO₂, glycolysis, energetics, and intracellular PO₂. *J. Appl. Physiol.* 68:833–842, 1990.

44. Connett R.J. and K. Sahlin. Control of glycolysis and glycogen metabolism. In: *Handbook of Physiology. Exercise: Regulation and Integration of Multiple Systems*, edited by Loring B. Rowell and John T. Shepherd, American Physiological Society. New York: Oxford University Press, 1996, pp. 870–911.

45. Daff S., W.J. Ingledew, G.A. Reid, and S.K. Chapman. New insights into the catalytic cycle of flavocytochrome b2. *Biochemistry* 35:6351–6357, 1996.

46. Daum G., P.C. Bohni, and G. Schatz. Import protein into mitochodria. Cytochrome b2 and cytochrome c peroxidase are located in the intermembrane space of yeast mitochondria. *J. Biol. Chem.* 252:13028–13033, 1982.

47. Depocas F., Y. Minaire, and J. Chatonnet. Rates of formation and oxidation of lactic acid in dogs at rest and during moderate exercise. *Can. J. Physiol. Pharmacol.* 47:603–610, 1969.

48. di Prampero P.E. Energetics of muscular exercise. *Rev. Physiol. Biochem. Pharmacol.* 89: 143–222, 1981.

49. di Prampero P.E. and R. Margaria. Relationship between O₂ consumption, high energy phosphates and the kinetics of the O₂ debt in exercise. *Pflugers Arch* 304:11–19, 1968.

50. Donovan C.M. and G.A. Brooks. Endurance training affects lactate clearance, not lactate production. *Am. J. Physiol.* 244 (*Endocrinol. Metab.* 7):E83–E92, 1983.

51. Drury D.R., A.N. Wick, and T.N. Morita. Metabolism of lactic acid in extrahepatic tissues. *Am. J. Physiol.* 180:345–349, 1955.

52. Dubouchaud H., G.E. Butterfield, E.E. Wolfel, B.C. Bergman, and G.A. Brooks. Endurance training, expression, and physiology of LDH, MCT1, and MCT4 in human skeletal muscle. *Am. J. Physiol. Endocrinol. Metab.* 278:E571–E579, 2000.

53. Duke T. Dysoxia and lactate. *Arch. Dis. Child.* 81:343–350, 1999.

54. Eggleton P. and G.P. Eggleton. XXV. The inorganic phosphate and a labile form of organic phosphate in the gastrocnemius of the frog. *Biochem. J.* 21:190–195, 1927.

55. Embden G., K. Baldes, and E. Schmitz. Über den Chemismus der Milchsaurebildung aus Traubenzucker im Teirkorper. *Biochem. Zeit.* 45:108–133, 1912.

56. Embden G. and F. Laquer. Über die Chemi des Lactocidogens. I. Isolierungsversuche. *Hoppe-Seyler's Z. Physiol.* 93:94–123, 1914.

57. Embden G. and F. Laquer. Über die Chemie des Lactacidogens. II. *Hoppe-Seyler's Z. Physiol.* 98:181, 1917.

58. Embden G. and G. Schmidt. Über Muskeladenylsaure und Hefeadenylsaure. *Hoppe-Seyler's Z. Physiol.* 181:130–139, 1929.

59. Embden G. and M. Zimmermann. Über die Bedeutung der Adenylsaure fur Muskelfunktion. I. Mitteilung: Das Vorkommen von Adenylsaure in der Skelettmuskulatur. *Hoppe-Seyler's Z. Physiol.* 167:137, 1927.

60. Englehardt W.A. and M.N. Ljubimowa. Myosine and adenosinetriphosphatase. *Nature* 144:668–669, 1939.

61. Fiske C.H. and Y. SubbaRow. The nature of the "inorganic phosphate" in voluntary muscle. *Science* 65:401–403, 1927.

62. Fiske C.H. and Y. SubbaRow. Phosphorus compounds of muscle and liver. I. Muscle. *Science* 70:381–382, 1929.

63. Fletcher W.M. and F.G. Hopkins. Lactic acid in amphibian muscle. *J. Physiol.* (*Lond.*) 35:247–309, 1907.

64. Gaesser G.A. and G.A. Brooks. Metabolic bases of excess post-exercise oxygen consumption: a review. *Med. Sci. Sports Exerc.* 16:29–43, 1984.

65. Garcia C.K., M.S. Brown, R.K. Pathak, and J.L. Goldstein. cDNA cloning of MCT2, a second monocarboxylate transporter expressed in different cells than MCT1. *J. Biol. Chem.* 270:1843–1849, 1995.

66. Garcia C.K., J.L. Goldstein, R.K. Pathak, R.G. Anderson, and M.S. Brown. Molecular characterization of a membrane transporter for lactate, pyruvate, and other monocarboxylates: implications for the Cori cycle. *Cell* 76:865–873, 1994.

67. Gladden L.B. Lactate transport and exchange. In: *Handbook of Physiology. Exercise: Regulation and Integration of Multiple Systems*, edited by L.B. Rowell and J.T. Shepherd, American Physiological Society. New York: Oxford University Press, 1996, pp. 614–648.

68. Gladden L.B., R.E. Crawford, and M.J. Webster. Effect of lactate concentration and metabolic rate on net lactate uptake by canine skeletal muscle. *Am. J. Physiol.* 266 (*Regulatory Integrative Comp. Physiol.* 35):R1095–R1101, 1994.

69. Gleeson T.T. Metabolic recovery from exhaustive activity by a large lizard. *J. Appl. Physiol.* 48:689–694, 1980.

70. Gollnick P.D., and L. Hermansen. Biochemical adaptations to exercise: anaerobic metabolism. *Exerc. Sport. Sci. Rev.* 1:1–43, 1973.

71. Halestrap A.P. and R.M. Denton. Specific inhibition of pyruvate transport in rat liver mitochondria and human erythrocytes by alpha-cyano-4-hydroxycinnamate. *Biochem. J.* 138:313–316, 1974.

72. Hamann J.J., K.M. Kelley, and L.B. Gladden. Effect of epinephrine on net lactate uptake by contracting skeletal muscle. *J. Appl. Physiol.* 91:2635–2641, 2001.

73. Harris, P. Lactic acid and the phlogiston debt. *Card. Res.* 3:381–390, 1969.

74. Harrison T.R. and C. Pilcher. Studies in congestive heart failure. I. The effect of edema on oxygen utilization. *J. Clin. Invest.* 8:259–290, 1929.

75. Harrison T.R. and C. Pilcher. Studies in congestive heart failure. II. The respiratory exchange during and after exercise. *J. Clin. Invest.* 8:291–315, 1929.

76. Hermansen L. and O. Vaage. Lactate disappearance and glycogen synthesis in human muscle after maximal exercise. *Am. J. Physiol.* 233 (*Endocrinol. Metab.* 2):E422–E429, 1977.

77. Hetenyi G. Jr., G. Perez, and M. Vranic. Turnover and precursor-product relationships of nonlipid metabolites. *Physiol. Rev.* 63:606–667, 1983.

78. Hill A. The energy degraded in recovery processes of stimulated muscles. *J. Physiol. (Lond.)* 46:28–80, 1913.

79. Hill A. The oxidative removal of lactic acid. *J. Physiol. (Lond.)* 48:x–xi, 1914.

80. Hill A.V. A challenge to biochemists. *Biochim. Biophys. Acta* 4:4–11, 1950.

81. Hill A.V. The heat produced in contracture and muscle tone. *J. Physiol. (Lond.)* 40:389–403, 1910.

82. Hill A.V. The revolution in muscle physiology. *Physiol. Rev.* 12:56–67, 1932.

83. Hill A.V., C.N.H. Long, and H. Lupton. Muscular exercise, lactic acid and the supply and utilisation of oxygen. Pt I–III. *Proc. R. Soc. B* 96:438–475, 1924.

84. Hill A.V., C.N.H. Long, and H. Lupton. Muscular exercise, lactic acid and the supply and utilisation of oxygen. Pt IV–VI. *Proc. R. Soc. B* 97:84–138, 1924.

85. Hill A.V., C.N.H. Long, and H. Lupton. Muscular exercise, lactic acid and the supply and utilisation of oxygen. Pt VII–VIII. *Proc. R. Soc. B* 97:155–176, 1924.

86. Hill A.V. and H. Lupton. Muscular exercise, lactic acid and the supply and utilization of oxygen. *Q. J. Med.* 16:135–171, 1923.

87. Hofmann P., V. Bunc, H. Leitner, R. Pokan, and G. Gaisl. Heart rate threshold related to lactate turn point and steady-state exercise on a cycle ergometer. *Eur. J. Appl. Physiol.* 69:132–139, 1994.

88. Hollmann W. Historical remarks on the development of the aerobic-anaerobic threshold up to 1966. *Int. J. Sports Med.* 6:109–116, 1985.

89. Hultman E. Studies on muscle metabolism of glycogen and active phosphate in man with special reference to exercise and diet. *Scand. J. Clin. Lab. Invest. Suppl.* 94:1–63, 1967.

90. Hultman E., J. Bergström, and N.M. Anderson. Breakdown and resynthesis of phosphorylcreatine and adenosine triphosphate in connection with muscular work in man. *Scand. J. Clin. Lab. Invest.* 19:56–66, 1967.

91. Hultman E. and H. Sjoholm. Energy metabolism and contraction force of human skeletal muscle in situ during electrical stimulation. *J. Physiol. (Lond.)* 345:525–532, 1983.

92. Issekutz B. Jr. Effect of beta-adrenergic bockade on lactate turnover in exercising dogs. *J. Appl. Physiol.* 57:1754–1759, 1984.

93. Jervell O. Investigation of the concentration of lactic acid in blood and urine under physiologic and pathologic conditions. *Acta Med. Scand.* 24:1–135, 1928.

94. Jöbsis F.F. and W.N. Stainsby. Oxidation of NADH during contractions of circulated mammalian skeletal muscle. *Respir. Physiol.* 4:292–300, 1968.

95. Jones N.L., N. McCartney, T. Graham, L.L. Spriet, J.M. Kowalchuk, G.J. Heigenhauser, and J.R. Sutton. Muscle performance and metabolism in maximal isokinetic cycling at slow and fast speeds. *J. Appl. Physiol.* 59:132–136, 1985.

96. Juel C. and H. Pilegaard. Lactate/H+ transport kinetics in rat skeletal muscle related to fibre type and changes in transport capacity. *Pflugers Arch. Eur. J. Physiol.* 436:560–564, 1998.

97. Kalckar H. The role of myokinase in transphosphorylations. II. The enzymatic action of myokinase on adenine nucleotides. *J. Biol. Chem.* 148:127–137, 1943.

98. Kayne H.L. and N.R. Alpert. Oxygen consumption following exercise in the anesthetized dog. *Am. J. Physiol.* 206:51–56, 1964.

99. Keilin D. *The History of Cell Respiration and Cytochrome.* Cambridge: Cambridge U. Press, 1966.

100. Kindermann W., G. Simon, and J. Keul. The significance of the aerobic-anaerobic transition for the determination of work load intensities during endurance training. *Eur. J. Appl. Physiol. Occup. Physiol.* 42:25–34, 1979.

101. Krebs H.A. and M. Woodford. Fructose 1,6-diphosphatase in striated muscle. *Biochem. J.* 94:436–445, 1965.

102. Krogh A. and J. Lindhard. The changes in respiration at the transition from work to rest. *J. Physiol. (Lond.)* 53:431–437, 1920.

103. Krogh A. and J. Lindhard. The relative value of fat and carbohydrate as sources of muscular energy. *Biochem. J.* 14:290–363, 1920.

104. Lehninger A.L. *Biochemistry; the Molecular Basis of Cell Structure and Function.* New York: Worth Publishers, 1970.

105. Leicester H.M. *Development of Biochemical Concepts from Ancient to Modern Times.* Cambridge: Harvard University Press, 1974, p. 286.

106. Lipmann F. Metabolic generation and utilization of phosphate bond energy. *Adv. Enzymol.* 1:99–162, 1941.

107. Lohmann K. Über das Vorkommen und den Umsatz von Pyrophosphat in Zellen. II. Mitteilung: Das Menge der leicht hydrolysierbaren P-Verbindung in tierischen und pflanzlichen Zellen. *Biochem. Zeit.* 203:164–171, 1928.

108. Lohmann K. Über das Vorkommen und den Umsatz von Pyrophosphat in Zellen. III. Mitteilung: Das physiologische Verhalten des Pyrophosphats. *Biochem. Zeit.* 203:172–207, 1928.

109. Lohmann K. Über die enzymatische Augspaltung der Kreatinphosphorsaure: zugleich ein Beitrag zum Chemismus der Muskelkontraktion. *Biochem. Zeit.* 271:264–277, 1934.

110. Lohmann K. Über die Pyrophosphatfraktion im Muskel. *Naturwissenschaften* 17:624–625, 1929.

111. Lundsgaard E. Betydningen af faenomenet maelkesyrefrie muskelkontraktioner fer opfattelsen af muskelkontraktiones kemi. *Danske. Hosp.* 75:84–95, 1932.
112. Lundsgaard E. Untersuchungen über Muskel–Kontraktion ohne Milschsaurebildung. *Biochem. Zeit.* 217:162–177, 1930.
113. Lundsgaard E. Weitere Untersuchungen über Muskelkontraktionen ohne Milchsaurebildung. *Biochem. Zeit.* 227:51–83, 1930.
114. Margaria R., R.H.T. Edwards, and D.B. Dill. The possible mechanisms of contracting and paying the oxygen debt and the role of lactic acid in muscular contraction. *Am. J. Physiol.* 106:689–715, 1933.
115. Mazzeo R.S., G.A. Brooks, D.A. Schoeller, and T.F. Budinger. Disposal of blood [1-^{13}C]-lactate in humans during rest and exercise. *J. Appl. Physiol.* 60:232–241, 1986.
116. McFarland E.W., M.J. Kushmerick, and T.S. Moerland. Activity of creatine kinase in a contracting mammalian muscle of uniform fiber type. *Biophys. J.* 67:1912–1924, 1994.
117. Meyer R.A., T.R. Brown, B.L. Krilowicz, and M.J. Kushmerick. Phosphagen and intracellular pH changes during contraction of creatine-depleted rat muscle. *Am. J. Physiol.* 250 (*Cell. Physiol.* 19):C264–C274, 1986.
118. Meyerhof O. Die Energieumwandlungen im Muskel I. Über die Beziehungen der Milchsaure zur Warmebildung and Arbeitsleistung des Muskels in der Anaerobiose. *Pflugers Arch. Physiol. Mensch. Tiere.* 182:232–283, 1920.
119. Meyerhof O. Die Energieumwandlungen im Muskel II. Das Schicksal der Milchsaure in der Erholungsperiode des Muskels. *Pflugers. Arch.. Physiol. Mensch. Tiere.* 182:284–317, 1920.
120. Meyerhof O. Die Energieumwandlungen im Muskel III. Kohlenhydrat und Milchsaureumsatz im Froschmuskel. *Pflugers Arch. Physiol. Mensch. Tiere.* 185:11–32, 1920.
121. Meyerhof O., University of Chicago, and University of Wisconsin. *A Symposium on Respiratory Enzymes.* Madison: The University of Wisconsin Press, 1942.
122. Molé P.A., Y. Chung, T.K. Tran, N. Sailasuta, R. Hurd, and T. Jue. Myoglobin desaturation with exercise intensity in human gastrocnemius muscle. *Am. J. Physiol.* 277 (*Regulatory Integrative Comp. Physiol.* 46):R173–R180, 1999.
123. Mommaerts W.F.H.M. *Muscular Contraction; a Topic in Molecular Physiology.* New York: Interscience Publishers, 1950.
124. Needham J. and R. Hill. *The Chemistry of Life; Eight Lectures on the History of Biochemistry.* Cambridge, England: University Press, 1970.
125. Newman E.V., D.B. Dill, H.T. Edwards, and F.A. Webster. The rate of lactic acid removal in exercise. *Am. J. Physiol.* 118:457–462, 1937.
126. Omachi A. and N. Lifson. Metabolism of isotopic lactate by the isolated perfused dog gastrocnnemius. *Am. J. Physiol.* 185:35–40, 1956.
127. Owles W.H. Alterations in the lactic acid content of the blood as a result of light exercise, and associated changes in the CO_2-combining power of the blood and in the alveolar CO_2 pressure. *J. Physiol.* (*Lond.*) 69:214–237, 1930.
128. Pascoe D.D. and L.B. Gladden. Muscle glycogen resynthesis after short term, high intensity exercise and resistance exercise. *Sports Med.* 21:98–118, 1996.
129. Pasteur L. Mémorie sur la fermentation appelée lactique. *Ann. Chim. Phys.* 52:404–418, 1858.
130. Pasteur L. Recherches sur putrefaction. *C. R. Acad. Sci.* (*Paris*) 56:1189–1194, 1863.
131. Pettenkofer M.v. and C. Voit. Untersuchungen über den Stoffverbrauch des normalen Menschen. *Zeit. Biol.* 2:459–573, 1866.
132. Piiper J., P.E. di Prampero, and P. Cerretelli. Oxygen debt and high-energy phosphates in gastrocnemius muscle of the dog. *Am. J. Physiol.* 215:523–531, 1968.
133. Pilegaard H., J. Bangsbo, E.A. Richter, and C. Juel. Lactate transport studied in sarcolemmal giant vesicles from human muscle biopsies: relation to training status. *J. Appl. Physiol.* 77:1858–1862, 1994.

134. Pilegaard H., G. Terzis, A. Halestrap, and C. Jeul. Distribution of the lactate/H⁺ transporter isoforms MCT1 and MCT4 in human skeletal muscle. *Am. J. Physiol.* 276 (*Endo-crinol. Metab.* 39):E843–E848, 1999.

135. Price N.T., V.N. Jackson, and A.P. Halestrap. Cloning and sequencing of four new mammalian monocarboxylate transporter (MCT) homologues confirms the existence of a transporter family with an ancient past. *Biochem. J.* 329:321–328, 1998.

136. Reeves J.T., E.E. Wolfel, H.J. Green, R.S. Mazzeo, A.J. Young, J.R. Sutton, and G.A. Brooks. Oxygen transport during exercise at altitude and the lactate paradox: lessons from Operation Everest II and Pikes Peak. *Exerc. Sport. Sci. Rev.* 20:275–296, 1992.

137. Richardson R.S., E.A. Noyszewski, K.F. Kendrick. J.S. Leigh, and P.D. Wagner. Myoglobin O_2 desaturation during exercise. Evidence of limited O_2 transport. *J. Clin. Invest.* 96:1916–1926, 1995.

138. Richardson R.S., E.A. Noyszewski, J.S. Leigh, and P.D. Wagner. Lactate efflux from exercising human skeletal muscle: role of intracellular PO_2. *J. Appl. Physiol.* 85:627–634, 1998.

139. Roth D.A. and G.A. Brooks. Lactate and pyruvate transport is dominated by a pH gradient- sensitive carrier in rat skeletal muscle sarcolemmal vesicles. *Arch. Biochem. Biophys.* 279:386–394, 1990.

140. Roth D.A. and G.A. Brooks. Lactate transport is mediated by a membrane-bound carrier in rat skeletal muscle sarcolemmal vesicles. *Arch. Biochem. Biophys.* 279:377–385, 1990.

141. Rothschuh K.E. *History of Physiology.* Huntington, NY: R.E. Krieger, 1973.

142. Rumsey W.L., C. Schlosser, E.M. Nuutinen, M. Robiolio, and D.F. Wilson. Cellular energetics and the oxygen dependence of respiration in cardiac myocytes isolated from adult rat. *J. Biol. Chem.* 265:15392–15402, 1990.

143. Saltin B., P.D. Gollnick, B.O. Eriksson, and K. Piehl. Metabolic and circulatory adjustments at onset of maximal work. In: *Onset of Exercise*, edited by A. Gilbert and P. Guille. Toulouse: University of Toulouse, 1971, pp. 63–76.

144. Schoenheimer R. *The Dynamic State of Body Constituents.* New York: Hafner, 1964.

145. Schwartz A. and A. Oschmann. Contribution au probleme du Mecanisme des contractures musculaires. Le taux de l'acide phosphorique musculair libre dans les contractures des animaux emposonnes par l'acide monobromecetique. *Social Biology* (*Paris*) 91:275–277, 1924.

146. Spriet L.L. Anaerobic metabolism during high-intensity exercise. In: *Exercise Metabolism,* edited by M. Hargreaves. Champaign, IL: Human Kinetics, 1995.

147. Stainsby W.N. and H.G. Welch. Lactate metabolism of contracting dog skeletal muscle in situ. *Am. J. Physiol.* 211:177–183, 1966.

148. Stanley W.C., E.W. Gertz, J.A. Wisneski, D.L. Morris, R.A. Neese, and G.A. Brooks. Systemic lactate kinetics during graded exercise in man. *Am. J. Physiol.* 249 (*Endocrinol. Metab.* 12):E595–E602, 1985.

149. Stegmann H., W. Kindermann, and A. Schnabel. Lactate kinetics and individual anaerobic threshold. *Int. J. Sports Med.* 2:160–165, 1981.

150. Straub F. B. Actin. *Studies from the Institute of Medical Chemistry, University Szeged.* II:3–16, 1942.

151. Szczesna-Kaczmarek A. L-lactate oxidation by skeletal muscle mitochondria. *Int. J. Biochem.* 22:617–620, 1990.

152. The Nobel Prize. The First 100 Years, edited by Agneta Wallin Levinovitz and Nils Ringertz. Singapore: World Scientific, London: Imperial College Press, 2001.

153. von Muralt A. The development of muscle chemistry, a lesson in neurophysiology. *Biochim. Biophys. Acta* 4:126–129, 1950.

154. Wallimann T., M. Wyss, D. Brdiczka, K. Nicolay, and H.M. Eppenberger. Intracellular compartmentation, structure and function of creatine kinase isoenzymes in tissues

with high and fluctuating energy demands: the 'phosphocreatine circuit' for cellular energy homeostatis. *Biochem. J.* 281:21–40, 1992.

155. Warburg O. The origin of cancer cells. *Science* 124:269–270, 1956.

156. Wasserman K. Coupling of external to cellular respiration during exercise: the wisdom of the body revisited. *Am. J. Physiol.* 266 (*Endocrinol. Metab.* 29):E519–E539, 1994.

157. Wasserman K. and M.B. McIlroy. Detecting the threshold of anaerobic metabolism in cardiac patients during exercise. *Am. J. Cardiol.* 14:844–852, 1964.

158. Wasserman K., B.J. Whipp, S.N. Koyal, and W.L. Beaver. Anaerobic threshold and respiratory gas exchange during exercise. *J. Appl. Physiol.* 35:236–243, 1973.

159. Watchko J.F., M.J. Daood, B. Wieringa, and A.P. Koretsky. Myofibrillar or mitochondrial creatine kinase deficiency alone does not impair mouse diaphragm isotonic function. *J. Appl. Physiol.* 88:973–980, 2000.

160. Watt M.J., K.F. Howlett, M.A. Febbraio, L.L. Spriet, and M. Hargreaves. Adrenaline increases skeletal muscle glycogenolysis, pyruvate dehydrogenase activation and carbohydrate oxidation during moderate exercise in humans. *J. Physiol. (Lond.)* 534:269–278, 2001.

161. Weber J.M., W.S. Parkhouse, G.P. Dobson, J.C. Harman, D.H. Snow, and P.W. Hochachka. Lactate kinetics in exercising Thoroughbred horses: regulation of turnover rate in plasma. *Am. J. Physiol.* 253 (*Endocrinol. Metab.* 22):R896–R903, 1987.

162. Welch H.G. and W.N. Stainsby. Oxygen debt in contracting dog skeletal muscle in situ. *Respir. Physiol.* 3:229–242, 1967.

163. Wickler S.J. and T.T. Gleeson. Lactate and glucose metabolism in mouse (Mus musculus) and reptile (Anolis carolinensis) skeletal muscle. *Am. J. Physiol.* 264 (*Regulatory Integrative Comp. Physiol.* 33):R487–R491, 1993.

164. Williams T.I. *A Biographical Dictionary of Scientists.* New York: Wiley, 1974.

165. Wilson M.C., V.N. Jackson, C. Heddle, N.T. Price, H. Pilegaard, C. Juel, A. Bonen, I. Montgomery, O.F. Hutter, and A.P. Halestrap. Lactic acid efflux from white skeletal muscle is catalyzed by the monocarboxylate transporter isoform MCT3. *J. Biol. Chem.* 273: 15920–15926, 1998.

166. Wiseman R.W. and M.J. Kushmerick. Creatine kinase equilibration follows solution thermodynamics in skeletal muscle. ^{31}P NMR studies using creatine analogs. *J. Biol. Chem.* 270:12428–12438, 1995.

167. Zuntz N. and W. Loeb. Über die Bedeutung der verschiedene Nahrstoff als Energiequelle der Muskelkraft. *Arch. Anat. Physiol.* 18:541–543, 1894.

168. Zuntz V.N. Umsatz der nahrstoffe. XI. Betrachtungen über die beziehungen zwischen nahrstoffen und leistungen des korpers. In: *Handbuch der Biochemie des Menschen und der Tiere*, edited by K. Oppenheimer, G. Fischer, Jena, 1911, pp. 826–863.

chapter 9

THE ENDOCRINE SYSTEM: METABOLIC EFFECTS OF THE PANCREATIC, ADRENAL, THYROIDAL, AND GROWTH HORMONES

Michael C. Riddell, Neil B. Ruderman,
Evangelia Tsiani, and Mladen Vranic

T HE purpose of this chapter is to review, from a historical perspective, the role of the key endocrine glands in modulating the metabolic responses to acute and chronic exercise. Its scope is limited to the metabolic effects of hormones released by the pancreas (insulin and glucagon), adrenal glands (catecholamines and glucocorticoids), anterior pituitary (growth hormone), and thyroid (thyroxine and triiodothyronine) that play a critical role in regulating substrate mobilization and utilization during muscular work. Although other hormones participate in the physiological response to exercise, their role is beyond the scope of this chapter. Although the study of endocrine diseases has enhanced our knowledge of hormonal actions, the chapter's coverage is restricted to physiological responses in healthy individuals. We also provide a brief history of some of the important studies conducted on exercise and diabetes, as they have expanded our knowledge of the influence of the pancreatic hormones on metabolism. Information on the historical developments of the study of the endocrine system is also found in Chapters 6 and 8. Together, these three chapters cover a considerable portion of the broad field of endocrinology and metabolism in exercise physiology.

The first connections between the endocrine functions and muscular activity were made in the nineteenth century by physiologists who observed that individuals with endocrine disorders have a decreased capacity for physical work. Later, in the late 1960s and beyond, there were systematic investigations of the effects of exercise and physical training on hormones and metabolism. Much of what is described in this chapter reflects methodological advances over the last 30 years in the study of endocrinology and metabolism. As will become evident, many of the investigations conducted before 1970 were inferential in nature, using mostly secondary measures of hormonal effects on metabolism. These experiments set the stage, however, for the recent explosion of knowledge concerning endocrine and metabolic responses to exercise. The development of new methods to assay hormones accurately in blood was an important contributor to these advances, as were the refinement and validation of tracer techniques that make it possible to measure substrate turnover more precisely. The introduction of glucose and hormonal clamps and the use of sophisticated techniques for measuring blood flow and arteriovenous (A–V) differences across liver and muscle in the 1960s and 1970s permitted informative studies to be conducted in animals and in humans. These methodological advances were complemented in the late 1970s by the development and characterization of isolated perfused rat hindquarter preparations, which allowed for specific hormonal manipulation of muscle during electrical stimulation. Since the 1990s, use of nuclear magnetic resonance spectroscopy (NMRS) has made it possible to obtain noninvasive, real-time measurements of intracellular pH, substrate fluxes, and energy metabolism in both liver and muscle tissues in humans. Investigators in the field of endocrinology and exercise have contributed considerably to the understanding of complicated hormone signalling pathways and have provided new insights into how tissues sense and respond to various energy needs. Their research has made many conceptual and methodological contributions to our understanding of human physiology in health and disease.

PANCREATIC HORMONES

Discovery and Characterization of Insulin and Glucagon

The quest for the internal pancreatic secretion that could correct diabetes is among the most impressive stories of modern physiology and medicine. Although several scientists had the discovery of insulin within their grasp at the turn of the twentieth century, Frederick Banting (1891–1941) and Charles Best (1899–1978), with the help of John Mcleod and James Collip, working in the Departments of Physiology and Medicine at the University of Toronto, reported the finding that secretions from the pancreas lowered blood glucose at the American Physiological meeting in New Haven, Connecticut, in the fall of 1921. Banting, a surgeon working in London, Ontario, had developed the idea that pancreatic internal secretions could be isolated and used to treat diabetes. This idea came to him after giving a physiology lecture to undergraduate students on the metabolic effects of the pancreas. He proposed his idea to Professor Macleod at the University of Toronto, an expert in the field of carbohydrate metabolism, who allowed him to perform experiments with Best, then a summer research associate at the university. Although it was Banting's idea, Best was particularly useful because of his analytical skills in measuring blood glucose using as little as 0.2 ml of blood with the recently introduced Lewis–Benedict method. With the help of Collip, a visiting biochemist who could further purify the extract, they worked on these experiments, which were frustrating at times, but they were ultimately successful in isolating and purifying a substance (insulin) that lowered blood glucose levels in diabetic dogs and eventually in humans with diabetes. The proof that the hormone could influence liver glycogen levels and correct excessive ketone production was also developed by these investigators. The group was so successful in their efforts, despite some internal friction, that clinical trials for the treatment of diabetes were initiated shortly after their first public announcement of the discovery in 1921. Those interested in the fascinating account of this highly emotional story, including the tremendous toll on the scientists involved and the immediate impact of the discovery of the hormone on the scientific and medical community, are directed to the definitive scholarly account on the subject, *The Discovery of Insulin*, by Michael Bliss (28).

When compared with insulin, the discovery of glucagon drew considerably less attention. Soon after the isolation of insulin, scientists observed that a pancreatic principle existed that had an effect on blood glucose that was opposite to that of insulin. The unknown factor, named initially "hyperglycemic glycogenolytic factor," was eventually purified and identified by Otto K. Behrens and colleagues at the Lilly Research Laboratories in Indianapolis, Indiana, in 1953 (213). The factor, now known as glucagon, was shown by Nobel laureate Earl W. Sutherland (1915–1974) to stimulate hepatic glucose production through the activation of adenylate cyclase by cAMP in the 1950s, a finding that revolutionized the hormone receptor concept (216). Later, in the 1970s, Mladen Vranic's laboratory found that glucagon could be produced in significant quantities outside of the pancreas in dogs and that there was a significant metabolic role for this hormone during exercise (59). A more detailed version of the discovery of the pancreatic hormones is found elsewhere in the People and Ideas series by Jay Tepperman (223). Within a few years of the discovery of each of these hormones, scientists began studying their effects on metabolism during exercise.

Insulin and Acute Exercise

The chance discovery of the radioimmunoassay technique for insulin by the Nobel laureate Rosalyn S. Yalow (1921–) and her colleague Solomon A. Berson in the late 1950s was revolutionary, as it allowed fairly accurate minute-by-minute measures of insulin in human plasma samples. Burt Cochran et al. (49) from Grant Gwinup's laboratory at the Department of Medicine, University of California, Irvine, appear to have been the first to show, in 1966, that serum immunoractive insulin levels decrease with muscular exercise. Gwinup, an endocrinologist with an interest in mechanisms of hypoglycemia, had noted from the studies by Ingle (123), Goldstein (93) and others that glucose is an important fuel during exercise, even in the absence of insulin (see below). The effects of exercise on plasma insulin levels were relatively unknown at the time and others such as David B. Dill (1891–1986), scientific director of the Harvard Fatigue Laboratory, had even proposed that insulin secretion might increase during exercise to account for the elevation in carbohydrate utilization (58). To clarify the effects of exercise on circulating insulin concentrations in humans, Gwinup's group measured glucose, insulin, lactate, and pyruvate concentrations before and after 25 minutes of stair climbing in lean nondiabetic and obese diabetic and nondiabetic men. Plasma insulin levels decreased in 10 of 17 subjects, and the drop was more pronounced in obese diabetic subjects than in the other groups (49). The investigators correctly speculated that the decrease in insulin is an important mechanism in maintaining blood glucose concentration during prolonged exercise and that it is a result of increased sympathetic activity.

A few years after the initial observations, Rasio (184), Hunter (119), and others confirmed that a decrease in insulin concentration occurs during muscular work in healthy men. Building upon these studies, Esther D. R. Pruett, from the Institute of Work Physiology in Oslo, Norway, systematically investigated the changes in plasma immunoreactive insulin levels during exercise in a comprehensive set of metabolic studies starting in 1970 (181, 182). The objectives of the Norwegian studies were to confirm that plasma insulin levels decrease with exercise of varying intensities (ranging from 20% to 90% of $\dot{V}O_2$max) and to determine if changes in pre-exercise diet could influence insulin levels during the activity. Pruett was perhaps the first to illustrate that a number of experimental factors make the finding of a drop in insulin levels somewhat elusive. For example, she noted that the timing of exercise, carbohydrate intake, intensity of exercise, and the duration of exercise all appeared to influence the magnitude of change in insulin concentration (181, 182). Indeed, increases, decreases, or no change in insulin levels could be observed depending on any of these factors. The timing of the blood sample taken was also shown to be crucial, as Pruett found that insulin levels could increase immediately and dramatically upon cessation of prolonged, or maximal intensity, exercise (182). These factors, which were seldom standardized in many of the earlier studies, explained some of the inconsistent findings in publications from other laboratories prior to the 1970s.

In the early 1970s, John Wahren, Gunvor Ahlborg, and their colleagues at the Huddinge Hospital and the Karolinska Institute in Stockholm, Sweden, began a series of human metabolic studies designed to illustrate the effects of exercise on metabolism. They too confirmed that plasma insulin levels decrease with prolonged moderate exercise (240). Based on what was known about the metabolic effects of insulin on the liver, the decrease in insulin concentration during exercise was proposed by Pruett, Wahren, Vranic, and others to facilitate hepatic glucose production to maintain blood glucose concentrations during physical activity. This proposal was consistent with the observations that insulin administration dramatically alters glucose homeostasis in humans and in animal models with diabetes (see below). That insulin concentration decreased with prolonged exercise fit with the concept that splanchnic glucose production, which was shown to increase during exercise in the mid-1960s by Loring B. Rowell et al. (198) at the University of Washington School of Medicine in Seattle, and Jonus Bergström and Erik Hultman (22) at St. Ericks Sjukhus in Stockholm, was likely under the influence of circulating insulin concentrations. It was also thought, at the time, that the reduction in insulin may limit peripheral glucose uptake into muscle and thereby preserve blood glucose levels for the central nervous system as the exercise progressed. However, the realization that muscle glucose uptake was accelerated during exercise even though plasma insulin levels were suppressed in normal individuals (240) was perplexing. Several theories were generated to explain this phenomenon, which are highlighted elsewhere in this chapter, although none was sufficiently substantiated. These theories were mostly proven incorrect with time, and ideas needed to be modified in subsequent years when it was found that muscle takes up glucose via insulin-independent mechanisms (see below).

By the mid-1970s, it was clear to a growing number of investigators including Hartley et al. (105), Bloom et al. (30), Jarhult and Holst (132), Ahlborg et al. (3), and Galbo et al. (89) that peripheral plasma insulin levels decreased with prolonged moderate-intensity exercise in humans. Although these studies were critical in developing our knowledge of exercise metabolism and the hormonal response to physical activity, they were limited to describing the effects of exercise on pancreatic hormone concentrations in peripheral plasma rather than providing information about hormone turnover rate and concentration near the target organs (i.e., liver and muscle). Techniques needed to be developed to determine the effects of exercise on pancreatic hormone secretions and the influence of portal concentrations of these hormones on substrate turnover (see Roles of the Pancreatic Hormones on Exercise Metabolism In Vivo: Evolution of Ideas).

The mechanism causing the decrease in plasma insulin concentration during acute exercise was another area of active investigation during the 1960s and 1970s. It was unknown initially if the decrease in insulin concentration during exercise was a result of reduced insulin secretion or increased insulin elimination. In 1968, Nikkila and associates used ^{131}I-insulin and dilution methods to determine the influence of exercise on insulin secretion in humans performing mild exercise (173). Surprisingly, the group found no changes in insulin secretion rate with exercise using this methodology, possibly because the infusion of the hormone altered its endogenous secretion in their subjects. A subsequent study in 1971 by Franckson et al. (73), using a constant infusion of ^{125}I-insulin, once again in humans, showed that the decrease in insulin concentration was the result of a decrease in insulin secretion. These studies were not regarded as conclusive since the use of iodinated insulin, which has a relatively high molecular weight, may have caused undesirable isotopic effects. In 1978, Michael Berger and colleagues, working in Renold's laboratory in Geneva, confirmed that the secretion of insulin decreased with exercise in rats (20) by using tritiated insulin that was synthesized by Halban, a lighter isotope very precise for kinetic analysis.

To further investigate the influence of exercise on insulin secretion in humans, other investigators in the early 1980s, such as Henrik Galbo's group from Denmark (112) and Pierre Lefebvre's team from Belgium (151), used C-peptide measurements to demonstrate decreased insulin secretion. The C-peptide molecule was known to be secreted in equimolar amounts with insulin, but unlike insulin, it is not metabolised to a great extent by liver or muscle. Thus, the decrease in C-peptide levels during exercise found by these investigators was taken as further evidence that insulin secretion is reduced during exercise. Because Galbo's contributions were pivotal in stimulating research in the area of hormonal regulation during exercise, and because he has mentored the careers of a number of investigators including Erik Richter, Thorkil Ploug, Bente Sonne, Michael Kjaer, and others, a brief background is warranted.

Henrik Galbo (1946–) graduated from the University of Copenhagen with a degree in internal medicine and rheumatology. His interest in the effects of exercise on the endocrine system and the role of hormones in the metabolic response to exercise was stimulated by studying the works of August Krogh (1874–1949) from Denmark and Per-Olof Astrand (1922–) from Sweden. As Galbo states, his inspiration was drawn from "the expansive Scandinavian tradition for the studies of those physiological and biochemical processes making physical exercise possible." His book published in 1983 (79) was the most comprehensive account of endocrinology and exercise physiology to date and remains a valuable resource for those interested in this field. Galbo appears to be the first to suggest that the complex integrated control of the hormonal response to exercise was determined by feed-forward and feedback control similar to the accepted mechanism for the control of respiration and circulation. Critical to his hypothesis underlying the control of the endocrine system during exercise was the role of impulses from working muscle and motor centers that modulate the activity in higher centers in the central nervous system according to the relative workload. Also critical in modulating the hormonal response were hormone–hormone interactions and feedback signals sensing glucose, temperature, oxygen tension, and intravascular volume. This complex and only partially unveiled schema forms the modern working hypothesis that is currently used to describe the normal endocrine response to exercise.

Once investigators knew with certainty that insulin secretion was lowered during exercise, the quest began to identify the mechanism by which this occurred. The decrease in insulin secretion was proposed to be a result of increased adrenergic stimulation, which is graded to the intensity and duration of exercise, as well as to the training status of the subjects. This hypothesis fit with the observations that suppression of insulin secretion depended on a sympathetic feed-forward mechanism that was, in turn, dependent on the intensity of exercise and fitness status of the organism. This concept, developed primarily by Galbo, was supported by investigations in the late 1970s by Bloom et al. (29, 30) and by Jarhult and Holst (132), who used adrenergic blockade in exercising humans and experimental animals to influence pancreatic insulin secretion. The group from Denmark tested the hypothesis that the decrease in insulin secretion with exercise was a result of an inhibitory effect of elevated catecholamines during the activity by using α- and β-blockade and by destroying the sympathoadrenal system and assessing the effect of replacement hormones in exercising rats (88). They were the first to demonstrate convincingly that the control of insulin secretion was under the influence of α-adrenergic activity via both sympathetic nerves to the pancreas and enhanced epinephrine secretion from the adrenal medulla (88, 192).

Glucagon and Acute Exercise
A number of the researchers who were studying the effects of exercise on insulin kinetics also found that glucagon concentrations increased with prolonged exercise. The first examples of this in human studies were provided in the early 1970s by the group from Sweden led by Wahren and Ahlborg in collaboration with Philip Felig from Yale University. They found that glucagon levels increased in plasma only after about 60 minutes of moderate exercise (3, 67). To complicate things initially, Galbo's (84, 85) and Bloom's (30) laboratories observed a decrease, rather than an increase, in glucagon concentration in peripheral plasma samples of humans during exercise that lasted less than 1 hour. Their studies suggested that if glucagon played a role in increasing glucose production in humans, it was not until the later stages of prolonged exercise. This hypothesis, which was primarily championed by the group from Denmark, was contested by others such as Vranic (237) and Issekutz (128), who conducted studies in catheterized dogs in which relatively rapid elevations in portal glucagon concentrations were seen (see below). The finding that glucagon secretion was increased soon after the start of exercise in dogs was supported in the late 1970s by studies by William W. Winder (1942–) and colleagues from Washington University in St. Louis, which demonstrated that glucagon levels increase almost immediately in rats during exercise (252). The discrepancy in these animal models and the relevance of glucagon in glucose homeostasis in humans during exercise has caused considerable debate. It also resulted in a vast number of elegant investigations to elucidate the influence of pancreatic hormones on metabolism during exercise (see below).

Similar to insulin, the control of glucagon secretion during exercise was thought to be under sympathetic control, which in turn was clearly influenced by the intensity and duration of exercise. However, Erik A. Richter, then a postdoctoral fellow with Galbo, and his co-workers in Denmark favored the idea that the rise in glucagon, which was often observed late in exercise in humans, could be attributed to increases in circulating epinephrine rather than to the sympathetic nervous system per se (192, 196). Since glucagon levels increase even when sympathetic drive is minimal and when adrenergic blockade is introduced (85), they proposed that other factors must be involved in regulating its secretion (80). These researchers favored the hypothesis that the stimulus for increased glucagon secretion is a decrease in blood glucose concentration rather than control by the autonomic nervous system (89). The idea that glucagon secretion was under a feedback mechanism was attractive to the group from Denmark because it fit with their observation of the delay in glucagon response after the onset of exercise when blood glucose levels were unchanged. In keeping with this conclusion were the findings of Ahlborg et al. (2) and of Galbo et al. in 1977 (81) that glucose administration reduces the glucagon response to exercise. This led them to propose that the pancreatic α cells were sensitized to small changes in plasma glucose during exercise compared with at rest. They found, however, that a variety of substances produced during exercise, such as alanine, free fatty acids (FFA), and lactate did not influence pancreatic α-cell sensitivity. This theory of feedback control did not fit with the earlier finding of the Wahren group that glucagon levels could increase during exercise without a decrease in glycemia (241). In addition, glucose clamp studies in the 1980s by Issekutz and Vranic (128) and by David Wasserman et al. (245) working in Vranic's laboratory showed conclusively that glucagon levels increase during exercise without any changes in glucose concentration. The control of glucagon secretion during exercise remains an active area of investigation and it appears that several factors can influence the exercise response (see Summary below).

Pancreatic Hormones and Training
The influence of physical training on pancreatic hormones during exercise has been a prime area of research interest for nearly 30 years. During the 1970s, Hartley (104, 105), Bloom (30), Gyntelberg (99), Rennie (188), Sutton (218), and their coworkers and others found that the decrease in insulin during exercise was less pronounced in trained versus untrained individuals. These findings led to the proposal that the training-induced attenuation in the decrease in insulin concentration may be caused by reduced catecholamine levels or by improvements in glucose homeostasis during exercise.

The attenuating effect of training on glucagon secretion was first suggested in a study using dogs by Bottger in Roger H. Unger's laboratory in Dallas in 1972 (34). The influence of fitness level and training on glucagon levels in humans was also demonstrated in the 1970s in publications by Bloom et al. (30), Gyntelberg et al. (99), and Winder et al. (259). In all of these investigations, the increase in glucagon concentration was less pronounced in trained compared with untrained individuals, even when exercise was performed at the same relative intensity. Because both sympathoadrenal activity and the tendency to a decrease in plasma glucose (and insulin) were shown to be reduced during exercise in trained compared to untrained individuals, the attenuated glucagon response in trained individuals was compatible with both the feedback and feed-forward theories of control.

The alterations in pancreatic hormones with training were not immediately tied in with alterations in substrate utilization caused by training, possibly because these effects had not yet been clearly illustrated. It was recognized in the early 1970s, however, by investigators such as L. Howard Hartley and colleagues (104, 105) that the hormonal adaptations caused by training could alter the provision of free fatty acids and glucose to the muscles during muscular work. It is also important to note that the discovery that regular exercise decreased basal insulin levels was critical since it was the first evidence that improvements in sensitivity to the hormone occurred (see Exercise and Insulin Sensitivity). The influence of training on pancreatic hormone kinetics and glucose metabolism are of continued interest to researchers. For example, a study published in 2001 in Lavoie's laboratory provided evidence that training increases liver glucagon receptor density (156), which may be important in facilitating training-associated increases in hepatic glucose production (62).

Roles of the Pancreatic Hormones in Exercise Metabolism In Vivo: Evolution of Ideas and Techniques

The catheterization technique, which was used extensively in the 1960s and 1970s by Rowell and others in the United States and by Keul and Doll in Germany, and by Pernow in Sweden, was instrumental in illustrating the relative importance of carbohydrate and lipid in working muscle and the provision of these substrates from the blood. The muscle and liver biopsy methods developed and used by Jonus Bergstöm and Erik Hultman from the Karolinska Institute in Sweden allowed these investigators, and their collaborators, to determine the importance of muscle and liver glycogen stores during anaerobic and aerobic exercise. As mentioned above, the A–V difference method, used by Wahren and coworkers in Sweden, was useful in describing splanchnic glucose production in human subjects during exercise. These investigations also gave some insight into insulin and glucagon secretions during exercise in humans, although measurements were somewhat clouded by hepatic degradation of these hormones.

The use of dog models and tracer methodology since the late 1960s has also been critical in developing our understanding of the effect of exercise on pancreatic hormone secretion and the role of these hormones in energy mobilization. These experimental techniques developed by Steele (214), first used in exercise by Bella Issekutz (1912–1999) from Dalhousie University in Halifax and by Mladen Vranic (1930–) and coworkers in Toronto, provided the opportunity for the accurate measurement and manipulation of pancreatic hormone concentrations during acute stresses such as exercise. These investigations were the first to show that the glucagon/insulin ratio is the main determinant of glucose production during moderate-intensity exercise in dogs. These experimental models, which were adopted by Alan Cherrington, David Wasserman, and others, were crucial in quantifying the changes in insulin and glucagon secretion that primarily determine hepatic glucose production rates during rest and exercise in diabetic and nondiabetic dogs. A more detailed discussion of the history of these studies is provided in the metabolic summary at the end of this chapter. Because the advent of tracer techniques was crucial in developing our knowledge of exercise metabolism, a brief history of the development of these techniques is warranted.

The development of tracer techniques for non–steady-state glucose turnover measurements in exercise began in the 1950s. Initially, Robert Steele and his collaborators from the Brookhaven laboratory in the United States developed a continuous tracer infusion method to repeatedly measure non–steady-state glucose turnover in vivo in 1956 (214). Gerry Wrenshall, a contemporary of Steele's, and a professor in the Department of Physiology at the University of Toronto, with doctoral degrees in physics and physiology, together with Geza Hetenyi, were instrumental in developing techniques to study turnover of various metabolites, including glucose, in vivo. Together with Vranic, a postdoctoral fellow of Charles H. Best, they were the first to be able to measure non–steady-state glucose turnover during exercise by applying three injections of radioactive glucose into exercising dogs—one before, one during, and one after the exercise period (239). It was clear from their results that exercise, in the complete absence of insulin (depancreatized dogs were used), increased glucose production without changing glucose metabolic clearance rate. This study was critical at the time because it clearly showed that some insulin is needed in vivo to increase glucose transport during exercise, possibly because insulin could counteract the inhibitory effects of elevated catecholamines. Uncertainty existed about the three-injection tracer technique since only one measure of glucose turnover could be extrapolated from the exponential decline in specific activity after each injection. The methods were further refined and validated in subsequent years (71) and have been widely adopted to assess glucose and FFA turnover during exercise with and without hormonal manipulation.

In addition to the use of non–steady-state tracer techniques that could be used on exercising animals, other developments were necessary to elucidate the role of the pancreatic hormones in glucose turnover. Depancreatized dogs could be used to investigate the role of the hormones in metabolism, although it was soon discovered that the dogs maintained glucagon levels via secretions from their stomach mucosa (59). This finding was useful, however, since exercise did not stimulate extrapancreatic glucagon, and the role of the drop in insulin could be determined independent from increases in glucagon concentrations (237). The discovery of somatostatin by the Nobel laureate Guillemin in 1973 was also instrumental in determining the effects of insulin and glucagon on glucose turnover, since it allowed for the complete suppression of insulin and glucagon secretion during exercise and permitted the concentration of these hormones to be manipulated by exogenous hormone infusions. Later, the use of glucose clamp techniques also allowed the role of these hormones to be investigated independent of changes in glucose concentration (245). The development of these techniques has been instrumental in determining the influence of the pancreatic hormones on substrate metabolism during exercise, and they are widely used in both human and animal studies of exercise and metabolism.

Regulation of Glucose Uptake during Exercise

The history of the study of the effects of exercise on muscle glucose uptake and of the requirement for insulin is among the most interesting in exercise metabolism. This story starts with a concept formed in the mid-1900s—that insulin is crucial in facilitating glucose transport into skeletal muscle, but ends with the realization that insulin has an additive effect to that of exercise, and that glucose transport during muscle contractions may be enhanced even in the absence of insulin.

In the late 1800s, Jean B. August Chauveau (1827–1917) of Paris and his colleague Kaufmann were attempting to illustrate that exercise altered the mixture of fuels used by the working muscles. More precisely, they felt that exercise causes a shift in fuel use from predominantly fat to mostly carbohydrate. In contrast, the

German Nathan Zuntz (1847–1920) and his followers thought that muscular work is produced by using a mixture of nutrients (i.e., lipid and carbohydrate) similar to that used at rest. Both groups of investigators used indirect measures of fuel utilization (i.e., respiratory exchange ratio) to support their theories. In 1887, Chauveau and Kaufmann gathered more definitive evidence that blood glucose was an important fuel during exercise (47). They had direct measures of oxygen, carbon dioxide, blood glucose, and blood flow in an artery and vein supplying the facial muscles of a conscious mare. In this remarkable experiment, when the mare was fed hay, they observed a large increase in glucose uptake and O_2 consumption in the working muscle (47).

Following the turn of the century, others became interested in measuring the mixture of fuels utilized during exercise. In North America, Francis G. Benedict (1870–1959), and his colleagues from Wesleyan University in Connecticut and later the Nutrition Laboratory in Boston—established by the Carnegie Institute of Washington—studied the effects of diet and cycling at different rates (on the Benedict ergometer) on the respiratory exchange ratio (16). They were the first to find that the RQ during exercise was lower when subjects were fed a diet poor in carbohydrate, presumably because their glycogen stores were low. In Denmark, Krogh and Johannes Lindhard (1870–1947) conducted carefully designed experiments which showed that both fat and carbohydrates can be used as fuel during exercise and that their relative use was determined by the availability of these substrates (150). Incorrectly, however, they supported Chauveau's earlier notion that fat must first be transformed into carbohydrate prior to its use, a process that required a loss in caloric value. After Word War I, Erik H. Christensen (1904–1996) and his coworkers, students of Krogh and Lindhard, continued with these studies and found that the ratio of carbohydrate to fat depends on a number of factors including the exercise intensity, duration, and antecedent diet (48). These observations, which have been highlighted in a well-written historical chapter on muscle metabolism by Asmussen (7), set the stage for future investigations into the regulation of glucose uptake during exercise—a topic that is now under intense investigation at a molecular level (see below).

Muscle Glucose Uptake and the Requirement of Insulin

Initially, it was thought that increased rates of enzymatic degradation of glucose to CO_2, water, heat, and energy during muscular work created a "vacuum" into which glucose "flowed." The development of the "glucose transport theory" of insulin action left researchers looking for the mechanism that facilitated glucose uptake into contracting muscle, something which was known to occur even in muscle of diabetic subjects. A major question was: how does muscular exercise stimulate glucose transport, independently of insulin?

One of the first theories to gather widespread attention was the "muscle activity factor" hypothesis proposed in 1953 by Maurice Goldstein and colleagues in Rachmiel Levine's (1910–1999) laboratory at Michael Reese Hospital in Chicago (93). These investigators felt that a substance released by working muscle "acts upon some cell surface barrier that is a specific determining factor for the entry of sugars." The group reported that exercising muscles release such a factor and that it apparently stimulates glucose uptake in both contracting and noncontracting muscles. In a follow-up study (92), using parabiotic eviscerated–nephrectomized dogs, these researchers found that when blood from an active dog was transfused into a resting dog, the resting dog had a dramatic increase in glucose uptake. This second study gathered considerable attention and left the investigators confident that a mysterious humoral factor was produced by the working muscle that circulated in the blood to facilitate glucose uptake into nonworking muscles. Their theory was strengthened by later studies of Havivi and Wertheimer in rat diaphragm muscle (107) and by a study from Bihler and colleagues (23) in isolated heart muscle, although no factor was isolated and purified in these investigations.

The muscle activity factor theory also had its detractors. Criticism of it occurred early from researchers such as Carl F. Cori from Washington University, who felt that no such factor existed (110). In the mid-1970s, the question was examined by Berger and colleagues (19) in the laboratory of Neil Ruderman (1937–) in Boston. Using the isolated perfused rat hindquarter preparation developed by Ruderman in 1969 while a fellow in Sir Hans Krebs's laboratory in Oxford (201), they observed that glucose uptake was not augmented in resting rat muscle when it was perfused with a medium taken from a similar preparation in which glucose uptake had been increased by electrical stimulation. The notion that a humoral factor influenced glucose uptake in nonworking muscle was also challenged by studies from the same laboratory by Richter et al. (194), in which they showed that electrical stimulation enhances glucose uptake only in muscles that were contracting or had previously contracted. Studies by Ahlborg et al. (4) in human subjects who performed one-legged exercise led to a similar conclusion. Based on these investigations and others, the theory that exercising muscles release a factor into blood that increases glucose uptake by distant muscles (working or nonworking) has been largely dismissed. Ironically, however, recent evidence (77) provides some support for the view that local factors such as nitric oxide may be responsible, or at least may contribute, to the higher rate of glucose transport in exercising and possibly adjacent muscle, a theory that is not too far removed from that proposed by Goldstein.

The Discovery of Insulin-Independent Glucose Uptake

As described above, evidence from Cochran in 1966 (49), Rasio in 1966 (184), and Hunter in 1968 (119) and their associates indicated that insulin levels decreased in exercising humans even though glucose uptake and utilization were known to be enhanced. The observation that exercise increases glucose utilization in the face of falling circulating insulin levels seemed somewhat paradoxical during this time period. This raised a question: How could increases in glucose transport be facilitated during exercise in the face of a falling insulin concentration? As a result, considerable interest grew in the roles of insulin-dependent and insulin-independent regulation of glucose uptake during exercise. Early kinetic studies in the 1960s by John

Holloszy (1933–), a research fellow at the time in the laboratory of Narahara at Washington University School of Medicine in St. Louis, showed that glucose uptake increases during contraction in isolated frog muscle in a medium devoid of insulin (116). This classic study, and the investigations conducted in diabetic animals by the Goldstein group in Chicago (see above), raised the critical question of whether insulin is required for exercise to increase glucose uptake by contracting muscle. Since that time, other investigators have examined this important question extensively and systematically using in vivo, in vitro and in situ animal models, with somewhat differing results (see below).

Although glucose uptake could be shown to increase in the absence of insulin in situ, the question remained concerning the need for insulin to normalize glucose uptake during exercise in vivo. Using an isolated, perfused rat hindquarter preparation, Ruderman's laboratory in Boston reported that the stimulation of glucose uptake during muscle contraction is markedly diminished in severely diabetic and ketotic rats (19). This group also showed that the addition of a small amount of insulin to the perfusate at a concentration that had no effect at rest restored the exercise-induced increase in glucose uptake to control values. They concluded from this and other experiments that a "permissive" amount of insulin (not present in rats with diabetic ketoacidosis) is required for contraction to stimulate glucose uptake. At the same time, investigators at the University of Toronto were studying how insulin influences glucose utilization in an in vivo dog model. These experiments addressed the requirement for insulin in contraction-stimulated glucose uptake in depancreatized dogs that were totally deprived of insulin but which had normal glucagon levels secreted by the stomach mucosa (238). In these dogs, exercise caused glucose production to increase normally but the glucose metabolic clearance rate did not increase above resting values and hyperglycemia ensued (239). Subsequent in vivo studies conducted in dogs (25, 247) and in humans (55) also indicated that at least a small amount of insulin was necessary to normalize glucose uptake during exercise. These findings supported the hypothesis that insulin has a "permissive" role in glucose uptake into muscle in vivo. Studies using ^{31}P-NMR also indicated that only a small amount of insulin was needed to maintain bioenergetic changes during musclular contraction and recovery in rat muscle (45). Based on these findings, it was thought that exercise must either increase insulin and glucose delivery to the skeletal muscle or increase insulin action or stimulate insulin-independent glucose uptake, or some combination of these, to facilitate glucose uptake. The "permissive" effect of insulin on glucose uptake may, in fact, be critical in counteracting the inhibitory effects of elevated catecholamines on glucose transport in vivo.

In contrast to the above finding that insulin plays a role in exercise-mediated glucose uptake in vivo, a number of investigators during the mid 1980s, including Ploug, Galbo, and Richter working in Copenhagen (178), Nesher and colleagues at Washington University in St. Louis (171), and Walberg-Henriksson working with Holloszy (243), used in situ experiments (isolated and perfused skeletal muscle) to show that muscle contraction stimulates glucose uptake in the total or near-total absence of insulin. These seminal studies were key in shaping our present view that, at least in vitro, contraction-stimulated glucose transport is essentially insulin-independent. Later in the 1990s, studies using the novel NMR ^{13}C glucose method in Gerald Shulman's laboratory at Yale demonstrated the existence of insulin-dependent and insulin-independent mechanisms of glycogen resynthesis in human skeletal muscle postexercise (180).

Over the last several decades, other factors have been identified that influence glucose uptake into exercising muscle and a brief history of these factors is warranted. In addition to insulin, catecholamine, FFA, and muscular contractions per se, glucose concentration also influences its own metabolic clearance rate (MCR). Indeed, the rate of glucose extraction is inversely proportional to glucose concentration so that glucose utilization (MCR × glucose concentration) remains relatively constant throughout a wide range of concentrations. This mechanism was found to be important in preventing excessive glucose entry into muscle during periods of hyperglycemia and allows for the maintenance of glucose uptake during periods of hypoglycemia (72), both of which may occur during exercise. The importance of the muscle's "metabolic state" in the regulation of glucose uptake is evident from the excessive increments in glucose uptake that appear to occur under conditions in which oxygen availability is limited. For example, Cooper, Karl and David Wasserman and colleagues collaborated with Vranic to show that the high rate of glucose uptake during exercise in anemic dogs (246) or in human subjects breathing a hypoxic gas mixture (51) occurs even though insulin levels are unaffected and catecholamine levels, which were known to be antagonistic to glucose uptake, are excessively elevated. A large body of evidence has long suggested that the increase in glucose uptake in these situations is related to changes in the energy state of the muscle cells, as reflected by decreases in concentrations of creatine phosphate and ATP (138). Recent studies from Winder's group at Brigham Young University in Utah (164), Goodyear's group at Harvard (109), and Shulman's group at Yale (21) suggest that this altered energy status may be reflected in changes in the activity of an AMP-activated protein kinase (AMPK). These and other researchers have shown that the activation of glucose transport during exercise may be stimulated by changes in AMPK which are distinct from an insulin signaling pathway (see below).

The findings of Ploug et al. (178), Nesher et al. (171), and others and the development of the AMPK-activated glucose transport theory clearly support the notion that insulin is not needed to mediate the enhancement of glucose transport into exercising muscle. Questions remain on the "permissive" effects of insulin in vivo, however, based on the observations by Vranic and Wrenshall (239), Berger et al. (18), Defronzo et al. (55), Bjorkman et al. (25), Wasserman et al. (247), and others that glucose uptake is not normalized unless some circulating insulin is available. Thus, there has been considerable progress in our knowledge about the role of insulin in glucose transport during exercise. The focus has now shifted toward the identification of the contraction-mediated signaling cascade for glucose transport. Questions also remain about the interaction between insulin-mediated and contraction-mediated signaling transduction during exercise.

Exercise and Molecular Biology of Glucose Transport

The search for the molecular mechanisms for increased glucose transport during exercise was fueled by near-simultaneous discoveries of the insulin-sensitive facilitative glucose transporter (GLUT-4) by David E. James et al. from Boston University (131) and by Maurice J. Birnbaum et al. from Harvard (24) during the late 1980s. The GLUT-4 transporter was shown to be localized in separate and distinct locations either within the cytosol, where it was considered to be inactive, or, on the plasma membrane where it could facilitate the transport of glucose into the cell. It was also shown that insulin caused the translocation of the GLUT-4 from the cytosol to the plasma membrane. The development of a procedure for separating skeletal muscle plasma membrane fractions, together with the use of polyclonal antibodies raised against the purified erythrocyte GLUT, was critical in this discovery. During the late 1980s and 1990s, several investigators looked to see if exercise had an effect on GLUT-4 translocation similar to insulin. An exercise-mediated glucose transport system could explain the earlier observations that exercise increases glucose uptake despite falling or completely absent insulin levels.

Studies published by Holloszy's group in 1989 (60) and by Laurie J. Goodyear, then in Edward Horton's laboratory in 1990 (97), appear to be the first to show that, in rat muscle, exercise causes an increase in GLUT-4 translocation in the absence of insulin. In 1990, Douen and collaborators that included John Holloszy and Amira Klip found that the exercise-induced recruitment of GLUT-4 occurs from a different muscle transporter pool than the insulin-sensitive intracellular pool (61). This observation explained why others had found that exercise and insulin have an additive effect on glucose transport. Exercise-induced GLUT-4 translocation was first described in humans in 1996 by Kristiansen, Hargreaves, and Richter by using sarcolemmal giant vesicles from human muscle biopsy samples obtained before and after exercise (149). The discovery of exercise-induced translocation revolutionized the concept of exercise-stimulated glucose uptake in contracting muscle. Now exercise physiologists had a plausible explanation for increased glucose uptake even if insulin delivery to the muscle is decreased during exercise.

To further develop an understanding of the exercise-activated pathway leading to an increase in glucose transport, investigators have recently focused on 5'AMP-activated protein kinase (AMPK). The first indication that AMPK could be involved in regulation of muscle metabolism came form a study published by Winder and Hardie in 1996 (256). AMPK activity was elevated by two- to threefold within 5 minutes of the start in the quadriceps muscle of running rats. It was hypothesized soon after by Winder's group that AMPK activation by muscle contraction stimulated both an increase in fatty acid oxidation and an increase in glucose uptake to meet the energy demands of working muscle (164). Later publications by Winder and his colleagues showed that the increase in AMP activity was dependent on work rate and that the activity remained elevated for several minutes following exercise. Using their gastrocnemius in situ model, Ruderman's group (231) first demonstrated that muscle contraction maximally activated the α-2 isoform of AMPK within 30 seconds of the beginning of electrical stimulation. Later collaborations between

Winder and Goodyear and their colleagues (152) showed that the perfusion of rat hindlimb skeletal muscle with 5-aminoimidazole-4-carboxamide ribonucleoside (AICAR), a cell-permeable precursor of the AMPK activator, increased glucose uptake due to an increase in the translocation of GLUT-4. A combination of maximal AICAR and insulin was partially additive, whereas no additive effect on glucose transport was seen with the combination of AICAR and contraction, suggesting they act by a common mechanism. Further evidence for AMPK as a potential trigger for exercise-induced GLUT-4 translocation has come from near-simultaneous publications from Goodyear's (108) and Shulman's (21) laboratories in 1999. Recent studies indicate that the α-2 isoform of AMPK is activated in human muscle during heavy exercise (70% $\dot{V}O_2$max) when muscle creatine phosphate and glycogen are significantly decreased (78). Ongoing studies will likely elucidate the mechanism by which AMPK during exercise increases glucose transport within the decade.

The Glucose Fatty Acid Cycle

The identification of the glucose–fatty acid cycle by Philip Randle (1926–) and colleagues from Cambridge in 1963 was instrumental in developing the understanding of the interaction between carbohydrate and lipid metabolism during exercise. Randle's idea was actually initiated by attending a CIBA Foundation conference in 1952 in which Drury and Wick presented a paper showing that acetate and 3-hydroxybutyrate decreased $^{14}CO_2$ production from ^{14}C-glucose in eviscerated–nephrectomized rabbits despite the presence of insulin. Randle, then an intern interested in diabetes and carbohydrate metabolism, was invited to join the Department of Biochemistry at Cambridge by the chair of the department, Frank Young, who was also at the conference. Studies initiated by Randle at Cambridge in 1955 initially showed that acetoacetate (albumin-bound nonesterified fatty acids [NEFA] were not yet discovered) decreased glucose uptake by 50% in the presence of insulin, but not in the absence of the hormone, in rat diaphragm in vitro. Although the study was published in the proceedings of a conference in Carmel, California, in 1956, it did not get widespread attention because the proceedings were only circulated among the conferees. Randle did not follow up this line of research until the early 1960s, soon after albumin-bound NEFAs in the circulation were discovered. In 1963, Randle et al. (183) showed, in their widely cited study, that fat availability significantly inhibits glucose uptake and oxidation acutely (within minutes) in resting heart and diaphragm muscle (i.e., glucose–FFA cycle). The decrease in glucose oxidation was attributed to inhibition of pyruvate dehydrogenase by an increase in mitochondrial acetyl CoA. It was proposed that the decrease in glucose uptake by the tissues was caused by inhibition of glycolysis at phosphofructokinase due to the accumulation of citrate in the cytosol. The interaction of glucose and free fatty acids in skeletal muscle may be different from what was proposed for heart and diaphragm tissues and the interaction between glucose and fat during exercise has been the topic of much debate over the past 40 years.

In 1977, Michael Rennie et al. (187) in Holloszy's laboratory reported that the free fatty acid oleate had an inhibitory effect on glucose uptake and glycogen break-down at rest and during electrically induced contraction in a rat hindquarter preparation. Others, using similar experimental preparations had failed to observe this effect (19). Since the mid-1970s, other animal experiments have both confirmed and refuted the finding that FFAs decrease glucose uptake in skeletal muscle during exercise. After discussions of the glucose–fatty acid cycle with Randle, during his time at Oxford, Vranic tested the hypothesis that the impaired glucose uptake during exercise in insulin deficiency was the result of elevated FFA oxidation by using an inhibitor of FFA oxidation (methylpalmoxirate) in insulin-deficient dogs. Although inhibition of FFA oxidation in these dogs caused a reduction in glucose production during exercise, it did not enhance glucose utilization (262). In a follow-up experiment, however, in the presence of small amounts of insulin, inhibition of FFA oxidation increased glucose uptake and clearance in exercising dogs (207), thereby suggesting that in the presence of small amounts of insulin, increased FFA oxidation lowered glucose utilization, likely as a result of the glucose–fatty acid cycle.

In humans, although FFAs appear to influence carbohydrate utilization, the role of the glucose–fatty acid cycle remains controversial. In 1991, Boden et al. (31) reported that FFA elevation diminishes muscle glucose uptake during a euglycemic hyperinsulinemic clamp in human subjects at rest, but they also showed that this was not associated with an increase in citrate or with the stimulation of glycogen synthesis, as was originally described by Randle. A study published in 1986 by Horton's group showed a sparing effect of elevated FFA levels on muscle glycogen and a reduced rate of total carbohydrate oxidation over 30 minutes of moderate-intensity exercise in human subjects (185). More recently, David Dyck and collaborators, including Lawrence Spriet at the University of Guelph and Erik Hutman from the Karolinska Institute in Stockholm, provided evidence that the classic Randle cycle does not occur in humans, at least during intense exercise (63). They found that although high FFA provision increases lipid utilization, it does not alter muscle acetyl-CoA or citrate levels or influence pyruvate dehydrogenase activity (63). Collectively, these findings, which have been confirmed by others in human and in rat models, suggest that an increase in plasma FFA can cause an inhibition of insulin-stimulated glucose uptake in skeletal muscle both at rest and during exercise, although the proposed mechanisms may not be as originally proposed by Randle.

The Malonyl CoA Fuel-Sensing and -Signaling Mechanisms

Beginning in the early 1990s, interest in how insulin and exercise regulate fatty acid oxidation in muscle was renewed, in great measure, by studies of the molecule malonyl CoA. Malonyl CoA is an intermediate in the de novo synthesis of fatty acids and, as originally demonstrated by Denis McGarry and coworkers in 1978 (163), an inhibitor of carnitine palmityl transferase (CPT1), the enzyme that controls the transfer of long-chain fatty acyl CoA (LCFA CoA) from the cytosol into mitochondria.

Thus, decreases in malonyl CoA concentration should enhance fatty acid oxidation while increases in this molecule should have the opposite effect. Studies by Winder and coworkers (256) in 1996 were the first to demonstrate that malonyl CoA levels in rat muscle decrease acutely (within minutes) during exercise, a finding consistent with its increase in fatty acid oxidation. One year later, investigations in the laboratories of Winder (164) and Ruderman (231) established that this decrease was due to the phosphorylation and inhibition of the acetyl CoA carboxylase isoform present in muscle (ACCβ) by AMPK. At the same time, Ruderman's group demonstrated that increases in insulin and glucose increase the concentration of malonyl CoA in muscle in vivo (200). Interestingly, this group found that this was not due to a decrease in the phosphorylation of ACCβ but was attributable to an apparent increase in the cytosolic concentration of citrate, an allosteric activator of ACC and a precursor of its substrate, cytosolic acetyl CoA. They also found that activation by citrate is substantially dampened, although not completely lost, when ACCβ is phosphorylated by AMPK during muscle contraction (231). From these investigations, the metabolic events initiated by exercise appeared to take precedence over, or at least diminish, those due to enhanced glucose availability. The finding that glucose deprivation, like exercise, acutely lowers the concentration of malonyl CoA and that inactivity (muscle denervation), like insulin and glucose, increases it has led to the hypothesis that malonyl CoA is a component of a fuel-sensing and -signaling mechanism in which ACCβ is the sensor (in response to changes in citrate or AMP) and malonyl CoA is the effector.

In contrast to the studies conducted on rats, the role of changes in malonyl CoA in regulating fatty acid oxidation during exercise in human muscle is somewhat unclear. For example, in response to low- and moderate-intensity exercise, human skeletal muscle malonyl CoA content was shown by Odland et al. (176) in 1996 to be largely unchanged in the face of an increase in fatty acid oxidation. In contrast, small but significant decreases in malonyl CoA, accompanied by a marked decrease in acetyl CoA carboxylase activity attributable to AMPK activation, was observed in a more recent study jointly carried out by Ruderman and Richter and their coworkers (54). Further research on malonyl CoA regulation in human muscle will provide further insight into this issue.

METABOLIC LESSONS LEARNED FROM STUDIES CONDUCTED ON EXERCISE AND DIABETES

Our existing knowledge of the role of the pancreatic hormones in glucoregulation and substrate utilization during exercise has evolved primarily from a number of pioneering studies investigating the acute responses to exercise in human and animal models of type 1 diabetes mellitus. This fact should not be surprising, since much of what is understood regarding the endocrine glands and metabolism has been gleaned from investigations in which the endocrine gland has been removed or in which the influence of the secreted hormone is blocked pharmacologically. Not long after the discovery of insulin, diabetologists found that exercise, together with subcutaneous insulin administration, causes a dramatic drop in blood glucose levels resulting in hypoglycemia and even coma (155). Exercise physiologists were subsequently interested in the role of the pancreatic hormones in glucose homeostasis and substrate flux during physical activity.

Hypoglycemic Effects of Exercise

In 1919, the Harvard-trained diabetologist Fredrick Allen (1879–1964) and his colleagues at the Rockefeller Institute in New York City demonstrated that exercise could cause either an increase or a decrease in blood glucose levels in patients with diabetes mellitus (5). Based primarily on the observations of patients with diabetes during the turn of the nineteenth century, the concept developed that a secretion from the pancreas could influence glucose production rate and glucose uptake during exercise. After insulin was discovered in 1921, many researchers were interested in how the interaction of insulin and exercise affected blood glucose concentration. Arguably, the most dramatic of these observations was that of the diabetologist Robin D. Lawrence from King's College Hospital, London, in 1926 (155). Having diabetes himself, Lawrence observed that exercise not only enhanced the lowering of his blood glucose levels by insulin but also could induce severe hypoglycemia. He concluded that exercise enhances the effect of subcutaneously injected insulin and suggested measures to avoid low blood glucose during exercise (e.g., prior carbohydrate administration, diminution of insulin dosage) that are still the foundation of clinical management for patients with type 1 diabetes. Interestingly, exercise-induced hypoglycemia in nondiabetic men who had just run 26 miles in the Boston Marathon had been described 2 years earlier by Samuel A. Levine and his colleagues in Boston (154). They found a close correlation between the condition of the runner at the finish of the race and the level of his blood glucose. In particular, they noted that those individuals with extremely low glucose presented a picture of shock not unlike that produced by an overdose of insulin. The group from Boston were perhaps the first to suggest that adequate ingestion of carbohydrate before and during prolonged and "violent" muscular effort would be of considerable benefit in preventing the hypoglycemia and accompanying development of the symptoms of exhaustion.

The 1978 Conference on Diabetes and Exercise

The 1970s were a renaissance period in the study of endocrinology, diabetes metabolism, and exercise. In particular, major strides were made in studying the effect of the pancreatic hormones on metabolism during exercise and the role of physical activity in the management of diabetes. It was in this setting that the first conference on diabetes and exercise (236) was held to review the current state of knowledge. We will briefly highlight the main findings of the first conference related to exercise and metabolism, which were followed by important advances by several of the participants and their colleagues over the next 20 years.

Erik Newsholme from Oxford, a colleague of Randle's, presented the following questions, related to exercise and control of fuel metabolism, at the conference: (172): (1) How is fuel mobilization from the body's stores regulated to tally with fuel utilization? (2) How is fuel oxidation in muscle regulated precisely to fit the energy requirements of the contractile system? and (3) How is the utilization of glucose and alternative fuels by exercising muscle coordinated to both the availability of these fuels and the fuel requirement of other tissues? In addition to these questions, theoretical aspects of the precision of metabolic control, in the hormonal milieu of exercise, and the importance of the sensitivity of near-equilibrium reactions in controlling substrate flux were discussed. Setting the stage for discussion of the regulation of substrate utilization during exercise was the "glucose–fatty acid cycle" concept (183). As described above, the glucose–fatty acid cycle hypothesis has generated a large number of studies attempting to elucidate the mechanisms involved in carbohydrate and fat interaction during exercise.

At the time of the 1978 conference, there was no information about insulin- or exercise-induced signaling. It was proposed, however, by Ruderman and Berger and their colleagues that pyruvate dehydrogenase played an important role in regulating glucose oxidation in response to insulin and exercise (19). It was also emphasized at the time by Ruderman et al. (199) and by Saltin et al. (202) that physical training in type 2 diabetic subjects enhances their ability to dispose of an intravenous glucose load. This was perhaps a key stimulus for subsequent investigations on the role of exercise training in insulin sensitivity and diabetes management. An important pioneering methodological advance in the assessment of substrate turnover during exercise was presented at the 1978 conference by Bella Issekutz from Dalhousie University. Issekutz and colleagues had shown that by simultaneously using different tracers, it was possible to study glucose and free fatty acid turnover and muscle glycogenolysis in exercising dogs (127). As described above, these techniques eventually led to our present knowledge of the interaction between insulin, glucagon, and catecholamines in regulating hepatic glucose production and peripheral glucose utilization during exercise. Based on their early experiments, Vranic and Kawamori presented their hypothesis detailing the relationship of *hypo-* and *hyper*insulinemia in inducing differential changes of glucose clearance and production in diabetes (139, 263). Those studies had shown, for the first time, that exercise-induced hypoglycemia, which occurs primarily in individuals with type 1 diabetes, is related to enhanced absorption of subcutaneous insulin and consequently to a reduction of glucose production and a further enhancement of glucose utilization. Lars Hagenfeldt from the Karolinska Institute in Sweden (100) presented a paper showing that exercise increases the net uptake of FFA by muscle in nondiabetic and diabetic subjects, primarily by increasing blood flow, but also by increasing the plasma FFA concentration, a finding that was suggested in his collaborative studies with Wahren (242). This overview by Hagenfelt was important in stimulating research on the mechanisms for increased FFA delivery and uptake during exercise, another area of current investigation.

The pioneering investigations illustrating that glucose uptake occurs in the face of extremely low insulin levels (see Discovery of Insulin-independent Glucose Uptake), were a focus of discussion at the conference and were undoubtedly an important impetus for subsequent investigations of the mechanisms regulating glucose transport in contracting muscle. More importantly, we believe that this conference was the first to unite study of the endocrine system with that of fuel metabolism during exercise and undoubtedly has helped to form the current explosion of studies on exercise metabolism in health and disease. We also feel that the first conference on exercise and diabetes was a major stimulus for clinical investigations demonstrating that exercise and lifestyle modifications can not only improve type 2 diabetes control but can prevent or delay the onset of the disease.

Exercise and Insulin Sensitivity

The theory that chronic activity enhances insulin action emanated from the studies of Per Björntorp (1931–) and his coworkers from the University of Göteborg, Sweden, in the early 1970s. In a seminal study published in 1972 (27), they compared glucose tolerance and plasma insulin levels in sedentary and active Swedish men. The results of glucose tolerance tests were clearly better in the active men, despite considerably lower plasma insulin levels, strongly suggesting that regular exercise elevates insulin sensitivity. It is interesting to note that abdominal fat, whose relationship to insulin resistance had not yet been appreciated, was not measured in the subjects; however, fat cell size, which tends to reflect abdominal fat mass, was more than twice as great in the sedentary group. Thus, to what degree these observations were attributable to differences in adiposity rather than to differences in physical activity could not be ascertained.

To some extent, the independent effects of exercise on insulin sensitivity were addressed in other studies by Björntorp and coworkers in which the same individuals were evaluated before and after a brief period of physical training. In the earliest of these investigations, published in 1970 (26), they observed a substantial decrease in plasma insulin concentration in severely obese nondiabetic subjects following a 6-week exercise program; however, no difference in glucose tolerance was observed. It was concluded that physical training "exerts a lowering effect on insulin concentrations in plasma in obesity, probably reflecting an increase in insulin sensitivity." Björntorp's investigations, in turn, have led to more recent efforts to examine how exercise increases insulin sensitivity.

After preliminary studies in the 1970s suggesting that regular exercise improves glucose tolerance in individuals with impaired glucose tolerance or type 2 diabetes (199, 202); but that the effect lasted no more than a few days (199), Ruderman's laboratory carried out a series of investigations that explored the effect of a single bout of exercise on insulin sensitivity in skeletal muscle. In a seminal study published in 1982, Richter, then a postdoctoral fellow in this group, reported that in rats, prior running (30 minutes at a moderate intensity) allowed a physiological concentration of insulin (75 μU/ml) to cause a fourfold-greater stimulation of glucose uptake into glycogen-depleted type IIA muscle (193). Later, studies from this group established that following more intense exercise in rats, glycogen depletion and increased insulin sensitivity occurred in both type IIA and IIB muscle (91). These effects of prior exercise, which lasted 4–24 hours, could be reproduced in muscles that were made to contract by electrical stimulation but not in resting contralateral muscles, indicating that the effects were mediated by local, rather than systemic, factors (194). As noted elsewhere in this chapter, the increase in insulin-independent glucose transport caused by muscle contraction is now attributed to AMPK activation. The basis for the increase in insulin-stimulated glucose uptake after a single bout of exercise is still unknown, as neither alterations in insulin receptor tyrosine kinase activation (229) nor later steps in the insulin signaling cascade have been consistently observed.

Although not directly germane to a historical review of exercise and the endocrine system, we would be remiss in not noting the implications of the above studies of acute and chronic effects of regular exercise on insulin sensitivity. In brief, they have provided the physiological underpinning for the use of exercise in the treatment and prevention of type 2 diabetes and other diseases characterized by insulin resistance. Of particular note is the recent demonstration that lifestyle changes that include exercise or exercise and diet, which also increase insulin sensitivity, prevent or at least substantially delay the progression of individuals with impaired glucose tolerance to overt type 2 diabetes (230).

ADRENAL HORMONES

Catecholamines

Catecholamines and Acute Exercise

The activation of the sympathoadrenal system during muscular exercise was first established in the early part of the last century. Indeed, a rise in catecholamine activity with exercise was first suggested in 1922 by Frank A. Hartman et al. (106) from the University of Buffalo as a result of examination of the pupils in exercising cats. Hartman, a former teaching fellow from Harvard who had worked with Walter B. Cannon (1871–1945), was a world-renowned physiologist with expertise in stress responses. Together, Hartman and Cannon were primarily interested in the roles of the adrenal medulla and sympathetic nervous system in the stress response. Based on the French physiologist Claude Bernard's concept of consistency in the "milieu interieur," Cannon created the term "homeostasis," which he defined as the integrated sum of the body's adaptations to maintain or restore consistency in the internal milieu. He and his colleagues appreciated that the sympathoadrenal system was critical in homeostatic regulation, particularly during an emergency reactions such as intense muscular exercise. Two years after Hartman's study, Cannon included exercise along with pain, hunger, fear, and rage into his theory of the emergency function of the adrenal medulla (41).

In the 1950s, research by the Nobel laureate Von Euler and Hellner demonstrated that urinary excretion of noradrenaline (norepinephrine) and adrenaline (epinephrine) increase during muscular work (235). Later, when technological advances made it possible to obtain fluorimetric assays for the hormones, measurements of increased catecholamines in blood were made by a number of investigators such as Gray and Beethan in 1957 (98). They were the first to show that exercise (running) increases catecholamine levels in the blood of humans within minutes of the onset of the activity. A more detailed description of the history of measurements of the catecholamines in humans and in animal models can found in Chapter 5.

Roles of the Catecholamines in Exercise Metabolism

Although the catecholamines were known to play critical roles in cardiovascular and respiratory responses to exercise and were thought to be under the influence of a number of factors such as oxygen availability, body temperature, hydration status, etc., the influence of the catecholamines on metabolism during exercise has been of considerable debate. From studies in Dill's Harvard Fatigue laboratory in the 1930s, it was known that epinephrine elevates blood glucose concentration during exercise in men (57). In the 1970s, Banister and Griffiths postulated that epinephrine augments cyclic AMP levels in liver and in skeletal muscle during exercise and, as a result of this, it has a regulatory influence on hepatic glycogenolysis, hepatic gluconeogenesis, and adipose tissue lipolysis (14). These functions were proposed to be important since they served to maintain or elevate plasma levels of glucose and free fatty acids, both of which were known to be key substrates for exercising muscle.

In 1975, using the recently developed radioenzymatic method (144), Galbo, Holst, and Christensen (83) showed that during short-term graded exercise both plasma norepinephrine and epinephrine increase in parallel with changes in oxygen uptake, while blood glucose remains unchanged. Based on these data and their failure to observe changes in insulin or glucagon during short-term exercise, they proposed that the rise in catecholamines was instrumental in maintaining the blood glucose concentration. The link between blood glucose and catecholamines was also strengthened by reports that if the drop in blood glucose was made more pronounced during exercise by having subjects exercise after carbohydrate stores had been depleted or during pharmacological inhibition of lipolysis, the magnitude of the increase in adrenaline was exaggerated (85). Convincingly, this group, which included Niels J. Christensen, who carried out extensive research related to the causes of hypoglycemia, also demonstrated hat the maintenance of blood glucose levels by a glucose infusion significantly attenuated the increase in plasma concentration of epinephrine and, to some extent, that of norepinephrine during exercise (81).

The concept that catecholamines contribute to hepatic glycogenolysis was further developed and supported in subsequent investigations, primarily by the group in Copenhagen. Some questions arose in connection with their theory because the activation of adrenaline was evident without any detectable change in glucose concentration. The group thought, however, that the blood glucose trigger for epinephrine secretion was far more sensitive to decreases in blood glucose levels during exercise than it was at rest (79). Winder and others later demonstrated that epinephrine was unessential for hepatic glucose production during moderate-intensity exercise (260) (see Summary), although it may play a role in inducing and increase in cAMP and fructose-2,6-bisphosphate (F2,6P2) in nonexercising muscle, which allows for lactate production and increased gluconeogenesis by the liver (253). Other hormonal factors regulating blood glucose levels, such as insulin and glucagon, were also thought to be important and were being investigated by researchers in North America. Thus, the battle of the role of the circulating catecholamines and hepatic nerves on hepatic glucose production during exercise began, a topic that is still currently debated (see below).

The role of catecholamines in stimulating muscle glycogenolysis and generating lactate has been investigated since the 1920s. Early studies in Dill's laboratory with dogs (57) and humans (8) showed that epinephrine administration enhances glycogenolysis and increases plasma lactate concentration. Indeed, as pointed out by Galbo (79), activation of the sympathoadrenal system parallels increased utilization of muscle glycogen during incremental exercise, both of which may be modulated by such factors as hypoxia, training, and the administration of glucose. In support of this hypothesis, Struck and Tipton (215) in 1974 showed that rat adrenodemedulation combined with chemical sympathectomy decreases exercise-induced muscular glycogen depletion. Using isolated muscle preparations and β-adrenergic blockade, Nesher et al. (170) provided additional evidence in 1980 that catecholamines facilitate muscle glycogenolysis during muscle contraction. Stemming from these observations, a series of experiments led in the 1980s by Richter et al. (191, 192), using in vivo and isolated perfused rat hindquarter preparations, has shown persuasively that adrenolmedullary control via the catecholamines influences glycogenolysis in exercising muscle by increasing phosphorylase A activity through increased cyclic AMP production.

Since the 1950s, catecholamines were known to activate adipose tissue lipolysis through accumulation of cAMP and activation of hormone-sensitive lipase. It was conceivable to many physiologists, therefore, that increases in catecholamines during exercise resulted in an increase in lipolytic rate, which would be important in energy provision. In the mid-1960s at the Karolinska Institute in Sweden, Sune Rosell and Kathryn Ballard conducted investigations using anesthetized dogs and electrical stimulation to show that sympathetic nerve activity is effective in producing lipolysis in subcutaneous and omental adipose tissue, but not in mesenteric adipose tissue, at activity levels thought to occur under normal exercising conditions (197). In a follow-up experiment, they found that the infusion of physiological levels of catecholamines did not produce significant lipolysis in subcutaneous or mesenteric adipose tissue but did mobilize lipid from the omental region (11). These investigations were important in indicating that regional differences exist in the mechanisms for lipid mobilization during exercise.

The importance of circulating catecholamines in mobilizing lipid was also illustrated in experiments conducted by Maling and colleagues at the National Institutes of Health in Bethesda. They reported in 1966 (160) that the increase in FFA levels during exercise is delayed in exercising adrenaldemedullated rats. Philip Gollnick's laboratory in Pullman, Washington, in studies published in the 1970s (96, 205), also found that compared with sham-treated animals, adrenodemedullated or chemically sympathectomized rats had lower concentrations of plasma FFA during exercise. Lefebvre's laboratory in Belgium had similar findings in immunosympathectomized animals (159) as did Nazar and colleagues in 1971 (168) and Issekutz in 1978 (126) using β-adrenergic receptor blockade in exercising dogs. In Issekutz's study (126), β-receptor blockade dramatically reduced the working capacity of dogs, which he attributed to the drop in blood FFA levels and a more rapid reduction in blood glucose levels due to enhanced glucose utilization. Similar effects on FFA mobilization were observed in humans given β-blockade in studies published in the 1970s by Hansen (102), Galbo et al. (85), and others. Adrenal catecholamines were not shown to be mandatory for increased lipolysis, however, as adrenalectomized humans were shown to have increases in FFA levels during prolonged exercise (15).

Catecholamines have also been proposed to play a role in the activation of muscle triglyceride lipase, as was first demonstrated in swimming rats in 1978 by Gorski's laboratory (212). Karlsson's group in Sweden (157) confirmed a few years later that catecholamines activate lipolysis in muscle tissue in humans, which is thought to be important in maintaining fuel provision.

Lundborg et al. (158), working with Smith and others from the University of Göteborg and Sahlgren's Hospital in Sweden, suggested that the decrease in exercise tolerance associated with propranolol, or with metoprolol (a selective β-adrenoceptor antagonist) was related to their influence on carbohydrate and lipid metabolism. In support of this proposition, they showed that plasma free fatty acid levels were greatly diminished by propranolol, making the body more dependent on carbohydrate as a fuel. More details of the development of our knowledge on the effects of adrenoceptor blockade and catecholamines on metabolism are found in the Summary at the end of the chapter.

Catecholamines and Physical Training

A more detailed discussion of training-induced alterations in the catecholamine response to exercise is found elsewhere in this text (see Chapter 5). As already noted, it had been demonstrated in the 1970s that physical training modifies the catecholamine response to exercise (90, 104, 105, 255) and that this may play a role in the altered metabolic response to a given workload in trained individuals. Winder, who had been a postdoctoral fellow of Holloszy's from 1971 to 1974, proposed in 1982, when he was at the University of South Dakota, that the attenuation in catecholamine response to exercise caused by training may be a result of higher liver glycogen levels in the trained animals (250). However, differences in the stress hormone responses (glucocorticoids, epinephrine, norepinephrine) to exercise between trained and untrained rats persisted when both groups exercised in the fasted state when liver glycogen levels were presumed to be depleted. Thus, differences in cate-cholamine responses between trained and untrained organisms were not the result of differences in liver glycogen levels. Their findings were important, however, since they also showed that training lowered the reliance on blood glucose as a fuel and re-sulted in improved blood glucose homeostasis during exercise (250). In addition, the attenuation in stress hormone response with training could be attributed to the higher blood glucose levels in the trained compared with untrained state (250). An-other key study published in 1983, after Winder moved back to Brigham Young Uni-versity in Provo, Utah, illustrated that liver norepinephrine content decreased sig-nificantly in response to acute exercise in untrained rats but failed to decrease in trained rats (251). From this, it was concluded that the liver sympathetics are acti-vated in response to exercise and hypoglycemia and that endurance training causes a reduction in the degree of exercise-induced activation of these neurons.

Glucocorticoids

Physiologists have been interested in the metabolic role of the adrenals during mus-cular work since the physician Thomas Addison (1793–1860) from Edinburgh con-ducted his pioneering investigations on the adrenal glands in the nineteenth century (1). There was little doubt that the adrenal cortex was important for general metab-olism since laboratory animals often failed to survive after adrenalectomy. Indeed, it was initially difficult to determine the specific role of the secretions from the adrenal cortex since the mortality rate was so high in animals who underwent adrenalec-tomy. Although the glucocorticoids were known to be critical in increasing physical working capacity, the specific role of these hormones was largely speculative until the nineteenth century (169). It was first suggested by physiologists such as Charles-Edouard Brown-Séquard (1817–1894) of London that the function of the adrenal cortex was related to detoxification of toxic metabolites of muscular activity (37). Clues regarding the metabolic implications of the extracts did exist, however, since Addison and others found that patients with adrenocortical insufficiency had pro-found muscular weakness and were intolerant to exercise. At the turn of the nine-teenth century, determining and isolating the full complement of the adrenal cortical extracts, which appeared to have effects on the electrolyte balance (i.e., mineralocor-ticoids) and metabolism (i.e., glucocorticoids), remained difficult. By the 1930s, how-ever, isolation and separation of the cortical extracts made it possible to differentiate the metabolic effects of these hormones (169).

A series of eloquent studies initiated by Dwight J. Ingle (1907–1978) in the 1930s (Mayo Foundation in Rochester, Minnesota) and 1940s (University of Penn-sylvania and Upjohn Company, Michigan) established that the exercise capacity of an adrenalectomized rat is dramatically reduced, whereas partial adrenalectomy or destruction of the adrenal medulla has almost no effect (120). Ingle, who was inter-ested in both animal behavior and endocrinology, acquired a number of relatively novel surgical techniques such as adrenalectomy, thyroparathyroidectomy, and hy-pophysectomy during his master's studies in psychology at the University of Idaho.

After moving to the University of Minnesota in the early 1930s, he began using a protocol of faradic stimulation of the load-bearing gastrocnemius muscle of rats to study the effects of a number of endocrine and metabolic variables on fatigue. The protocol, which developed into the Ingle "work test," was critical in studying the effects of adrenalectomy on the loss of muscular work capacity in rats. In a series of convincing studies, Ingle used autotransplants of adrenal cortical tissue to illustrate that work fatigue was a result of an absence of cortical tissue per se and not a consequence of surgical shock or absence of the adrenal medulla (120). He showed persuasively that glucocorticoid replacement can fully restore physical working capacity in adrenalectomized animals, whereas mineralocorticoid or adrenaline replacement has no effect on exercising rats. Ingle also noted that liver glycogen and blood glucose levels were lowered in adrenalectomized laboratory rats who had survived surgery. During his 4 year stay at the Mayo Clinic between 1934 and 1938, Ingle published more than 20 papers on this topic. Ingle's interest in the adrenal expanded when he went to the University of Chicago in 1953, and he made important contributions concerning the role of adrenal hormones in such areas as "steroid diabetes," cancer biology, and the response to severe stress.

The endocrinologist Hans Selye (1907–1982) was another major contributor to our knowledge of the role of the adrenals in exercise. Selye, who was originally from Prague, worked at McGill University in the 1930s and then at the University of Montreal from 1945 until his retirement in 1976. In his textbook *Stress* published in 1950 (204), he wrote that "intense muscular exercise has been one of the agents routinely employed in the earliest experiments concerned with the production of (stress-related) changes in the adrenal." In support of his theory, Selye cited evidence—from a vast number of histological experiments—that adrenal hyperemia of the principal glucocorticoid secretory portion of the gland, the zona fasciculate, had depleted cortical lipids and cholesterol content as a result of muscular exercise. He also noted that physical training causes hypertrophy of this area, with increases in both the number of cortical sinusoids and their blood content. Even more impressive, Selye realized that systemic effects of muscular exercise had a "great similarity to the alarm reaction" but that "the systemic consequences of exercise differ from those produced by other alarming stimuli." Since Selye's early papers, exercise physiologists, with the help of more direct measures of adrenal activation, have attempted to identify the effect of the adrenal hormones on the exercise response.

After Selye and Ingle found that cortisol, along with catecholamines, prepares an animal to confront a given stressful situation, the influence of exercise on activation of the hypothalamo–hypophysial–adrenal axis became of major interest. In studies conducted in the 1950s, changes in the levels of circulating lymphocytes or eosinophils, as a result of physical activity, were taken as indicators of adrenocortical function. Results with respect to plasma and urine levels of adrenal metabolites were inconsistent and did not lead to any conclusions regarding the effects of exercise on the corticosteroids (226). The development of techniques that allowed measurement of free cortisol levels in urine in the 1960s was an important advancement since they were considered to be an index of adrenocorticoid secretion rate and reflective of the biologically active plasma fraction. Better still was the measurement of plasma cortisol turnover during exercise in humans, which became possible by the 1970s, as will be discussed later. Investigators such as Tharp (226), Thorn (111), Davies and Few (53), and Sutton (219) reconciled many of the earlier inconsistencies by establishing that exercise intensity and anxiety levels influenced blood cortisol response and that long-term training often produces an initial increase and a later decrease in cortisol levels at a given exercise intensity (see below).

Glucocorticoids and Endurance Performance

As mentioned above, Ingle was a pioneer in examining the role of the glucocorticoids in exercise performance. Using his work test protocol, he showed that adrenalectomized rats had greatly diminished work times compared to control rats and that adrenalectomized rats who were given injections of cortical extract had significant delays in fatigue rates (120). The effect of glucocorticoids in delaying fatigue was impressive, with increased work times that ranged anywhere from 15 to 145 hours, depending on the amount of extract given. Later, Ingle's laboratory reported that injections of either cortisone or glucose, and less so aldosterone and epinephrine, had restorative power on work performance (121). From these studies, Ingle hypothesized that glucocorticoids influence performance by acting on substrate mobilization.

In 1974, Struck and Tipton (215) published an investigation showing that, compared with control rats, adrenalectomized rats had lower muscle glycogen utilization and reduced endurance performance. Also of historical importance, Sellers et al., in Winder's laboratory, reported in 1988 (203) that adrenalectomized rats have a reduced exercise capacity that could be improved by glucocorticoid treatment. Observations have not always been consistent in this regard, however, particularly during comparatively light-intensity exercise (228). It would appear therefore that the glucocorticoids influence exercise capacity, possibly through substrate mobilization, at some but not all work intensities.

Glucocorticoids and Acute Exercise

Eleanor H. Venning and her colleague V. Kazmin from McGill University Clinic at the Royal Victoria Hopital in Montreal appear to have been the first to demonstrate (in 1946) that urinary corticosteroid excretion increases during muscular exercise (232). An apparent conflict arose soon after, however, when George W. Thorn and his colleagues (227) found no change in urinary 17-hydroxycorticosteroids excretion in untrained men as a result of running (12 miles in 4 hours). These discrepancies likely occurred because indirect and inappropriate markers of adrenocortical activity were used. S. Richardson Hill Jr., working with Thorn in Boston, published in 1956 "Studies on Adrenocortical and Psychological Responses to Stress in Man" (111), which showed that glucocorticoid levels increase significantly in race and time trial days compared to practice days in oarsmen of a college crew even though the muscular work accomplished is comparable. This important publication inspired a number of later investigators, including Connell and Redfearn (1950s), Steadman and Sharkey (1960s), and Sutton and Casey (1970s), to attempt to separate the effects of emotional and physical stress on glucocorticoid secretion. The results of these studies are consistent in showing that the emotional stress of competition appears to be responsible, at least in part, for the increase in glucocorticoid levels during the activity.

The glucocorticoid response to acute exercise continued to be a major focus of investigation in a number of laboratories in the 1960s and 1970s. The Belgian researcher A. Cornil in 1965, and others (52) from Brussels University, appear to have been the first to measure plasma cortisol levels during muscular exercise in humans. Their work was stimulated by the publications from Thorn's laboratory reporting that emotional factors increased adrenal cortex activity. Unfortunately, they found that in untrained men, cycle ergometry exercise (100 W for 20 minutes) caused a significant fall in plasma cortisol levels, which may have been particularly misleading at the time. Other groups such as Davies and Few (53) and Sutton and Casey (219), however, soon found that a rise in cortisol concentration could occur within minutes of cycling at slightly higher intensities. The increase in concentration during exercise was first attributed to an elevated secretion rate in man in 1974 by Few (69), a finding that had been reported in dogs a few years earlier (222). In a follow-up publication (53), Davies and Few showed that the relative intensity was a critical factor in determining the glucocorticoid response to exercise.

Research continued to clarify the role of exercise intensity and duration on glucocorticoid concentration in plasma during the mid-1970s. For example, Jan Häggendal from the University of Goteborg, Sweden, in a study published with Saltin (101), confirmed that a threshold intensity exists for cortisol response and that relative exercise intensity is important in comparing individuals. Davies and Few also suggested that a workload of 60% $\dot{V}O_2$max was the critical value, above which a rise in plasma cortisol occurred in humans performing treadmill running. Others such as Viru (233), Bloom et al. (30), and Sutton and coworkers (221) supported Davies and Few's theory that moderate-to-severe exercise was necessary to increase blood cortisol levels in humans and that the activity of the pituitary–adrenocortical system is a good indicator of effort during exercise. It was clear, therefore, that short-term exercise at a low or moderate intensities did not increase pituitary–adrenocortical activity and blood cortisol levels, observations that had been made in previous years by others (226). An important observation was made in 1976 when Bonen found that urinary free cortisol excretion rate, which depends upon the concentration of free cortisol in the plasma, is increased after heavy exercise but not after mild exercise (33). In that study, he also showed that cortisol excretion correlates positively with % $\dot{V}O_2$max once an intensity threshold has been surpassed.

A number of researchers in the 1960s and 1970s, including Atko Viru and his colleague Akke (234) from Tartu State University in Estonia, reported that although plasma glucocorticoid concentrations increase during exercise, levels may be decreased at exhaustion. The decrease at exhaustion reported both in humans and in rats was initially interpreted by Frenkl and Csalay (74) as being caused by a limitation in adrenal secretion through the exhaustion of the gland itself. However, a study published in 1969 by Viru and Akke (234), in which adrenocorticotropic hormone (ACTH) injected at the end of exhaustive exercise caused increased glucocorticoid levels in guinea pigs was inconsistent with this interpretation. They proposed that ACTH release by the pituitary is suppressed prior to exhaustion in heavy exercise, which would reduce glucocorticoid release from the adrenals. This hypothesis was subsequently refuted by Peter Farrell and colleagues (65) in Milwaukee, however, when ACTH levels were shown to dramatically increase in humans as the exercise intensity increased to a point of exhaustion. The mechanisms for the reported decrease in glucocorticoid concentrations at exhaustion, and the reduction in glucocorticoids near the end of prolonged exercise, remain a mystery.

The influence of exercise on cortisol turnover was established primarily by Davies and Few (53) in the 1970s by using ^{3}H-cortisol infusion. Initially, they were interested in glucocorticoid responses to changes in body temperature in humans. In a series of subsequent studies, the group identified glucocorticoid turnover rates to various exercise intensities in their subjects. They also showed that moderate-to-heavy exercise often causes cortisol levels to rise progressively because of an increase in secretion rate rather than a decrease in cortisol elimination (44). Compared with rest, light exercise was associated with an increase in the rate of removal of cortisol from plasma and a reduced secretion rate, which would explain the commonly observed decrease in plasma concentrations. During exercise at heavier intensities, they noted that the cortisol removal rate is even more markedly increased but that it is exceeded by an increase in the rate of cortisol secretion (44, 69).

The influence of the timing of exercise on cortisol response was initially proposed by Gabrielle Brandenberger and Marguerite Follenius from the Centre d'Etudes Bioclimatiques in Strasbourg, France, during the 1970s (35). In another publication in 1982, they showed that diurnal variation in cortisol secretion "clearly influenced the exercise response" of the hormone (36). In that study, exercise performed during normally "quiescent" periods of cortisol secretion (10:00, 14:00, 17:00 and 21:30 hours) produced exaggerated increases in plasma cortisol levels but levels for exercise in the evening or during meal times were somewhat dampened. These findings were important to understanding the influence of timing of exercise and meals on glucocorticoid response, and to understanding that feedback inhibition occurs from transitory diurnal peaks.

Since it was generally accepted during the 1970s that exercise stimulated glucocorticoid secretion, the search immediately began to determine the source of the stimulus. Blood glucose concentration was demonstrated as a likely mediator, since Winder and colleagues found that the increase in corticosterone concentration was enhanced in overnight-fasted rats, who had decreases in blood glucose compared to control rats in which glucose levels were maintained (252). In addition, Nazar established that the exercise-induced increase in plasma 17-hydroxycorticoid concentrations in dogs is eliminated when hypoglycemia is prevented by intravenous infusion of glucose (167). She proposed that glucoreceptors in brain and liver activate the pituitary–adrenocortical axis, which could sense small changes in glucose concentration in the carotid artery or portal vein. Indeed, by infusing small amounts of glucose into these vascular regions, Nazar eliminated the usual exercise-induced glucocortocoid response to exercise. Galbo and colleagues later tested Nazar's theory that glucose-sensitive receptors control glucocorticoid output during exercise since they felt that several of the hormonal responses to exercise were driven by glycemia. They, too, found that a rapid decrease in blood glucose levels, caused by dietary manipulation, is accompanied by a rapid increase in cortisol concentration in humans (84). Again, however, as with other hormones described in this chapter, increases in cortisol were also observed during exercise performed with high circulating glucose levels (30, 104). More importantly, the group from Copenhagen found that the restoration of blood glucose late in exercise by glucose infusion did not influence cortisol concentrations (84). Euglycemic clamp in dogs also failed to completely prevent the exercise-induced increase in cortisol levels (245). Clearly, glucose concentration was not the only stimulus for glucocorticoid secretion.

Farrell and colleagues from the University of Wisconsin, Milwaukee, first suggested that it was a stimulus from the muscle itself, such as lactate production or altered pH, which stimulates ACTH secretion by the pituitary gland and, in turn, increases glucocorticoid secretion by the adrenal cortex. This hypothesis was similar to that for the increase in growth hormone secretion during exercise that was also being proposed at the time (see the section called Growth Hormone). Farrell and colleagues reported that a positive relationship exists between ACTH and the change in lactate concentration in untrained subjects (65). The theory that plasma ACTH levels increase in humans, particularly during exercise at heavy intensities, gained further support from observations by Few and (70) and by Farrell (65) and their colleagues. Thus, it appeared that an increase in ACTH was the stimulus for glucocorticoid secretion rather than direct sympathetic regulation of the adrenal cortex. The questioned remained, however, as to whether lactate triggered ACTH secretion during exercise. In Few's study, four subjects held 20–25 kg weights in one hand for 5 minutes and plasma ACTH levels doubled from resting values while lactate concentration did not change significantly (70). This finding indicated that isometric exercise increases plasma ACTH secretion independent of increases in lactate concentration in the plasma. As pointed out by Farrell and others, it may not be the lactate anion per se that triggers ACTH secretion but, rather, an associated physiological change such as pH, or another messenger from exercising muscle. Over the years, other factors such as decreased levels of circulating insulin and increases in body temperature (86) have also been proposed as stimulating glucocorticoid secretion during exercise, although no one factor has been found to be critical.

Glucocorticoids and Training

Studies conducted in the late 1960s, following the development of sensitive hormone assays for glucocorticoids in plasma, have shown that plasma glucocorticoid levels tend to increase less following exercise training (56, 75), although this observation has not always been consistent. A key study published early on by Frenkl indicated that the hypothalamic–pituitary axis may undergo exercise training adaptation, similar to other stressors, that lowers the amount of ACTH released (75). This training adaptation was proposed to conserve adrenal cortical function. It was also thought that the adrenal cortex could have an elevated sensitivity to circulating ACTH.

Interestingly, however, exogenous ACTH treatment caused similar increases in plasma glucocorticoid levels in trained rats compared with untrained rats, showing that endogenous secretion of ACTH may be lowered with training. In 1973, Song et al. (210) in Gollnick's laboratory reported that exercise training results in adrenal gland hypertrophy and hyperplasia in rats, possibly as an adaptive response to repeated ACTH stimuli.

In 1978, John Sutton (1941–1996) from McMaster University in Hamilton, Ontario, published a study indicating that plasma cortisol levels tend to increase less, or may even decrease, during exercise in fit compared with unfit subjects (218). In an earlier study published in 1969, however, Sutton had found that the cortisol pattern was similar in both fit and unfit subjects who pedalled a cycle ergometer with increasing power outputs designed to produce complete exhaustion by 30 minutes (221). Confusion arose, therefore, with respect to whether training modified the cortisol response to exercise performed at the same relative intensity. Indeed, both Shephard et al. (206) and Hartley et al. (104) found no differences in cortisol response between trained and untrained humans. In contrast, Bloom et al. (30) reported that cortisol values were significantly higher in six well-trained cyclists than in six untrained subjects. These conflicting reports were attributed to the widely differing experimental conditions that make valid comparisons nearly impossible. It was logical to many at the time, however, that lower corticosteroid levels in the trained state during the same absolute exercise intensity were a function of a reduction in physiological and psychological stress caused by an adaptive response to the training. The ability of trained individuals to dramatically increase glucocorticoid levels during maximal effort was evident (226), however, as were the repeated observations of hypertrophy of the adrenal glands caused by training (210).

Roles of the Glucocorticoids in Exercise Metabolism

Although the glucocorticoids were known since the mid-1800s to play a role in various stress responses, the role of glucocorticoids in the normal metabolic response to exercise has only been an area of considerable interest since the 1970s. It was first proposed, based on the early studies by Krystyna Nazar from the Polish Academy of Sciences, Warsaw, Poland (167), and others, that glucocorticoids participate in regulating glucose homeostasis, either by acting directly on the liver or by sensitizing the liver to other glucoregulatory hormones such as glucagon and epinephrine. In addition, both cortisol and, possibly, ACTH were proposed to play a possible role in augmenting epinephrine-induced lipolysis. The use of an adrenalectomized rat model has been instrumental in elucidating the significance of glucocorticoids for substrate mobilization. Galbo's group, for example, showed that during exercise, FFA concentrations are lower, and the decline of blood glucose concentrations are more marked, in adrenalectomized rats compared to control rats (89). The mobilization of fatty acids from adipose tissue and, perhaps, hepatic glucogenogenesis during the later stages of prolonged exercise is suggested by Friedman et al. (76) to be related to glucocorticoid secretion. It would appear, therefore, that ACTH secretion is stimulated during prolonged exercise to mobilize fuels for oxidation, possibly in response to a fall in glucose level.

Winder and colleagues also investigated the effect of the increase in glucocorticoids during exercise on metabolism and endurance capacity. In a study published in 1988, they found that although adrenalectomy markedly reduced endurance capacity in rats, it had little influence on carbohydrate metabolism (203). In contrast, Winder's laboratory proposed that glucocorticoids could play a role in lipid mobilization since adrenalectomized rats had markedly lower FFA levels at the onset of exhaustion than control rats (203).

GROWTH HORMONE

GH and Acute Exercise

Elevations of the blood concentration of growth hormone (GH) during exercise were established by W. M. Hunter from Edinburgh, Scotland, and his colleague Greenwood after they developed a radioimmunoelectrophoretic assay for the hormone in 1962. Soon after, Hunter, working with Fonseka and Passmore, was instrumental in developing the concept that GH has an important impact on fuel mobilization during exercise (118). In 1969, Sutton and colleagues (221) confirmed that GH levels increase in humans during exercise. They also proposed that GH plays a role in metabolism. The consistency of the blood GH response to exercise, shown by these and other researchers, led to the clinical use of the exercise test as a screening test for GH deficiency in the late 1960s. In the mid-1970s, Sutton and colleagues best illustrated that the GH response to exercise depends on the intensity and duration of the activity as well as on the physical fitness of the individual (217, 220). Sutton concluded that an intensity threshold must be met before significant increases in GH occur since levels did not increase appreciably during cycling at or below 300 kpm/min. It was also clear from Sutton's studies that once the threshold intensity is met, increases in power outputs dramatically increase GH concentration. Similar findings of an intensity threshold were reported by Hartley and colleagues (105) at Harvard in 1972 and by Karagiorgos et al. (137) in Brooks's laboratory in Berkley in 1979. As discussed earlier, this concept of a threshold was similar to what was reported for the glucocorticoids.

In the mid-1970s, several investigators verified that the GH response to exercise was dependent on both the duration and the intensity of the activity performed (153, 165, 174). In their studies, increases in both intensity and duration were associated with increases in GH levels. Together, these studies suggested that GH concentration peaks between 5 and 15 minutes after the start of moderate-intensity exercise (~50% $\dot{V}O_2$max) of 15–40 minute duration and that the concentration decreases after the end of exercise. It was also apparent from these investigations that a lag time existed between the onset of exercise and the increase in GH concentration, which is inversely proportional to the intensity of exercise.

Interesting findings published by both Hartley et al. (104) in 1972 and by Sutton's group in 1976 (217) indicated that plasma GH levels may be lower at maximal effort than at rest and at submaximal effort. These observations paralleled the finding by Sutton et al. (221) that the increase in GH concentration during prolonged exercise may be followed by a decrease if the exercise was of a long duration. No definitive explanation of this phenomenon has been given, but Hartley and Sutton suggested that the suppression in GH levels may be due to some factor released during heavy or exhaustive exercise. Plausible factors such as blood lactate and pH were proposed but ruled out in their later publications (see below).

Stimulus for GH Secretion

A number of hypotheses have been advanced by exercise physiologists to explain the trigger for the increase in GH levels during exercise. Buckler, one of the pioneers in developing the exercise GH test in the early 1970s, suggested that GH output was dependent on the cumulative effects of exercise rather than on one particular stimulus (39). This notion was only supported by the widely cited observations that the speed at which GH increases are detected are dependent on the intensity of the activity, and thus the magnitude of the cumulative effects. Hansen (102) suggested in 1971 that there were at least two possible mechanisms for GH release during exercise: (1) a humoral factor liberated from the working muscle that was carried by the blood to the hypothalamus or (2) a neural reflex mechanism from the working muscles. The first proposal fit with his observations that a lag time exists following the onset of moderate exercise before plasma GH levels increase. In his studies, Hansen had not observed any changes in blood GH levels during 20 minutes of cycling at 450 kpm/min (102). A significant rise was detected 10 minutes after the cessation of exercise, however, and he proposed that an elevation in catecholamine or cortisol levels may be the stimulus for GH release. Hansen also suggested, based on his studies of insulin-dependent diabetes with autonomic neuropathy, that the adrenergic control of GH secretion, if present, is likely mediated by catecholamines released locally from monoaminergic fibres in the hypothalamus (102).

The finding that the GH response was graded to exercise intensity once a threshold was met, similar to the activation of anaerobic metabolism, led researchers to propose in the 1970s that some anabolic metabolite may trigger GH release. In an attempt to provide the metabolic link to hypoxia that could stimulate GH secretion, Sutton and Lazarus suggested in 1976 that an increase in blood lactate, or a change in acid–base balance, may be the stimulus for release of GH (217). This hypothesis was similar to what they proposed for the glucocorticoids (see above). These investigators were also instrumental in showing that exercise under hypoxic conditions, such as is found at high altitude, could dramatically elevate GH levels. Similarly, Mètivier and colleagues reported in 1978 that the stimulus could be localized hypoxia within the contracting muscle (165). The failure of GH levels to increase during moderate, but not severe, exercise appeared to be inconsistent with this observation, however, since one would expect that intense exercise would produce an effect similar to hypoxia. It was also problematic because the intensity threshold, based on Sutton's investigations, was about 30% $\dot{V}O_2$max, an intensity well below a lactate threshold. More importantly, Brooks's laboratory found that although larger elevations in lactate occurred during intermittent, compared with continuous, exercise, elevations in GH concentration were similar in the two types of activities (137). Thus, it appeared that lactate was not a primary stimulus for GH secretion. In a more definitive publication, Sutton, in collaboration with his colleagues at McMaster University, rejected their hypothesis in 1976 by showing that artificially changing blood pH and lactate concentrations with NH_4Cl and $NaHCO_3$ capsules did not influence GH secretion over a range of work loads (220). Some aspect of oxygen debt may prove to be important in triggering GH secretion, however, since anaerobic exercise is thought to be a potential moderator of GH response (68).

Some investigators proposed in the 1970s that the reduction in insulin levels with exercise may trigger the rise in GH levels, although this theory was also questioned when insulin infusion late in exercise did not suppress GH levels (84). The initial decrease in plasma FFA was also proposed as the trigger for GH secretion during exercise, but the infusion of neither triglyceride emulsion nor heparin diminished the GH response to exercise (118). Body temperature was also proposed as a potential trigger for GH hormone secretion during exercise since Buckler and others (40) found that an increase of a subject's body temperature elevates GH levels. Increase in body temperature per se was not the only factor in the exercise response, since Buckler also showed in 1973 that increases in body temperature by hot baths, to values similar to what is experienced during exercise, did not induce the same rise in GH levels (40). More recently, muscarinic cholineric and α-adrenergic mechanisms are thought to play roles in GH secretion during exercise, either by decreasing somatostatin levels or by increasing GH-releasing hormone (103). Thus, a number of factors have been proposed to trigger and modulate the GH response to exercise and the knowledge of the GH secretion during exercise is still an area of active investigation.

Roles of GH in Exercise Metabolism

Research in the early 1960s on isolated fat cells of the rat indicated that GH had lipolytic actions. In addition, in vitro experiments on the diaphragm of hypophysectomized rats showed that GH reduced the triglyceride content in muscle and increased lipid oxidation. Because increases in GH levels during prolonged exercise paralleled increases in fatty acid mobilization, investigators were interested in identifying whether a causal relationship existed.

Several researchers investigating GH response to exercise attempted to show a link between GH secretion and the shift from carbohydrate to fat mobilization. Indeed, in the mid-1960s, Hunter, Fonseka, and Passmore proposed that GH plays a role in fat mobilization during exercise. In their 1965 paper published in *Science* (118), they suggested that GH plays a critical role in FFA mobilization and utilization during exercise, although little data were provided to back up their hypothesis.

They felt that an increase in energy expenditure is an important factor in stimulating GH secretion and that ingested fats must first be taken up by adipose tissue and subsequently released as FFA, a process that was determined by GH. In their seminal publication, they reported that during exercise the increase in the blood level of GH is accompanied by an elevated concentration in free fatty acids (118). Since both responses were eliminated by glucose administration, they concluded that GH plays an important role in the mobilization of FFAs. Studies of hypophysectomized rats in Gollnick's laboratory during the 1970s supported the concept initially, since rats that had undergone the surgery had a depression in the rise of fatty acids during exercise (95). In contrast, Lassarre et al. (153) and Vigas's group (142) found in the mid-1970s that changes in lipid and carbohydrate metabolism were not associated with elevations in GH secretion. As mentioned above, these authors favored the hypothesis that peak GH values were instead related to O_2 deficit during exercise, which could be manipulated by hypoxic conditions such as high altitude. Subsequent investigations of exercising humans with hypopituitarism or Cushing's disease who were both adrenalectomized and hypophysectomized also indicated that an increase in GH was not essential in increasing FFA levels (15). Thus, other factors seemed to be involved in regulating lipid oxidation (see below).

In addition to potentially having a role in lipolysis, GH was proposed to influence carbohydrate mobilization. Investigators who felt that glucose concentration was the primary hormonal trigger for a number of hormones cited evidence that hypoglycemia augments GH secretion during exercise (118). The rise in GH levels was, in fact, was shown to be prevented with glucose administration in 1965 in the seminal paper by Hunter et al. (118). Evidence that glucose concentration was not the only stimulus for GH secretion came from studies published a decade later indicating that high GH and high glucose levels can occur concurrently (84, 188, 218). It was unclear, therefore, whether GH secretion was a stimulus for glucose production by a feed-forward mechanism or if small decreases in glycemia were the trigger for GH secretion. The variety of experimental approaches made it difficult to draw this distinction (see below). At least during intense exercise, increases in GH levels do not appear to play a significant role in glucose metabolism. For example, during intense exercise (~80% $\dot{V}O_2max$), increases in glucose production were similar in males when GH was clamped, along with insulin and glucagon, at basal levels, compared to when these hormones were allowed to change with exercise (208).

The influence of GH on physical working capacity was initially determined from Ingle's studies using hypophysectomized rats. These studies showed that, unlike corticotropin or adrenocortical extract, which dramatically improved exercise performance, GH replacement did little to restore the rats' capacity for exercise (122).

GH and Physical Training

Several studies have found that training has an attenuating effect on GH levels during exercise. In 1969, Sutton and colleagues (221), then in Sydney, reported that GH release was higher during maximal effort in unfit ($\dot{V}O_2$max 24–32 ml·kg·min^{-1}) than in fit ($\dot{V}O_2$max 58–78 ml·kg·min^{-1}) subjects. They also reported that cycling at 40% $\dot{V}O_2$max elevated blood GH levels only in the unfit subjects, while GH levels remained unchanged by exercise in the fit subjects. If the exercise was at a higher intensity, GH increased during short-term exercise in both fit and unfit subjects, but the increase was less pronounced and returned to baseline faster in fit persons. Similarly, in Bloom's study published in 1976 (30), trained cyclists had an attenuated increase in GH levels compared with untrained cyclists exercising at a similar relative work load. Within a few years of Bloom's study, Mètivier (165) demonstrated that highly trained athletes do not show any significant changes in GH levels from rest while performing exercise on a bicycle at 50% $\dot{V}O_2$max, which was considerably higher than the intensity normally required to induce an increase in GH levels in untrained individuals. When the exercise was increased to 66% $\dot{V}O_2$max, a significant rise in GH was noted even in the trained subjects. In agreement at the time, Hartley et al. (104), Koivisto et al. (143), Rennie and Johnson (188), and Sutton et al. (218) also confirmed that GH levels are lower in trained compared with untrained subjects during exercise at either the same relative or absolute intensity. These findings were not what were expected if GH is a major determinant in the increase in lipid mobilization associated with training. If GH is a major contributor to FFA mobilization, then one would expect that trained individuals would have higher GH levels during exercise to facilitate a greater FFA mobilization. Nonetheless, major differences between fit and unfit individuals were proposed to exist in the exercise-intensity threshold and recovery of blood GH response to exercise, the significance of which is currently unknown.

In a landmark study published in 1972 that was designed to eliminate the influence of pituitary hormones on training adaptations in rats, Gollnick and Ianuzzo showed that increases in the activities of oxidative enzymes and in the concentration of mitochondrial protein in skeletal muscle occur with training in the absence of GH via hypophysectomy (94). Ten years later, Edwards et al. (64) also reported that trained hypophysectomized rats had normal metabolic adaptations to endurance training, which included higher muscle cytochrome oxidase activity and increased exercise capacity. It was clear from these, and other studies, that GH had little influence on training-induced adaptations in metabolism.

THYROID HORMONES

The thyroid gland has long been known to influence metabolic rate, growth, and development in mammals. Therefore the role of the thyroid hormones—thyroxine (T_4) and triiodothyronine (T_3)—in regulating the acute and chronic responses to exercise has been a focus of investigation for over a century. Similar to the investigations of other endocrine glands, such as the pancreas, adrenals, and pituitary, studies on the role of thyroid hormones during exercise have frequently involved either the removal of the gland (i.e., thyroidectomy) or exogenous administration of the hormones or assessments of the effects of clinical conditions causing hypo- or hyperthyroidism. In general, early experiments were inconclusive and somewhat speculative while later studies were more informative due to advances in methods used to investigate the influence of these hormones during exercise. As with glucose tracers, the more recent use of radioactive tracers for T_4 and T_3 hormones has led to a clearer understanding of the influence of acute and chronic exercise on thyroid hormone turnover rates. The roles of exercise training on thyroid function and of thyroid hormones in mediating the training response have been investigated extensively over the past several decades.

Thyroid Hormones and Acute Exercise

In the early part of the twentieth century, the relationship between the thyroid gland and acute exercise was unclear and largely speculative. The simple question concerning how exercise influences thyroid activity is extremely controversial. Many of the apparent contradictions in early observations likely resulted from limitations in the indirect methods used to estimate T_4 metabolism during exercise. Prior to the 1940s, a number of investigators, such as Roy G. Hoskins (117), a former student of Cannon's from Ohio State University, and Curt P. Richter (190) from Johns Hopkins in Baltimore, examined the effect of thyroid gland removal on spontaneous running activity in rats. The results were contradictory, however, as Hoskins (117) found that thyroidectomy did not alter spontaneous activity patterns, while Richter observed that the complete removal of the thyroid gland dramatically reduced spontaneous activity (190). Richter's studies were more convincing than Hoskins' results because he was able to show that partial removal of the gland did not curtail spontaneous activity and that feedings of thyroid extract appeared to increase activity. These observations led Richter to propose that the thyroid gland is important in regulating acute exercise responses, a topic of debate that still lives today.

By the 1950s and 1960s, the influence of acute exercise on indirect measures of thyroid hormone metabolism were being assessed by thyroid histology in rodents (43), iodide uptake or content in the thyroid gland (189), and blood protein-bound iodine levels (32). These initial studies were the first attempts to determine the influence of exercise on thyroid activity and hormone turnover. Of particular note, using an isotope equilibrium method, Rhodes reported in *Nature* in 1967 that rats which exercised for several weeks store about 50% less radioactive iodine in the thyroid gland than do sedentary rats (189). He proposed that his data could be interpreted in one of two ways. Either thyroid activity is suppressed by physical exercise, or the utilization of the hormone is increased so that more dietary iodine is converted to circulating hormonal iodine and less of it is stored. As cited in his publication, Rhodes preferred the latter explanation (189).

Based on the observations that acute exercise increased T_4-derived iodine elimination in animals, Winder and Heninger proposed in 1973 that either T_4 provision to tissues is enhanced by exercise or that exercise increases tissue T_4 degradation. To investigate these possibilities, they examined T_4-deiodinating activity in homogenates of muscle, liver, and kidney from trained and untrained rats and found that training did not alter the deiodinating capacities of these tissues (258). Since exercise was shown not to increase T_4 degradation, attention was directed toward events associated with T_4 supply to the tissues. Because it is only the circulating free T_4 (i.e., that which is unbound with plasma proteins) which is considered to be available for uptake by tissues, focus on the role of exercise on free T_4 levels emerged. It was postulated that increased T_4 degradation associated with exercise might result from decreased circulating levels of T_4 binding globulin or T_4 binding prealbumin, which would be expected to increase the free T_4 removal rate constant. To examine this possibility, Terjung and Tipton (225) assessed the binding capacity of thyroid binding globulin (TBG) and thyroid binding prealbumin (TBPA) before, during, and after 30 minutes of cycle ergometry exercise at 900 kpm/min. They found that plasma T_4 binding actually increased with exercise, thereby making it impossible to attribute the increased T_4 turnover to a reduction in the binding capacity of T_4 binding proteins. Although a decrease in T_4 binding was not found during exercise, the authors noted that maximal free T_4 concentrations increased by 20%–35% during exercise and recovery above baseline, possibly because of increases in hemoconcentration (225). Other laboratories also observed that free T_4 levels increased with exercise in humans although, again, this finding was not always consistent. For example, Caralis et al. (42) from Paul J. Davis' laboratory in Buffalo found that subjects with some prior training had an increase in free T_4 after a treadmill exercise test, while the subjects who were untrained had a decrease. In addition, Johannessen et al. (133) were unable to detect any change in serum concentrations of thyroxine and triiodothyronine within 80 minutes of exercise leading to exhaustion, although they did observe an increase in thyroid-stimulating hormone (TSH). Others interested in the field during the late 1970s and early 1980s, including O'Connell (175), Premachandra and Holloszy (179), and Refsum and Stromme (186), found that levels of T_4 and T_3 are essentially unchanged during a single bout of exercise. In general, no consensus could be made. The long turnover time of the thyroid hormones and the delay in the stimulus (i.e., TSH) secretion coupling as well as other factors may explain the failure to observe increases even though it was expected that exercise stimulated TSH secretion and increases the turnover rate of the hormones.

To explain the increased secretion and degradation of T_4 hormone during acute exercise, thyroid stimulating hormone (TSH) levels were also investigated in a number of laboratories. The mechanism causing the increase in free T_4 levels during exercise in some of the studies may have been related to increased FFA levels, since elevations in these lipids were known to enhance thyroxine disassociation with the binding globulins (115). It was also hypothesized by a number of investigators that increases of thyroid hormone levels during exercise are caused by an increase in TSH secretion, possibly through stimulation by the central nervous system. A number of investigators failed to show that TSH levels increase with exercise. In a seminal study conducted in 1971 in humans by Terjung and Tipton, TSH concentration in males remained unchanged before, during, and after 30 minutes of exercise at 60% $\dot{V}O_2$max (225). Although TSH levels were apparently unaffected by exercise in their study, inferences on TSH turn-over rate could not be made, because of methodological limitations, and a definitive conclusion concerning the effects of TSH on T_4 metabolism during exercise was not possible. It was proposed, therefore, that any increase in TSH release would be rapidly cleared from the plasma, making it difficult to view an increase in TSH concentration. As emphasized earlier, however, Galbo and colleagues found that TSH levels were increased after repeated 20 minute sets of cycling at high intensity in their study published in 1977 (87). In addition, levels were shown to increase with increasing exercise intensities to a maximum of 107% above resting values after maximal exercise (87). They concluded, therefore, that the TSH response to exercise was duration and intensity specific (87) and that the levels may be influenced by carbohydrate intake (133). Once again, the influence of blood glucose levels on hormone secretion was proposed (133).

Other laboratories provided some evidence that TSH levels increase with exercise. For example, Refsum and Strömme (186) reported in 1979 that a modest increase in plasma TSH levels in male athletes after they completed a 70 km cross-country ski race. This group also found an increase in plasma T_4 and free T_4 levels immediately following a 90 km race although the hormones declined below pre-race values by the next day. More importantly, perhaps, was their observation that TSH levels rose to a peak value of 175% of control after the race and did not return to prerace levels for several days after the end of the competition even though the T_4 and free T_4 levels were suppressed. These findings were attributed to a decrease in thyroid hormone feedback, which was normalized after approximately 4 days. It was suggested by Refsum and Stromme (186), and others, that the increase in TSH levels during exercise contributed to the increase in thyroid hormone secretion rate, which had been observed by a number of the previous investigators. Mason and colleagues even reported that TSH levels could be increased even in anticipation of exercise (162).

Roles of Thyroid Hormones in Exercise Metabolism

Although considerable attention had been focused on the influence of acute and chronic exercise on thyroid function, it was difficult to establish the role of thyroid hormones in the normal response to acute exercise.

The influence of thyroid hormones on substrate metabolism has been investigated in only a limited number of studies, mostly from Kaciuba-Uscilko's laboratory at the Polish Academy of Science in Warsaw. Using an exercising dog model, Kaciuba-Uscilko confirmed in the mid-1970s that treadmill exercise increased serum T_4 concentration considerably (84%) and that exogenous and exercise-induced elevations of thyroid hormones enhanced free fatty acid mobilization (136). Although free T_4 and free T_3 levels were not measured in their studies, it is expected that both the rise in body temperature and the FFA levels associated with the exercise would have caused a decrease in thyroid hormone binding to plasma proteins and increases in the levels of unbound hormones. A subsequent investigation published in 1979 by this group (38) revealed that T_3 and T_4 treatment in dogs lowered glycogen levels in resting muscle and in the liver and that the rate of glycogen degradation during prolonged exercise was also suppressed. The rate of postexercise muscle and liver glycogen resynthesis also appeared to be impaired in hormone-treated dogs (38). In collaboration with Terjung two years later, Kacibuta-Uscilko et al. (135) reported that hypothyroidism causes an increase in muscle lipoprotein lipase activity and enhanced muscle and plasma TG utilization during exercise when compared with euthyroid rats. These studies have been instrumental in illustrating that the thyroid hormones have at least some influence on substrate mobilization and utilization during exercise.

About the same time period as the studies conducted in Kaciuba-Uscilko's laboratory, O'Connell et al. at the University of Vermont (175) showed that in humans, prolonged moderate cycling increased reverse triiodothyronine (rT_3) and T_4 concentrations and lowered T_3 concentrations. Interestingly, thyroidal responses to exercise were attenuated by glucose infusion, suggesting an interaction between carbohydrate metabolism and thyroid function during exercise. In addition, in this same study (175), rT_3 levels correlated with plasma FFA ($r = 0.95$) and glucose ($r = -0.87$) concentrations, further suggesting that thyroid hormones may have some role in the mobilization or utilization of these substrate.

Thyroid Hormones and Exercise Capacity

Using a rodent model in a study published in 1980, Baldwin et al. (10) at the University of California showed that hypothyroidism reduces both high-intensity and prolonged endurance exercise capacity, likely by reducing oxidative capacity of skeletal muscle. This finding of reduced exercise tolerance is consistent with C. P. Richter's 1933 study of thyroid hormones and basal metabolic rate (BMR) in rats (190) and by Kaciuba-Uscilko's observations in the late 1970s in thyroidectomized dogs (134). To explain this lowered exercise tolerance, and to examine the potential role of the thyroid hormones in energy status during exercise, Argov et al. (6) at the University of Pennsylvania used in vivo phosphorus-31 NMR to show in the late 1980s that hypothyroid patients ($n = 2$) and rats ($n = 8$) had significantly lower PCr/Pi ratios during exercise and delayed postexercise recovery of PCr/Pi. These alterations were shown to be normalized with thyroxine therapy. In addition, hyperthyroidism was associated with normal PCr metabolism at rest and during exercise and postexercise PCr/Pi recovery was somewhat enhanced (6). These bioenergetic alterations suggest that thyroid hormones may play a role in mitochondrial bioenergetics during exercise and recovery, which may help to explain exercise intolerance associated with hypothyroidism.

Thyroid Hormones and Physical Training

As noted above, Rhodes reported in 1967 that the thyroids of spontaneously active rats contain about half as much iodine as nonexercising controls and that the amount of iodine in the gland was negatively correlated with the amount of exercise performed by each animal (189). The influence of training on T_4 metabolism was first illustrated by Cliff H. G. Irvine from Lincon College in Canterbury, New Zealand, in the 1960s. Irvine, who was trained as a veterinary surgeon in Sydney, reported in 1967 (124) the influence of acute exercise and training on T_4 turnover in a racehorse by using T_4-^{125}I injection and measuring T_4 degradation over time. He found that 12 weeks of training increased T_4 secretion by 65%, indicating that the exercise caused an increase in thyroid gland stimulation. In addition, Irvine found that as the training progressed, the rate of T_4 degradation increased to a greater extent than its secretion rate, thereby causing a reduction in circulating T_4 levels (124). Thus, the major influence of regular exercise on T_4 metabolism was an increase in T_4 disposal and, as a result, an increase in T_4 production. In the following year, Irvine showed that track athletes had a 75% higher T_4 turnover rate at the end of their training season than nonathletes (125). He calculated that T_4 secretion must be increased by approximately 75% in athletes to maintain T_4 levels at a constant level. Thus, the greater T_4 removal rate was matched by enhanced T_4 production in athletes, thereby causing plasma levels to remain relatively unchanged during exercise. This is a classic illustration of the necessity to use tracer technology to determine turnover rates of hormones and substrates rather than plasma concentrations alone. Irvine also showed that when three of the athletes ceased training for 3 days T_4 degradation rates tended to fall. In the same study, 12 nonathletes were assigned to two equal groups; one group exercised daily for 6 days and one group did not. The exercising nonathletes increased their T_4 turnover by 33% and their urinary excretion of iodide by 34%, suggesting that exercise per se increased T_4 turnover, presumably by an increase in tissue deiodination.

Using a technique similar to that used by Irvine, Winder and Heninger in 1973 (258) and Balsam and Leppo in 1974 (12) reported enhanced T_4 metabolic clearance rate (MCR) in trained rats compared with untrained rats. These findings led the investigators to assess in what tissues T_4 MCR was enhanced during exercise training. Winder and Heninger addressed this question by systematically evaluating the influence of exercise on tissue levels of thyroid hormones and found that the liver was the primary tissue responsible for the training-induced increase in MCR (257). Balsam and Leppo (13) also observed that MCR of T_4 and T_3 increases in humans after a 6-week track-running program. Collectively, these studies also indicated that exercise training enhances thyroid hormone turnover in humans and in rats by causing an increase in peripheral tissue deiodination.

Kraus and coworkers (145) suggested in 1969 that thyroxine mediated the training-induced increase in mitochonidrial oxidative enzymes, although the observation in 1971 by Winder and Heninger (257) that muscle T_4 concentration was not influenced by exercise seemed inconsistent with other data at the time. In 1972, Gollnick and Ianuzzo (94) showed that T_4 turnover plays a small role, if any, in skeletal muscle mitochondrial adaptations to training, since thyroidectomized rats showed similar exercise-induced adaptations in mitochondrial density and succinate dehydrogenase activity as did control rats. This finding was particularly important, and somewhat controversial at the time, since thyroidectomy was shown to exert considerable depression in mitochondrial content and oxidative capacity of muscle prior to the training regimen (94).

To further characterize the role of thyroid hormones in the biochemical adaptations to training, Terjung and Koerner (224) examined cytochrome c concentrations of different muscle fiber types in normal and thyroidectomized rats exposed to 12 weeks of treadmill training. They found a 50% reduction in baseline oxidative capacity of all fiber types after thyroidectomy; however, the absolute increase in cytochrome c in the muscles of these rats after training was the same or even higher than in control trained rats (224). Winder et al. also found in 1981 (254) that the activity of liver α-glycerophosphate dehydrogenase, which is a sensitive index of tissue thyroid activity, does not increase in trained animals. Thus, it was proposed that greater T_4 turnover, found with training, seemed to occur through a process that was distinct from the "normal" actions of T_4, which would normally cause hyperthyroidism. Winder and colleagues (254) measured T_4 turnover in exercise-trained and sedentary rats under conditions of controlled food intake and body weight to further characterize the role of the thyroid in training. They showed that although relative heart weight and plantaris muscle citrate synthase activity were both significantly increased in trained animals, thyroxine secretion and degradation rates were not affected by training. In addition, markers of thyroid status at the tissue level, including liver mitochondrial α-glycerophosphate dehydrogenase, were unchanged by training (254). Thus, it appeared from these data that increased thyroid activity is not essential to inducing and maintaining the adaptations in muscle mitochondrial enzymes that occur in response to prolonged endurance training.

The influence of the thyroid hormones on increased mitochonidria number, elevated content of mitochondrial protein, and augmented activity of mitochondrial enzymes resulting in enhanced oxidative capacity has been of particular interest for a number of investigators. Thyroxine administration was shown by Winder et al. (249) in Holloszy's laboratory to elicit biogenesis in skeletal muscle mitochondria. However, Gollnick et al. showed that the increased content of muscle mitochondrial proteins and augmented activity of succinate dehydrogenase with training could not be avoided by hypophysectomy or thyroidectomy (94, 95). Terjung and Koerner (224) showed that an increase in cytochrome c content in normal and thryroidectomized rats due to training in all three types of muscle (types I, IIa, and IIb). Thus, it appears that the adaptive response to training of an increased oxidative capacity in the skeletal muscles occurs in the absence of normal thyroid function.

In summary, it appears that thyroid hormones play an important role in normal metabolic process, both at rest and during exercise, and that thyroid hormones exert a majority of biological effects by influencing gene expression. Interestingly, it appears that although thyroid hormones are increased with acute and chronic exercise, many of the normal training-induced alterations within skeletal muscle are not caused by these hormonal adaptations alone.

A HISTORICAL SUMMARY OF THE METABOLIC INFLUENCES OF PANCREATIC, ADRENAL, THYROIDAL, AND GROWTH HORMONES DURING EXERCISE

Over the past 70 years, considerable progress has been made in determining the function of the endocrine glands in regulating metabolism during exercise. The development of our knowledge has evolved from contributions from a number of investigators who were either interested in using exercise as a tool to understand basic physiology and metabolism or who wished to elucidate the importance of the endocrine system in the metabolic response to exercise. This summary is our integrative perspective on the control of carbohydrate and lipid metabolism based on a great number of these studies.

Carbohydrate Metabolism

During the 1920–1950s endocrinologists and diabetologists began developing an understanding of the role of insulin in the control of blood glucose concentration during rest and exercise. The findings that excessive insulin causes hypoglycemia, while too little insulin aggravates hyperglycemia and results in ketosis during exercise, lead physiologists to propose that insulin plays a critical role in hepatic and muscle glucose flux during exercise. The problem was, however, that measurements of insulin concentrations in the plasma were not available until the 1960s. Later, in the 1970s and 1980s, a number of studies indicated that the drop in insulin sensitizes the liver to glucagon and thereby facilitates glucose production to match glucose utilization during exercise. In addition, other hormones and hepatic nerves have been shown to modulate glucose production and utilization during muscular work.

Glucose Production

Quantification of glucose A–V differences across the liver and measurement of arterial insulin and glucagon concentrations during exercise in humans by Wahren and colleagues in the early 1970s were important in developing the concept that glucose production was greatly enhanced during exercise (240). In addition, technical breakthroughs in measuring portal and arterial insulin and glucagon concentrations to quantify hepatic glucose output using tracer technology in dogs, initially by Issekutz et al. (130), in 1967 and by Vranic and colleagues in 1969 (239), have been instrumental in determining the role of these hormones in glucose production. At the time, however, the mechanisms for regulating glucose production were very unclear. Subsequent collaborations by Issekutz and Vranic in 1980 (128) showed that somatostatin-induced glucagon suppression reduced hepatic glucose production, which resulted in a fall in blood glucose that was normalized by glucagon replacement. Although these studies demonstrated that basal glucagon levels are required for normal glucose production during exercise, compensation by other neuroendocrine regulators in response to a decrease in blood glucose prevented a quantitative assessment of its role. Clamping the arterial glucose concentration, in the presence of glucagon suppression with somatostatin in dogs, in a subsequent investigation published in 1984 by Wasserman et al. (245) permitted the role of this hormone to be assessed without the confounding effects of hypoglycemic counter-regulation. Under these controlled conditions, the increase in glucagon is necessary for over 60% of the total glucose production during moderate exercise in dogs. Furthermore, Wasserman demonstrated that the counterregulatory responses (i.e., catecholamines and cortisol increases) that occurred when plasma glucose levels were allowed to fall can compensate for ~40% of the deficiency in hepatic glucose production created by glucagon deficiency. In a classic study published in 1989 using the dog model once again, Wasserman and colleagues (248) reported that a rise in glucagon is necessary for the full increment in both hepatic glycogenolysis and gluconeogenesis during prolonged moderate-intensity exercise. This stimulatory effect of glucagon on gluconeogenesis was shown to be due to an accelerated rate of gluconeogenic precursor extraction by the liver and enhanced channeling of precursor to glucose within the liver. These observations in the dog model were consistent with the later finding published in 1991—that a rise in glucagon is necessary for normal glucose homeostasis during moderate exercise in humans (113).

In addition to glucagon, other hormones were known to play a role in facilitating glucose production during exercise. Issekutz was the first to test the hypothesis that the decrease in plasma insulin has a major role in glucose production and is important in the transition from carbohydrate to fat utilization. He used steady-state exercise and the infusion of manno-heptulose in to lower the already-reduced insulin levels in dogs (129). As Issekutz hypothesized, this treatment was associated with a marked increase in hepatic glucose production, which could be lowered to normal by insulin infusion. Wasserman and colleagues (244) followed up this line of thinking in 1989 by infusing insulin into the portal vein at a rate that prevented the usual fall in its concentration. It was calculated from the results of their study that the fall in insulin accounts for ~55% of the increase in hepatic glucose production largely based on insulin's effect on hepatic glycogenolysis (244).

The importance of increases in glucagon and decreases in insulin in humans has been explored over the last 20 years. The studies of Issekutz, Vranic, and Wasserman and their associates from North America clearly illustrate that an increase in glucagon/insulin ratio are the key regulators of hepatic glucose production during prolonged exercise in dogs. The use of dogs allowed these investigators to obtain arterial, portal, and hepatic vein blood samples and to observe changes in glucagon, insulin and other hormones reaching the liver. Later, the use of dogs with pancreatic hormone suppression and with portal insulin and glucagon infusion under euglycemia conditions was instrumental in illustrating that the rise in glucagon is important in stimulating glucose production. As mentioned previously, however, others have noted that glucose production increases in humans during exercise without apparent changes in peripheral insulin concentrations. Investigations by Wolfe and colleagues in 1986 (261), Holloszy's group in 1991 (113), and Galbo's laboratory in 1993 (140) are all important in showing that glucagon and insulin changes are important in glucoregulation during prolonged exercise in humans although both the timing of these hormonal changes and the concentrations of these hormones in the portal circulation remain a topic of debate. It is also possible that other, yet to be determined, factors link glucose production with glucose utilization, as has recently been suggested by Coker, Wasserman, Kjaer, and others (50). The investigations in humans and in dogs are in agreement that increases in other counterregulatory hormones, particularly the catecholamines, can partially compensate for the absence of exercise-induced changes in the pancreatic hormones in minimizing or preventing the fall in plasma glucose during exercise.

The failure of changes in the pancreatic hormones to be associated with increases in hepatic glucose production in the 1975 publication of Wahren et al. (242) led others to investigate the role of the catecholamines in facilitating glucose production. Circulating norepinephrine and catecholamines or norepinephrine released from sympathetic nerves to the liver were attractive candidates since all were known to increase glucose production, at least during rest, and because catecholamine levels had been shown to increase during muscular activity. By measuring hepatic cAMP in rats, Winders group reported in 1979 that the stimulation of hepatic β-adrenergic receptors varies directly with both the relative intensity and the duration of exercise (252). Advancing this concept that catecholamines increase glucose production during exercise, experiments with β-adrenergic blockade showed that the rate of decline in blood glucose was exaggerated in humans (80) and in dogs (126, 168). Using a rat model in a study published in 1978, Galbo and colleagues also showed that the decrease in hepatic glycogen content during exercise was less dramatic in chemically sympathectomized, adrenodemedullated rats than in sham-treated rats (88).

The role of catecholamines in regulating glucose production has also been investigated in a number of other laboratories (144, 166, 209) and, in general, few studies have shown that the catecholamines play a significant role during low or moderate-intensity exercise of <120 minutes duration. These studies have clashed somewhat with the earlier studies from the group from Denmark and their finding that adrenodemodulation in the rat reduces hepatic glycogenolysis and glucose production in the rat (191). Later studies by Kjaer and his coworkers (140), in which human subjects exercised during celiac ganglion blockade and epinephrine infusion, have confirmed that epinephrine is not a major stimulus of glucose production during moderate-intensity exercise. A role for the catecholamines during moderate exercise has not yet ruled out, however, since Moates and colleagues (166) found that epinephrine plays a significant role in stimulating glucose production late in exercise, possibly by mobilizing gluconeogenic substrates from the periphery.

Studies published by Marliss's laboratory provide strong evidence that the catecholamines may have a more significant role in glucose production during high-intensity exercise. Using glucose tracer technology to study glucose turnover during exercise in humans, the group found that hepatic glucose production is considerably higher during heavy exercise (80%–90% $\dot{V}O_2max$) compared with exercise performed at lesser intensities (161). This maximal exaggeration in glucose production during heavy exercise results in an increase in arterial glucose levels that extends into the postexercise state. It was hypothesized in this classic study published in 1991 that during high-intensity exercise there may be a shift in the control of glucose production away from the pancreatic hormones to a dependency on catecholamines (161). This hypothesis is based on the observation that norepinephrine and epinephrine levels can increase severalfold during exercise performed during heavy exercise, whereas the increase in the glucagon/insulin ratio is very small at these intensities. In support of Marliss's hypothesis was his observation that glucose production is not affected by greatly divergent responses of glucagon, insulin, and growth hormones induced during heavy exercise in humans using a trihormonal infusion clamp (208). In addition, intense exercise, performed either during glucose infusion or following a meal, both of which were associated with elevations in insulin secretion and glucagon suppression, caused increased endogenous glucose production, likely because of elevations in the catecholamines (147). This was in contrast to moderate exercise when glucose infusion limits endogenous glucose production. Stronger evidence still are the recent observations that infusion of epinephrine (148) and norepinephrine (146) during moderate exercise, to physiological concentrations seen during intense exercise, increase glucose production rates to those measured during intense exercise. Thus, although the role of the catecholamines during moderate exercise is questionable, they appear to have a major role during more severe muscular activities. It is also important to mention that this concept appears to favor the feed-forward control of glucose production rather than feedback control, since glucose rate of appearance (Ra) exceeds rate of disappearance (Rd) during the activity and blood glucose levels are elevated.

Muscle Glycogen Utilization

Based on studies conducted since the late 1970s, it has appeared that muscle glycogen utilization during exercise is influenced by the catecholamines. It was demonstrated in Holloszy's laboratory in the early 1970s (9) that contraction increases phosphorylase activity and that catecholamines increase muscle glycogen utilization and lactate production. In addition, in 1982, Chasiostis et al. (46) in Hultman's laboratory, using muscle biopsies from subjects studied during β-blockade, demonstrated that the catecholamines stimulate glycogen utilization during exercise. In the early 1980s, Richter and colleagues also demonstrated that adrenomedullary hormones enhanced muscular glycogenolysis, glucagon secretion, and hepatic glycogenolysis but inhibited insulin secretion in exercising rats (191). Studies published a year later by Richter et al. (195) using the isolated perfused rat hindquarter and electrical stimulation indicate that epinephrine and contractions can have an additive effect on activating phosphorylase. Increases in muscle glycogen utilization with infusions of epinephrine have also been observed in human subjects by Spriet and coworkers in 1988 (211) and by Febbraio and colleagues in 1998 (66). More recent findings from Kjaer and colleagues, however, indicate that although epinephrine increases the glycogen phosphorylase activity it is not essential for muscle glycogen breakdown, although it may play a role in activating muscle triglyceride lipolysis in humans (141).

Lipid Metabolism

It was well known by the mid-1970s that low insulin availability and increased sympathetic activity were the likely candidates for increased lipolysis during exercise.

Issekutz suggested in 1967 that during prolonged exercise the decrease in blood insulin was a tool for the transition from carbohydrate to lipid oxidation. The decrease in insulin availability would reverse its dampening role on cAMP in adipose tissue. Pavle Paul (177) from the Lankenau Hospital in Philadelphia used isotope dilution techniques, which he had learned from Issekutz, to show that the turnover rates of FFA and glycerol in plasma are linearly related to their respective concentrations in plasma, which were ultimately influenced by circulating insulin concentration. In addition, studies conducted in the mid-1970s in patients with diabetes by investigators such as Berger et al. (17) and Wahren et al. (242) indicated that hypoinsulinemia increases lipolysis during exercise. Also noted by these and others at that time was that respiratory quotients were lower, and FFA uptake across the working limb higher, in diabetic subjects compared with non–insulin-deficient subjects. From this, it was proposed that the drop in insulin levels during exercise in healthy humans is, at least in part, responsible for the increased rate of FFA utilization during exercise. Galbo's investigations with normal fasted (82) and fat-fed (84) subjects provided further support for the concept that the reduction in insulin has a major role in stimulating lipolysis. The decrease in insulin was not, however, considered to be mandatory for the increase in lipolysis, since it was often observed that enhanced lipolysis occurs without a concomitant decrease in the plasma insulin concentration below preexercise concentration. Other factors such as the catecholamines, GH, glucagon, and the thyroid hormones were also considered.

Studies conducted in the 1970s in adrenaldemedullated rats by Gollnick and others showed that the catecholamines were important in stimulating lipolysis during exercise (96). Later, investigations conducted by Gorski (212) and Karlsson (157) and others confirmed that the catecholamines regulate lipolysis during exercise, both in adipose tissue and in skeletal muscle. In addition to insulin and catecholamines, Hunter and colleagues reported in 1965 (118) that GH may also be important in stimulating lipolysis during muscular activity. These and other studies conducted by Gollnick et al. (95) using hypophysectomized rats were instrumental in showing that GH is important in fat mobilization. Studies in patients with hypopituitarism or hypophysectomy, however, indicate that a normal rise in FFA may be expected without changes in GH concentration, likely because of other compensatory responses. Evidence from studies conducted in the 1970s by Kaciuba-Uscilko and coworkers (38, 136) on exercising dogs administered thyroxine or triiodothyronine indicated that the thyroid hormones may have a role in lipid mobilization. This evidence was backed by their observations that humans with hyperthyroidism had exaggerated lipid levels during exercise. Results from Maling et al. (160) from the mid-1960s, using adrenalectomized cortisone-treated rats, also suggest that the glucocorticoids may play a role in lipolysis, probably by suppressing the rate of degradation of cAMP. Once again, however, results from Gollnick's studies using adrenalectomized rats (96) indicate that the glucocorticoids are not essential in facilitating lipolysis, likely because of higher levels of GH, noradrenaline, and glucagon.

Summary

In summary, considerable progress in the field of endocrinology and metabolism has been made, to a large extent because of visionary contributions by a number of pre-eminent researchers described in this chapter. Strides have been made primarily because of interests in the effects of hormones on physical work capacity and because of the large number of studies conducted on the hormonal control of metabolism in health and disease. This vast interest in exercise physiology and biochemistry of animals and humans has developed our knowledge of the complex integrated control of glycemia and the supply of the energy of substrates to working muscle. These investigations have shown that there is broad, and sometimes redundant, neural–endocrine control of metabolism during exercise. In addition, these studies have shown us that a complex feed-forward and feedback control of hormones and metabolism exists during exercise, which fits well with what we have learned about the cardiorespiratory and neuromuscular systems during muscular work. Knowledge of the endocrine system during and after exercise is incomplete, however, and the role of many hormones still needs further exploration.

We thank Professors Michael Berger and Erik Richter for reading this chapter. In addition, we appreciate the time given by others to provide information about the stimuli for their research over the years. Finally, we acknowledge the assistance of Elizabeth Moore for her invaluable help in editing this manuscript.

References

1. Addison T. On the Constitutional and Local Effects of Disease of the Suprarenal Capsules. London: Highley. 1855.
2. Ahlborg G. and P. Felig. Influence of glucose ingestion on fuel-hormone response during prolonged exercise. *J. Appl. Physiol.* 41:683–688, 1976.
3. Ahlborg G., P. Felig, L. Hagenfeldt, R. Hendler, and J. Wahren. Substrate turnover during prolonged exercise in man: splanchnic and leg metabolism of glucose, free fatty acids and amino acids. *J. Clin. Invest.* 53D:1080–1090, 1974.
4. Ahlborg G., L. Hagenfeldt, and J. Wahren. Substrate utilization by the inactive leg during one-leg or arm exercise. *J. Appl. Physiol.* 39:718–723, 1975.
5. Allen F. M., E. Stillman, and R. Fitz. Total dietary regulation in the treatment of diabetes. In: *Exercise*, edited by F. M. Allen, E. Stillman, and R. Fitz. New York: Rockefeller Institute. 1919, pp. 468–499.
6. Argov Z., P. F. Renshaw, B. Boden, A. Winokur, and W. J. Bank. Effects of thyroid hormones on skeletal muscle bioenergetics. In vivo phosphorus-31 magnetic resonance spectroscopy study of humans and rats. *J. Clin. Invest.* 81:1695–1701, 1988.
7. Asmussen E. Muscle metabolism during exercise in man: a historical survey. In: *Muscle Metabolism During Exercise*, edited by B. Pernow and B. Saltin. New York: Plenum Press. 1971, pp. 1–12.
8. Asmussen E., J. W. Wilson, and D. B. Dill. Hormonal influences on carbohydrate metabolism during work. *Am. J. Physiol.* 130:600–607, 1940.

9. Baldwin K. M., W. W. Winder, R. L. Terjung, and J. O. Holloszy. Glycolytic enzymes in different types of skeletal muscle: adaptation to exercise. *Am. J. Physiol.* 225:962–966, 1973.

10. Baldwin K. M., A. M. Hooker, R. E. Herrick, and L. F. Schrader. Respiratory capacity and glycogen depletion in thyroid-deficient muscle. *J. Appl. Physiol.* 49:102–106, 1980.

11. Ballard K., C. A. Cobb, and S. Rosell. Vascular and lipolytic responses in canine subcutaneous adipose tissue following infusion of catecholamines. *Acta Physiol. Scand.* 81: 246–253, 1971.

12. Balsam A. and L. E. Leppo. Stimulation of the peripheral metabolism of L-thyroxine and and 3,5,3'-L-triiodothyronine in the physically trained rat. *Endocrinology* 95:299–302, 1974.

13. Balsam A. and L. E. Leppo. Effect of physical training on the metabolism of thyroid hormones in man. *J. Appl. Physiol.* 38:212–215, 1975.

14. Banister E. W. and J. Griffiths. Blood levels of adrenergic amines during exercise. *J. Appl. Physiol.* 33:674–676, 1972.

15. Barwich D., H. Hagele, M. Weiss, and H. Weiker. Hormonal and metabolic adjustments in patients with central Cushing's disease after adrenalectomy. *Int. J. Sports Med.* 2:220– 227, 1980.

16. Benedict F. G. and E. P. Cathcart. Muscular Work. A Metabolic Study, with Special Reference to the Efficiency of the Human Body as a Machine. *Carnegie Institute of Washington, Pub 187,* 1913.

17. Berger M., P. Berchtold, H. J. Cuppers, H. Drost, H. K. Kley, W. A. Muller, W. Wiegelmann, H. Zimmermann-Telschow, F. A. Gries, H. L. Kruskemper, and H. Zimmermann. Metabolic and hormonal effects of muscular exercise in juvenile type diabetics. *Diabetologia* 13:355–365, 1977.

18. Berger M., S. A. Hagg, M. N. Goodman, and N. B. Ruderman. Glucose metabolism in perfused skeletal muscle. Effects of starvation, diabetes, fatty acids, acetoacetate, insulin and exercise on glucose uptake and disposition. *Biochem. J.* 158:191–202, 1976.

19. Berger M., S. A. Hagg, and N. B. Ruderman. Glucose metabolism in perfused skeletal muscle. Interaction of insulin and exercise on glucose uptake. *Biochem. J.* 146:231–238, 1975.

20. Berger M., P. A. Halban, W. A. Muller, R. E. Offord, A. E. Renold, and M. Vranic. Mobilization of subcutaneously injected tritiated insulin in rats: effect of muscular exercise. *Diabetologia* 15:113–140, 1978.

21. Bergeron R., R. R. Russell, III, L. H. Young, J. M. Ren, M. Marcucci, A. Lee, and G. I. Shulman. Effect of AMPK activation on muscle glucose metabolism in conscious rats. *Am. J. Physiol.* 276 (Endocrinol. Metab. 39):E938–E944, 1999.

22. Bergstrom J. and E. Hultman. A study of the glycogen metabolism during exercise in man. *Scand. J. Clin. Lab. Invest.* 19:218–228, 1967.

23. Bihler I., M. Hollands, and P. E. Dresel. Stimulation of sugar transport by a factor released from gas-perfused hearts. *Can. J. Physiol. Pharmacol.* 48:327–332, 1970.

24. Birnbaum M. J. Identification of a novel gene encoding an insulin-responsive glucose transporter protein. *Cell* 57:305–315, 1989.

25. Bjorkman O., P. Miles, D. Wasserman, L. Lickley, and M. Vranic. Regulation of glucose turnover during exercise in pancreatectomized, totally insulin deficient dogs: effects of beta-adrenergic blockade. *J. Clin. Invest.* 81:1759–1767, 1988.

26. Björntorp P., K. De Jounge, L. Sjostrom, and L. Sullivan. The effect of physical training on insulin production in obesity. *Metabolism* 19:631–638, 1970.

27. Björntorp P., M. Fahlen, G. Grimby, A. Gustafson, J. Holm, P. Renstrom, and T. Schersten. Carbohydrate and lipid metabolism in middle aged physically well-trained men. *Metabolism* 21:1037–1042, 1972.

28. Bliss, M. *The Discovery of Insulin.* Chicago, IL: University of Chicago Press. 1982.

29. Bloom S. R. and A. V. Edwards. The release of pancreatic glucagon and inhibition of insulin in response to stimulation of the sympathetic innervation. *J. Physiol. (Lond.)* 253: 157–173, 1975.

30. Bloom S. R., R. H. Johnson, D. M. Park, M. J. Rennie, and W. R. Sulaiman. Differences in the metabolic and hormonal response to exercise between racing cyclists and untrained individuals. *J. Physiol. (Lond.)* 258:1–18, 1976.

31. Boden G., F. Jadali, J. White, Y. Liang, M. Mozzoli, X. Chen, E. Coleman, and C. Smith. Effects of fat on insulin-stimulated carbohydrate metabolism in normal men. *J. Clin. Invest.* 88:960–966, 1991.

32. Bondy P. K. and M. A. Hagewood. Effects of stress and cortisone on plasma-bound iodine and thyroxine metabolism in rats. *Proc. Soc. Exp. Biol. Med.* 81:328–331, 1952.

33. Bonen A. Effects of exercise on excretion rates of urinary free cortisol. *J. Appl. Physiol.* 40:155–158, 1976.

34. Bottger I., E. M. Schlein, G. R. Faloona, J. P. Knochel, and R. H. Unger. The effect of exercise on glucagon secretion. *J. Clin. Endocrinol. Metab.* 35:117–125, 1972.

35. Brandenberger G. and M. Follenius. Influence of timing and intensity of musclar exercise on temporal patterns of plasma cortisol levels. *J. Clin. Endocrinol. Metab.* 40:845–849, 1975.

36. Brandenberger G., M. Follenius, and B. Hietter. Feedback from meal-related peaks determines diurnal changes in cortisol response to exercise. *J. Clin. Endocrinol. Metab.* 54: 592–596, 1982.

37. Brown-Séquard C. E. Recherches experimentales sur la physiologie et la pathologie des capsules surrenals. *C. R. Acad. Sci.* 43:422–425, 1856.

38. Brzezinska Z. and H. Kaciuba-Uscilko. Low muscle and liver glycogen contents in dogs treated with thyroid hormones. *Horm. Metab. Res.* 11:675–678, 1979.

39. Buckler J. M. Exercise as a screening test for growth hormone release. *Acta Endocrinol. (Copenh.)* 69:219–229, 1972.

40. Buckler J. M. The relationship between changes in plasma growth hormone levels and body temperature occurring with exercise in man. *Biomedicine* 19:193–197, 1973.

41. Cannon W. B., J. R. Linton, and R. R. Linton. The effects of muscle metabolites on adrenal secretion. *Am. J. Physiol.* 71:153–162, 1924.

42. Caralis D. G., L. Edwards, and P. J. Davis. Serum total and free thyroxine and triiodothyronine during dynamic muscular exercise in man. *Am. J. Physiol.* 233 (Endocrinol. Metab. Gastrointestinal 182):E115–E118, 1977.

43. Carriere R. and H. Isler. Effects of frequent housing changes and muscular exercise on the thyroid gland of mice. *Endocrinology* 64:414–418, 1959.

44. Cashmore G. C., C. T. Davies, and J. D. Few. Relationship between increases in plasma cortisol concentration and rate of cortisol secretion during exercise in man. *J. Endocrinol.* 72:109–110, 1977.

45. Challiss R. A., M. Vranic, and G. K. Radda. Bioenergetic changes during contraction and recovery in diabetic rat skeletal muscle. *Am. J. Physiol.* 256 (Endocrinol. Metab. 19): E129–E137, 1989.

46. Chasiostis D., D. Sahlin, and E. Hultman. Regulation of glycogenolysis in human muscle at rest and during exercise. *J. Appl. Physiol.* 53:708–715, 1982.

47. Chauveau M. A. and M. Kaufmann. Experiences pour la determination du coefficient de l'activite nutritive et respiratoire des muscles in repos et en travail. *C. R. Acad. Sci. (Paris)* 104:1126–1132, 1887.

48. Christensen E. H. and O. Hansen III. Arbeitsfähigkeit und ernährung. *Scand. Arch. Physiol.* 81:160–171, 1939.

49. Cochran B. Jr., E. P. Marbach, R. Poucher, T. Steinberg, and G. Gwinup. Effect of acute muscular exercise on serum immunoreactive insulin concentration. *Diabetes* 15:838–841, 1966.

50. Coker R. H., L. Simonsen, J. Bulow, D. H. Wasserman, and M. Kjaer. Stimulation of splanchnic glucose production during exercise in humans contains a glucagon-independent component. *Am. J. Physiol. Endocrinol. Metab.* 280:E918–E927, 2001.

51. Cooper D. M., D. H. Wasserman, M. Vranic, and K. Wasserman. Glucose turnover in response to exercise during high- and low-FIO$_2$ in humans. *Am. J. Physiol.* 251 (Endocrinol. Metab. 14):E209–E214, 1986.

52. Cornil A., A. Decoster, G. Copinschi, and J. R. M. Franckson. Effect of muscular exercise on the plasma level of cortisol in man. *Acta Endocrinol. (Copenh.)* 48:163–168, 1965.

53. Davies C. T. and J. D. Few. Effects of exercise on adrenocortical function. *J. Appl. Physiol.* 35:887–891, 1973.

54. Dean D., J. R. Daugaard, M. E. Young, A. Saha, D. Vavvas, S. Asp, B. Kiens, K. H. Kim, L. Witters, E. A. Richter, and N. Ruderman. Exercise diminishes the activity of acetyl-CoA carboxylase in human muscle. *Diabetes* 49:1295–1300, 2000.

55. Defronzo R. A., E. Ferrannini, Y. Sato, P. Felig, and J. Wahren. Synergistic interaction between exercise and insulin on peripheral glucose uptake. *J. Clin. Invest.* 68:1468–1474, 1981.

56. Dieter M. P. Glucose metabolism in rat lymphatic tissues: effects of acute and chronic exercise. *Life Sci.* 8:459–468, 1969.

57. Dill D. B., H. T. Edwards, and R. H. de Meio. Effects of adrenalin injection in moderate work. *Am. J. Physiol.* 111:9–20, 1935.

58. Dill D. B., H. T. Edwards, and S. Mead. Blood sugar regulation in exercise. *Am. J. Physiol.* 111:21–30, 1935.

59. Doi K., M. Prentki, C. Yip, W. Muller, B. Jeanrenaud, and M. Vranic. Identical biological effects of pancreatic glucagon and a purified moiety of canine gastric glucagon. *J. Clin. Invest.* 63:525–531, 1979.

60. Douen A. G., T. Ramlal, A. Klip, D. A. Young, D. Cartee, and J. O. Holloszy. Exercise induced increase in glucose transporters in plasma membrane of rat skeletal muscle. *Endocrinology* 124:449–454, 1989.

61. Douen A. G., T. Ramlal, S. Rastogi, P. J. Bilan, G. D. Cartee, M. Vranic, J. O. Holloszy, and A. Klip. Exercise induces recruitment of the "insulin-responsive glucose transporter." Evidence for distinct intracellular insulin- and exercise-recruitable transporter pools in skeletal muscle. *J. Biol. Chem.* 265:13427–13430, 1990.

62. Drouin R., C. Lavoie, J. Bourque, F. Ducros, D. Poisson, and J. L. Chiasson. Increased hepatic glucose production response to glucagon in trained subjects. *Am. J. Physiol.* 274 (Endocrinol. Metab. 37): E23–E28, 1998.

63. Dyck D. J., C. T. Putman, G. J. F. Heigenhauser, E. Hultman, and L. L. Spriet. Regulation of fat-carbohydrate interaction in skeletal muscle during intense aerobic cycling. *Am. J. Physiol.* 265 (Endocrinol. Metab. 28):E852–E859, 1993.

64. Edwards J. G., D. D. Lund, T. G. Bedford, C. M. Tipton, R. D. Matthes, and P. G. Schmid. Metabolic and cardiovascular adaptations in trained hypophysectomized rats. *J. Appl. Physiol.* 53:448–454, 1982.

65. Farrell P. A., T. L. Garthwaite, and A. B. Gustafson. Plasma adrenocorticotropin and cortisol responses to submaximal and exhaustive exercise. *J. Appl. Physiol.* 55:1441–1444, 1983.

66. Febbraio M. A., D. L. Lambert, R. L. Starkie, J. Proietto, and M. Hargreaves. Effect of epinephrine on muscle glycogenolysis during exercise in trained men. *J. Appl. Physiol.* 84:465–470, 1998.

67. Felig P., J. Wahren, R. Hendler, and G. Ahlborg. Plasma glucagon levels in exercising man. *N. Engl. J. Med.* 287:184–185, 1972.

68. Felsing N. E., J. A. Brasel, and D. M. Cooper. Effect of low and high intensity exercise on circulating growth hormone in men. *J. Clin. Endocrinol. Metab.* 75:157–162, 1992.

69. Few J. D. Effect of exercise on the secretion and metabolism of cortisol in man. *J. Endocrinol.* 62:341–353, 1974.

70. Few J. D., F. J. Imms, and J. S. Weiner. Pituitary-adrenal response to static exercise in man. *Clin. Sci. Mol. Med.* 49:201–206, 1975.

71. Finegood D. T., P. D. Miles, H. L. Lickley, and M. Vranic. Estimation of glucose production during exercise with a one-compartment variable-volume model. *J. Appl. Physiol.* 72: 2501–2509, 1992.

72. Fisher S. J., M. Lekas, Z. Q. Shi, D. Bilinski, G. Carvalho, A. Giacca, and M. Vranic. Insulin-independent acute restoration of euglycemia normalizes the impaired glucose clearance during exercise in diabetic dogs. *Diabetes* 46:1805–1812, 1997.

73. Franckson J. R., R. Vanroux, R. Leclercq, H. Brunengraber, and H. A. Ooms. Labelled insulin catabolism and pancreatic responsiveness during long-term exercise in man. *Horm. Metab. Res.* 3:366–373, 1971.

74. Frenkl R. and L. Csalay. Effects of regular muscular activity on adrenocorticoid function in rats. *J. Sport. Med.* 2:207–211, 1962.

75. Frenkl R., L. Csalay, G. Csakvary, and T. Zelles. Effect of muscular exertion on the reaction of the pituitary-adrenocortical axis in trained and untrained rats. *Acta Physiol. Acad. Sci. Hung.* 33:435–438, 1968.

76. Friedman J. E. Role of glucocorticoids in activation of hepatic PEPCK gene transcription during exercise. *Am. J. Physiol.* 266 (Endocrinol. Metab. 29):E560–E566, 1994.

77. Fryer L. G., E. Hajduch, F. Rencurel, I. P. Salt, H. S. Hundal, D. G. Hardie, and D. Carling. Activation of glucose transport by AMP-activated protein kinase via stimulation of nitric oxide synthase. *Diabetes* 49:1978–1985, 2000.

78. Fujii N., T. Hayashi, M. F. Hirshman, J. T. Smith, S. A. Habinowski, L. Kaijser, J. Mu, O. Ljungqvist, M. J. Birnbaum, L. A. Witters, A. Thorell, and L. J. Goodyear. Exercise induces isoform-specific increase in 5'AMP-activated protein kinase activity in human skeletal muscle. *Biochem. Biophys. Res. Commun.* 273:1150–1155, 2000.

79. Galbo H. Hormonal and Metabolic Adaptation to Exercise. New York: Thieme-Stratton. 1983, pp. 5–40.

80. Galbo H., N. J. Christensen, and J. J. Holst. Catecholamines and pancreatic hormones during autonomic blockade in exercising man. *Acta Physiol. Scand.* 101:428–437, 1977.

81. Galbo H., N. J. Christensen, and J. J. Holst. Glucose-induced decrease in glucagon and epinephrine response to exercise in man. *J. Appl. Physiol.* 42:525–530, 1977.

82. Galbo H., N. J. Christensen, K. J. Mikines, B. Sonne, J. Hilsted, C. Hagen, and J. Fahrenkrug. The effect of fasting on the hormonal response to graded exercise. *J. Clin. Endocrinol. Metab.* 52:1106–1112, 1981.

83. Galbo H., J. J. Holst, and N. J. Christensen. Glucagon and plasma catecholamine responses to graded and prolonged exercise in man. *J. Appl. Physiol.* 38:70–76, 1975.

84. Galbo H., J. J. Holst, and N. J. Christensen. The effect of different diets and of insulin on the hormonal response to prolonged exercise. *Acta Physiol. Scand.* 107:19–32, 1979.

85. Galbo H., J. J. Holst, N. J. Christensen, and J. Hilsted. Glucagon and plasma catecholamines during beta-receptor blockade in exercising man. *J. Appl. Physiol.* 40:855–863, 1976.

86. Galbo H., M. E. Houston, N. J. Christensen, J. J. Holst, B. Nielsen, E. Nygaard, and J. Suzuki. The effect of water temperature on the hormonal response to prolonged swimming. *Acta Physiol. Scand.* 105:326–337, 1979.

87. Galbo H., L. Hummer, I. B. Peterson, N. J. Christensen, and N. Bie. Thyroid and testicular hormone responses to graded and prolonged exercise in man. *Eur. J. Appl. Physiol.* 36: 101–106, 1977.

88. Galbo H., E. A. Richter, N. J. Christensen, and J. J. Holst. Sympathetic control of metabolic and hormonal responses to exercise in rats. *Acta Physiol. Scand.* 102:441–449, 1978.

89. Galbo H., E. A. Richter, J. Hilsted, J. J. Holst, N. J. Christensen, and J. Henriksson. Hormonal regulation during prolonged exercise. *Ann. N. Y. Acad. Sci.* 301:72–80, 1977.

90. Galbo H., E. A. Richter, J. J. Holst, and N. J. Christensen. Diminished hormonal responses to exercise in trained rats. *J. Appl. Physiol.* 43:953–958, 1977.

91. Garetto L. P., E. A. Richter, M. N. Goodman, and N. B. Ruderman. Enhanced muscle glucose metabolism after exercise in the rat:the two phases. *Am. J. Physiol.* 246 (Endocrinol. Metab. 9):E471–E475, 1984.

92. Goldstein M. S. Humoral nature of hypoglycemia in muscular exercise. *Am. J. Physiol.* 200:67–70, 1961.

93. Goldstein M. S., V. Mullick, B. Huddlestun, and R. Levine. Action of muscular work on transfer of sugar across cell barriers: comparison with the action of insulin. *Am. J. Physiol.* 173:212–216, 1953.

94. Gollnick P. D. and C. D. Ianuzzo. Hormonal deficiencies and the metabolic adaptations of rats to training. *Am. J. Physiol.* 223:278–282, 1972.

95. Gollnick P. D. and C. D. Ianuzzo. Acute and chronic adaptations to exercise in hormone deficient rats. *Med. Sci. Sports* 7:12–19, 1975.

96. Gollnick P. D., R. G. Soule, A. W. Taylor, C. Williams, and C. D. Ianuzzo. Exercise-induced glycogenolysis and lipolysis in the rat: hormonal influence. *Am. J. Physiol.* 219:729–733, 1970.

97. Goodyear L. J., P. A. King, M. F. Hirshman, C. M. Thompson, E. D. Horton, and E. S. Horton. Contractile activity increases plasma membrane glucose transporters in absence of insulin. *Am. J. Physiol.* 258 (Endocrinol. Metab. 21):E667–E672, 1990.

98. Gray I. and W. P. Beetham. Changes in plasma concentration of epinephrine and norepinephrine with muscular work. *Proc. Soc. Exp. Biol. Med.* 96:636–638, 1957.

99. Gyntelberg F., M. J. Rennie, R. C. Hickson, and J. O. Holloszy. Effect of training on the response of plasma glucagon to exercise. *J. Appl. Physiol.* 43:302–305, 1977.

100. Hagenfeldt L. Metabolism of free fatty acids and ketone bodies during exercise in normal and diabetic man. *Diabetes* 28:66–70, 1979.

101. Haggendal J., L. H. Hartley, and B. Saltin. Arterial noradrenaline concentration during exercise in relation to the relative work levels. *Scand. J. Clin. Lab. Invest.* 26:337–342, 1970.

102. Hansen A. P. The effect of adrenergic receptor blockade on the exercise-induced serum growth hormone rise in normals and juvenile diabetics. *J. Clin. Endocrinol. Metab.* 33:807–812, 1971.

103. Harada T., T. Yamauchi, A. Tsukanaka, Y. Matsumura, M. Kurono, A. Honda, and N. Matsui. Involvement of muscarinic cholinergic and alpha2-adrenergic mechanisms in growth hormone secretion during exercise in humans. *Eur. J. Appl. Physiol.* 83:268–273, 2000.

104. Hartley L. H., J. W. Mason, R. P. Hogan, L. G. Jones, T. A. Kotchen, E. H. Mougey, F. E. Wherry, L. L. Pennington, and P. T. Ricketts. Multiple hormonal responses to graded exercise in relation to physical training. *J. Appl. Physiol.* 33:602–606, 1972.

105. Hartley L. H., J. W. Mason, R. P. Hogan, L. G. Jones, T. A. Kotchen, E. H. Mougey, F. E. Wherry, L. L. Pennington, and P. T. Ricketts. Multiple hormonal responses to prolonged exercise in relation to physical training. *J. Appl. Physiol.* 33:607–610, 1972.

106. Hartman F. A., R. A. Waite, and H. A. McCordock. The liberation of epinephrine during muscular work. *Am. J. Physiol.* 62:225–241, 1922.

107. Havivi E. and H. E. Wertheimer. A muscle activity factor increasing sugar uptake by rat diaphragms in vitro. *J. Physiol. (Lond.)* 172:342–352, 1964.

108. Hayashi T, M. F. Hirshman, S. D. Dufresne, and L. J. Goodyear. Skeletal muscle contractile activity in vitro stimulates mitogen-activated protein kinase signalling. *Am. J. Physiol.* 277 (Cell Physiol. 46):C701–C707, 1999.

109. Hayashi T., M. F. Hirshman, E. J. Kurth, W. W. Winder, and L. J. Goodyear. Evidence for 5′ AMP-activated protein kinase mediation of the effect of muscle contraction on glucose transport. *Diabetes* 47:1369–1373, 1998.

110. Helmreich E. and C. F. Cori. Studies of tissue permeability. II. Distribution of pentoses between plasma and muscle. *J. Biol. Chem.* 224:663–679, 1957.

111. Hill S., F. Goetz, H. Fox, B. Murawski, L. Krakauer, R. Reifenstein, S. Gray, W. Reddy, S. Hedberg, J. St Marc, and G. Thorn. Studies on adrenocortical and psychological responses to stress in man. *Arch. Int. Med.* 97:269–298, 1956.

112. Hilsted J., H. Galbo, T. Sonne, T. Schwartz, O. Fahrenkrug, K. Schaffalitzky De Muckadell, K. Lauritsen, and B. Tronier. Gastroenteropancreatic hormonal changes during exercise. *Am. J. Physiol.* 239 (Gastrointest. Liver Physiol. 2):G136–G140, 1980.

113. Hirsch I. B., J. C. Marker, L. J. Smith, R. J. Spina, C. A. Parvin, J. O. Holloszy, and P. E. Cryer. Insulin and glucagon in prevention of hypoglycemia during exercise in humans. *Am. J. Physiol.* 260:E695–E704, 1991.

114. Hoelzer D., G. Dalsky, W. Clutter, S. D. Shah, J. O. Holloszy, and P. E. Cryer. Glucoregulation during exercise: hypoglycemia is prevented by redundant glucoregulatory systems during exercise: sympathochromaffin activation, and changes in hormone secretion. *J. Clin. Invest.* 77:212–221, 1986.

115. Hollander C. S., R. L. Scott, J. A. Burgess, D. Rabinowitz, T. J. Merimee, and J. H. Oppenheimer. Free fatty acids: a possible regulator of free thyroid hormone levels in man. *J. Clin. Endocrinol. Metab.* 27:1219–1223, 1967.

116. Holloszy J. O. and H. T. Narahara. Studies of tissue permeabilty. X. Changes in permeabilty to 3-methylglucose associated with contraction of isolated frog muscle. *J. Biol. Chem.* 240:3493–3500, 1965.

117. Hoskins R. G. Studies on vigor. XVI. Endocrine factors in vigor. *Endocrinology* 11:97–105, 1927.

118. Hunter W. M., C. C. Fonseka, and R. Passmore. Growth hormone: important role in muscular exercise in adults. *Science* 150:1051–1053, 1965.

119. Hunter W. M. and M. Y. Sukkar. Changes in plasma insulin levels during muscular exercise. *J. Physiol. (Lond.)* 196:110P–112P, 1968.

120. Ingle D. The time for the work capacity of the adrenalectomized rats treated with cortin. *Am. J. Physiol.* 116:622–625, 1934.

121. Ingle D., J. Nezamis, and E. Morley. The comparative value of cortisone, 17-hydroxycorticosteroids and adrenal cortical extract given by continuous intravenous injection in sustaining the ability of the adrenalectomized rat to work. *Endocrinology* 50:1–4, 1952.

122. Ingle D. J., H. D. Moon, and H. M. Evans. Work performance of hypophysectomized rats treated with anterior pituitary extracts. *Am. J. Physiol.* 123:620, 1938.

123. Ingle D. J., J. E. Nezamis, and E. H. Morley. Work output and blood glucose values in severely diabetic rats with and without insulin. *Am. J. Physiol.* 165:469–472, 1951.

124. Irvine C. H. Thyroxine secretion rate in the horse in various physiological states. *J. Endocrinol.* 39:313–320, 1967.

125. Irvine C. H. Effect of exercise on thyroxine degradation in athletes and non-athletes. *J. Clin. Endocrinol. Metab.* 28:942–948, 1968.

126. Issekutz B. Role of beta-adrenergic receptors in mobilization of energy sources in exercising dogs. *J. Appl. Physiol.* 44:869–876, 1978.

127. Issekutz B. Energy mobilization in exercising dogs. *Diabetes* 28:39–44, 1979.

128. Issekutz B. and M. Vranic. Role of glucagon in regulation of glucose production in exercising dogs. *Am. J. Physiol.* 238 (Endocrinol. Metab. 1):E13–E20, 1980.

129. Issekutz B. Jr. The role of hypoinsulinemia in exercise metabolism. *Diabetes* 29:629–635, 1980.

130. Issekutz B. Jr., P. Paul, and H. I. Miller. Metabolism in normal and pancreatectomized dogs during steady-state exercise. *Am. J. Physiol.* 213:857–862, 1967.

131. James D. E., R. Brown, J. Navarro, and P. F. Pilch. Insulin-regulatable tissues express a unique insulin-sensitive glucose transport protein. *Nature* 333:183–185, 1988.

132. Jarhult J. and J. Holst. The role of the adrenergic innervation to the pancreatic islets in the control of insulin release during exercise in man. *Pflugers Arch.* 383: 41–45, 1979.

133. Johannessen A., C. Hagen, and H. Galbo. Prolactin, growth hormone, thyrotropin, 3,5,3′-triiodothyronine, and thyroxine responses to exercise after fat- and carbohydrate-enriched diet. *J. Clin. Endocrinol. Metab.* 52:56–61, 1981.

134. Kaciuba-Uscilko H., Z. Brzezinska, and A. Kobryn. Metabolic and temperature responses to physical exercise in thyroidectomized dogs. *Eur. J. Appl. Physiol.* 40:219–226, 1979.

135. Kaciuba-Uscilko H., G. A. Dudley, and R. L. Terjung. Muscle LPL activity, plasma and muscle triglycerides in trained thyroidectomized rats. *Horm. Metab. Res.* 13:688–690, 1981.

136. Kaciuba-Uscilko H., J. E. Greenleaf, S. Kozlowski, Z. Brzezinska, K. Nzar, and A. Ziemba. Thyroid hormone-induced changes in body temperature and metabolism during exercise in dogs. *Am. J. Physiol.* 229:260–264, 1975.

137. Karagiorgos A., J. F. Garcia, and G. A. Brooks. Growth hormone response to continuous and intermittent exercise. *Med. Sci. Sports* 11:302–307, 1979.

138. Katz A., S. Brobert, K. Sahlin, and J. Wahren. Leg glucose uptake during maximal dynamic exercise in humans. *Am. J. Physiol.* 251 (Endocrinol. Metab. 14):E65–E70, 1986.

139. Kawamori R. and M. Vranic. Mechanism of exercise-induced hypoglycemia in depancreatized dogs maintained on long-acting insulin. *J. Clin. Invest.* 59:331–337, 1977.

140. Kjaer M., K. Engfred, A. Fernandez, and H. Galbo. Regulation of hepatic glucose production during exercise in humans: role of sympathoadrenergic activity. *Am. J. Physiol.* 265 (Endocrinol. Metab. 28):E275–E283, 1993.

141. Kjaer M., K. Howlett, J. Langfort, T. Zimmerman-Belsing, J. Lorentsen, J. Bulow, J. Ihlemann, U. Feldt-Rasmussen, and H. Galbo. Adrenaline and glycogenolysis in skeletal muscle during exercise: a study in adrenalectomised humans. *J. Physiol. (Lond.)* 528 Pt 2:371–378, 2000.

142. Klimes I., M. Vigas, J. Jurcovicova, and S. Nemeth. Lack of effect of acid-base alterations on growth hormone secretion in man. *Endocrinol. Exp.* 11:155–162, 1977.

143. Koivisto V., R. Hendler, E. Nadel, and P. Felig. Influence of physical training on the fuel-hormone response to prolonged low intensity exercise. *Metabolism* 31:192–197, 1982.

144. Kotchen T. A., L. H. Hartley, T. W. Rice, E. H. Mougey, L. G. Jones, and J. W. Mason. Renin, norepinephrine, and epinephrine responses to graded exercise. *J. Appl. Physiol.* 31:178–184, 1971.

145. Kraus H., R. Kirsten, and J. R. Wolff. Effect of swimming and running exercise on the cellular function and structure of muscle. *Pflugers Arch.* 308:57–79, 1969.

146. Kreisman S. H., M. N. Ah, J. B. Halter, M. Vranic, and E. B. Marliss. Norepinephrine infusion during moderate-intensity exercise increases glucose production and uptake. *J. Clin. Endocrinol. Metab.* 86:2118–2124, 2001.

147. Kreisman S. H., A. Manzon, S. J. Nessim, J. A. Morais, R. Gougeon, S. J. Fisher, M. Vranic, and E. B. Marliss. Glucoregulatory responses to intense exercise performed in the postprandial state. *Am. J. Physiol. (Endocrinol. Metab.)* 278:E786–E793, 2000.

148. Kreisman S. H., N. A. Mew, M. Arsenault, S. J. Nessim, J. B. Halter, M. Vranic, and E. B. Marliss. Epinephrine infusion during moderate intensity exercise increases glucose production and uptake. *Am. J. Physiol. (Endocrinol Metab.)* 278:E949–E957, 2000.

149. Kristiansen S., M. Hargreaves, and E. A. Richter. Exercise-induced increase in glucose transport, GLUT-4, and VAMP-2 in plasma membrane from human muscle. *Am. J. Physiol.* 270 (Endocrinol. Metab. 33):E197–E201, 1996.

150. Krogh A. and J. Lindhard. The relative values of fat and carbohydrate as sources of muscular energy. *Biochem. J.* 14:290, 1920.

151. Krzentowski G., F. Pirnay, N. Pallikarakis, A. S. Luyckx, M. Lacroix, F. Mosora, and P. J. Lefèbvre. Glucose utilization during exercise in normal and diabetic subjects. The role of insulin. *Diabetes* 30:983–989, 1981.

152. Kurth-Kraczek E. J., M. F. Hirshman, L. J. Goodyear, and W. W. Winder. 5' AMP-activated protein kinase activation causes GLUT 4 translocation in skeletal muscle. *Diabetes* 48: 1667–1671, 1999.

153. Lassarre C., F. Girard, J. Durand, and J. Raynaud. Kinetics of human growth hormone during submaximal exercise. *J. Appl. Physiol.* 37:826–830, 1974.

154. Lavine S. A., B. Gordon, and C. L. Derick. Some changes in the chemical constituents of the blood following a marathon race. *JAMA* 82:1778–1779, 1924.

155. Lawrence R. D. The effects of exercise on insulin action in diabetes. *Br. Med. J.* 1:648–652, 1926.

156. Legare A., R. Drouin, M. Milot, D. Massicotte, F. Peronnet, G. Massicotte, and C. Lavoie. Increased density of glucagon receptors in liver from endurance-trained rats. *Am. J. Physiol. Endocrinol. Metab.* 280:E193–E196, 2001.

157. Lithell H., M. Cedermark, J. Froberg, P. Tesch, and J. Karlsson. Increase of lipoprotein-lipase activity in skeletal muscle during heavy exercise. Relation to epinephrine excretion. *Metabolism* 30:1130–1134, 1981.

158. Lundborg P., H. Astrom, C. Bengtsson, E. Fellenius, H. von Schenck, L. Svensson, and U. Smith. Effect of beta-adrenoceptor blockade on exercise performance and metabolism. *Clin. Sci. (Colch.)* 61:299–305, 1981.

159. Luyckx A. S., A. Dresse, A. Cession-Fossion, and P. J. Lefebvre. Catecholamines and exercise-induced glucagon and fatty acid mobilization in the rat. *Am. J. Physiol.* 229:376–383, 1975.

160. Maling H. M., D. N. Stern, P. D. Altland, B. Highman, and B. B. Brodie. The physiologic role of the sympathetic nervous system in exercise. *J. Pharmacol. Exp. Ther.* 154:35–45, 1966.

161. Marliss E. B., E. Simantirakis, P. D. G. Miles, C. Purdon, R. Gougeon-Reygburn, C. J. Field, J. B. Halter, and M. Vranic. Glucoregulatory and hormonal responses to repeated bouts of intense exercise in normal male subjects. *J. Appl. Physiol.* 71(3):924–933, 1991.

162. Mason J. W., L. H. Hartley, T. A. Kotchen, F. E. Wherry, L. L. Pennington, and L. G. Jones. Plasma thyroid-stimulating hormone response in anticipation of muscular exercise in the human. *J. Clin. Endocrinol. Metab.* 37:403–406, 1973.

163. McGarry J. D., G. F. Leatherman, and D. W. Foster. Carnitine palmitoyltransferase I. The site of inhibition of hepatic fatty acid oxidation by malonylCoA. *J. Biol. Chem.* 253: 4128–4136, 1978.

164. Merrill G. F., E. J. Kurth, D. G. Hardie, and W. W. Winder. AICA riboside increases AMP-activated protein kinase, fatty acid oxidation, and glucose uptake in rat muscle. *Am. J. Physiol.* 273 (Endocrinol. Metab. 36):E1107–E1112, 1997.

165. Métivier G., J. Poortmans, and R. Vanroux. Metabolic controls of human growth hormone in trained athletes performing various workloads. In: *Third International Symposium on Biochemistry of Exercise*, edited by W. A. R. Orban. Miami: Symposia Specialists. 1978, pp. 209–218.

166. Moates J. M., D. B. Lacy, R. E. Goldstein, A. D. Cherrington, and D. H. Wasserman. Metabolic role of the exercise-induced increment in epinephrine in the dog. *Am. J. Physiol.* 255 (Endocrinol. Metab. 18):E428–E436, 1988.

167. Nazar K. Adrenocortical activation during long-term exercise in dogs: evidence for a glucostatic mechanism. *Pflugers Arch.* 329:156–166, 1971.

168. Nazar K., Z. Brzezinska, J. Lyszczarz, and A. Danielewicz-Kotowicz. Sympathetic control of the utilization of energy substrates during long-term exercise in dogs. *Arch. Int. Physiol. Biochim.* 79:873–879, 1971.

169. Nelson D. H. Pituitary-adrenal system. In: *Endocrinology*, edited by S. M. McCann. Bethesda: American Physiological Society, 1988, pp. 87–115.

170. Nesher R., I. E. Karl, and D. M. Kipnis. Epitrochlearis muscle. II. Metabolic effects of contraction and catecholamines. *Am. J. Physiol.* 239 (Endocrinol. Metab. 7):E461–E467, 1980.

171. Nesher R., I. E. Karl, and K. M. Kipnis. Dissociation of the effects of insulin and contraction on glucose transport in rat epitrochlearis muscle. *Am. J. Physiol.* 249 (Cell Physiol. 18):C226–C232, 1985.

172. Newsholme E. A. The control of fuel utilization during exercise and starvation. *Diabetes* 28 (Suppl. 1):1–7, 1979.

173. Nikkila E. A., M. R. Taskinen, T. A. Miettinen, R. Pelkonen, and H. Poppius. Effect of muscular exercise on insulin secretion. *Diabetes* 17:209–218, 1968.

174. Nilsson K. O., L. G. Heding, and B. Hokfelt. The influence of short term submaximal work on the plasma concentrations of catecholamines, pancreatic glucagon and growth hormone in man. *Acta Endocrinol. (Copenh.)* 79:286–294, 1975.

175. O'Connell M., D. C. Robbins, E. S. Horton, E. A. Sims, and E. Danforth Jr. Changes in serum concentrations of 3,3',5'-triiodothyronine and 3,5,3'-triiodothyronine during prolonged moderate exercise. *J. Clin. Endocrinol. Metab.* 49:242–246, 1979.

176. Odland L. M., G. J. Heigenhauser, G. D. Lopaschuk, and L. L. Spriet. Human skeletal muscle malonyl-CoA at rest and during prolonged submaximal exercise. *Am. J. Physiol.* 270 (Endocrinol. Metab. 33):E541–E544, 1996.

177. Paul P. FFA metabolism of normal dogs during steady-state exercise at different work loads. *J. Appl. Physiol.* 28:127–132, 1970.

178. Ploug T., H. Galbo, J. Vinten, M. Jorgensen, and E. Richter. Increased muscle glucose uptake during contraction: no need for insulin. *Am. J. Physiol.* 247 (Endocrinol. Metab. 10):E726–E731, 1984.

179. Premachandra B. N., W. W. Winder, R. Hickson, S. Lang, and J. O. Holloszy. Circulating reverse triiodothyronine in humans during exercise. *Eur. J. Appl. Physiol.* 47:281–288, 1981.

180. Price T. B., D. L. Rothman, R. Taylor, M. J. Avison, G. I. Shulman, and R. G. Shulman. Human muscle glycogen after exercise: insulin-dependent and independent phases. *J. Appl. Physiol.* 76:104–111, 1994.

181. Pruett E. D. Glucose and insulin during prolonged work stress in men living on different diets. *J. Appl. Physiol.* 28:199–208, 1970.

182. Pruett E. D. Plasma insulin concentrations during prolonged work at near maximal oxygen uptake. *J. Appl. Physiol.* 29:155–158, 1970.

183. Randle P. J., P. B. Garland, C. N. Hales, and E. A. Newsholme. The glucose-fatty acid cycle: its role in insulin sensitivity and the metabolic disturbances of diabetes mellitus. *Lancet* 1:785–789, 1963.

184. Rasio E., W. Malaisse, J. R. Franckson, and V. Conard. Serum insulin during acute muscular exercise in normal man. *Arch. Int. Pharmacodyn. Ther.* 160:485–491, 1966.

185. Ravussin E., C. Bogardus, K. Scheidegger, B. LaGrange, E. D. Horton, and E. S. Horton. Effect of elevated FFA on carbohydrate and lipid oxidation during prolonged exercise in humans. *J. Appl. Physiol.* 60:893–900, 1986.

186. Refsum H. E. and S. B. Stromme. Serum thyroxine, triiodothyronine and thyroid stimulating hormone after prolonged heavy exercise. *Scand. J. Clin. Lab. Invest.* 39:455–459, 1979.

187. Rennie M. J. and J. O. Holloszy. Inhibition of glucose uptake and glycogenolysis by availability of oleate in well-oxygenated perfused skeletal muscle. *Biochem. J.* 168:161–170, 1977.

188. Rennie M. J. and R. H. Johnson. Alteration of metabolic and hormonal responses to exercise by physical training. *Eur. J. Appl. Physiol. Occup. Physiol.* 33:215–226, 1974.

189. Rhodes B. A. Effects of exercise on the thyroid gland. *Nature* 216:918–919, 1967.

190. Richter C. P. The role played by the thyroid gland in the production of gross body activity. *Endocrinology* 17:73–87, 1933.

191. Richter E. A., H. Galbo, and N. J. Christensen. Control of exercise-induced muscular glycogenolysis by adrenal medullary hormones in rats. *J. Appl. Physiol.* 50:21–26, 1981.

192. Richter E. A., H. Galbo, B. Sonne, J. J. Holst, and N. J. Christensen. Adrenal medullary control of muscular and hepatic glycogenolysis and of pancreatic hormonal secretion in exercising rats. *Acta Physiol. Scand.* 108:235–242, 1980.

193. Richter E. A., L. P. Garetto, M. N. Goodman, and N. B. Ruderman. Muscle glucose metabolism following exercise in the rat. Increased sensitivity to insulin. *J. Clin. Invest.* 69:785–789, 1982.

194. Richter E. A., L. P. Garetto, M. N. Goodman, and N. B. Ruderman. Enhanced muscle glucose metabolism following exercise in the rat: modulation by local factors. *Am. J. Physiol.* 246 (Endocrinol. Metab. 9):E476–E482, 1984.

195. Richter E. A., N. B. Ruderman, H. Gavras, E. Belur, and H. Galbo. Muscle glycogenolysis during exercise: dual control by epinephrine and contractions. *Am. J. Physiol.* 242 (Endocrinol. Metab. 5):E25–E32, 1982.

196. Richter E. A., B. Sonne, N. J. Christensen, and H. Galbo. Role of epinephrine for muscular glycogenolysis and pancreatic hormonal secretion in running rats. *Am. J. Physiol.* 240 (Endocrinol. Metab. 3):E526–E532, 1981.

197. Rosell S. Release of free fatty acids from subcutaneous adipose tissue in dogs following sympathetic nerve stimulation. *Acta Physiol. Scand.* 67:343–351, 1966.

198. Rowell L. B., E. J. Masoro, and M. J. Spencer. Splanchnic metabolism in exercising man. *J. Appl. Physiol.* 20:1032–1037, 1965.

199. Ruderman N. B., O. P. Ganda, and K. Johansen. Effects of physical training on glucose tolerance and plasma lipids in maturity onset diabetes mellitus. *Diabetes* 28 (Suppl.):89, 1979.

200. Ruderman N. B., Saha A. K., Vavvas D., and L. A. Witters. Malonyl CoA, fuel sensing and insulin resistance. *Am. J. Physiol.* 276 (Endocrinol. Metab. 39):E1–E18, 1999.

201. Ruderman N. B., C. R. Houghton, and R. Hems. Evaluation of the isolated perfused rat hindquarter for the study of muscle metabolism. *Biochem. J.* 124:639–651, 1971.

202. Saltin B., F. Lindgarde, M. Houston, R. Horlin, E. Nygaard, and P. Gad. Physical training and glucose tolerance in middle-aged men with chemical diabetes. *Diabetes* 28 (Suppl. 1):30–32, 1979.

203. Sellers T. L., A. W. Jaussi, H. T. Yang, R. W. Heninger, and W. W. Winder. Effect of the exercise-induced increase in glucocorticoids on endurance in the rat. *J. Appl. Physiol.* 65:173–178, 1988.

204. Selye, H. *Stress.* Montreal: Acta. 1950, p. 37.

205. Sembrowich W. L., C. D. Ianuzzo, C. W. Saubert, R. E. Sheperd, and P. D. Gollnick. Substrate mobilization during prolonged exercise in 6-hydroxydopamine treated rats. *Pflugers Arch.* 349:57–62, 1974.

206. Shephard R. J. and K. H. Sidney. Effects of physical exercise on plasma growth hormone and cortisol levels in human subjects. *Exerc. Sport Sci. Rev.* 3:1–30, 1975.

207. Shi Z. Q., A. Giacca, K. Yamatani, S. J. Fisher, H. L. Lickley, and M. Vranic. Effects of subbasal insulin infusion on resting and exercise-induced glucose turnover in depancreatized dogs. *Am. J. Physiol.* 264 (Endocrinol. Metab. 27):E334–E341, 1993.

208. Sigal R. J., S. Fisher, J. B. Halter, M. Vranic, and E. B. Marliss. The roles of catecholamines in glucoregulation in intense exercise as defined by the islet cell clamp technique. *Diabetes* 45:148–156, 1996.

209. Simonson D. C., V. Koivisto, R. S. Sherwin, E. Ferrannini, R. Hendler, J. Juhlin-Dannfeldt, and R. DeFronzo. Adrenergic blockade alters glucose kinetics during exercise in insulin-dependent diabetics. *J. Clin. Invest.* 73:1648–1658, 1984.

210. Song M. K., C. D. Ianuzzo, C. W. Saubert, and P. D. Gollnick. The mode of adrenal gland enlargement in the rat in response to exercise training. *Pflugers Arch.* 339:59–68, 1973.

211. Spriet L. L., J. M. Ren, and E. Hultman. Epinephrine infusion enhances muscle glycogenolysis during prolonged electrical stimulation. *J. Appl. Physiol.* 64:1439–1444, 1988.

212. Stankiewicz-Choroszucha B. and J. Gorski. Effect of beta-adrenergic blockade on intramuscular triglyceride mobilization during exercise. *Experientia* 34:357–358, 1978.

213. Staub A., L. Sinn, and O. B. Behrens. Purification and crystallization of hyperglycemic-glycogenolytic factor (HGF). *Science* 117:628–629, 1953.

214. Steele R., J. S. Wall, R. C. deBodo, and N. Altszuler. Measurement of size and turnover rate of body glucose pool by isotope dilution method. *Am. J. Physiol.* 187:15–24, 1956.

215. Struck P. J. and C. M. Tipton. Effect of acute exercise on glycogen levels in adrenalectomized rats. *Endocrinology* 95:1385–1391, 1974.

216. Sutherland E. W. Studies on the mechanism of hormone action. *Science* 177:401–408, 1972.

217. Sutton J. and Lazarus L. Growth hormone in exercise: comparison of physiological and pharmacological stimuli. *J. Appl. Physiol.* 41:523–527, 1976.

218. Sutton J. R. Hormonal and metabolic responses to exercise in subject of high and low work capacities. *Med. Sci. Sports* 10:1–6, 1978.

219. Sutton J. R. and J. H. Casey. The adrenocortical response to competitive athletics in veteran athletes. *J. Clin. Endocrinol. Metab.* 40:135–138, 1975.

220. Sutton J. R., N. L. Jones, and C. J. Toews. Growth hormone secretion in acid-base alterations at rest and during exercise. *Clin. Sci. Mol. Med.* 50:241–247, 1976.

221. Sutton J. R., J. D. Young, L. Lazarus, J. B. Hickie, and J. Maksvytis. The hormonal response to physical exercise. *Australas. Ann. Med.* 18:84–90, 1969.

222. Suzuki T., K. Otsuka, H. Matsui, S. Ohukuzi, K. Sakai, and Y. Harada. Effect of muscular exercise on adrenal 17-hydroxycorticosteroid secretion in the dog. *Endocrinology* 80:1148–1151, 1967.

223. Tepperman J. A view of the history of biology from an islet of langerhans. In: *Endocrinology*, edited by S. M. McCann. Bethesda: American Physiological Society, 1988, pp. 285–333.

224. Terjung R. L. and J. E. Koerner. Biochemical adaptations in skeletal muscle of trained thyroidectomized rats. *Am. J. Physiol.* 230:1194–1197, 1976.

225. Terjung R. L. and C. M. Tipton. Plasma thyroxine and thyroid-stimulating hormone levels during submaximal exercise in humans. *Am. J. Physiol.* 220:1840–1845, 1971.

226. Tharp G. D. The role of glucocorticoids in exercise. *Med. Sci. Sports* 7:6–11, 1975.

227. Thorn G. W., D. Jenkins, and J. C. Laidlaw. The adrenal response to stress in man. *Recent Progr. Horm. Res.* 7:171–215, 1953.

228. Tipton C. M., P. J. Struck, K. M. Baldwin, R. D. Matthes, and R. T. Dowell. Response of adrenalectomized rats to chronic exercise. *Endocrinology* 91:573–579, 1972.

229. Treadway J. L., D. E. James, E. Burcel, and N. B. Ruderman. Effect of exercise on insulin receptor binding and kinase activity in skeletal muscle. *Am. J. Physiol.* 256 (Endocrinol. Metab. 19):E138–E144, 1989.

230. Tuomilehto J., J. Lindstrom, J. G. Eriksson, T. T. Valle, H. Hamalainen, P. Ilanne-Parikka, S. Keinanen-Kiukaanniemi, M. Laakso, A. Louheranta, M. Rastas, V. Salminen, and M. Uusitupa. Prevention of type 2 diabetes mellitus by changes in lifestyle among subjects with impaired glucose tolerance. *N. Engl. J. Med.* 344:1343–1350, 2001.

231. Vavvas D., A. Apazidis, A. K. Saha, J. Gamble, A. Patel, B. E. Kemp, L. A. Witters, and N. B. Ruderman. Contraction-induced changes in acetyl-CoA carboxylase and 5′-AMP-activated kinase in skeletal muscle. *J. Biol. Chem.* 272:13255–13261, 1997.

232. Venning E. H. and V. Kazmin. Excretion of urinary corticoids and 17-ketosteroids in the normal individual. *Endocrinology* 39:131–133, 1946.

233. Viru A. Dynamics of blood corticoid content during and after short term exercise. *Endokrinologie* 59:61–68, 1972.

234. Viru A. and H. Akke. Effects of muscular work on cortisol and corticosterone content in the blood and adrenals of guinea pigs. *Acta Endocrinol. (Copenh.)* 62:385–390, 1969.

235. Von Euler U. S. and S. Hellner. Excretion of noradrenaline and adrenaline in muscular work. *Acta Physiol. Scand.* 26:183–191, 1952.

236. Vranic M., S. Horvath, and J. Wahren. Proceedings of a conference on diabetes and exercise. Sponsored by the Kroc Foundation, Santa Ynez Valley, California. *Diabetes* 28 (Suppl. 1):1–113, 1979.

237. Vranic M., R. Kawamori, S. Pek, N. Kovacevic, and G. Wrenshall. The essentiality of insulin and the role of glucagon in regulating glucose utilization and production during strenuous exercise in dogs. *J. Clin. Invest.* 57:245–256, 1976.

238. Vranic M., S. Pek, and B. Kawamori. Increased "glucagon immunoreactivity" in plasma of totally depancreatized dogs. *Diabetes* 23:905–912, 1974.

239. Vranic M. and G. A. Wrenshall. Exercise, insulin and glucose turnover in dogs. *Endocrinology* 85:165–171, 1969.

240. Wahren J., P. Felig, G. Ahlborg, and L. Jorfeldt. Glucose metabolism during leg exercise in man. *J. Clin. Invest.* 50:2715–2725, 1971.

241. Wahren J., L. Hagenfeldt, and P. Felig. Splanchnic and leg exchange of glucose, amino acids and free fatty acids and ketones in insulin-dependent diabetics during exercise. *J. Clin. Invest.* 55:1303–1314, 1975.

242. Wahren J., L. Hagenfeldt, and P. Felig. Splanchnic and leg exchange of glucose, amino acids, and free fatty acids during exercise in diabetes mellitus. *J. Clin. Invest.* 55:1303–1314, 1975.

243. Wallberg-Henriksson H. and J. O. Holloszy. Activation of glucose transport in the diabetic muscles: responses to contraction and insulin. *Am. J. Physiol.* 249 (Cell Physiol. 18):C233–C237, 1985.

244. Wasserman D., P. E. Williams, and D. B. Lacy. Exercise-induced fall in insulin and hepatic carbohydrate metabolism during muscular work. *Am. J. Physiol.* 256 (Endocrinol. Metab. 19):E500–509, 1989.

245. Wasserman D. H., H. L. A. Lickley, and M. Vranic. Interactions between glucagon and other counterregulatory hormones during normoglycemic and hypoglycemic exercise. *J. Clin. Invest.* 74:1404–1413, 1984.

246. Wasserman D. H., H. L. A. Lickley, and M. Vranic. Effect of hematocrit reduction on hormonal and metabolic responses to exercise. *J. Appl. Physiol.* 58:1257–1262, 1985.

247. Wasserman D. H., T. Mohr, P. Kelly, D. B. Lacy, and D. Bracy. Impact of insulin deficiency on glucose fluxes and muscle glucose metabolism during exercise. *Diabetes* 41:1229–1238, 1992.

248. Wasserman D. H., J. S. Spalding, D. B. Lacy, C. A. Colburn, R. E. Goldstein, and A. D. Cherrington. Glucagon is a primary controller of hepatic glycogenolysis and gluconeogenesis during muscular work. *Am. J. Physiol.* 257 (Endocrinol. Metab. 20):E108–E117, 1989.

249. Winder W. W., K. M. Baldwin, R. L. Terjung, and J. O. Holloszy. Effects of thyroid hormone administration on skeletal muscle mitochondria. *Am. J. Physiol.* 228:1341–1345, 1975.

250. Winder W. W., M. A. Beattie, and R. T. Holman. Endurance training attenuates stress hormone responses to exercise in fasted rats. *Am. J. Physiol.* 243 (Regulatory Integrative Comp. Physiol. 12):R179–R184, 1982.

251. Winder W. W., M. A. Beattie, C. Picquette, and R. T. Holman. Decrease in liver norepinephrine in response to exercise and hypoglycemia. *Am. J. Physiol.* 244 (Regulatory Integrative Comp. Physiol. 13):R845–R849, 1983.

252. Winder W. W., J. Boullier, and R. D. Fell. Liver glycogenolysis during exercise without a significant increase in cAMP. *Am. J. Physiol.* 237 (Regulatory Integrative Comp. Physiol. 5):R147–R152, 1979.

253. Winder W. W., S. R. Fisher, S. P. Gygi, J. A. Mitchell, E. Ojuka, and D. A. Weidman. Divergence of muscle and liver fructose 2,6-diphosphate in fasted exercising rats. *Am. J. Physiol.* 260 (Endocrinol. Metab. 23):E756–E761, 1991.

254. Winder W. W., S. J. Garhart, and B. N. Premachandra. Peripheral markers of thyroid status unaffected by endurance training in rats. *Pflugers Arch.* 389:195–198, 1981.

255. Winder W. W., J. M. Hagberg, R. C. Hickson, A. A. Ehsani, and J. A. McLane. Time course of sympathoadrenal adaptation to endurance exercise training in man. *J. Appl. Physiol.* 45:370–374, 1978.

256. Winder W. W. and Hardie DG. Inactivation of acetyl-CoA carboxylase and activation of AMP-activated protein kinase in muscle during exercise. *Am. J. Physiol.* 270 (Endocrinol. Metab. 33):E299–E304, 1996.

257. Winder W. W. and R. W. Heninger. Effect of exercise on tissue levels of thyroid hormones in the rat. *Am. J. Physiol.* 221:1139–1143, 1971.

258. Winder W. W. and R. W. Heninger. Effect of exercise on degradation of thyroxine in the rat. *Am. J. Physiol.* 224:572–575, 1973.

259. Winder W. W., R. C. Hickson, J. M. Hagberg, A. A. Ehsani, and J. A. McLane. Training-induced changes in hormonal and metabolic responses to submaximal exercise. *J. Appl. Physiol.* 46:766–771, 1979.

260. Winder W. W., M. L. Terry, and V. M. Mitchell. Role of plasma epinephrine in fasted exercising rats. *Am. J. Physiol.* 248 (Regulatory Integrative Comp. Physiol. 17):R302–R307, 1985.

261. Wolfe R. R., E. R. Nadel, J. H. F. Shaw, L. A. Stephenson, and M. Wolfe. Role of changes in insulin and glucagon in glucose homeostasis in exercise. *J. Clin. Invest.* 77:900–907, 1986.

262. Yamatani K., Z. Shi, A. Giacca, R. Gupta, S. Fisher, L. Lickley, and M. Vranic. Role of FFA-glucose cycle in glucoregulation during exercise in total absence of insulin. *Am. J. Physiol.* 263 (Endocrinol. Metab. 26):E646–E653, 1992.

263. Zinman B., F. T. Murray, M. Vranic, A. M. Albisser, B. S. Leibel, P. A. McClean, and E. B. Marliss. Glucoregulation during moderate exercise in insulin treated diabetics. *J. Clin. Endocrinol. Metab.* 45:641–652, 1977.

chapter 10

THE TEMPERATURE REGULATORY SYSTEM

Elsworth R. Buskirk

TEMPERATURE regulation has been demonstrated to occur through two processes: (1) physiological and (2) behavioral. Behavioral temperature regulation plays its greatest role in coping with cold environments, whereas both physiological and behavioral temperature regulation operate in hot environments, with physiological regulation perhaps playing the dominant role. Exercise induces both physiological and behavioral regulation by altering heat production, blood flow redistribution for convective heat transfer, and sweating. In this chapter little attention will be paid to behavioral regulation, but many of those who have contributed to our current understanding of physiological regulation and processes of heat exchange will be identified. Where possible those who first proposed a concept will be featured as well as some of those who either broadened the concept or forcibly brought it to the attention of other investigators through definitive experimentation. Since the literature on temperature regulation and heat exchange is so vast, many investigators who have made significant contributions will not be identified because of space limitations. An apology is hereby made for their exclusion. Focusing on the early research means that many relatively recent and current studies will not be covered; the reader is referred to the excellent reviews that have appeared such as the recent (1996) *Handbook of Physiology*, Section 4, *Environmental Physiology* edited by Fregley and Blatteis (40).

One of the very earliest investigators of body temperature regulation was Charles Blagden (12) (1748–1820), who performed a variety of investigations in heated rooms using thermometry. His observations during the period of 1774–1777 are discussed in L. L. Langley's book (72). Not only did Blagden describe both internal and body-surface temperature differences with different types of heat exposure, but he also documented the essential function of evaporative cooling, the differential effects of dry and moist heat, as well as the modification of heat exchange that clothing provides. His commentary regarding dry and moist heat is revealing (72):

> Dr. Fordyce has since had occasion, in making other experiments, to go frequently into a much greater heat, where the air was dry, and to stay there a much longer time without being affected nearly so much, for which he assigns the reasons that dry air does not communicate its heat like air saturated with moisture and that the evaporation from the body, which takes place when the air is dry, assists its living powers in producing cold. (William Fordyce, 1724–1792)

Blagden was apparently aware of the added thermal stress induced by exercise, but his only exercise experiments involved random walking in heated rooms. He made no comment about the specific impact of exercise but dutifully reported the elevated body temperatures and pulse rates he observed. Nevertheless, his delineation of the differential effects of dry and moist heat was an important observation on the conduct of exercise in environments varying in humidity.

The early story of insensible perspiration and its role in heat exchange has been interestingly set forth by Renbourn (93). Among others he points to Galen (ca. 129–200), who not only established the term, but cited the process as one that continued both day and night and was differentiated from liquid sweat. Renbourn also cited Santorio Santorio (1561–1636), who in 1614 quantified insensible perspiration by constructing a sensitive beam balance and weighing himself, his food, and various excretions during both day and night. He literally lived on the balance for protracted periods. Various techniques were applied to clarify insensible perspiration such as magnification devices to examine the skin, condensation surfaces applied near the skin, illumination of the vapor given off by contact on a white wall or light background, and improved balances. Renbourn credits Antoine Lavoisier (1743–1794) and Armand Sequin (1767–1835) for showing in 1790 that the essential function of insensible perspiration was the control of animal heat. Nevertheless, it was not until the twentieth century that insensible perspiration was better understood and applied to the calculation of basal metabolism (Benedict and Root: 8) and became a standard, still utilized today as an independent variable and as a base for the calculation of the metabolic severity of exercise. The concentrated effort on insensible water loss thus set the stage for baseline usage and clarification.

Although early work in thermometry was focused on disease, there were limited reports dealing with exercise. One of the first was John Davy (28) (1791–1860); he gave special attention to the temperatures of healthy persons, including the temperature elevation associated with passive exercise. Davy reported that physical activity of any type elevated heart rate and body temperature and that these elevations were augmented in warm environments—particularly in the tropics in which he lived for an extended period. Thermometry did not really become established until Karl A. Wunderlich of Liepzig (1815–1877), who in 1851 formulated the laws of thermometry in disease. He thought every disease had a different temperature profile. The use of the thermometer on a universal basis was soon established, but it involved the use of a version of the John Aitkin (d. 1790) self-registering mercury thermometer by Allbutt—the prototype of the thermometer used today (41).

In terms of sensing equipment, our knowledge of body temperatures and heat exchange during rest and exercise has depended on the use of ever-more-sophisticated devices that started with the mercury thermometer and progressed to thermocouples, thermistors, radiometers, infrared imaging, scanning devices, and calorimeters. Although of considerable interest to those interested in the history of temperature regulation/heat-exchange investigations, here the interest is focused on the use of these measuring devices in the pursuit of knowledge about mechanisms invoked by exercise.

We tend to forget the contributions made by early twentieth-century workers in a given field of endeavor and concentrate on the recent "hot" observations. Doing so is critical for moving the research frontier ever forward, but the basic information provided by these early workers still serves as a foundation. In this connection some early workers might be considered "giants in the field"—those who have not only contributed original meaningful data but who have also provided interpretations and synthesis that have expanded our perspectives. Among those in the area of exercise, temperature regulation, and heat exchange who have since passed from the scene, some are listed in Table 10.1. Presumably there are many more who should be added to this list because of their initiatives, and some of these are listed in Table 10.2. All of the investigators have come to mind as having made outstanding contributions that have stood the test of time.

Table 10.1. Some Deceased Investigators Who Have Contributed Measurably to the General Topic of Temperature Regulation and Heat Exchange with Exercise

Investigator	Contribution
Adolph, Edward (1895–1986)	Heat stress, acclimatization, and adaptability
Asmussen, Erling (1907–1991)	Temperature regulation during exercise
Bazett, H. C. (1885–1950)	Reflex control of temperature regulation
Belding, Harwood (1909–1973)	Performance limits in the heat
Brouha, Lucian (1899–1968)	Industrial performance limits in the heat
Burton, Alan C. (1904–1979)	Cold exposures and heat exchange calculations
Carlson, Loren (1915–1972)	Heat loss and physiological response in the cold
Dill, David B. (1891–1986)	Physiological responses to exercise in the heat
Gagge, Pharo (1908–1993)	Mathematics and schematics of heat exchange
Hardy, James D. (1918–1985)	Modeling of temperature regulation and heat exchange
Hertzman, Alrick B. (1898–1991)	Cutaneous blood flow and thermoregulation
Kuno, Yas (1882–1977)	Sweating mechanisms and responses
Nielsen, Marius (1903–2000)	Regulation of body temperature with exercise
Randall, Walter (1916–1993)	Sweating and peripheral control of responses to heat
Robinson, Sid (1902–1982)	Physiological adjustments to exercise in the heat
Scholander, Peter (1905–1980)	Chronic adaptation to cold stress and countercurrent heat exchange

Table 10.2. Other Elderly or Deceased Investigators Who Have Contributed
Measurably to Our Understanding of Temperature Regulation and Heat Exchange
during Exercise

Investigator	Contribution
Bass, David	Circulatory and body fluid adjustments with exercise in the heat
Hammel, Harold T.	Control systems in thermoregulation
Hatch, Theodor	Assessment of heat stress
Hensel, Herbert	Neural processes in thermoregulation
Ladell, William S. S.	Heat induced disorders and sweating mechanisms
Thauer, Rudolf	Central and peripheral mechanisms in thermoregulation
Weiner, J. S.	Body temperature and the sweating mechanism
Wyndham, Cyril	Heat stress, exercise and acclimatization to heat

We are indebted to Theodore H. Benzinger (10) for unearthing and thoroughly
reviewing the seminal early papers on temperature regulation and heat exchange.
His review was published as a two-volume work entitled *Temperature, Part I: Arts
and Concepts and Temperature, Part II: Thermal Homeostasis*. Both the original papers and Benzinger's interpretation of them constitute not only interesting reading
but a valuable base for the consideration of subsequent findings.

EARLY ENDEAVORS

Marius Nielsen (88) is responsible for the concept that core temperature (T_c) (rectal)
not only increases during exercise but also is controlled at a higher value. He demonstrated that this value, relatively independent of temperatures between 5° and 30°C,
was related to the severity of work. Although core temperature was linearly increased with exercise, body-surface temperature changed little. A rectal temperature
(T_{re}) increase of 1°–2°C is produced by a 900–1000 kgm·min^{-1} work rate despite body
heat-loss mechanisms that are more than adequate for dissipating such heat gain.
Thus, except at high environmental temperatures, core temperature was regulated
more closely during exercise than at rest despite different exercise intensities. In contrast, surface temperatures changed with the environmental conditions, but exercise
did not effect a substantial change in the observed values.

In reviewing Nielsen's work and that of Robinson as published in Newburgh
(86), Bazett (6) was led to believe that there are reflex responses to both heat and cold
and that the dermal receptors mediating these reflex responses were stimulated by
temperature gradients. Bazett also believed that another heat receptor lay deeper in
tissue, perhaps at what he termed the "subcutaneous junction." Adjacent arterial and
venous plexuses would serve the sensing of temperature differences between arterial and venous blood. Bazett concluded that superficial heat receptors and their adjacent vessels were sensing external heat and that the deeper ones sensed heat
induced by working muscle. Bazett further rationalized that the superficial receptor
was involved with peripheral vascular readjustments and the deeper receptor was af-

fected by internal heat and was concerned with sweating. Thermal gradients, he thought, were responsible for the apparent relationship of sweating to skin temperature. These views have subsequently been challenged but formed the basis for further comparisons and conjecture.

Nielsen (88) interpreted the response he observed as controlled physiologically and not a reflection of inadequacy of the thermoregulatory system. That the control could be overwhelmed in hot environments with further elevations in body temperature was first discovered by Lind (75). This led Lind to describe a "prescriptive climate or zone" within which thermoregulation remained controlled and steady for a given rate of exercise. Later investigations (Lind, 76) extended the duration of exercise and work/rest cycles and found little change in such physiological responses as core temperature, pulse rate, or body weight loss. His conclusion was that within "prescriptive climates" thermoregulation remains dependent on the rate of exercise —a confirmation of Nielsen's earlier results.

Studies in Scandinavia following those of Nielsen involved study of the effects of elevated muscle and core temperature on subsequent performance (cycle ergometry). Erling Asmussen and O. Bøje (4) increased core and muscle temperature by both active exercise and passive heating. More intense active warm-up (30 minute ride) and passive warm-up (radiodiathermy) or a 10 minute hot shower (47°C) all increased work capacity. Time to perform a short-term fixed cycling task decreased steadily over the range of 36.5° to 40.8°C in the vastus lateralis muscle. Muscle temperature elevation was regarded as more important than that of core temperature. Although such factors as decreased muscle blood viscosity, increased muscle elastic properties, increased metabolic turnover, and increased dissociation of oxyhemoglobin were alluded to, specific investigation of mechanistic import remained for others to explain.

TEMPERATURE REGULATION

In 1972 James D. Hardy (51) (Fig. 10.1) reviewed models of temperature regulation and found that only one or two were developed in each decade between 1880 and 1940. Hardy chose 1880 as a starting date because of the discovery that the brain was important in temperature regulation (Ricket 1885 [b. 1816], Ott 1887 [1847–1916]). After 1950, there was an escalating increase in the number of published models, so in the decade 1960–1970, more than 20 appeared. Based on his analysis, Hardy (51) proposed a modified servosystem designed to facilitate research. It involved the use of experimental observations modified by intuitive judgment for the formation of conceptual models. From these models, hypothetical implications were drawn about feedback through analytical or intuitive test loops to modify judgments or to provide experimental challenges. The scheme was an adaptation of earlier work by Grodins (47), who set forth principles for model development in regulatory biology. Hardy described a variety of models including verbal (to which he gave Claude Bernard [1813–1878] credit for his "fixite du milieu interieur," 1865), pictorial, ana-

Fig. 10.1. James D. Hardy, Ph.D. (1918–1985) of the John B. Pierce Foundation Laboratory in New Haven, Connecticut. Photo was graciously provided by John Stitt, a fellow of the John B. Pierce Foundation Laboratory.

logue, mathematical, neuronal, and chemical (Table 10.3). Although these models do not represent current integrative perspectives, they provided a background for studying the specific effects of exercise on temperature regulation.

TEMPERATURE REGULATION: SET POINT

The concept of "set-point" theory, perhaps originally proposed by Hellon (53), was promoted by Benzinger (9) (Fig. 10.2) who pioneered the utilization of tympanic membrane temperature (as a surrogate measurement of brain temperature), gradient layer calorimetry (partitioning of heat loss), and the rapid response indirect calorimetry (heat production). Based on his studies and those of his collaborators, Benzinger proposed that the anterior hypothalamus is thermally sensitive and controls heat loss mechanisms, whereas the posterior hypothalamus was insensitive to temperature but relayed cold receptor impulses from the periphery and other sources, thereby stimulating heat production, particularly shivering. Thus the concept involved anterior excitation and posterior inhibition. To quote Benzinger (9):

> Moreover, the 'set-point' of the regulatory system, previously a theoretical concept, was defined experimentally, determined individually, and reproduced with high precision as a characteristic of the human thermostat with significant differences between individuals.

Benzinger also demonstrated on–off, proportional, and rate control characteristics, reportedly for the first time, with respect to the processes of central warm reception and peripheral cold reception. Experiments involving exercise—that is, augmented

Table 10.3. Prototype Temperature Regulation Models

Type	Date	Author	Concept
Verbal	1865	Bernard, C.	Preserving the milieu interieur
	1885	Richet, C.	Regulation in brain
	1887	Ott, I.	Thermotoxic centers
	1913	Meyer, H.	Thermogenic and heat-loss centers
Pictorial	1948	DuBois, E. F.	Nervous system control
	1952	Hensel, H.	Stimulatory and inhibitory controls
	1959	Benzinger, T. H.	Central warmth, peripheral cold receptors
	1965	Chatonnet, J. Cabanac, M.	Common origins for physiological and behavioral regulation
Analogue	1950	MacDonald, R. K. C. Wyndham, C. H.	Heat transfer and control for vasomotion and sweating
	1958	Aschoff, J. Wever, R.	Hydrodynamic countercurrent heat exchange
Mathematical	1934	Burton, A. C.	Cylinder with uniform properties
	1961	Wissler, E.	Six-cylinder, 15-compartment system
Analogue: closed loop	1964	James, E. W. Smith, P. E.	Feedback loops from skin and hypothalamic multification
Neuronal	1965	Hammel, H. T.	Peripheral afferents modify central signal error
	1971	Hardy, J. D. Guieu, J. D.	Networks of hypo- and hyperthermic activity
Chemical	1970	Myers, R. D.	Peripheral control: posterior hypothalamic "ionic" set point
	1970	Carlson, L. D. Hsieh, A. C. L.	Sympathetic peripheral vasomotor control
	1970	Bruck, K. Wunnenberg, W.	Simulation of brown fat for metabolic heat

Source: Adapted from Hardy (51).

heat production, elevated tympanic temperature, sweating rate, and vasodilation, showed that both set-point and proportional control exist (See Fig. 10.3 for a composite illustration using data derived from experiments involving both rest and work.) Both sweating rate and peripheral vasodilation (conductance, as determined by the calorimetry assessment) were distinctly related to tympanic membrane temperature. Benzinger recognized local temperature effects on vascular smooth muscle (i.e., vasoconstriction with cooling and vasodilation with warming), but he relegated these local effects to moderating influences with central control being dominant.

A broader view of temperature regulation was advanced by Hensel (55) who contended that the control actions of thermoregulation involving sweating, vasomotion, and fluid volume redistribution were functions of multiple thermal and other inputs and that integration occurred in the CNS, most notably in the hypothalamus.

Fig. 10.2. Theodore H. Benzinger, M.D. (1905–1999), former physiologist and director of the German Air Forces Testing Center (1934–1944), physiologist and director of the bioenergetics division of the U.S. Naval Medical Research Institute (1947–1970), and scientist at the National Institute of Standards and Technology (1970–1974). Photo courtesy of Mrs. Theodore Benzinger.

Such integration involving central and peripheral signals was construed to be multiplicative rather than additive. Central drive for sweating, for example, was viewed as being modified by local temperature effects on the sweat glands themselves. Thus, a dynamic concept of a set point was based on a comparison of warm- and cold-sensitive feedback signals. Hensel also advanced the concept that set temperature does not change with exercise and that voluntary physical activity is possible because the body relies to an important extent on physiological thermoregulation as contrasted with behavioral thermoregulation.

Fox and Macpherson (39) provided evidence for proportional control of temperature regulation during exercise above an elevated set point induced by fever. In the presence of fever, a young man undertook an exercise regimen for 6 weeks. His pattern of body temperature elevation with exercise was of normal magnitude, but it was superimposed upon an elevated resting temperature. Peripheral receptors were also involved with regulation during fever so that the fibrile state could be maintained despite warm or cold exposure (Hardy, 50).

Evidence that any elevation in T_c during exercise is sensed by thermodetectors located both in the anterior hypothalamus and the spinal cord was provided by Thauer (113), and for receptors in limb muscles by Jessen et al. (64). These receptors

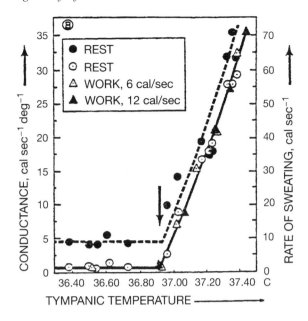

Fig. 10.3. Vasodilation and central temperature. Conductances of resting man measured by gradient calorimetry and by tympanic and skin thermometry at a variety of environmental temperatures were plotted against tympanic temperatures together with sweat rates. Powerful vasodilation in response to central warming as a stimulus; set-point and proportional-control characteristics were evident. Over the range of less than 0.5°C, peripheral blood flow increased by seven times its normal rate. From Benzinger (9) (1969) with permission.

provided effective signals for augmenting skin blood flow and sweating. A balance was reached when heat loss equaled heat production and T_c reached a stable value for a particular intensity of exercise. That the elevation in T_c with exercise is passive was demonstrated by Nielsen and Nielsen (87) via diathermy heating experiments at rest that duplicated T_c elevation with exercise. This finding argued that heat-loss mechanisms were not substantially modified by exercise and that the set point is also unaltered by exercise. Irma Astrand (5) first proposed that the relative metabolic rate is more important than the absolute rate in determining T_c.

Chronic processes such as heat acclimatization (70) and endurance training (89) appear to shift the set point and reduce thresholds for peripheral vasodilation (110) and sweating (85). Another shift in set point appears to be related to the daily rhythm of change in core temperature. For example, Stephenson et al. (unpublished, as cited by Gisolfi and Wenger, 43) found that thresholds for torso sweating and forearm vasodilation tracked core temperature during a 20 hour observation period. One major finding in the alteration of set point was clearly demonstrated by Stitt (109), who induced fever and compared changes in set point, heat production, and heat loss with those occurring during exercise. Fever unequivocally raised the set point, whereas exercise did not.

The interaction of effector responses for dissipation of heat with respect to core temperature and skin temperature has been described by several investigators and perhaps best put in perspective by Wenger et al. (117) and Gisolfi and Wenger (43). They provided equations to illustrate the impact of the concept of set point and the dominance of central temperature sensing in the control of effector responses, with peripheral sensing providing a secondary or modulating influence. They also emphasized a central thermoregulatory controller rather than a temperature sensor. This central controller would generate a thermal command signal sent to integrators controlling effector responses with perhaps different weighting factors for each integrator. Modification of signaling from other receptors such as baroreceptors would also affect the integrators. Gisolfi and Wenger's (43) concept of a schematic model of this system is presented in Figure 10.4. They interpret the figure in terms of a "load error" wherein $T_{ws} - T_{set}$ sets forth the thermal state of the body. As T_{ws} increases, the load error increases and the separate effector systems are activated when the load error reaches a characteristic value for that response. T_{ws} is regarded as the weighted sum of core temperature (T_c) and mean skin temperature ($\bar{T}_{sk}$).

TEMPERATURE REGULATION: ZONES OF CONTROL

Hardy (50) in his review settled on the concept of zones of control, a concept set forth by Hellon and Lind (54). The zones of regulation were characterized as the neutral zone (vasomotor control), the hot zone (evaporative and vasomotor control), and the cold zone (control of metabolic rate). Body temperature regulation in the cold and neutral zones was described as largely effected by peripheral receptor involvement,

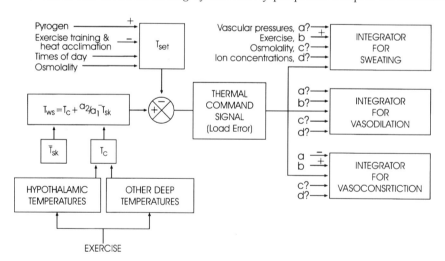

Fig. 10.4. Modified model of the control of thermoregulatory effector responses with emphasis on exercise. Known effectors having postive (+) or negative (–) impacts are identified. T_{ws}, a function of core (T_c) and mean skin ($\bar{T}_{sk}$) temperature, is displaced from set-point temperature (T_{set}) and has been identified as the load error. From Gisolfi and Wenger (43) with permission.

whereas during exercise and in heat the regulation was effected by central receptors; coordinated or reciprocal control across zones was possible. Hardy concluded that protection against overheating constitutes the major function of physiological thermoregulation. Receptors of the skin, hypothalamus, and other body areas participate in regulation. The respective receptors coordinate in an additive fashion producing thermoregulatory action with the properties of proportional and rate control, but not integral control.

Since the review by Burton (18) we have known that the elevation in core temperature (T_c) during steady-state exercise responds essentially as a proportional control system for effector responses. That this proportional control can be altered by regular exercise (training) has been demonstrated by an increase in the slope of the sweat rate response; that is, a smaller increase in T_c gives a greater sweat rate response to effectively eliminate the added internal heat load (85).

TEMPERATURE REGULATION: IONIC CONTROL

Feldberg et al. (37) and Myers and Veale (83) set forth the ionic set-point concept involving principally sodium and calcium ions. Follow-up studies on exercise have been reviewed by Gisolfi and Wenger (43), who concluded that increased cerebral calcium increases heat loss and decreased calcium increases heat gain. It was further concluded that the $[Na^+]/[Ca^{2+}]$ ratio in the posterior hypothalamus may establish and maintain the thermoregulatory set point. Through experiments on monkeys, Gisolfi et al. (42), using calcium injections into the cerebral ventricles, argued against such a conclusion based on a failure to initiate or enhance sweating or lower core temperature with heat exposure. In contrast, Sobocinska and Greenleaf (108) reported that increasing calcium concentration either in plasma or cerebrospinal fluid increased heat loss and lowered core temperature during steady-state exercise. The key experiment involved infusing a hypercalcic solution into the cerebral ventricle of resting and exercising dogs and observing weight loss and core temperature. No doubt specific ions may be involved in effector mechanism control, but their role in set-point alteration is unresolved and perhaps unlikely.

Feldberg and Myers (36) set forth the amine theory for body-temperature regulation involving the effect of three amines on the hypothalamus. They described heat gain and loss via a "balanced release" of norepinephrine and 5-hydroxytryptamine within the preoptic region of the anterior hypothalamus. Subsequently it has been concluded that the role of such amines in the respective heat gain and loss pathways has more credence than the ionic set-point theory.

TEMPERATURE REGULATION: CURRENT CONCEPTS

Following a series of pivotal experiments in the 1950s and 60s, Hammel (49) provided us with a more comprehensive scheme for temperature regulation, which is depicted in Figure 10.5. He should be given credit for building on the models re-

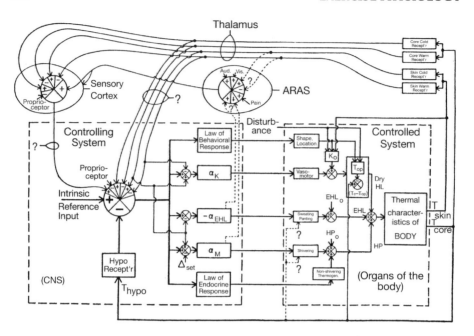

Fig. 10.5. Hammel's interpretation of the controlling and controlled systems for the regulation of body temperature emphasizing a feedback control system. The neural links are depicted. ARAS = ascending reticular activating system. From Hammel (49) with permission.

viewed and proposed by Hardy and for involving feedback neuronal control. He also raised the issue of a possible decline in set point with exercise. Boulant (14) has built on Hammel's model and pursued the various interactions that allow thermoregulation to operate effectively. He postulates three sets of neurons in the preoptic–anterior hypothalamus, as well as input from receptors in the skin, muscle, and spinal cord. These neurons operate by integrating control and peripheral information. Thus, warm-sensitive neurons, cold-sensitive neurons, and insensitive neurons not only react to thermal stimuli but also respond to changes in many metabolites including glucose; in endocrines, including testosterone, estrogen, and progesterone; as well as to changes in osmolality. Since exercise not only places a thermal load on the body but also changes metabolites, hormones, and osmolality, it is clear that understanding this in respect to temperature regulation with different types, intensities, and durations of exercise in different environments is immensely complicated. Boulant (14) gives credit in his review to the early work of many others including Brück, Carpenter, Hensel, Hori, Magoun, Nakayama, Pierau, and Wit.

HEAT ACCLIMATIZATION

The terms "acclimation" and "acclimatization" have frequently been used to describe the same phenomena, but as here used acclimation refers to repeated exposures in a chamber or with induced conditions whereas acclimatization refers to that adaptation that takes place under natural conditions. Classical descriptive studies of heat acclimatization during hard rock-mining were made by Dresoti (31), who preceded Wyndham, Strydom, and colleagues (119), the investigators who vastly extended such studies in South Africa. Dresoti developed a rock-shoveling test to evaluate the rate of heat acclimatization and found that heart rate was lessened and sweat rate augmented so that work performance was enhanced within 14 days. Progressive beneficial effects proceeded thereafter for approximately 1 month. Subsequently, Robinson et al. (98) (Robinson 1902–1982) brought further attention to what might be regarded as a heat acclimatization model by following heart rate and skin and rectal temperatures for a period of 24 days of regular work in the heat. Their conclusions: (1) rapid changes in all three variables occurred within the first 7 days, (2) regular exercise of 1.0–1.5 hours per day was necessary for effective acclimatization, (3) substantial heat acclimatization persisted postexposure for about 3 weeks, and (4) following 7 days of acclimatization the heat balance was satisfactorily obtained during work in a hot environment about as well as in a cool environment.

The picture in regard to the physiological impact of heat acclimatization was extended by two offshoots from the Harvard Fatigue Laboratory—namely, the Fort Knox Laboratory in Kentucky and the Physiological Hygiene Laboratory at the University of Minnesota. According to Eichna et al. (33) the former extended the investigations to hot, humid climates and elucidated many other factors that could alter the acclimatization process, including: lack of sleep, alcohol ingestion, season of the year, physical fitness, and hydration level. Horvath and Shelley (61) as well as Eichna et al. (34) extended the research to hot, dry environments and found that acclimatization was effective in any lesser stressful environment. Taylor et al. (112) at Minnesota (Henry L. Taylor 1912–1983) focused on cardiovascular adjustments to heat acclimatization including the increase in stroke volume that accompanied the decrease in heart rate. They documented that about 50% of improved performance during work in the heat occurred within the first 4 days and that the cardiovascular responses were important for successful acclimatization.

Mention of heat acclimatization with respect to work in a desert environment should not overlook the classical studies of Edward Adolph and his colleagues (1). They studied various facets of heat/exercise exposures and the acclimatization process including the roles of fluid and electrolyte balance as well as the potential for thermal injury.

In a classical study from the Harvard Fatigue Laboratory, Robinson et al. (95) studied the effects of a hot, humid climate (29°–32°C, 85% RH [relative humidity]) on work performance entailing a sevenfold increase in metabolism. Two-hour work periods were undertaken by white and black sharecroppers and members of the research team and significant differences in work performance were found. All sharecroppers successfully completed the work bouts, whereas members of the research team were exhausted at the end of 2 hours. The black sharecroppers finished their stints with a rectal temperature (T_{re}) of 38.3°C and a pulse rate of 150, whereas the white sharecroppers, although lean and fit, reached a T_{re} of 38.9°C and a pulse rate of 170. The black sharecroppers sweated less but had lower skin temperature, drank more water, and achieved greater overall efficiency. Superior cardiovascular function was stated as the reason for the black sharecroppers' practical superiority. The data emphasized not only ethnic differences, but also effects of heat acclimatization and physical fitness.

Investigation of the heat acclimation process among women were initiated by Hertig and Sargent (57). They compared the responses of men and women to walks at 4.8 km·h^{-1} for 2 hours under thermally neutral conditions ($T_a = 21°C$) and 10 exposures to heat 45°C/25.5°C (T_{db}/T_{wb}) plus one at 50°C/26.5°C. All subjects were minimally clothed. The women, aged 20–43 years, some of whom were physically quite active, reached limits of endurance easily tolerated by the men and showed comparable physiological adjustments with acclimation: reduced pulse rate, reduced body core and skin temperature elevations, onset of sweating at a lower skin temperature, and lessened discomfort with the regimen. In addition, it was suggested that possible comparative disadvantages for the women were a lower thermal gradient (core to skin) for metabolic heat removal and lesser reserve capacity to move blood to the skin. Precise assessments of these variables were not made.

Cleland et al. (24) extended such observations on women over 9 days of heat exposure, with effective acclimatization reached within that period in a pattern comparable to that of men. Horvath (60), in reviewing historical perspectives, indicated that the relative workload concept—that is, percentage of maximal oxygen intake—was not utilized for gender comparative purposes until much later and cites Kupprat et al. (1980) for initiating this leveling consideration, although the concept had been discussed among investigators since the 1950s.

REGIONAL BLOOD FLOW

With respect to assessment of skin blood flow and forearm blood flow, pioneering investigations involving supine leg exercise were performed by Bevegard and Shepherd (11). They observed an initial dilation followed by constriction of vessels in forearm muscle and a gradual dilation of vessels in the skin. Capacity vessels constricted with exercise; the constriction persisted throughout exercise and was graded to workload similar to that of resistance vessels. They attributed the vessel reactions to sympathetic fiber mediation and suggested that the venomotor reflex could have been elicited by muscle contractions. This work was expanded substantially by Rowell and colleagues (100), who undertook a variety of experiments to delineate such variables as posture, environmental conditions, and intensity of workload. For example, Johnson et al. (65) evaluated upright exercise and the skin blood flow–body temperature relationship. They found that both upright posture and exercise reduced forearm skin blood flow at a given core temperature when compared to that during supine rest. Skin temperature was maintained constant at 38°C. It was concluded that exercise in heat or upright posture in heat causes relative vasoconstriction in the skin and that these effects are greater in the upright as compared to the supine position.

Altered control of skin blood flow during exercise involving high core temperatures was investigated by Brengelmann et al. (16). They found that forearm blood flow (determined plethysmographically) did not increase in direct relation to core temperature but leveled off at approximately 38°C T_c. Only local heating of the forearm could increase the flow. Thus, under these conditions when core (esophageal) temperature exceeded 39°C, forearm skin blood flow remained at the level attained at 38°C core temperature. Clearly the competition for blood flow supplying working muscle and that providing conductive/convective heat transfer in the skin favored support of the working muscle, at least until the regulating mechanisms were overwhelmed. Nevertheless, flow in cutaneous vascular beds apparently responds differently from beds in resting muscle and the splanchnic area, with the two latter bed-flows graded with respect to exercise intensity and the former reaching a threshold (99).

Gisolfi and Wenger (43) cautioned that it is misleading to consider muscle perfusion as dominating thermoregulation during exercise in the heat. Modifying influences and interindividual variability moderate the ultimate responses.

Rowell (99), reporting on experiments using direct heating of the skin, found that the predominant acute effect was relocation of a portion of the blood volume into the compliant venous system near the body surface. Despite this relocation, maximum skin vasodilation was averted if sufficient increases in cardiac output or splanchnic vasoconstriction did not occur, thus avoiding a precipitous fall in mean pressure. Accurate quantitative assessment of the true diversion of blood from working muscle to skin during exercise in heat could not be assessed, nor could the precise responsible mechanisms be illuminated. Still, the relocation of blood volume was apparent and was reversed by lowering skin temperature during heavy exertion in heat.

An interesting feature of heat exchange is the role of superficial veins which traverse subcutaneous tissue. Dilation of these superficial veins facilitates heat conduction and convection from core to skin via blood flow. That superficial veins are innervated and subject to both constriction and dilation was demonstrated by Webb-Peploe and Shepherd (115). Dilation of the superficial veins with heating and exercise increases peripheral blood pooling with reduction in central blood volume that limits cardiac filling, stroke volume, and cardiac output until compensatory factors intervene. A fraction of the heat produced by working muscle was brought to the skin both by passive conduction and by venous drainage through communicating veins to superficial veins (25). Thus, some heat generated by muscle was probably dissipated from overlying skin, effectively attenuating elevation in core temperature.

Cardiovascular readjustments during exercise in heat were thoroughly reviewed by Rowell (99). Bradley (15) is credited with first measuring reduction in splanchnic blood flow with exercise (supine) in men. Wade and Bishop (114) extended these observations to heavy supine exercise. Although the question of whether posture effected these changes was unresolved, Rowell et al. (100) clarified the issue by studying the effects of upright exercise and found that reduction in splanchnic blood flow was inversely proportional to intensity of exercise

($\%\dot{V}O_2$max). Thus, posture had no major influence on splanchnic blood flow reduction during exercise. About the same time Grimby (46) found similar proportional reductions in renal blood flow with incremental exercise ($\%\dot{V}O_2$max). Grimby's work extended that of White and Rolf (118) who had found that reduction in renal blood flow was also proportional to the severity of exercise. Rowell (99) should be credited for elevating understanding of the impact on regional circulations of exercise, heat stress, and hypohydration in humans and for pointing out the important effects of augmented sympathetic vasomotor outflow on the control of splanchnic and renal blood flow. He did not rule out action of local mediators and hormonal mechanisms in regional blood flow control. Rowell (99) explained—

> The extent to which unheated skin and non-working muscle respond to such stresses is unknown. If they do not respond then the question is whether sympathetic vasomotor outflow in man is highly discrete or whether local conditions in skin and muscle can effectively counteract the stimulus.

The convective transfer of heat via the circulation from the interior of the body to the surface comprises an important homeostatic function facilitated by exercise. Consideration of this transfer has involved construction of equations that are derived from environmental, skin, and internal body temperatures. Burton (17) was one of the first to prepare appropriate equations. Heat flow was considered to go in only one direction, but his simplification was modified when the concept of countercurrent thermal exchange was elucidated by Scholander (104). This concept, involving heat flow between adjacent blood vessels parallel to the skin surface, makes the relationship between blood flow and thermal exchange efficient but more complicated. Thus, the pattern of blood flow becomes important in the heat-exchange process. With countercurrent exchange operating effectively, a relatively large blood flow between the warm interior of the body and a cool skin results in attenuated heat loss.

Cooper et al. (25) studied heat transfer from working muscle to skin and found an abrupt rise of 1.5°–3°C in skin temperature following onset of exercise. This increased temperature appeared restricted to the palpable borders of the working muscle where venous occlusion delayed and total blood flow occlusion abolished the rise. Epinephrine iontophoresis did not prevent the rise nor was it associated with arterial dilation. Thus, direct venous convection from working muscle to the skin was responsible for the increased heat transfer. Countercurrent heat exchange was not involved.

SWEATING

Perhaps the ancient Greeks were the first to believe that pores exist in skin, but it was left to Steno in 1664 to describe the *fontes sudores* (sweat glands) through which both insensible water and sweat pass (92). Five years later, Marcello Malpighi (1628–1694) described the ducts of the *glandulae sudores* and mentioned that they resembled those of salivary glands (92). The early history of the development and understanding of sweating and the sweat rash has been prepared by Renbourn (92), who followed early writings in Greek, Latin, and Arabic through more recent entries in English, German, French, and Dutch. From early times it was evident that intense exercise is inextricably linked with the sweating process.

Additional early history of sweating and sweat gland function has been presented by Kuno in his original 1934 monograph entitled *The Physiology of Human Perspiration*, which was later updated as *Human Perspiration* (69) because of contacts with Eugene DuBois and Otto Edholm. Yas Kuno cites Johann Purkinje (1787–1869) as the discoverer of the sweat glands in 1833. Kuno and associates deserve credit for focusing on the sweating mechanism as a topic for continual study in the area. Iwataki (1936) is credited by Kuno (only selected references are provided in Kuno's revision, but not Iwataki) with providing the "exact features of the sweating during muscular exercise." Several experiments were conducted at various exercise intensities; one example involved 20 minutes of running (presumably under comfortable environmental conditions) at a treadmill speed of 160 m·min^{-1} (Fig. 10.6). The time course of sweating on the chest as well as the palm of the hand is depicted during pre-exercise, running, and recovery. Conclusions were: (1) the rate of sweating increase and profuseness depended on running speed; (2) palmar sweating was augmented but to a lesser extent than sweating on the chest; (3) sweating decreased rapidly during recovery even though rectal temperature remained elevated, the duration of which was dependent on exercise intensity; and (4) sweating during exercise involved a combination of thermal and mental sweating with the latter contributing relatively little to heat loss via evaporation of the sweat. Innervation of the palmar eccrine glands was regarded as neurogenic and cholinergic under most circumstances.

Kuno (69) also reviewed the various constituents of sweat and credits Kiltsteiner who, in 1911, observed that sweat chloride concentration increased progressively with the rate of sweating, an observation confirmed many times. In summarizing the Japanese work as well as that of others, Kuno reported the high variability of sweat constituent concentrations found in the active investigative period from 1930 to 1950 for most inorganic and organic substances. This variability was ascribed to different experimental exposures, dietary balances, rates and duration of exercise, sweat collection procedures, and contamination from the skin surface. He asserted that collection of "pure" sweat is a virtual impossibility—a fact now well known to most investigators, with the possible exception of Sato et al. (102, 103), who managed to microcannulate individual sweat glands.

Robinson and Gerking (96) observed that the average rate of sweating declined by 10%–30% when men worked continuously for 6 hour periods in extreme heat. The decline was greater in humid than in dry heat and was somewhat dependent on mean skin temperature; that is, the higher the skin temperature the greater the decline. Calling this phenomenon "sweat gland fatigue" is inappropriate because the sweat gland can be reactivated to secrete sweat at higher levels with appropriate intervention stimulation or treatment such as stripping of the stratum cornium with plastic tape.

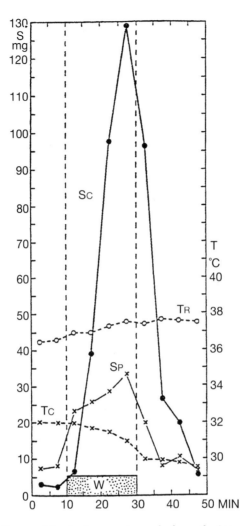

Fig. 10.6. Sweating and body temperatures before, during, and after running on a treadmill at 160 m·min⁻¹. Palmar sweating (S_p), chest sweating (S_c), as well as rectal temperature (T_r) and chest surface temperature (T_c) are indicated. From Iwataki as reported in Kuno (69) with permission from the publisher, C. C. Thomas.

The heat required for evaporation of sweat was speculative because of such factors as ambient temperature and humidity, as well as the solutes contained in sweat, which would change the heat requirements. Wenger (116) thoroughly reviewed the problem and concluded that for all practical situations the heat of evaporation was 2,426 J·g⁻¹ (0.580 kcal·g⁻¹)—the value for the latent heat of vaporization of water at 30°C. Having confidence in this value for use with a variety of environmental conditions and exercise intensities facilitated calculations of evaporative heat exchange.

Sid Robinson and Robinson (97) (see Fig. 6.13) found a reduction of salt content in both urine and sweat with heat acclimatization. In the unacclimatized individual salt losses of 20 g·d⁻¹ were found during exercise in the heat. Associated with such considerable losses of salt, Talbott documented dangers of heat-induced cramps and disturbances of cardiovascular function that compromise exercise performance (111).

Hypohydration

It has been well known since the observations of Pitts et al. in the 1940s that hypohydration leads to elevated core temperature during exercise in heat (91). In contrast, Moroff and Bass (81) demonstrated that hyperhydration lowered core temperature and increased sweating during heat exposure. In exploring the reasons for these changes, Greenleaf and Castle (45) found that at least 77% of the excessive increase in body temperature could be accounted for by sweating "deficits" but that either hypohydration "modifies" the postulated core–skin temperature control mechanisms or an essential component of the basic control mechanism resides elsewhere. Based on their data, Greenleaf and Castle implicated inhibition in the transfer of interstitial fluid (ISF) to the sweat glands during hypohydration to increased ISF osmotic pressure as well as to a decrement in plasma volume (45). Senay and Christensen (106) earlier had suggested that the set point for osmoreceptors is changed, thereby modifying neural impulse traffic from the hypothalamus to the sweat glands, resulting in sweating reduction with hypohydration. Senay (105) found that changes in serum [Na⁺] and osmolality were both significantly correlated with evaporative weight loss (r = –0.63 and –0.62, respectively), whereas changes in serum [K⁺] or changes in ([Na⁺]/[K⁺] were not. Ekblom et al. (35) confirmed these observations.

Dill (30) provided evidence that the continuously rising heart rate during walks in desert heat relates to the diminished blood (plasma) volume. Not only did his own measurements support this concept, but so did those of Gosselin and Adolph (1). In the latter experiments men sweated at a rate of about 1% of body weight per hour when exposed to a T_a of 50°C over a 6 hour period and there was no reduction in sweat rate as body water was depleted. From measurements of plasma volume it was concluded that plasma contributed the major portion of the water lost via sweat secretion. Myhre and Robinson (84), in similar studies, found that men walking in the heat for 2 or more hours at 50°C T_{db} and 26°C T_{wb} lost 4.2% of their body weight and 17% of their plasma volume. During recovery 45% of plasma volume was regained although the men remained hypohydrated at the 4.5% body-weight-loss level, suggesting internal fluid shifts. Thus plasma and consequently blood volume decreased in proportion to the level of hypohydration whether it occurred under resting or exercise conditions in the heat.

Pearcy et al. (90) observed that sweat rate during exercise in heat among hypohydrated men was consistently below that for those fully hydrated. Under opposite circumstances, hyperhydration produced an increase in sweat rate but no changes in sweat composition; that is, Na$^+$, K$^+$, Cl$^-$ (Cage et al., 22). Interpretation at that time revealed no significant deviation in sweat electrolyte concentrations as long as surfeit or deficiencies were avoided.

In regard to compensation for the hypohydration that can occur with exercise in heat, Hunt (63) in 1912 recommended water ingestion to restore better water balance. He recognized evaporative cooling and the processes of conductive and radiative heat exchange. Although he set no limits for sweat evaporation, he found that men exposed to 109°F consumed 13.6 l of water per day and attributed such consumption as compensation for evaporative loss.

In assessing the effects of heat acclimatization, Moss in 1922 (cited by Kuno, 69) studied trained miners who secreted 4.5–8.5 l of sweat per 5 hour work shift in a hot mine, whereas new workers secreted only 2.5 liters per shift. Szakall in 1937 (cited by Kuno, 69), found that sweating rate increased with repeated exercise in a hot chamber and the amount of sweat per unit of work was also elevated.

Further demonstration of this acclimatization effect on sweating was described by those working in the Harvard Fatigue Laboratory just prior to and during World War II. Forbes, Dill, and Hall (38) found changes in the volume of blood and tissue fluids when traveling from a northern to southern climate. Blood volume increased by about 5%, whereas interstitial fluid decreased in the warmer environment by about 11%. Fairly large interindividual differences were noted. Asmussen (3), reporting on the same subjects, suggested that elevated blood volume and pulse rate compensated effectively for the physiological demands of the hot climate on cardiac output among the various subjects.

HEAT EXCHANGE CAVEATS

The first law of thermodynamics is basic for studies of thermoregulation. A standard equation prepared by Bligh and Johnson describes the net rate at which heat is generated and exchanged with the environment (13):

$$S = M \pm E = (\pm W \pm R \pm C \pm K), \text{ units [W] or [W·m}^{-2}]$$

where:

S = rate of STORAGE OF BODY HEAT (+ for net gain by body)

M = METABOLIC FREE ENERGY PRODUCTION (always +)

E = EVAPORATIVE HEAT TRANSFER (− for net loss)

W = WORK (+ for POSITIVE WORK against external forces)

R = RADIANT HEAT EXCHANGE (+ for net gain)

C = CONVECTIVE HEAT TRANSFER (+ for net gain)

K = CONDUCTIVE HEAT TRANSFER (+ for net gain)

Perhaps the explanation prepared by Mitchell (80) provides the essential caveats for those working in this field in that the impact of exercise requires major considera- tion of M, W, and E. With respect to heat storage, Mitchell pointed out that there is no accurate thermometric method for calculating mean body temperature or body- heat storage because of the variation in the many possible temperatures throughout and on the surface of the body. Approximations can be made, but precision is elusive even when excluding the multiplicity of body composition variables.

Heat-Stress Indices

Consideration of heat-stress indices applicable for ascertaining heat strain during ex- ercise has been fraught with problems because of the large number of variables. Nev- ertheless, one relatively simple index has found application in several settings, such as the military, athletics, recreational activities, industry, and special events. Origi- nally proposed by Minard et al. (79), the wet-bulb globe temperature index (WBGT) has not only survived, but prevails.

$$WBGT = 0.7T_{wb} + T_g + 0.1T_a$$

where T_{wb} = ambient wet-bulb temperature, T_g = block globe temperature, and, T_a = ambient dry-bulb temperature.

Various criticisms of and modifications of the WBGT have arisen over the years including erroneous readings associated with different air movement and movement of the subject in relation to air movement. It has also been demonstrated that strains associated with a WBGT index of 85 as compared to 89 do not differ under dry con- ditions. Other factors affecting the WBGT are the physical fitness and acclimatiza- tion status of the subjects, amount and type of clothing worn, body composition of subjects (obesity) and gender.

Although other simple indices have been proposed, such as the "probable or predicted four-hour sweat rate" (P4SR) proposed by McArdle et al. (78), they have not enjoyed sustained usage. Another heat stress index, formulated by Harwood Belding and Theodore Hatch (7), was more sophisticated and involved use of psy- chometric charts, metabolic rate, air speed, and consideration of the maximum evap- orative capacity of a known environment. A single value was established for overall heat stress in a given exposure and this index was widely used, particularly by the military. Since these early indices used to characterize environmental heat stress, a wide variety of indices have appeared that were even more comprehensive. Since then several investigators have contributed measurably to our understanding of the physical environmental factors that contribute to heat stress and how they interact with man's physiological responses to evoke heat-induced strain.

HEAT INJURY

Attention was focused on heat injury when Dill (29) published a summary in his book (1) entitled *Heat, Life and Altitude* (1938). Additional attention emanated from

Adolph et al.'s book (1) *Physiology of Man in the Desert* (1947). Thereafter compendia appeared by Ladell (71), Leithead (73), Leithead and Lind (74), and by Buskirk and Grasley (19) that detailed the experimental and clinical observations of heat injury which were categorized into heat stroke, heat exhaustion, and heat cramps.

Knochel and Schlein provided further categorization to include special conditions such as rhabdomyolysis (68). Knochel (67) recently updated his analysis of the three entities mentioned above as well as his consideration of such other conditions as hypernatremia. Similarly, Hubbard (62) and Hales et al. (48) have provided new insights into our thinking about heat injury.

> There are three phases during progressive heat stress: (1) below 39°C, changes in body fluids, electrolytes, and cardiovascular activity are the major contributors to morbidity; (2) between about 40° and 42.5°C, LPS toxicity and overwhelming cardiovascular demands cause morbidity and mortality; (3) above about 43°C, thermal damage becomes critical and there may also be a significant contribution of LPS (lipopolysaccharide) toxicity. Critical changes in cell membrane leaks, pumps, and energy metabolism probably underpin essentially the entire cascade of events. (48)

Thus, there is need for further clarification of the etiology of heat injury.

Moseley and Gisolfi (82) have explored the development of circulatory shock and heat stroke and have hypothesized that this complex condition is related to modified immune function and to multiorgan system failure. The process would involve combined thermal and ischemic injury to the gut leading to increased epithelial permeability and subsequent endotoxemia within the portal circulation. If the liver's clearance mechanisms are overwhelmed, cytokines (tumor necrosis factor) and interleukins are released, causing circulatory shock and heat stroke. Review of the available supporting literature on human and animal studies, as well as in vitro cellular and molecular work, suggests that the hypothesis is tenable. Locke et al. and Ryan et al. provided demonstrations of the synthesis and appearance of heat-shock protein with severe exercise in the heat, which supports this general hypothesis (77, 101). Further work in this area of heat-induced injury and thermal intolerance should help reveal operational mechanisms and possibilities for protective action.

COLD

Hardy (50) stated that it was impossible to separate temperature regulation from the physiological responses to heat and cold. In regard to cold, he cited data from the Russell Sage calorimeter maintained at 23°C in which heat storage efficiency was calculated for both shivering and exercise. That for shivering was 48%, whereas exercise (moving the arms and legs) elicited only 20% because much of the heat produced by the exercise was lost by convection due to limb movement. Carlson et al. (23), in confirming this conclusion, studied the relative value of exercise on a cycle ergometer with that of shivering for maintaining body temperature in the cold (10°C, men in shorts). Exercise at threefold the shivering metabolism could not prevent the large convective heat losses from the legs during pedaling. They also suggested that increased leg blood flow contributed to convective heat loss. They concluded that shivering was more efficient in preserving body heat because of peripheral vasoconstriction and alteration in posture—for example, curling up, which reduces exposed body surface area. From such data it was also concluded that despite the observed differences in heat conservation between exercise and shivering in the cold, peripheral vasoconstriction assists in maintaining some degree of thermal insulation in both situations.

An important concept was proposed by Aschoff and Wever concerning a body core and shell, and it has affected thinking about heat exchange, particularly in cool or cold environments (2). The core represents that portion of the body maintained at relatively constant temperature, whereas the variable shell changes with environmental circumstances. The shell responds to ambient temperature and becomes thicker and larger as ambient temperature decreases but is modified by clothing and the type, intensity, and duration of exercise.

In regard to the cool or cold environment and its interaction with exercise, a number of influential investigators have contributed and only a few can be cited. We recognize the early contributions of Beckman, Burton, Carlson, Hammel, Hayward, Hong, Keatinge, Irving, and Molnar. Such contributions have ranged from elucidation of the physical components of heat exchange (18) to adaptations to exercise in cold water (52, 59). In one significant paper Craig and Dvorak (27) studied thermoregulation during graded exercise while the subjects were immersed in water of different temperatures. They established that increasing heat production by exercise countered a decrease in heat stores, whereas vasomotor responses provided protection for the resting individual. Temperature gradients from body core to water were attenuated with greater workloads. Others have extended such studies to water temperatures that overwhelm the body's ability to preserve relative thermal equilibrium for reasonable periods of time whether for athletic competition or survival.

Existing data indicate that it is not only water temperature and exercise intensity that are important considerations regarding preservation of thermal balance; type of exercise, body fatness (particularly subcutaneous fatness), and fitness (ability to maintain high heat production) are important as well. In regard to thermal protection of subcutaneous adipose tissue, Keatinge (66) as well as Buskirk and Kollias (20) have described this insulative component and its effectiveness. Similar observations were observed by Buskirk et al. when men and women were exposed to cold air (21).

The question of whether regular exercise training affects cold acclimatization has been explored, but the conclusions are tenuous. Dressendorfer et al., Hong, and Skresler and Aarefjord have undertaken such studies on swimmers and underwater divers and the most prominent finding is blunted or delayed shivering during cold water immersion (32, 58, 107). In addition there was a pattern of minimization of thermal conductance through the skin while maintaining blood flow to underlying muscle indicating an insulative form of cold acclimatization that perhaps involved some muscle as well as subcutaneous adipose tissue. A variety of other studies reviewed by Young have been done on indigenous populations and habitually cold-exposed workers (120).

SUMMARY

While this review focuses to a major extent on early contributions to understanding temperature regulation and heat exchange during exercise, many worthy contributors were not mentioned because of space limitations for which I apologize. Early investigators set the stage for significant following work that has occurred and we are indebted to them for their enterprise. As our knowledge broadens and specific mechanisms are revealed more precisely, we still must acknowledge our predecessors and thank them for insights that provided stimuli for building on their efforts.

REFERENCES

1. Adolph E. F. and associates. *Physiology of Man in the Desert*. New York: Interscience, 1947.
2. Aschoff J. and R. Wever. Kern und schale im warmehauschalt des menschen. *Naturwissenschaften* 45:477–485, 1958.
3. Asmussen E. The cardiac output in rest and work in humid heat. *Am. J. Physiol.* 131:54–59, 1940.
4. Asmussen E. and O. Bøje. Body temperature and capacity for work. *Acta Physiol. Scand.* 10:1–22, 1945.
5. Astrand I. Aerobic work capacity in men and women with special reference to age. *Acta Physiol. Scand.* 49(Suppl. 169):1–92, 1960.
6. Bazett H. C. Theory of reflex control to explain regulation of body temperature at rest and during exercise. *J. Appl. Physiol.* 4:245–262, 1951.
7. Belding H. S. and T. F. Hatch. Index for evaluating heat stress in terms of resulting physiological strain. ASHRAE 27:129–136, 1955.
8. Benedict F. G. and H. F. Root. Insensible perspiration; its relation to human physiology and pathology. *Arch. Int. Med.* 34:1–35, 1926.
9. Benzinger T. H. Heat regulation: homeostasis of central temperature in man. *Physiol. Rev.* 49:671–759, 1969.
10. Benzinger T. H. (ed.). In: *Temperature*. Stroudsburg, PA: Dowden, Hutchinson, and Ross, 1977.
11. Bevegard B. S. and J. T. Shepherd. Reaction in man of resistance and capacity vessels in forearm and hand to leg exercise. *J. Appl. Physiol.* 21:123–132, 1966.
12. Blagden C. Experiments and observations in a heated room. *Philos. Trans. R. Soc. (Lond.)* 65:111–123, 1775.
13. Bligh J. and K. G. Johnson. Glossary of terms for thermal physiology. *J. Appl. Physiol.* 35:941–961, 1973.
14. Boulant J. A. Hypothalamic neurons regulating body temperature. In: *Handbook of Physiology*, edited by M. J. Fregley and C. M. Blatteis. New York: Oxford University Press, 1996, Vol. 1, Section 4, pp. 105–126.
15. Bradley S. E. Hepatic blood flow: effect of posture and exercise upon blood flow through the liver. In: *Transactions of the Seventh Conference on Liver Injury*, edited by F. W. Hoffbauer. New York: Josiah Macy Jr. Foundation, 1948, pp. 53–56.

16. Brengelmann G. L., J. M. Johnson, L. Hermansen, and L. B. Rowell. Altered control of skin blood flow during exercise at high internal temperatures. *J. Appl. Physiol.* 43:790–794, 1977.

17. Burton A. C. The application of the theory of heat flow to the study of energy metabolism. *J. Nutr.* 7:497–533, 1934.

18. Burton A. C. The operating characteristics of the human thermoregulatory mechanisms. In: *Temperature, Its Measurement and Control in Science and Industry.* New York: Reinhold, 1941, Vol. 7, American Institute of Physics, National Bureau of Standards, Research: Instrument Society of America, pp. 522–528.

19. Buskirk E. R. and W. C. Grasley. Heat injury and conduct of athletics. In: *Science and Medicine of Exercise and Sport* (2nd ed.), edited by W. R. Johnson and E. R. Buskirk. New York: Harper and Row, 1974, pp. 206–210.

20. Buskirk E. R. and J. Kollias. Total body metabolism in the cold. *The Bulletin (N.J. Acad. Sci.)* March:17–25, 1969.

21. Buskirk E. R., R. H. Thompson, and G. D. Whedon. Metabolic response to cold air in men and women in relation to total fat content. *J. Appl. Physiol.* 18:603–612, 1963.

22. Cage G. W., S. M. Wolfe, R. H. Thompson, and R. S. Gordon, Jr. Effects of water intake on composition of thermal sweat in normal human volunteers. *J. Appl. Physiol.* 29:687– 690, 1970.

23. Carlson L. D., D. Pearl, and W. Scheyer. Effects of temperature and work on metabolism and heat loss in man. *Am. J. Physiol.* 179:625, 1954 (Abstract).

24. Cleland T. S., S. M. Horvath, and M. Phillips. Acclimatization of women to heat after training. *Int. Z. Angew. Physiol.* 27:15–24, 1964.

25. Cooper T., W. C. Randall, and A. B. Hertzman. The vascular convection of heat from active muscle to overlying skin. *Wright Air Development Center (WADC) Tech. Report,* (*Dayton, Ohio*) Pt. 7:6680, 1951.

26. Cooper T., W. C. Randall, and A. B. Hertzman. Vascular convection of heat from active muscle to overlying skin. *J. Appl. Physiol.* 14:207–211, 1959.

27. Craig A. B. and M. Dvorak. Thermal regulation of man exercising during water immersion. *J. Appl. Physiol.* 25:28–35, 1968.

28. Davy J. On the temperature of man within the tropics. *Proc. R. Soc. (Lond.)* 437–466, 1850.

29. Dill D. B. *Life, Heat and Altitude.* Cambridge: Harvard University Press, 1938, pp. 1–211.

30. Dill D. B. Desert sweat rates. In: *Advances in Climatic Physiology,* edited by S. Itoh, K. Ogata, and H. Yoshimura. New York: Springer-Verlay, 1972, pp. 134–143.

31. Dreosti A. O. The results of some investigations into the medical aspect of deep mining on the Witwatersrand. *J. Chem. Metal Mining Soc. S. Africa* 36:102–129, 1935.

32. Dressendorfer R. H., R. M. Smith, D. G. Baker, and S. K. Hong. Cold intolerance of long-distance runners and swimmers in Hawaii. *Int. J. Biometeorol.* 21:51–63, 1977.

33. Eichna L. W., W. B. Bean, W. F. Ashe, and N. Nelson. Performance in relation to environmental temperature: Reactions of normal young men to hot, humid (simulated jungle) environment. *Bull. Johns Hopkins Hosp.* 76:25–58, 1945.

34. Eichna L. W., C. R. Park, N. Nelson, S. M. Horvath, and E. D. Palmes. Thermal regulation during acclimatization in a hot dry (desert type) environment. *Am. J. Physiol.* 163:585–597, 1950.

35. Ekblom B., C. J. Greenleaf, J. E. Greenleaf, and L. Hermansen. Temperature regulation during exercise dehydration in man. *Acta Physiol. Scand.* 79:475–483, 1970.

36. Feldberg, W. and R. D. Myers. Effects on temperature of amines injected into the cerebral ventricles. A new concept of temperature regulation. *J. Physiol. (Lond.)* 173:226–227, 1964.

37. Feldberg W., R. D. Myers, and W. L. Veale. Perfusion from cerebral ventricle to cisterna magna in the unanesthetized cat. Effect of calcium on body temperature. *J. Physiol. (Lond.)* 207:403–416, 1970.

38. Forbes W. H., D. B. Dill, and F. G. Hall. The effects of climate upon the volumes of blood and of tissue. *Am. J. Physiol.* 130:739–746, 1940.

39. Fox R. H. and R. K. Macpherson. The regulation of body temperature during fever. *J. Physiol. (Lond.)* 125:21p, 1954.

40. Fregley M. J. and C. M. Blatteis. II. Thermal environment. In: *Handbook of Physiology,* edited by M. J. Fregley and C. M. Blatteis. New York: Oxford University Press, 1996, Vol. 1 and 2, Section 4, pp. 45–626.

41. Gershon-Cohen J. A short history of medical thermometry. *Ann. NY Acad. Sci.* 121:1–11, 1964.

42. Gisolfi C. V., P. T. Wall, and W. R. Mitchell. Thermoregulatory responses to central injections of excess calcium in monkeys. *Am. J. Physiol.* 245:R76–R83, 1983.

43. Gisolfi C. V. and C. B. Wenger. Temperature regulation during exercise: old concepts, new ideas. In: *Exercise and Sports Science Reviews,* edited by R. L. Terjing. Lexington, MA: The Collamore Press, 1984, Vol. 12, pp. 339–372.

44. Greenleaf J. E. Hyperthermia and exercise. In: *International Review of Physiology, Environmental Physiology III,* edited by D. Robertshaw. Baltimore: University Park Press, 1979, Vol. 20, pp. 157–208.

45. Greenleaf J. E. and B. L. Castle. Exercise temperature regulation in man during hypohydration and hyperhydration. *J. Appl. Physiol.* 30:847–853, 1971.

46. Grimby G. Renal clearances during prolonged supine exercise at different loads. *J. Appl. Physiol.* 20:1294–1298, 1965.

47. Grodins F. Theories and models in regulatory biology. In: *Physiological and Behavioral Temperature Regulation,* edited by J. D. Hardy, P. Gagge, and J. A. J. Stolwijk. Springfield, IL: Chas C. Thomas, 1970, pp. 722–726.

48. Hales J. R. S., R. W. Hubbard, and S. L. Gaffin. Limitation of heat tolerance. In: *Handbook of Physiology,* edited by M. J. Fregley and C. M. Blatteis. New York: Oxford University Press, 1996, pp. 285–355.

49. Hammel H.T. Regulation of internal body temperature. *Ann. Rev. Physiol.* 30:641–710, 1968.

50. Hardy J. D. Physiology of temperature regulation. *Physiol. Rev.* 41:521–606, 1961.

51. Hardy J. D. Models of temperature regulation—a review. In: *Essays on Temperature Regulation,* edited by J. Bligh and R. E. Moore. New York: Elsevier, 1972, pp. 163–186.

52. Hayward J. S., J. Eckerson, and M. L. Collis. Effect of behavioral variables on cooling rate of man in cold water. *J. Appl. Physiol.* 38:1073–1077, 1975.

53. Hellon R. F. Thermal stimulation of hypothalamic neurons in unanesthetized rabbits. *J. Physiol. (Lond.)* 193:381–395, 1967.

54. Hellon R. F. and A. R. Lind. The influence of age on peripheral vasodilation in a hot environment. *J. Physiol. (Lond.)* 141:262, 272, 1958.

55. Hensel H. Neural processes in thermoregulation. *Physiol. Rev.* 53:948–1017, 1973.

56. Hertig B. A., H. S. Belding, K. K. Kraning, D. L. Batterton, C. R. Smith, and F. Sargent II. Artificial acclimatization of women to heat. *J. Appl. Physiol.* 13:383–386, 1963.

57. Hertig B. A. and F. Sargent II. Acclimatization of women during work in hot environments. *Fed. Proc.* 22:810–813, 1963.

58. Hong S. K. Comparison of diving and nondiving women of Korea. *Fed. Proc.* 22:831–833, 1963.

59. Hong S. K., C. K. Lee, J. K. Kim, S. H. Song, and D. W. Rennie. Peripheral blood flow and heat flux of Korean women divers. *Fed. Proc.* 28:1143–1148, 1969.

60. Horvath S. M. Historical perspectives of adaptation to heat. In: *Environmental Physiology: Aging, Heat and Altitude,* edited by S. M. Harvath and M. K. Yousef. New York: Elsevier/North Holland, 1981, pp. 11–26.

61. Horvath S. M. and W. B. Shelley. Acclimatization to extreme heat and its effects on the ability to work in less extreme environments. *Am. J. Physiol.* 146:336–343, 1946.

110. Stolwijk J. A. J., M. F. Roberts, C. B. Wenger, and E. R. Nadel. Changes in thermoregulatory and cardiovascular function with heat acclimation. In: *Problems with Temperature Regulation During Exercise*, edited by E. R. Nadel. New York: Academic Press, 1977, pp. 77–90.
111. Talbott J. H. Heat cramps. *Medicine* 14:323–376, 1935.
112. Taylor H. L., A. F. Henschel, and A. Keys. Cardiovascular adjustments of man in rest and work during exposure to dry heat. *Am. J. Physiol.* 139:583–591, 1943.
113. Thauer R. Thermosensitivity of the spinal cord. In: *Physiological and Behavioral Temperature Regulation*, edited by J. D. Hardy. Springfield, IL: Charles C. Thomas, 1970, pp. 472–492.
114. Wade O. L. and J. M. Bishop. *Cardiac Output and Regional Blood Flow.* Oxford: Blockwell, 1962, pp. 1–268.
115. Webb-Peploe M. M. and J. T. Shepherd. Response of large hindlimb veins of the dog to sympathetic nerve stimulation. *Am. J. Physiol.* 215:299–307, 1968.
116. Wenger C. B. Heat of evaporation of sweat: thermodynamic considerations. *J. Appl. Physiol.* 32:456–459, 1972.
117. Wenger C. B., M. F. Roberts, J. A. J. Stolwijk, and E. R. Nadel. Forearm blood flow during body temperature transients produced by leg exercise. *J. Appl. Physiol.* 38:58–63, 1975.
118. White H. L. and D. Rolf. Effects of exercise and of some other influences on the renal circulation in man. *Am. J. Physiol.* 152:505–516, 1948.
119. Wyndham C. H., N. B. Strydom, C. G. Williams, J. F. Morrison, and G. A. G. Bredell. The heat reaction of Bantu males in various states of acclimatization. *Int. Z. Angew. Physiol. Einschl. Arbeitsphysiol.* 23:79–92, 1966.
120. Young A. J. Homeostatic responses to prolonged cold exposure: human cold acclimatization. In: *Handbook of Physiology*, edited by M. J. Fregley and C. M. Blatteis. New York: Oxford University Press, 1996, pp. 419–438.

chapter 11

THE RENAL SYSTEM

Jacques R. Poortmans and Edward J. Zambraski

RENAL function is markedly altered in response to the stress of acute exercise. The changes include decreased renal blood flow, decreased glomerular filtration rate, decreased excretion of water, decreased excretion of sodium/chloride, the release of renin–angiotensin and norepinephrine, increased excretion of proteins and other macromolecules, and changes in metabolic functions.

Exercise alters renal hemodynamics, excretory function, and hormone release. All this has implications for total-body homeostasis. Thus, changes in cardiovascular function, hormonally mediated events, and water and electrolyte balance associated with acute exercise could be influenced or even primarily mediated by alterations in renal function. Many of these changes could affect exercise performance. Therefore, it is appropriate to review the developments that have contributed to our present understanding of the renal response to exercise.

Despite the essential role of the kidney in the homeostasis of mammalian organisms at rest, and the diversified and potentially important changes in renal function with exercise, it is surprising that the field of exercise physiology has placed so little emphasis on the issue of the kidneys and exercise. The first meaningful textbook, *The Physiology of Muscular Exercise*, was published in 1919 by F. A. Brainbridge (8). Out of 214 pages none was devoted to the kidney and its functions.

In the 1930s four textbooks on exercise physiology were published, of which three did contain some information about the kidney. The first was written by Adrian G. Gould and Joseph A. Dye and called *Exercise and Its Physiology* (42). Of their 312 pages, approximately one page was related to kidney functions.

The second text, written by Percy M. Dawson and published in 1935 (22), was likely used by most physical education departments. It had about seven pages on the kidney and its changes during exercise. The third textbook of that period, published by Edward C. Schneider in 1933, ran to 366 pages and had no information whatsoever on the kidney (104). However, the third edition of this book, published in 1948, with Peter V. Karpovich as a coauthor, had one page (out of 313), which noted that lactic acid is found in urine after exercise (105). The fourth book was a human physiology text published in 1936 by Evans that included a few sentences on the composition of the urine after severe exercise. Specifically, it stated that "small trace of albumin will often be found in the urine which is passed shortly after taking muscular exercise, but this has no pathognomonic significance" (32).

In the 1950s, Sarah R. Reidman wrote a textbook entitled *The Physiology of Work and Play* of which about five out of 565 pages were devoted to urinary changes with exercise (93). By the end of that decade (1959), Karpovich became the sole author of the fifth edition of Schneider's text but it still had only one page (out of 368) on the kidney (55). He wrote that "changes in the urine after excessive exertion cannot represent pathological conditions of the kidneys."

Information on renal function during exercise is almost absent from most exercise physiology textbooks published in the 1960s. However, a 13 page chapter called "Kidney Function and Exercise" emphasized the participation of the kidney in the general adjustments designed to cut down the renal plasma flow and glomerular filtration rate (123). Nevertheless, Ernst Simonson and Ariel Keys did not mention the word "kidney" in their 1971 textbook, *Physiology of Work Capacity and Fatigue* (108). In contrast, David Lamb devoted a whole chapter (11 pages) to renal responses and adaptations to exercise in his 1978 book, *Physiology of Exercise: Responses and Adaptations*. He noted the complex nature of the responses and the lack of research on the subject has contributed "to our poor understanding of the function of this system under exercise conditions" (59). Around this time, some other textbooks referred to the kidneys and exercise. In their book, *Exercise Physiology*, Phillip Rasch and I. Dodd Wilson wrote part of a chapter on the kidney (nine pages), mentioning postexercise proteinuria, hemoglobinuria, and myoglobinuria (91). In 1982, Roy Shephard included four pages on renal responses during exercise in his book (672 pages) *Physiology and Biochemistry of Exercise* (107).

Historically, sports medicine textbooks have also given little attention to the kidney and exercise. In 1933 there was a publication by Herbert Herxheimer, who related that "kidney function during exercise and in recovery has been explained in a very incomplete manner" (47). More recent sports medicine textbooks describe only some of the renal abnormalities during exercise—namely, hematuria and proteinuria observed in athletes (38, 40).

It is surprising that recent textbooks on exercise physiology have meager information on kidney functions, with the possible exception of a statement concerning the redistribution of blood flow from the kidney to skeletal muscle. Scott K. Powers and Edward T. Howley introduced one half-page (out of 552 pages) on regulation of acid–base balance via the kidneys (89), while no renal material at all was included in the 849 pages of the exercise physiology text by William D. McArdle, Frank I. Katch, and Victor L. Katch (67). Nevertheless, the second edition of *Physiology of Sport and Exercise* (1999) by Jack H. Wilmore and David L. Costill did introduce two and one-half pages on erythropoietin doping and a half-page on electrolyte loss in urine during exercise (out of 710 pages) (125). The most recent textbook by George A. Brooks and associates, *Physiology of Exercise*, provided two pages (out of 830 pages) on renal hemodynamics and fluid volume regulation by the kidney during acute exercise (10).

THE EARLY DAYS OF RENAL PHYSIOLOGY

When one examines the past, it is evident that the history of nephrology and renal physiology is limited compared to other fields of medicine. Through clinical observations, however, Hippocrates (460–370 B.C.) was well aware of the problems related to renal diseases. Hippocrates was convinced that no other system or organ of the human body provides so much diagnostic information via its excretion as does the urinary system (65). Around 330 B.C., Aristotle (384–322 B.C.) performed anatomic studies on the kidneys of aborted fetuses and attributed two functions to the kidney (64). The first was the anchoring of the blood vessels to the body and the second, and most important function, was the secretion of residual fluid into the bladder.

In the Bible (Psalm 16:7), according to Kopple (56), the kidneys were considered to be associated with the innermost part of personality. They were viewed as central to the soul and to morality, but no reference was made to the fact that the kidneys produce urine.

The Middle Ages saw the development of Arabic medicine, extending the analytical approaches of Hippocrates (460–370 B.C.) and Galen (ca. 129–200). The writings of Avicenna (980–1037) give numerous details on the color, odor, density, and sediments of urine, which are put forth as a reliable guide for the diagnosis of the illness (31).

The overall structural organization of the kidney, its vasculature, and its excretory units were described by Alexander Shumlyansky (1748–1795), a Russian obstetrician, in his doctoral dissertation in 1783 in Strasbourg (41). First, he confirmed the early work of Marcello Malpighi (1628–1694) from Bologna describing "the glands" (now called glomeruli) of the kidney and their arterial and venous connections. He also described the glomerular–tubular relationship, a description which was extended later, in 1842, by Sir William Bowman (1816–1892). Bowman noted that the malpighian corpuscle was suitable for the separation of water from plasma. He also assumed that all other urinary constituents were added to the urine by the tubular epithelium. That same year, Carl F. W. Ludwig (1816–1895), a German physician, presented a thesis "On the Physical Forces That Promote the Secretion of Urine" (21). Thus, the glomerular ultrafiltration theory was born. In the twentieth century, the theory of glomerular filtration was provided an unchallengeable foundation by several investigators, including Richards et al., Chambers, and Walker et al. (109). As a corollary of glomerular filtration, it followed that substances present in the filtrate, but absent from the bladder urine, must be reabsorbed during transit along the tubule. During the same period (1923–1925) the reality of tubular secretion was demonstrated by Marshall (66), who administered phenol red dye intravenously to dogs. The dye was bound to plasma protein and could not be filtered by the glomeruli. However, it accumulated in high concentration in the proximal tubule lumen. This observation was the first evidence of renal tubular secretion in mammals.

Urines from diseased kidneys also contain albumin and hyaline cylinders (casts). Richard Bright (1789–1858) reported in 1836, the presence of coagulable protein in the urine of patients (9). However, in 1764, Domenico Cotugno (1736–1822), an Italian physician, had described the presence of "a dense white body like albumen of boiled eggs" (actually albumin) in the urine of a patient with nephrotic syndrome (102). By 1843, a German physiologist, Johann Franz Simon (1807–1843), observed the presence of hyaline cylinders in pathological urine. In 1896, another German physician, Hermann Senator, said that "all cylinders in urine owe their shape to the transit through the tubules. They are derived from the kidney and are indicative of renal disease" (94). From several lines of reasoning Senator also concluded that "the epithelia through some kind of secretion provide the material for the formation of hyaline cylinders."

Despite these many early observations and insights, it is surprising that there have been so few descriptions of abnormal constituents in the urine of individuals engaged in strenuous physical exercise. We know that "bloody urine" can occur as a consequence of exercise. Perhaps one of the first papers to describe the collection of bloody urine from athletes was that of Barach in 1910, who studied the renal effects of the marathon run (4).

EXERCISE AND RENAL PLASMA/BLOOD FLOW

Up to 1940, the literature on renal plasma flow (RPF) changes during exercise is limited. This lack of information may be attributed to the absence of an adequate method for measuring renal blood flow (RBF) in humans. Nevertheless, in 1935 Edwards and associates suggested that a diversion of blood flow from the kidney during exercise could provide more blood flow for working muscles (27).

Observations in Humans

In the early 1940s the application of the Fick first law was extended to renal physiology for the measurement of RPF. It was found that the clearance of diodrast or p-aminohippurate (PAH) could be used to determine RPF or renal blood flow (RBF). For practical reasons (to minimize interference or trauma to the subjects), when the clearance technique was first applied to humans to measure RBF and RPF, it was performed almost exclusively under rest conditions.

The first report on the effects of exercise on RPF came from the Department of Physiology, University of Birmingham (England) (5). Barclay et al. measured diodrast clearance in 10 healthy students immediately after cessation of a quarter mile run at full speed. They found that this short-term exercise reduced RPF 30% from the resting value. Ten to 40 minutes was required for RPF to return to resting levels.

Barclay et al. also investigated the effect of 15 minutes of sustained bicycle exercise (intensity not specified) in two subjects. They found a steady decline in RPF during the exercise. There was no correlation between changes in RPF and in urine flow. One year later, in 1948, additional reports appeared on the effect of exercise on RPF. White and Rolf investigated the renal responses of five subjects subjected to relatively light exercise in the form of jogging (6.8–11.5 km/h for 13–22 minutes). They noted little change in RPF during this workload, but found an 80% fall during heavy exercise in two subjects (124). The authors postulated that the renal vascular bed constricts in response to the normal stimuli of exercise and that the degree of renal vasoconstriction is proportional to the intensity of exercise, the renal vascular resistance being increased at least fivefold during heavy exercise.

A more detailed and controlled study was made that same year by Chapman et al., who measured PAH clearance in nine young healthy male subjects. These tests involved constant intravenous infusion of PAH while the subjects were at rest, walking on a motor-driven treadmill, and during recovery (15). Each subject was investigated during two 20 minute walking exercise periods (5 km/h) at three different slopes on a motor-driven treadmill, resulting in exercise at different intensities. Their results were in accord with those obtained by other groups, but they demonstrated more clearly the relationship between the decline in RPF and the intensity of the exercise. Moreover, they emphasized that the recovery of RPF after exercise was considerably slower than that of the pulse rate and blood pressure. RPF remained depressed even after 40 minutes of rest following all three levels of exercise.

Since these studies had demonstrated the effects of exercise on RPF, the effects of various environmental conditions in addition to exercise were then examined in healthy subjects. Knowing that exposure to heat may cause profound circulatory stress, Radigan and Robinson in 1949 investigated the effects of moderate exercise (walking 5 km/h up a 5% grade for 80 minutes) on RBF in a "cool" (21°C) and hot (50°C) environment (90). The decreases in RPF were significantly greater when exercising in the heat (50°C) than in a cool environment (21°C). Together with Smith and Pearcy, Robinson extended his study of the adaptations of exercising men to dehydration (110). At 50°C temperature while in a dehydrated state, moderate exercise (walking 5 km/h up a 5% grade) induced a 56% decrease in RPF; the subjects became fatigued or syncopal and were unable to complete the exercise period. These results were later confirmed by German researchers in 1960 (112).

All previous investigators restricted their studies to the effects of absolute power on renal hemodynamics. The next step was to relate changes in RPF to exercise quantified as relative power outputs. Thus, Grimby from Göteborg exercised 15 healthy young males on a bicycle ergometer at different power requirements expressed as percentage of their maximal oxygen uptake ($\dot{V}O_2max$) (43). Using PAH clearance, it was shown that RPF decreased linearly as exercise intensity increased from 18% to 89% of VO_2max, the lowest value of RPF being 33% of the resting value.

These reports indicated that the degree of renal vasoconstriction is proportional to the intensity of exercise. The first experiment to attempt to alter this response was reported in 1967 by the Swede Jan Castenfors from the Karolinska Institute (14). He used dihydralazine, a vasodilator agent which causes smooth muscle relaxation. When dihydralazine was injected intravenously at the start and in the middle of a 45 minute bicycle exercise at approximately 170 beats per minute, the decrease in RPF was attenuated.

Both the afferent and efferent glomerular arterioles are densely innervated by sympathetic adrenergic nerves. Neurally induced renal vasoconstriction during exercise could also be supplemented by the effects of elevated circulating norepinephrine and epinephrine.

In the 1970–1980s sensitive assays became available for the measurement of plasma and urine catecholamines. Such measurements, however, represent only a generalized or global measure of sympathetic nervous system activation, both at rest and during perturbations such as exercise. As discussed in more detail in Chapter 5, the measurement of plasma norepinephrine (NE) spillover rates in the 1980s emerged as a more rigorous method to evaluate sympathetic responses from select organs. In 1991, Tidgren and associates (114) demonstrated that the differences in RPF at several exercise intensities were closely related to the renal NE spillover rate, which indicated that the renal vasoconstriction during exercise was related to increased renal sympathetic activity. In 1988 Hasking et al. measured renal NE spillover both at rest and during exercise in humans (44). Their work demonstrated that at rest renal NE spillover, as a contribution to total NE spillover, was 22%, an amount that was similar to the contribution from muscle. With exercise renal NE spillover was increased.

Animal Models for Human Responses

In contrast to studies in humans utilizing indirect clearance techniques, direct measurements of RPF/RBF have been made during exercise in dogs, miniature swine, and baboons. Measurements made in conscious healthy dogs indicate that RBF does not decrease during exercise, as it does in humans (6, 51, 117). In contrast, the renal hemodynamic response to exercise in miniature swine appear to be similar to what has been observed in humans. Bloor and associates in the late 1980s conducted experiments in miniature swine that were trained to run on a motor-driven treadmill at various power intensities (7). Radioactive microspheres were used to measure both total RBF as well as the distribution of RBF within the kidney (i.e., cortex versus medulla) (7, 100). Severe exercise reduced RBF to 30% of control, while the intrarenal blood flow distribution was unchanged throughout the rest and exercise period. These results clearly showed that exercising swine, like humans, increase blood flow to skeletal muscle in association with a decrease in RBF. In contrast, these changes have not been seen in the baboon. Vatner implanted Doppler ultrasonic flow probes on the left renal arteries of conscious baboons that were permitted to "exercise" (climbing, running, jumping) in a large outdoor enclosure (118). Moderate exercise (heart rate frequency of about 179 beats/min) did not decrease RBF.

Exercise and Renal Sympathetic Nerve Activity

An important contribution of the animal studies was quantifying changes in renal sympathetic nerve activity (RSNA) during exercise in conscious animals. To obtain renal nerve recordings from awake animals, chronic implantation of recording electrodes on the renal nerve bundle is required. Successful recording of RSNA in animals during exercise was achieved by Schad and Seller in 1975 using cats (101). These results have been confirmed by others using rabbits (25, 78). All three studies demonstrated that RSNA was increased by acute exercise. Also, the increase in RSNA was related to the intensity of exercise and RSNA remained elevated well into the postexercise recovery period. This direct evidence showing increased RSNA during acute exercise was extremely important and has profound implications regarding the mechanisms responsible for changes in various elements of renal function during exercise (i.e., changes in renal hemodynamics, water/electrolyte excretion, hormone release). A clear example is renin release. With acute exercise there is an increase in renal renin release. Although there are multiple mechanisms that could potentially be responsible for the increased renin secretion (i.e., change in perfusion pressure, prostaglandin release, macula densa feedback), Zambraski and associates showed that the increased renin release during exercise is mediated by renal sympathetic nerves (130).

THE INFLUENCE OF EXERCISE ON THE GLOMERULAR FILTRATION

Prior to the 1930s methodology had not been developed to measure glomerular filtration rate (GFR); consequently, the initial studies to examine the effects of exercise on renal excretion function were limited to the examination of excretion rates of water and solutes.

In 1921, James Campbell and Thomas Webster were the first to report creatinine excretion rates in subjects at rest, performing daily activities, and after exercise had been completed on a bicycle ergometer. Their results varied and no consistent conclusions could be drawn concerning the effects of exercise on glomerular filtration (11, 12). Twenty years later, Eggleton reported that in a group of 60 of his biochemistry class students, full-speed running over a quarter mile distance resulted in a significant decrease in creatinine excretion (29).

The first attempt to directly measure changes in GFR with exercise was made in 1936 by F. G. Covian and P. B. Rehberg on two healthy young men who performed bicycle exercise on a Krogh ergometer at power intensities varying from 120 W to 240 W during 12–40 minutes (19). They used endogenous creatinine clearance (creatinine rate excretion divided by plasma creatinine concentration) to evaluate GFR before and after exercise. The authors observed that exercise at low and moderate intensity did not affect GFR, while heavy or intense exercise reduced GFR by nearly 40%. This result was later confirmed by several authors using inulin clearance, a more accurate method to measure GFR (5, 70, 124). A further step in the evaluation of the effect of exercise was introduced by Radigan and Robinson (90) and by Smith et al. (110), who exercised their subjects in a hot environment (50°C) or when in a dehydrated state. Under these conditions GFR showed earlier and more exaggerated decreases.

Experiments conducted between 1965 and 1993 have confirmed the importance of exercise intensity and duration on GFR. Grimby (43), Kachadorian and Johnson (52), and Freund et al. (37) reported that GFR changes are exercise intensity dependent. GFR is constant, or even slightly elevated, when heart rates are under 150 beats/minute, an exercise intensity representing 50%–60% of the VO_2max. Above these values, there is a gradual decrease in GFR as the power output is increased. Thus, the kidney maintains GFR relatively unchanged over low to moderate power outputs, despite the fact that RPF is decreased. This indicates that filtration fraction (FF), the ratio of GFR to RPF, is increased during exercise (5, 14). Prolongation of exercise duration does not produce further changes. Severe exercise may cause a doubling of the FF as the decrease in RPF is greater than the decrease in GFR (14).

Both endocrine and neural factors influence GFR. In 1981 atrial natriuretic peptide (ANP) was discovered by de Bold and colleagues (23). This endogenous peptide is a potent renal vasodilator and increases both GFR and sodium excretion. Plasma levels of ANP increase during acute exercise, and it has been speculated that ANP may be important in maintaining or even increasing GFR during low-intensity exercise (37). In 1990 Zambraski (129) speculated that with moderate-to-severe exercise, the effects of ANP are likely overridden by elevated RSNA, angiotensin II, and circulating catecholamines—factors which probably are responsible for the observed decreased GFR.

The role of ANP as a determinant of renal function, both at rest and during exercise, however, is likely to be supplemented by the discovery of a similar peptide, urodilatin, in 1998 (106). Urodilatin is of renal origin. It has properties similar to ANP, and it has been recently shown that the excretion of urodilatin is increased with exercise (103). The exact roles that ANP and urodilatin play in mediating changes in kidney function with exercise will not be known until such time as pharmacological antagonists are available for each of these compounds to be used during an exercise test.

THE INFLUENCE OF EXERCISE ON WATER AND ELECTROLYTE EXCRETION

In 1862, Major-General R. A. Sabine, president and chair of the Royal Society (London), emphasized that there are large and frequent variations in the daily elimination of urea in a person of "the most regular habits" (exercise included) (96). However, there were no data provided to support this allegation. The scientific investigation of exercise and water/electrolyte excretion began in May 1870 when Edward Weston, a 31-year-old pedestrian, walked 100 miles in 21 hours and 39 minutes in New York City. All urine was collected in a single vessel by Professor Austin Flint in New York (35). The quantity of water in the urine was described as being "immensely greater" during the exercise (a 142% increase) and the author attributed this in part to the large quantity of liquids taken during the walk (beef essence, lemonade, champagne, brandy, and water). Since Weston also consumed 18 raw eggs he excreted "a considerable proportion of excess urea." Flint concluded that, besides larger amounts of "nitrogenized excrementitious matters" in the urine, chloride did not significantly change, while sulfates and phosphates increased during the walk.

Later, Mr. Weston underwent a 100 mile walk in England under the supervision of F. W. Pavy, a physician of Guy's Hospital in London. The urine analyses were extensively reported in 10 issues of *The Lancet* and the *British Medical Journal* (79). For example, his daily urine volume was reported as 1700 ml before the race and as an average of 1810 ml after 12 days walking. In summary, Pavy contradicted the conclusion reached by Flint. Indeed, he said that "although the elimination of urinary nitrogen is increased by muscular exercise, yet the increase is nothing nearly sufficient to give support to the proposition that the source of the power manifested in muscular action is due to the oxidation of muscular tissue" (79).

Historically, all of the studies that have evaluated the effects of exercise on creatinine excretion, or the renal clearance of compounds to assess changes in GFR or RPF, confirm that in the absence of excessive fluid loading, strenuous exercise results in decreased urine output. The urine excreted is reduced in volume and its composition is changed such that it is more concentrated (i.e., higher osmolarity) and more acidic. Resting urine flow rates in humans of approximately 1.0–1.5 ml/min are reduced by approximately 50% with strenuous exercise. One of the first papers to actually comment on the mechanisms of this exercise-induced reduction in urine volume was Castenfors in 1967 (14). He described the reduction in urine flow during exercise as largely a function of renal vasoconstriction and decreased GFR. Also, in the 1960–1970s, new assays to measure hormones such as aldosterone, renin-angiotensin, and vasopressin became available. Numerous groups observed that plasma concentrations of these hormones increased during moderate-to-heavy exercise, and these would be predicted to contribute to the conservation of fluid by the kidney during exercise.

In this regard, one interesting finding concerning urine output and exercise was first made by Refsum and Stromme in 1977 (92) and Wade and Claybaugh in 1980 (121). These two groups reported that during heavy exercise there was an impairment of the kidney's capacity to concentrate urine. Despite elevated levels of vasopressin there may actually be a decrease in urine osmolarity and an increase in free water clearance. This response is not seen in all cases and the consequences of this impairment in terms of fluid balance during exercise are minimal. The mechanism responsible for this renal resistance to vasopressin has not been determined.

Sodium Excretion

Because sodium is the major extracellular osmotically active electrolyte and because its homeostasis is tightly regulated to maintain plasma and total body tonicity at a fixed level, the renal excretion of sodium is a major factor in the regulation and control of extracellular volume (including plasma volume). Thus, during the stress of exercise, especially when coupled with dehydration due to excess water and sodium loss through sweating, the renal handling of sodium will be important in terms of maintaining or reestablishing body fluid homeostasis. In this exercise setting, renal conservation of sodium and water (i.e., decreased excretion) is essential.

It is quite interesting to observe that sodium and chloride excretion during exercise are concordant. Koriakina et al. (57) observed that running 3–28 km or playing soccer reduced chloride excretion. From the 1930s through 1947 these findings have been confirmed by several researchers (5, 30, 45, 62, 126). More recent studies have evaluated the effects of exercise on sodium excretion, both in the laboratory and during athletic competitions, such as the marathon (129). In general, with intense prolonged exercise, sodium excretion decreases by 30%–50% or more. In 1967 Castenfors (14) suggested that the decrease in sodium excretion could not be fully accounted for by changes in GFR or decreased filtered sodium load. Hence, with exercise there is an increase in renal tubular sodium reabsorption. In 1991 Freund et al. demonstrated that the antinatriuretic effect of exercise is dependent on the intensity of the exercise (37).

Because of the paramount importance of the control of sodium handling by the kidney, there are several different mechanisms that can influence sodium excretion during exercise stress. These include changes in GFR (filtered sodium load), the renal effects of renin–angiotensin and aldosterone, both of which may independently increase renal tubular sodium reabsorption, and ANP—which decreases renal tubular sodium reabsorption. In addition, it has been conclusively shown that renal sympathetic nerves directly increase renal tubular sodium reabsorption, an effect that is independent of changes in renal hemodynamics or other hormonal systems (24).

The identification of the mechanism(s) responsible for increased renal tubular sodium reabsorption during exercise is an important issue. With moderate-to-heavy exercise, since the increased renal tubular sodium reabsorption parallels changes in circulating renin–angiotensin and aldosterone, it has been widely presumed (and most textbooks espouse) that the antinatriuretic effect of exercise is due to increases in these hormones. These associations, however, are probably incorrect. This association between the changes in urinary sodium excretion and changes in renin–angiotensin and aldosterone during exercise was challenged by Wade et al. in 1987 (121). They demonstrated that the changes in urinary sodium excretion with exercise were not altered in subjects who were treated with drugs to block the angiotensin II production. This response was confirmed by Mittleman et al. in 1996 (72). In the exercised miniature swine, the antinatriuretic response to exercise is extremely fast, suggesting a neural effect. Also, the antnatriuretic response was unaltered by treatment with an aldosterone antagonist (129). It has been suggested that there is significant neurogenic control of tubular sodium reabsorption during exercise (129). However, experiments involving exercise in the presence of renal denervation have not been performed.

An important question concerning the renal excretory response to exercise involves the problem of "water intoxication" and the resultant hyponatremia (77). Subjects who drink copious amounts of water during a long-term exercise event, such as a marathon, may develop dilutional hyponatremia. Questions have arisen as to why the kidney fails to excrete the excess free water. In 1984 Poortmans (83), from the Université Libre de Bruxelles, demonstrated that in fluid-loaded subjects with urine flow rates of 10–12 ml/min, the antidiuretic effects of strenuous exercise were profound. In these subjects exercise reduced their urine flow rates to 2–3 ml/min despite continued fluid loading. It is likely that the observed renal hemodynamic changes, especially the decrease in GFR and filtered loads of sodium and water, substantially reduce the delivery of filtrate to the diluting segment of the nephron. The result is an antidiuresis rather than an anticipated diuresis.

THE URINE METABOLITES AS EVIDENCE OF MUSCLE ENERGETICS

Historically, the examination of urine metabolites was used to indirectly assess changes in muscle energetics during exercise. Considering urinary urea during exercise, Justus von Liebig (1803–1873), from the University of Giessen (Germany), was convinced that "the sum of the mechanical effects produced is proportional to the amount of nitrogen in the urine." In other words, muscle protein breakdown was necessarily involved in muscular contraction (60). Subsequent studies have shown that this does not occur to an appreciable extent. This issue was clarified when the concept of renal clearance was introduced by Moller, McIntosh, and Van Slyke in 1928 (75). In 1932, D. D. Van Slyke et al. from the Rockefeller Institute for Medical Research (116) reported a decrease of urea clearance during several types of exercise—tennis, weight lifting, "vigorous exercise," football, soccer, basketball, and running on a treadmill. The decrease was 25%–75% of the original clearance depending on the intensity and duration of the exercise. They concluded that during severe exercise it is probable that "some dehydration of the blood by sweating occurred, and perhaps also some diversion of blood flow from the kidneys to the heavily working muscles" (116).

With regard to the renal metabolism of lactic acid, small amounts of lactate have been found in the urine of humans at rest. Historically, in 1877, Karl A. Spiro isolated zinc lactate from the urine of tetanized animals and a small quantity of impure material from the urine of men after mild exercise (111). Ten years later, Colasanti and Moscatelli confirmed this observation in the urine of soldiers after a forced march (17). Insight into the origin of lactic acid excretion came from Sir Archibald V. Hill's (1886–1977) work, which showed that mild/moderate exercise did not lead to a continuous accumulation of lactate in the plasma. This may account for the fact that only small quantities of lactate have been found in the urine after long periods of exercise, such as mountain climbing (49) or hardwork for 5 hours (12). In 1925, Liljestrand and Wilson made a thorough study of urinary lactate excretion after muscular exercise (61). They were the first to actually isolate d(−)lactate ("sarcolactate") from exercise urine and to construct a complete curve for rates of lactate excretion during recovery from exercise.

In 1937, Johnson and Edwards addressed the problem of evaluating blood and urine lactates (50). They calculated that the total excretion of excess lactate and pyruvate accounts for about 2% of the total excess that the body has dissipated after short-term intense exercise. This figure is remarkably close to the present-day estimates. Meanwhile, Miller and Miller examined urinary lactate excretion in two subjects who were running on a treadmill for 2–15 min at speeds of 10–16 miles/h (71). A threshold plasma lactate of 5–6 mmol·l^{-1} was associated with increased rates of lactate clearance (20 ml/min versus 1 ml/min at rest). Dies et al. in 1969, concluded that lactate is actively reabsorbed in the renal proximal tubule (26).

What is the fate of the lactate reabsorbed by the kidney? Biochemical investigations by Sir Hans A. Krebs and T. Yoshida demonstrated that rat kidney cortex slices synthesize glucose from lactate under resting conditions (however, five times less than in liver). They also collected cortex slices 35 minutes after swimming (58). Gluconeogenesis from lactate increased by 55% immediately after exercise. This increase was interpreted as an adaptive reaction effecting a more rapid removal of lactate from blood. In order to understand this enhanced gluconeogenesis, Sanchez-Medina investigated the activities of the enzymes responsible for these biosynthetic pathways and showed that the activity of kidney phosphoenolpyruvate carboxykinase doubled when rats were forced to swim for 2 hours (97–99). In 1971, Wahren et al. determined arterial concentrations and net substrate exchange across the kidney in two human subjects who bicycled for 40 minutes at increasing intensities up to 200 W (122). There was inconsistent renal uptake and production of glucose during the exercise period. A limitation of this study was that the mean arterial lactate was less than 4 mmol·l^{-1}. Thus, the question of whether excess lactate produced during heavy exercise is oxidized versus converted to glucose has not as yet been answered.

THE HANDLING OF MACROMOLECULES BY THE KIDNEY DURING EXERCISE

The Nineteenth–Twentieth Century Transition

The presence of proteins in urine of healthy subjects who undergo physical exercise was first recognized in 1878 by W. von Leube from the medical clinic of Erlangen (120). He noted that out of 119 soldiers whose early morning urine had been free of protein, 12% developed albuminuria within 1–3 days after a strenuous battalion march. Von Leube believed that this phenomenon was a "physiologischen albuminurie," suspecting that its origin could be due to filtration through the pores of the glomerular membrane. A colleague of von Leube, Fleischer, published a case report 3 years later of a young soldier who excreted hemoglobin in his urine after strenuous exercise (without blood cell elements) (34). In 1906 a much more strenuous exercise, a 100 km march, revealed that 12 out of 30 young people had urines which contained protein, blood cells, and casts (3). The authors claimed that these findings were almost identical to an acute parenchymatous nephritis. A year later a very interesting report appeared in the *British Medical Journal* from a physician at the Radcliffe Infirmary, Oxford (18). Collier examined the urine of 156 young men involved in training for the University of Oxford Torpids, a famous rowing contest. Using the heat and acetic acid test, he demonstrated that 57% of the rowers had "functional albuminuria" which disappeared after a night's rest. But he also extended his conclusion to two important statements: (1) no one should any longer advise young men who pass large quantities of albumin in the urine after severe muscular exercise to give up all hard athletic competition, and (2) insurance companies should not continue to refuse to consider the acceptance of the lives of young men between the ages of 18 and 30 whose urines are found to contain albumin after exercise.

In 1910, J. H. Barach investigated the effects on the kidney of running a marathon race. He confirmed the presence of urinary protein, red cells, and casts after the race (4). In addition he pointed out that the subjects having the highest maximum blood pressures at the end of the race excreted the largest amounts of albumin.

While several studies confirmed the quantitative observation of postexercise proteinuria, which may amount to several grams of proteins per liter, other investigators tried to find an explanation for the occurrence of exercise proteinuria. Isaac Starr, from the University of Pennsylvania, introduced the concept of induced renal vasoconstriction (by emotional excitement in cats, stimulation of renal nerves in dogs, or subcutaneous injection of adrenaline in men) to explain a concomitant excretion of albumin in urine (113). He ended his summary by saying that "this concept may apply to those albuminurias in man which result from excessive muscular exercise."

In 1931, Edwards, Richards, and Dill from the Harvard Fatigue Laboratory examined protein in urine estimated by the degree of turbidimetry produced in the nitric acid test (28). Of the 42 football players studied after competition, 60% had proteinuria. They related these findings to the duration of exercise. In 1932, Frances A. Hellebrandt from the University of Wisconsin was convinced that albuminuria was related to the subnormal phase in blood pressure which follows intense exercise (46). She tested this relationship in 47 women after bicycling, running, or rowing exercise and concluded that the greatest amount of albumin occurred during the maximal depression in postexercise blood pressure. Moreover, Hellebrandt also claimed that strenuous exercise redirects blood flow to the working muscles and the skin, affecting the circulation of the kidney in such a way as to cause "asphyxiation of the renal cells beyond that compatible with normal function." She suspected that abnormal accumulation of acids in the renal tissue alters its permeability to the blood proteins and induces albuminuria. This conclusion was also reached in 1944 by Guy Viarnaud of Lyon, France, when he concluded that albuminuria after soccer and rugby games may indicate underlying renal disease (119).

The Rise of the "Athletic Pseudonephritis" Concept

As indicated previously, the concept of "pathological urine" induced by exercise originated with Baldes et al. from Frankfort (Germany) in 1906 (3). In 1950, Erik Hohwü Christensen and P. Högberg reported that postexercise proteinuria occurred in practically all their best Swedish skiers after horizontal and uphill skiing (16). In 1952,

Javitt and Miller suggested three possible causes of exercise proteinuria: increased glomerular filtration, increased acidity, or decreased tubular reabsorption of protein (48). The subject was revisited and extended by Kenneth D. Gardner 6 years later (39). Gardner collected urine specimens from 47 members of a football team. Of the 424 specimens, 44.8% met the criteria established for abnormal urine (red blood cell casts, broad granular casts). He also noted that the occurrence of protein was related to the estimated team activity. This study was intended to confirm and reemphasize the fact that formed elements heretofore considered almost pathognomonic of parenchymal renal disease may appear in the urine of players during the athletic season. However, Gardner stated that these elements usually do not indicate serious renal disease, for their disappearance from the urinary sediment is prompt when the athlete is withdrawn from daily exertion. The name "athletic pseudonephritis" was proposed for this phenomenon "with the hope that the sediment abnormalities that appear in the urine of an athlete will not lead to the incorrect diagnosis of nephritis in candidates for life insurance or induction into the armed forces."

Since 1956, the concept of "athletic pseudonephritis" has been widely used to characterize the urinary responses of healthy subjects to heavy exercise. It has been observed in many types of exercise (2, 20, 53), including nontraumatic exercise (1). A relation between postexercise proteinuria and the intensity of the exercise was mentioned in a few reports (54, 80, 86, 115) which hypothesized a link between the degree of proteinuria, heart rate, blood lactate, and the intensity of the exercise load imposed on the subjects.

The Identification of Macromolecules

In order to characterize urinary protein excretion one has to identify the origin of these macromolecules. Prior to 1958 incomplete information was available. After the introduction of paper electrophoresis, further insight was obtained by Nedbal and Seliger from Charles University (Prague). In their studies albumin remained the major constituent (67%) of postexercise proteinuria, but four globulin fractions were present as well (76). They concluded that the ratio of the different protein fractions in postexercise proteinuria is practically identical with the protein spectrum in blood serum. In 1961, two British scientists, Rowe and Soothill, studied exercise proteinuria, and using immune antisera, identified plasma albumin, transferrin, ceruloplasmin, and immunoglobulin G in urine (95). They suggested that "exercise" urine proteins might be the product of a selective filtration through the glomerular basement membrane. This heterogeneity of postexercise urinary proteins was confirmed by Poortmans in 1962 (82) and subsequently by other investigators (36, 68).

The introduction of sensitive and accurate immunochemical techniques allowed Poortmans and colleagues to suggest that postexercise proteinuria was in fact of both glomerular and tubular origin (81, 84, 87). In 1977 Mogensen and Solling (74) reported that certain amino acids (i.e., lysine) inhibit, either partially or completely, renal tubular protein reabsorption. They showed that those amino acids, positively charged, interfere with tubular protein reabsorption. Under those conditions, the results obtained under lysine perfusion may be considered as an index of glomerular passage of proteins. Using this information Poortmans administered lysine to healthy subjects submitted to severe exercise to confirm that exercise proteinuria involved both an increase in glomerular permeability as well as impaired renal tubular protein reabsorption (85).

There are probably several mechanisms for exercise proteinuria (88). As indicated in the studies cited, the magnitude of exercise proteinuria is related to exercise intensity and duration. Changes in renal hemodynamics (i.e., renal vasoconstriction) are also likely to be involved. One of the most definitive studies dealing with the mechanism of exercise proteinuria in man was published by Mittleman et al. in 1992 (73). In eight subjects undergoing 30 minutes of steady-state treadmill exercise at 75% of $\dot{V}O_2$max, the prostaglandin (PG) inhibitor indomethacin significantly attenuated the exercise proteinuria observed during the placebo tests. This effect of PG inhibition was not associated with any change in renal hemodynamics, excretory function, or systemic arterial pressure. In these same subjects, angiotensin-converting enzyme inhibition did not alter exercise proteinuria. These data suggest that PGs are contributing to an increase in glomerular membrane permeability to protein during exercise.

ACUTE RENAL FAILURE

Acute renal failure (ARF) is defined as a sudden decline in GFR. With regard to exercise, ARF could be the result of renal tubular obstruction, such as would be seen with exertional rhabdomyolysis. With rhabdomyolysis, the muscle breakdown and release of intracellular enzymes and myoglobin result in myoglobinuria and the possibility of renal tubular obstruction. ARF may also occur during exercise due to excessive renal vasoconstriction, which would decrease renal perfusion and GFR. When combined with exercise, environmental factors such as dehydration, sodium depletion and heat stress would be predicted to increase renal vasoconstriction and thereby the likelihood of ARF.

Historically, there has not been great research or research interest concerning the issue of exercise-induced ARF. This is mainly due to the fact that this complication of exercise is relatively rare. The first attempts to quantitate the frequency of ARF in an athletic event, such as a marathon, was made by MacSearraigh et al. (63). They evaluated the number of cases of ARF that occurred in the 90 km South African Comrades Marathon. In this race over a 9 year period, with more than 20,000 competitors, only 10 cases of ARF were reported. Besides the low frequency of exercise-induced ARF, another limitation in studying this condition is that it is difficult to predict who will get ARF in terms of the exercise intensity, individual's level of training, or other environmental factors that might contribute to this condition.

Rather than address the issue of what causes ARF during exercise, in a 1996 review Zambraski (128) approached this question from a different perspective; asking what prevents ARF during exercise. It was emphasized that during exercise high levels of RSNA, angiotensin II, endothelin, neuropeptide Y, and circulating catecholamines, would actually be predicted to cause excessive renal vasoconstriction and to potentially cause ARF. It was hypothesized that these renal vasoconstriction factors must be counterbalanced by some yet to be identified renal vasodilatory influence(s). Possibilities may include ANF, PG, and/or nitric oxide. Recent studies which have evaluated the effects of PG inhibtion during exercise in humans tested under conditions designed to mimic those occurring in an event such as a marathon suggest that PGs are not responsible for this renal vasodilatory protective effect (33). Moreover, PG inhibition does not prevent postexercise proteinuria following short-term high-intensity exercise (88). Definitive studies evaluating the role of nitric oxide on kidney function during exercise have yet to be conducted. Lastly, the suggestion has been made that the effects of exercise training on RSNA (see below) may be an important factor in preventing exercise-induced ARF (128).

ADAPTATIONS TO CHRONIC EXERCISE TRAINING

An area of research in which there is a lack of information is the effect of exercise training on intrinsic renal function. While a large effort has gone into elucidating the effect of training on the metabolic, structural, and functional properties of various tissues (i.e., skeletal muscle, bone, liver, pancreas, heart), to our knowledge, no studies have evaluated the effects of training on renal tissue(s). The ability to isolate, culture, and study individual renal cell types has only been refined over the last decade. As yet, the utility of these techniques has not been applied to the chronic exercise setting.

The various functions of the kidney are largely dictated by extrarenal regulating factors such as perfusion pressure, neural influences, and endocrine factors. With chronic exercise renal function would be predicted to change to the extent that these regulatory factors may be altered by exercise training. A clear example is RSNA. As discussed previously, during exercise the renal sympathetic nerves appear to influence renal hemodynamics, renin release, and possibly tubular sodium reabsorption and proteinuria. It has been stressed that, in humans who are exercise trained, the peripheral sympathetic nerve activity response to a given absolute exercise power output is decreased (127). In fact, an important human study by Meredith et al. in 1991 (69) showed that of the significant decrease in total body NE spillover (at rest) seen with chronic exercise training, two-thirds of this reduction is attributable to a decrease in renal NE spillover (RSNA). Thus, there is a selective and disproportionate reduction in RSNA with chronic exercise training. This may explain the reduction of postexercise proteinuria observed in trained sbjects submitted to submaximal exercise (13, 83). This training effect has profound implications, yet to be tested, not only for the acute renal response to exercise but also for changes in renal function that may be important in certain disease states, such as hypertension.

SUMMARY

The understanding of and interest in kidney function during exercise have lingered behind those for the other organ systems. There are two reasons. First, renal function during exercise is usually not considered to be essential or critical in terms of limiting or determining exercise capacity. Second, and more important, is the fact that renal function is extremely difficult to assess during exercise. While a urine analysis provides some information, it is extremely difficult to apply invasive techniques to assess kidney function in an exercising subject. Also, an array of noninvasive methodologies to evaluate various elements of renal function (e.g., hemodynamics, tubular transport, hormone release, etc.) simply are not available. This is particularly true with regard to the issue of possible exercise-induced renal damage or acute renal failure.

In the last decade or so there has been an increased interest in and/or acknowledgement of the importance of kidney function during exercise. Individuals studying exercise and environmental stressors and the associated changes in extracellular or plasma volumes have begun to look at alterations in kidney function as the mediator of some of these modifications. Changes in cardiovascular and thermoregulatory function with aging are pointing toward important alterations in kidney function with age. The issue of exercise- or exercise-and-dehydration-induced acute renal failure continues to be a problem. The fact that many athletes are using anti-inflammatory drugs or other compounds which may be nephrotoxic under certain conditions has generated interest in the potential deleterious combined effects of exercise and these agents on renal function. Lastly, the role of the kidneys as an endocrine organ, not only for the release of renin–angiotensin but also norepinephrine, has profound implications that may possibly explain the effects of chronic exercise on certain disease states. This may be particularly true for those diseases associated with elevated peripheral sympathetic nerve activity, such as hypertension, congestive heart failure, or obesity.

REFERENCES

1. Alyea E. P. and A. W. Boone. Urinary findings resulting from nontraumatic exercise. *Southern Med. J.* 50:905–910, 1957.
2. Bailey R. R., E. Dann, A. H. B. Gillies, K. L. Lynn, M. H. Abernethy, and T. J. Neale. What the urine contains following athletic competition. *New Z. Med. J.* 83:309–313, 1976.
3. Baldes, Heichelheim, and Metzger. Untersuchungen über den einfluss grosser köperanstrengungen auf zirkulationapparat, nieren und nervensystem. *Muenchen Med. Wschr.* 53:1865–1866, 1906.
4. Barach J. H. Physiological and pathological effects of severe exertion (the marathon race) on the circulatory and renal systems. *Arch. Intern. Med.* 5:382–405, 1910.
5. Barclay J., W. Cooke, R. Kenney, and M. Nutt. The effects of water diuresis and exercise on the volume and composition of the urine. *Am. J. Physiol.* 148:327–337, 1947.
6. Blake W. Effect of exercise and emotional stress on renal hemodynamics, water and sodium excretion in the dog. *Am. J. Physiol.* 165:149–157, 1951.

7. Bloor C., M. McKirnan, B. Guth, and F. White. Renal blood flow during exercise in trained and untrained swine. In: *Swine in Biomedical Research*, edited by M. Tumbleson. New York: Plenum Press, 1986, pp. 1789–1794.

8. Brainbridge F. A. *The Physiology of Muscular Exercise*. London: Longmans, Green and Co, 1919.

9. Bright R. Tabular view of the morbid appearances in 100 cases connected with albuminous urine. *Guy's Hosp. Rep.* 1:380–400, 1836.

10. Brooks G. A., T. D. Fahey, T. P. White, and K. M. Baldwin. *Exercise Physiology: Human Bioenergetics and Its Applications* (3rd ed.). Mountain View, CA: Mayfield, 2000, pp. 93–95.

11. Campbell J. and T. Webster. LXXX. Day and night urine during complete rest, laboratory routine, light muscular work and oxygen administration. *Biochem. J.* 15:660–664, 1921.

12. Campbell J. and T. Webster. XV. Effect of severe muscular work on the composition of the urine. *Biochem. J.* 16:106–110, 1922.

13. Cantone A. and P. Cerretelli. Effect of training on proteinuria following muscular exercise. *Int. Z. Angew. Physiol. Arbeitsphysiol.* 18:324–329, 1960.

14. Castenfors J. Renal function during exercise. *Acta Physiol. Scand.* 70 (Suppl. 293):1–43, 1967.

15. Chapman C., A. Henschel, J. Minckler, A. Forsgren, and A. Keys. The effect of exercise on renal plasma flow in normal male subjects. *J. Clin. Invest.* 27:639–644, 1948.

16. Christensen E. and P. Högberg. Physiology of skiing. *Arbeitsphysiologie.* 14:292–303, 1950.

17. Colasanti G. and R. Moscatelli. L'acido paralattico nell'orina dei soldati dopo le marcie di resistenza. *Bull. D.R. Acad. Med. Roma* viii:482–488, 1886–7.

18. Collier W. Functional albuminuria in athletes. *Br. Med. J.* I:4–5, 1907.

19. Covian F. and P. Rehberg. Uber die nierenfunktion während schwerer muskelarbeit. *Skand. Arch. Physiol.* 75:21–37, 1936.

20. Coye R. D. and R. R. Rosandich. Proteinuria during the 24-hour period following exercise. *J. Appl. Physiol.* 15:592–594, 1960.

21. Davis J., C. Gottschalk, D. Häberle, and K. Thurau. Foreword. Carl Ludwig's habilitation thesis. *Kidney Int.* 46 (Suppl. 46):i–iii, 1994.

22. Dawson P. M. *The Physiology of Physical Education*. Baltimore: Williams & Wilkins, 1935.

23. de Bold A. J., H. B. Borenstein, A. T. Veress, and H. Sonnenberg. A rapid and potent natriuretic response to intravenous injection of atrial myocardial extract in rats. *Life Sci.* 28: 89–94, 1981.

24. Dibona G. F. and U. C. Kopp. Neural control of renal function. *Physiol. Rev.* 77:75–197, 1997.

25. Dicarlo S. and V. Bishop. Onset of exercise shifts operating point of arterial baroreflex to high pressures. *Am. J. Physiol.* 262 (Heart Circ. Physiol. 31):H303–H307, 1992.

26. Dies E., G. Ramos, E. Avelar, and M. Lennhoff. Renal excretion of lactic acid in the dog. *Am. J. Physiol.* 216:106–111, 1969.

27. Edwards H., M. Cohen, and D. Dill. Renal function in exercise. *Arbeitsphysiologie.* 9: 610–618, 1935–1937.

28. Edwards H. T., T. K. Richards, and D. B. Dill. Blood sugar, urine sugar and urine protein in exercise. *Am. J. Physiol.* 98:352–356, 1931.

29. Eggleton M. A class experiment on urinary changes after exercise. *J. Physiol. (Lond.)* 101: 1P–2P, 1942.

30. Eggleton M. G. The effect of exercise on chloride excretion in man during water diuresis and during tea diuresis. *J. Physiol. (Lond.)* 102:140–154, 1943.

31. Eknoyan G. Arabic medicine and nephrology. *Am. J. Nephrol.* 14:270–278, 1994.

32. Evans L. *Starling's Principles of Human Physiology.* London: Churchill, 1936.
33. Farquhar W., A. Morgan, E. Zambraski, and W. Kenney. Effects of acetaminophen and ibuprofen on renal function in the stressed kidney. *J. Appl. Physiol.* 86:598–604, 1999.
34. Fleischer R. Ueber eine neue form von haemoglobinurie beim menschen. *Berliner Klin. Wschr.* 18:691–694, 1881.
35. Flint A. The influence of excessive and prolonged muscular exercise upon the elimination of effete matters by the kidneys; based on an analysis of the urine passed by Mr. Weston, while walking one hundred miles in twenty-one hours and thirty-nine minutes. *N.Y. Med. J.* 12:280–289, 1870.
36. Freedman M. H. and G. E. Connell. The heterogeneity of gamma-globulin in post-exercise urine. *Can. J. Biochem.* 42:1065–1097, 1964.
37. Freund B., E. Shizuru, G. Hashiro, and J. Claybaugh. Hormonal, electrolyte, and renal responses to exercise are intensity dependent. *J. Appl. Physiol.* 70:900–906, 1991.
38. Gardner K. Exercise and the kidney. In: *Sports Medicine*, edited by O. Appenzeller. Baltimore: Urban & Schwarzenberg, 1988, pp. 189–195.
39. Gardner K. D. "Athletic pseudonephritis": alteration of urine sediment by athletic competition. *JAMA* 161:1613–1617, 1956.
40. Goldszer R. and A. Siegel. Renal abnormalities during exercise. In: *Sports Medicine*, edited by R. Strauss. Philadelphia: W.B. Saunders, 1984, pp. 130–139.
41. Gottschalk C. Alexander Schumlansky's De structura renum. *Am. J. Nephrol.* 14:320–324, 1994.
42. Gould A. G. and J. A. Dye. *Exercise and Its Physiology.* New York: Barnes A.S., 1932.
43. Grimby G. Renal clearances during prolonged supine exercise at different loads. *J. Appl. Physiol.* 20:1294–1298, 1965.
44. Hasking G., M. Esler, G. Jennings, E. Dewar, and G. Lambert. Norepinephrine spillover to plasma during steady-state supine bicycle exercise. *Circulation* 78:516–521, 1988.
45. Havard R. E. and G. A. Reay. The influence of exercise on the inorganic phosphates of the blood and urine. *J. Physiol. (Lond.)* 61:35–48, 1926.
46. Hellebrandt F. A. Studies on albuminuria following exercise. I. Its incidence in women and its relationship to the negative phase in pulse pressure. *Am. J. Physiol.* 101:357–364, 1932.
47. Herxheimer H. *The Principles of Medicine in Sport for Physicians and Students.* Berlin: George Thieme (translated from German by W. W. Tuttle and G. C. Knowlton), 1933.
48. Javitt N. B. and A. T. Miller. Mechanisms of exercise proteinuria. *J. Appl. Physiol.* 4:834–839, 1952.
49. Jerusalem E. Ueber ein neuer verfahren zur quantitativen bestimmung der milchsäure in organen und tierischen flüssigkeiten. *Biochem. Z.* Berlin xii:361–389, 1908.
50. Johnson R. E. and H. T. Edwards. Lacatate and pyruvate in blood and urine after exercise. *J. Biol. Chem.* 118:427–432, 1937.
51. Joles J. A. Renal function and exercise in dogs. Ph.D. thesis, University of Utrecht, 1984.
52. Kachadorian W. and R. Johnson. Renal responses to various rates of exercise. *J. Appl. Physiol.* 28:748–752, 1970.
53. Kachadorian W. A. and R. E. Johnson. Athletic pseudonephritis in relation to rate of exercise. *Lancet* 1:472, 1970.
54. Kachadorian W. A., R. E. Johnson, R. E. Buffington, L. Lawler, J. J. Serbin, and T. Woodall. The regularity of "athletic pseudonephritis" after heavy exercise. *Med. Sci. Sports* 2:142–145, 1970.
55. Karpovich P. *Physiology of Muscular Activity* (5th ed.). Philadelphia: Saunders, 1959.
56. Kopple J. The Biblical view of the kidney. *Am. J. Nephrol.* 14:279–281, 1994.

57. Koriakina A. F., E. B. Kossowskaja, and A. N. Krestownikoff. Uber die schwankungen des chloridgehaltes im blut, harn und schweib bei muskeltätigkeit. *Arbeitsphysiologie.* 2:461–473, 1930.

58. Krebs H. A. and T. Yoshida. Muscular exercise and gluconeogenesis. *Biochem. Z.* 338: 241–244, 1963.

59. Lamb D. *Physiology of Exercise. Responses and Adaptations.* New York: Macmillan, 1978.

60. Liebig V. J. *Animal Chemistry, or Chemistry in Its Application to Physiology and Pathology.* London: Taylor and Walton, 1843.

61. Liljestrand S. H. and D. W. Wilson. The excretion of lactic acid in the urine after muscular exercise. *J. Biol. Chem.* 65:773–782, 1925.

62. MacKeith N. W., M. S. Pembrey, W. R. Spurell, E. C. Warner, and H. J. W. J. Westlake. Observations on the adjustment of the human body to muscular work. *Proc. R. Soc. B* 95: 413–439, 1923.

63. MacSearraigh E. T. M., J. C. Kallmeyer, and H. B. Schiff. Acute renal failure in marathon runners. *Nephron* 24:236–240, 1979.

64. Marandola P., S. Musitelli, H. Jallous, A. Speroni, and T. De Bastiani. The Aristotelian kidney. *Am. J. Nephrol.* 14:302–306, 1994.

65. Marketos S. Hippocratic medicine and nephrology. *Am. J. Nephrol.* 14:264–269, 1994.

66. Marshall E. K. The secretion of urine. *Physiol. Rev.* 6:440–484, 1926.

67. McArdle W. D., F. I. Katch, and V. L. Katch. *Exercise Physiology* (4th ed.). Baltimore: Williams & Wilkins, 1996.

68. McKay E. and R. J. Slater. Studies on human proteinuria. II. Some characteristics of the gamma globulins excreted in normal, exercise, postural, and nephrotic proteinuria. *J. Clin. Invest.* 41:1638–1652, 1962.

69. Meredith I., P. Friberg, G. Jennings, E. Dewar, V. Fazio, G. Lambert, and M. Esler. Exercise training lowers resting renal but not cardiac sympathetic activity in humans. *Hypertension* 18:575–582, 1991.

70. Merrill A. and W. Cargill. The effect of exercise on the renal plasma flow and filtration rate of normal and cardiac subjects. *J. Clin. Invest.* 27:272–277, 1948.

71. Miller A. T. and J. O. Miller. Renal excretion of lactic acid in exercise. *J. Appl. Physiol.* 1:614–618, 1949.

72. Mittelman K. Influence of angiotensin II blockade during exercise in the heat. *Eur. J. Appl. Physiol.* 72:542–547, 1996.

73. Mittleman K. D. and E. J. Zambraski. Exercise-induced proteinuria is attenuated by indomethacin. *Med. Sci. Sports Exerc.* 24:1069–1074, 1992.

74. Mogensen C. E. and K. Solling. Studies on renal tubular absorption: partial and near complete inhibition by certain amino acids. *Scand. J. Clin. Lab. Invest.* 37:477–486, 1977.

75. Moller E., J. F. McIntosh, and D. D. Van Slyke. Studies of urea excretion. *J. Clin. Invest.* 6:427–465, 1928.

76. Nedbal J. and V. Seliger. Electrophoretic analysis of exercise proteinuria. *J. Appl. Physiol.* 13:244–246, 1958.

77. Noakes T. D., N. Goodwin, B. L. Rayer, T. Branken, and R. K. N. Taylor. Water intoxication: a posssible complication during endurance exercise. *Med. Sci. Sports Exerc.* 17:370–375, 1985.

78. O'Hagan K., L. Bell, S. Mittelstadt, and P. Clifford. Effect of dynamic activity on renal sympathetic nerve activity in conscious rabbits. *J. Appl. Physiol.* 74:2099–2104, 1993.

79. Pavy F. W. Effect of prolonged muscular exercise upon the urine in relation to the source of muscular power. *Lancet* II:815–818, 1876.

80. Perlman L. V., D. Cunningham, H. Montoye, and B. Chiang. Exercise proteinuria. *Med. Sci. Sports* 2:20–23, 1970.

81. Poortmans J. and R. J. Jeanloz. Quantitative immunological determination of 12 plasma proteins excreted in human urine collected before and after exercise. *J. Clin. Invest.* 47:386–393, 1968.

82. Poortmans J., E. Van Kerchove, and P. Jaumain. Aspects physiologiques et biochimiques de la protéinurie d'effort. *Int. Z. Angew. Physiol. Einschl. Arbeitsphysiol.* 19:337–354, 1962.

83. Poortmans J. R. Exercise and renal function. *Sports Med.* 1:125–153, 1984.

84. Poortmans J. R., E. Blommaert, M. Baptista, M. E. De Broe, and E. J. Nouwen. Evidence of differential renal dysfunctions during exercise in men. *Eur. J. Appl. Physiol.* 76:88–91, 1997.

85. Poortmans J. R., H. Brauman, M. Staroukine, A. Verniory, C. Decaestecker, and R. Leclercq. Indirect evidence of glomerular mixed-type postexercise proteinuria in healthy humans. *Am. J. Physiol.* 254 (Renal Fluid Electrolyte Physiol. 23):F277–F283, 1988.

86. Poortmans J. R. and D. Labilloy. The influence of work intensity on postexercise proteinuria. *Eur. J. Appl. Physiol.* 57:260–263, 1988.

87. Poortmans J. R. and B. Vancalck. Renal glomerular and tubular impairment during strenuous exercise in young women. *Eur. J. Clin. Invest.* 8:175–178, 1978.

88. Poortmans J. R. and J. Vanderstraeten. Kidney function during exercise in healthy and diseased humans. *Sports Med.* 18:419–437, 1994.

89. Powers S. K. and E. T. Howley. *Exercise Physiology.* Boston: McGraw-Hill, 1996.

90. Radigan L. and S. Robinson. Effects of environmental heat stress and exercise on renal blood flow and filtration rate. *J. Appl. Physiol.* 2:185–191, 1949.

91. Rasch P. and I. Dodd Wilson. Other body systems and exercise. In: *Exercise Physiology,* edited by H. Falls. New York: Academic Press, 1978, pp. 129–151.

92. Refsum H. E. and S. B. Stromme. Renal osmol clearance during prolonged heavy exercise. *Scand. J. Clin. Lab. Invest.* 38:19–22, 1977.

93. Reidman S. R. *The Physiology of Work and Play*: Dryden Press, 1950.

94. Ritz E., S. Küster, and M. Zeier. Clinical nephrology in 19th century Germany. *Am. J. Nephrol.* 14:443–447, 1994.

95. Rowe D. S. and J. F. Soothill. The proteins of postural and exercise proteinuria. *Clin. Sci.* 21:87–91, 1961.

96. Sabine R. A. Remarks upon the most correct methods of inquiry in reference to pulsation, respiration, urinary products, weight of the body, and food. *Proc. R. Soc.* 9:561–575, 1860–1862.

97. Sanchez-Medina F., L. Sanchez-Urrutia, J. M. Medina, and F. Mayor. Effect of muscular exercise and glycogen depletion on rat liver and kidney phosphoenolpyruvate carboxykinase. *F.E.B.S. Lett.* 19:128–130, 1971.

98. Sanchez-Medina F., L. Sanchez-Urrutia, J. M. Medina, and F. Mayor. Effect of short-term exercise on gluconeogenesis by rat kidney cortex. *F.E.B.S. Lett.* 26:25–26, 1972.

99. Sanchez-Urrutia L. and J. P. Garcia-Ruiz. Lactic acidosis and renal phosphoenolpyruvate carboxykinase activity during exercise. *Biochem. Med.* 14:355–367, 1975.

100. Sanders M., S. Rasmussen, D. Cooper, and C. Bloor. Renal and intrarenal blood flow distribution in swine during severe exercise. *J. Appl. Physiol.* 40:932–935, 1976.

101. Schad H. and H. Seller. A method for recording autonimic activity in unanesthetized, freely moving cats. *Brain Res.* 100:425–430, 1975.

102. Schena F. Domenico Cotugna and his interest in proteinuria. *Am. J. Nephrol.* 14:325–329, 1994.

103. Schmidt W., A. Bub, M. Meyer, T. Weiss, G. Schneider, N. Maasen, and W. Forssmann. Is urodilatin the missing link in exercise-dependent renal sodium retention? *J. Appl. Physiol.* 84:123–128, 1998.

104. Schneider E. C. *Physiology of Muscular Exercise.* London: W.B. Saunders, 1933.

105. Schneider E. C. and P. V. Karpovich. *Physiology of Muscular Exercise* (3rd ed.). London, 1948.

106. Schulz-Knappe P., K. Forssman, F. Herbst, D. Hock, R. Pipkorn, and W. G. Forssman. Isolation and structural analysis of "urodilatin," a new peptide of the cardiodilatin-[ANP]-family, extracted from human urine. *Klin. Wochenschr.* 66:752–759, 1988.

107. Shephard R. *Physiology and Biochemistry of Exercise.* New York: Praeger, 1982.

108. Simonson E. and A. Keys. *Physiology of Work Capacity and Fatigue.* Springfield: Charles C. Thomas, 1971.

109. Smith H. *Principles of Renal Physiology.* New York: Oxford University Press, 1956.

110. Smith J., S. Robinson, and M. Pearcy. Renal responses to exercise, heat and dehydration. *J. Appl. Physiol.* 4:659–665, 1952.

111. Spiro P. Beiträge zur physiologie der milchsäure. *Z. Physiol. Chem.* 1:111–119, 1877–1878.

112. Starlinger H. and R. Bandino. Der einfluss von arbeit bei hoher temperatur auf die Inulin- und PAH-clearance sowie die elektrolyt- und 17-21-dihydroxy-20-ketosteroidausscheidung im harn. *Arbeitsphysiologie.* 18:285–305, 1960.

113. Starr I. The production of albuminuria by renal vasoconstriction in animals and in man. *J. Exp. Med.* 43:31–51, 1926.

114. Tidgren D., P. Hjemdahl, E. Theodorsson, and J. Nussberger. Renal neurohormonal and vascular responses to dynamic exercise in humans. *J. Appl. Physiol.* 70:2279–2286, 1991.

115. Todorovic B., B. Nikolic, R. Brdaric, V. Nikolic, B. Stojadinovic, and V. Pavlovic-Kentera. Proteinuria following submaximal work of constant and variable intensity. *Int. Z. Angew. Physiol.* 30:151–160, 1972.

116. Van Slyke D. D., A. Alving, and W. C. Rose. Studies of urea excretion. VII. The effects of posture and exercise on urea excretion. *J. Clin. Invest.* 11:1053–1064, 1932.

117. Vatner S. and M. Pagani. Cardiovascular adjustments to exercise: hemodynamics and mechanisms. *Progr. Cardiovasc. Dis.* 19:91–108, 1976.

118. Vatner S. F. Effects of exercise and excitement on mesenteric and renal dynamics in conscious, unrestrained baboons. *Am. J. Physiol* 243 (Heart Circ. Physiol. 3):H210–H214, 1978.

119. Viarnaud G. L'albuminurie d'effort. Université de Lyon, 1944.

120. von Leube W. Ueber die ausscheidung von eiweiss im harn des gesunden menschen. *Virchow Archiv Pathol. Anat. Physiol. Klin. Med.* 72:145–157, 1878.

121. Wade C., S. Ramee, M. Hunt, and C. White. Hormonal and renal responses to converting enzyme inhibition during maximal exercise. *J. Appl. Physiol.* 63:1796–1800, 1987.

122. Wahren J., P. Felig, G. Ahlborg, and L. Jorfeldt. Glucose metabolism during leg exercise in man. *J. Clin. Invest.* 50:2715–2725, 1971.

123. Wesson L. G. Kidney function in exercise. In: *Science and Medicine of Exercise and Sports,* edited by W. R. Johnson. New York: Harper & Brothers, 1960, pp. 270–284.

124. White H. and D. Rolf. Effects of exercise and of other influences on the renal circulation in man. *Am. J. Physiol.* 152:505–516, 1948.

125. Wilmore J. H. and D. L. Costill. *Physiology of Sport and Exercise.* Champaign: Human Kinetics, 1999.

126. Wilson D. W., H. C. Long, H. C. Thompson, and S. Thurlow. Changes in the composition of the urine after muscular exercise. *J. Biol. Chem.* 65:755–771, 1925.

127. Winder W. W., J. M. Hagberg, R. C. Hickson, A. A. Ebashi, and J. A. McLane. Time course of sympatho-adrenal adaptation to endurance exercise. *J. Appl. Physiol.* 45:370–374, 1978.

128. Zambraski E. The kidney and body fluid balance during exercise. In: *Body Fluid Balance Exercise and Sport,* edited by E. Buskirk and S. Puhl. Boca Raton: CRC Press, 1996, pp. 75–95.

129. Zambraski E. Renal regulation of fluid homeostasis during exercise. In: *Perspectives in Exercise Science and Sports Medicine,* edited by C. Gisolfi and D. Lamb. Carmel, Indiana: Benchmark Press, 1990, pp. 247–280.

130. Zambraski E., M. Tucker, C. Lakas, S. Grassl, and C. Scanes. Mechanism of renin release in exercising dog. *Am. J. Physiol.* 246:E71–E78, 1984.

chapter 12

THE GASTROINTESTINAL SYSTEM

Carl V. Gisolfi

D URING prolonged exercise, the loss of fluid and electrolytes in sweat, together with the depletion of energy stores in the liver and working skeletal muscles, can produce dehydration, hyperthermia, hyponatremia, hypoglycemia, and glycogen depletion. Any one of these factors could contribute to fatigue and limit physical performance. These effects are exacerbated when exercise is performed in the heat. The magnitude of the alterations can be substantial but they are highly variable among individuals. Sweat losses can range from 1 to 4 $l \cdot h^{-1}$ and sodium losses from 10 to 120 $mEq \cdot l^{-1}$ of sweat. Exogenous carbohydrate utilization can exceed 1 $g \cdot min^{-1}$.

The gastrointestinal (GI) system is essential in attenuating these losses and supplying energy, but its capacity to do so may be compromised by reduced blood flow, changes in hormone and neurotransmitter profiles, and disruption of intestinal barrier function. The latter impairment could produce circulatory and thermal dysfunction.

Despite the key role of the GI system in fluid homeostasis and the provision of nutrient during exercise, it is difficult to identify a gastroenterologist in the last 150 years who was devoted to understanding the effects of exercise on GI function. Moreover, it is equally difficult to identify an exercise physiologist who devoted his or her research career to understanding the role of the GI tract in an exercise response. As a consequence, knowledge of how exercise affects this system and how this system affects other organ systems during exercise was slow to develop. Early research focused primarily on how exercise affected the stomach. Little effort was made to understand the effects of exercise on the small intestine, virtually no studies were designed to understand how exercise might influence colonic function, and no studies investigated the effect of exercise on gut barrier function.

This chapter focuses on human studies, beginning with the classical work of William Beaumont (1785–1853) in 1833 (Fig. 12.1). The important observations that influenced exercise physiology since then were not necessarily made by exercise physiologists, nor were they necessarily made during exercise. Furthermore, in some cases, prevailing ideas cannot be attributed to a single researcher. I apologize in advance for offending investigators whose research is not included or not emphasized in this chapter.

GASTRIC EMPTYING

Perhaps the first report of how exercise influenced the gut was provided by William Beaumont (Fig. 12.1), the surgeon and pioneer physiologist who treated a gastric fistula secondary to a shotgun wound suffered by a French Canadian named Alexis St. Martin in 1833 (77). Beaumont concluded that moderate exercise was conducive to rapid and healthy digestion. He wrote (77),

> Exercise, sufficient to produce moderate perspiration, increases the secretions from the gastric cavity, and produces an accumulation of a limpid fluid, within the stomach, slightly acid, and possessing the solvent properties of the gastric juice in an inferior degree. This is probably a mixed fluid, a small proportion of which is gastric juice.

Beaumont concluded, however, that severe and fatiguing exercise retarded digestion (77).

In the next century, there was little change in thinking about the effects of exercise on GI function. During this period, exercise was described as mild, moderate, or severe, and interest focused on how such exercise influenced hunger pangs, gastric emptying, and gastric secretions. Anton J. Carlson (1875–1956) (12) demonstrated that gastric contractions associated with hunger were unaffected by walking but were inhibited by running. Campbell et al. (10) in 1928 were the first to show that severe exercise (running 1–4 miles) reduced gastric emptying, which was used as a measure of gastric digestive motility, following consumption of a solid test meal. They showed, however, that mild exercise could increase gastric emptying and that persons who were more fit could exercise at higher intensities without reducing gastric emptying. Hellebrandt and Tepper (40) showed that mild-to-moderate exercise had no effect on gastric emptying and may even have a stimulatory effect. They noted that exhaustive exercise, which inhibits gastric emptying, may be followed by a period of augmented gastric activity. Later, Hellen Brandt and Hoopes studied the effects of various grades of exercise on gastric secretion. Similar to gastric emptying, mild-to-moderate exercise had no effect or stimulated gastric secretion, whereas severe exercise was inhibitory (39). The concept emerged that the mechanisms controlling gastric motility and secretion were one and the same (38).

In their 1960 review of the effects of exercise on gastrointestinal function, Stickney and Van Liere (103) reached the same conclusion as Beaumont—that is, mild-to-moderate exercise does not alter normal gastric function. Although severe exercise

Fig. 12.1. A portrait of William Beaumont as published by Jessee S. Meyer (in ref. 66). Permission for use granted by W.B. Saunders Company for the C.V. Mosby Company.

reduced gastric motility and secretion, the stomach was considered to have a large margin of safety. Even in subjects performing severe and exhausting muscular work, these investigators could find no evidence of "permanent harm" to digestive function. An interesting observation in support of this concept was made by Mellinkoff and Machella (60). They found that treadmill walking 4 h/day at 200 m/h by rats did not interfere with liver regeneration after 70% of the liver was removed. Resting control and exercised animals reached 100% restoration in 10 days.

A pioneer in the field of gastric emptying under resting conditions was the British investigator John N. Hunt. Exercise experiments performed during the last four decades have not advanced our understanding of the mechanisms responsible for gastric emptying but have fine-tuned many of the observations originally made by Hunt (44–48) at rest. The ideas explored the most during exercise have been the inhibitory influences of high carbohydrate concentration and high osmolality and the stimulatory effect of high fluid volume on gastric emptying. A role for energy density in the regulation of gastric emptying was studied by Hunt and Stubbs (48) in 1975, and later by Brener et al. (4), but has garnered limited support.

The finding that mild exercise can enhance gastric emptying, first reported by Campbell et al. (10) in 1928, has also been observed in more recent years by Neufer et al. (68, 70) and Marzio et al. (58). Mild exercise in these three studies consisted of walking, running at 50%–70% $\dot{V}O_2$max, and running at 50% of predicted maximal

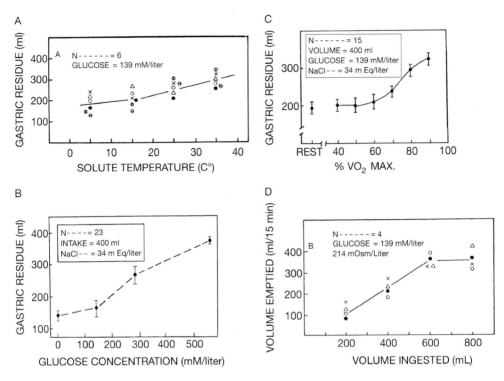

Fig. 12.2. Effects of solution temperature (*A*), glucose concentration (*B*), and exercise intensity expressed as %$\dot{V}O_2$max on the volume of fluid remaining in the stomach 15 minutes after a 400 ml feeding. Also shown is the influence of volume of ingested solution on the rate of gastric emptying (*D*). Figure obtained from reference 18. Permission for use granted by American Physiological Society.

heart rate, respectively. Taken together, these observations support the concept that mild-to-moderate exercise promotes digestion. Note that the "mode of exercise" in these studies involved walking or running. Whether or not submaximal cycling exercise or cross-country skiing also enhance gastric emptying is unclear (17).

One of the more influential papers regarding gastric emptying during exercise was published by David C. Costill (1932–) and Bengt Saltin (1935–) (18). This was a comprehensive study that evaluated the effects of exercise intensity and duration as well as the effects of glucose concentration, temperature, osmolality, and volume of a beverage on gastric emptying (Fig. 12.2). Although it did not generate any new ideas, it did quantify the effects of exercise on gastric emptying and it did influence the first published Position Stand on fluid replacement during prolonged exercise by the American College of Sports Medicine. It served as a "benchmark" paper showing that gastric emptying was markedly reduced when the glucose concentration of the ingested beverage exceeded 139 mmol·l^{-1} (~2.5%). This observation stimulated a marked increase in research on the gastric emptying of sport drinks containing a variety of different forms of carbohydrate (CHO) during exercise (52, 59, 64). However, subsequent studies have not supported the conclusions drawn about the effects of carbohydrate concentration and beverage temperature on the gastric emptying of fluids. For example, in 1986, Owen et al. (78) reported that 10% solutions of glucose or glucose polymers were emptied from the stomach as quickly as water during a 2 hour bout of treadmill exercise. Similar observations were reported by Neufer et al. (68), Rehrer et al. (85), and Mitchell et al. (62). The different results are in part attributed to when gastric emptying was measured. If the same volume of a dilute and concentrated CHO solution is ingested, the concentrated solution initially empties more slowly because of the inhibitory influence of the high CHO concentration. However, because gastric volume is also a major factor determining gastric emptying, the dilute solution empties more slowly than the concentrated solution as time progresses. This is because its volume is smaller following its greater initial gastric emptying than the concentrated solution. When gastric emptying was measured throughout its exponential time course, Vist and Maughan of Scotland (108) found that glucose solutions >4%–5% delayed gastric emptying. When glucose polymers or different forms of carbohydrates in varying concentrations are combined to formulate rehydration beverages, the results have been conflicting (17, 59). The reason for the disparity is attributed in part to the method employed to measure gastric emptying, when the measurement was made, and whether or not gastric secretions were measured and taken into account when calculating gastric emptying.

The emerging concept today is that gastric volume is a primary factor in determining gastric emptying during exercise and that beverages containing CHO concentrations up to 6%–8% (independent of CHO type, energy density, and to a large extent osmolality) do not significantly reduce gastric emptying. However, beverages formulated only with glucose seem to exert a greater inhibitory effect than beverages formulated with a combination of different (glucose, fructose, sucrose, maltodextrin, etc.) CHOs (65, 78).

Lambert and his Iowa colleagues (53) capitalized on the idea of promoting a high gastric emptying rate by maintaining a high gastric volume. They attached a nasogastric tube to a multilumen tube and fluoroscopically guided the former into the gastric antrum and the latter into the duodenojejunum. By combining repeated measures of gastric emptying with segmental sampling from the intestine (14, 31, 63), gastric emptying and intestinal absorption were simultaneously measured following the oral ingestion of beverages during exercise. They demonstrated that after ingesting a bolus (~300–500 ml) of fluid, repeat drinking during exercise produced a relatively constant gastric volume and gastric emptying rate over a prolonged period of time (90 minutes). Intestinal absorption of a solution orally ingested was the same as intestinal absorption of the same solution infused directly into the duodenum at the gastric emptying rate. This methodology subsequently served as a model with which to study the effect of various formulations of oral rehydration beverages on gastrointestinal function and fluid homeostasis during exercise.

Another important observation made in the last two decades involved the influence of environmental stress on gastric emptying. In 1986, Owen et al. (78) were the first to show that heat stress significantly reduced gastric emptying. This was confirmed by Neufer et al. (69), who further studied the effects of heat acclimation and dehydration on gastric emptying. Heat acclimation had no effect on gastric emptying during exercise in a 35°C environment, but hypohydration by 4% of body weight significantly reduced both gastric emptying and gastric secretion during exercise in 35°C heat. Rehrer et al. (86) also reported a significant decline in gastric emptying following dehydration by 4% of body weight, but Ryan et al. (91) did not find a decrease in gastric emptying in subjects hypohydrated by only 3% of body weight.

INTESTINAL/COLONIC MOTILITY

GI motility in humans, especially with regard to the colon, is a neglected area of investigation. Cammack and coworkers at the University of Sheffield (9) in 1982 were the first to measure small bowel transit (breadth hydrogen) in humans. They found that intermittent moderate cycle exercise had no effect on the passage of a solid meal. Subsequent studies of "small intestinal" transit supported this observation (75, 101), or found that exercise increased (61) or decreased (50) transit.

Exercise has been reported to decrease (16, 73), increase (55), or not to change colonic transit time (101) in humans. To identify the mechanism responsible for changes in colonic motility with exercise, Rao et al. (84) recorded colonic motility from six sites using a solid-state probe. An important aspect of this study is that it examined 75% of the colon compared with 25% in other studies. Moreover, the probe was placed the night before experiments were performed. Thus the colon was in a more normal state rather than having been emptied by enema or lavage immediately before testing. In this more normal state, Rao et al. (84) found an intensity-dependent decrease in colonic phasic activity during acute exercise. However, after exercise, the number and amplitude of propagated waves increased. It was concluded that colonic motility was significantly reduced by exercise, but increased in recovery. These findings support the concept of enhanced colonic transit because the reduction in motility during exercise reduces resistance to colonic flow, whereas enhanced motility in recovery increases propulsion.

The different observations made in these studies are, in part, attributed to the effects of exercise duration, exercise intensity, whether the subject was fasted or fed, whether the test meal was liquid or solid, the size and composition of the meal, and/or the mode of exercise. This is a crucial area of future research because many of the GI symptoms experienced during prolonged exercise have been attributed to colonic dysfunction.

INTESTINAL ABSORPTION

In the 1960s came the lure of endurance running. President John F. Kennedy emphasized the need for greater fitness, and Ken Cooper authored the best seller entitled *Aerobics* (15). Thousands of Americans took to the roads, jogging. However, in contrast to current practice, fluid replacement during team games (especially American football) was considered unnecessary and inappropriate. Even elite marathon runners abstained from drinking. In 1957, Jim Peters from Britain—one of the greatest distance runners of all times—and his coauthors wrote, "There is no need to take any solid food at all and every effort should also be made to do without liquid as the moment food or drink is taken, the body has to start dealing with its digestion and in so doing some discomfort will almost invariably be felt" (81).

The Olympic rule in 1968 was that you could not drink during the first 15 kilometers (9 miles) of the Olympic marathon (74). This handicapped numerous runners with high sweat rates who would dehydrate sufficiently in the early portion of the race to compromise their health and performance. The consequences of these erroneous ideas and rules led to many cases of heat exhaustion in football and in distance runners (102). The current Olympic policy (rule 165) is that: "In all events 10 km or longer, water shall be provided at intervals of no more than 5 km. In addition, race management may provide refreshments (other than water) and/or sponging stations at positions approximately midway between water station." (76).

In the 1960s, human studies of intestinal absorption were markedly influenced by a discovery unrelated to exercise, but which subsequently had a marked influence on exercise physiology. This discovery not only provided the stimulus for research on gastric emptying and intestinal absorption that continued for several decades but was responsible for the initiation of a multibillion dollar industry—*sport drinks*. The discovery was sodium–glucose cotransport in the small intestine, an unprecedented example of how biological brinkmanship can impact clinical medicine. The discovery of this mechanism is primarily credited to the in vitro studies of Robert K. Crane (19) (Fig. 12.3), who sketched the first model of sodium–glucose coupling in the sand on Stony Beach at Woods Hole in 1959 (20). A year later he presented his formal model of sodium–glucose cotransport at the International Conference on Membrane Transport and Metabolism in Prague.

Important in vitro observations regarding sodium–glucose transport were also made by Riklis and Quastel (87), Schultz and Zalusky (96), Curran (23), and Csaky and Thale (21). However, Schedl and Clifton (94) at the University of Iowa in 1963 were the first to demonstrate the link between sodium and glucose transport in the human small intestine. This discovery of sodium–glucose cotransport led to the development of oral rehydration solutions for the treatment of diarrheal disease in Third World countries.

In the 1970s the marathon became one of the most popular running events worldwide (personal communication, Hal Higdon, senior writer, *Runner's World* magazine). Running long slow distance (LSD) created a need to prevent the deleterious effects of dehydration, heat illness, and early fatigue by drinking during exercise. It was during this period that a research team headed by Robert Cade, M.D., at the University of Florida, formulated a carbohydrate–electrolyte solution which came to be known as Gatorade. When Stokely–Van Camp acquired the rights to produce and mass market Gatorade in 1967, the sport drink industry was born. Despite the widespread use of Gatorade and similar sport drinks, the prevailing idea among many practitioners, coaches, athletes, and laypersons for decades (and even today)

Fig. 12.3. A picture of Robert K. Crane taken from the text *Comprehensive Biochem-istry*, Vol. 35, 1983, pp. 44, "Selected Topics in the History of Biochemistry: Personal Recollections," that was provided to the author upon request.

was that the ideal fluid replacement beverage during exercise was water. This led to a marked increase in research to identify the ideal fluid replacement beverage before, during, and after exercise. In part, this was an outgrowth from the development of oral rehydration solutions (ORSs) for the treatment of diarrheal disease; however, the electrolyte composition of an ORS was fashioned based on electrolyte losses in the stool, whereas the electrolyte composition of a sport drink was based on electrolyte losses in the sweat. Sport drinks require only a fraction of the sodium and potassium typically included in a solution to treat diarrhea, but require considerably more carbohydrate.

In the early 1960s, much progress was made toward understanding solute and water transport in humans by using the technique of segmental perfusion. John Fordtran was a pioneer in this field (Fig. 12.4). He designed the triple-lumen tube for segmental perfusion (14) and performed many of the early studies in humans that elucidated the mechanism of fluid and electrolyte transport. Because this technique is considered the gold standard for evaluating intestinal absorption in humans (95), more importance is given to experiments performed during exercise using this methodology than to results derived from less invasive techniques.

Fig. 12.4. A picture of John S. Fordtran that was taken in 1998. Permission for use was provided to the author upon request.

In 1967, John S. Fordtran and B. Saltin at Dallas (27) published the results of a landmark study designed to determine if exercise limits intestinal absorption. They used segmental perfusion with a triple-lumen catheter and found that 1 hour of treadmill exercise at 64%–78% $\dot{V}O_2max$ had no consistent effect on jejunal or ileal absorption of water, Na^+, Cl^-, K^+, glucose, L-xylose, urea, or tritiated water. This study was limited to only five subjects and five different solutions tested only once or twice; nevertheless, the conclusion that intense exercise does not affect intestinal absorption was subsequently confirmed by experiments conducted by Gisolfi and associates (32).

This conclusion was not supported by the results of Barclay and Turnberg (2), who also used segmental perfusion, but who perfused an electrolyte solution without CHO. They found that exercise significantly reduced water, sodium, chloride, and potassium absorption, but the values for water and solute absorption were extremely low. Water absorption was only 1–2 $ml \cdot cm^{-1} \cdot h^{-1}$. However, adding 2%–6% CHO to saline increases water absorption to 10–23 $ml \cdot cm^{-1} \cdot h$, a 6- to 10-fold increase (32).

Thus, removing CHO from the perfusion solution unmasked an inhibitory effect of exercise on absorption, but the biological importance of this reduction is probably of little consequence provided CHO is included in the solution.

In the 1960s (and even today) water was generally believed to be absorbed faster than other beverages in normal healthy individuals. For sporting events lasting an hour or more and contests known to require high sweat rates it became important to formulate a beverage that would empty from the stomach and be rapidly absorbed from the intestine. In 1969, Sladen and Dawson (100) investigated the range of intestinal glucose and sodium concentrations required to maximize fluid absorption under resting conditions but did not include water as a test solution. They reported that a glucose:sodium molar ratio of 2:1 stimulated the greatest fluid absorption. Although the importance of this ratio has been emphasized in numerous studies, it had little impact on the formulation of sport drinks. The leading sport drinks contain 6%–8% carbohydrate and approximately 20 mM·l^{-1} Na^{+}, yielding a glucose:sodium molar ratio of about 17:1. In a subsequent segmental perfusion study on net fluid absorption that evaluated the effects of 2%–8% solutions of CHO monomers, dimers, and polymers, there were no significant differences in water absorption among 2%–6% solutions (99). Increasing the Na^{+} concentration of a fluid replacement beverage, at least up to 50 mEq·l^{-1}, does not increase intestinal fluid absorption (33). The effect of exercise on simultaneous anion/cation exchange has not been investigated.

The study that demonstrated that a CHO–electrolyte beverage was absorbed as rapidly as water during exercise was recently reported by Lambert and coworkers at the University of Iowa (54). The key features of this study were: (1) the subjects drank the beverages, (2) they exercised at 65% $\dot{V}O_2$max, and (3) both gastric emptying and intestinal absorption from three different segments of the intestine were measured simultaneously. The last aspect was important because water was shown to be absorbed primarily by the duodenum by flowing down an osmotic gradient, whereas the CHO–electrolyte beverage was absorbed primarily in the jejunum and absorption rate was related to total solute flux secondary to Na^{+}–glucose cotransport.

Water crosses the intestinal epithelium passively by moving through and between cells down an osmotic gradient. Fordtran et al. (26) were among the first investigators, in 1961, to provide evidence that a hypotonic solution was more efficacious than an isotonic solution in promoting water absorption. This idea was confirmed by Wapnir and Lifshitz (110), who suggested, using an animal model of diarrheal disease, that luminal osmolality is a major factor promoting water absorption. Thillaninayagam et al. (105) came to the same conclusion after reviewing the evidence gathered over the last decade, pointing out that hypotonic solutions promote greater water absorption and reduce stool volume and frequency. All of these studies were performed at rest and the vast majority of them involved patients or animal models of diarrheal disease. The importance of hypotonicity for hydration beverages during exercise is not as clear.

During 1986, Leiper and Maughan (56) studied the role of osmolality in normal healthy subjects with the idea of evaluating the efficacy of such beverages for use during exercise. These were segmental perfusion experiments in the normal human jejunum. It was concluded that net water absorption is determined primarily by active solute transport rather than the osmotic gradient across the mucosa and that an isotonic solution promotes the greatest absorption of water, electrolytes, and glucose. However, Hunt et al. (42, 43) found significantly greater water absorption from hypotonic solutions in humans using segmental perfusion of the jejunum at rest. When Shi et al. (98) evaluated absorption from the duodenojejunum by perfusing 6% CHO–electrolyte solutions at rest ranging in osmolality from 183 to 403 mOsm·kg⁻¹, they concluded that differences in osmolality among beverages were eliminated in the duodenum and that osmolality within the range studied had no effect on intestinal water absorption or fluid homeostasis (based on changes in plasma volume). Thus, the different conclusions drawn by Hunt et al. (42, 43) and by Shi et al. (98) are attributed to differences in the intestinal segment perfused.

Water movement occurs passively down an osmotic gradient, but this gradient is usually created by active and passive solute transport. Thus, water movement is usually secondary to total solute flux. Shi et al. (99), at Iowa, illustrated the interaction of net water movement, luminal osmolality, and solute transport. They evaluated nine different solutions ranging in osmolality from 165 to 477 mOsm·kg⁻¹ that were formulated with only one or multiple transportable solutes. Formulating a rehydration beverage with multiple carbohydrates that activated multiple transport mechanisms maximized water absorption at any given osmolality (99). They concluded that solute flux was more important than luminal osmolality in determining net water movement in humans.

Given this dependence of water movement on solute flux, the idea of creating a "super" rehydration beverage by adding amino acids to a CHO–electrolyte beverage to maximally activate independent solute transport mechanisms is theoretically sound, but, to date has not improved fluid absorption (22, 79, 93).

The idea that water is the ideal fluid replacement beverage during prolonged exercise is not supported by research gathered in the last two decades. Under no condition is water better than a CHO–electrolyte solution, but under numerous conditions a CHO–electrolyte solution is far superior to water. Moreover, the ingestion of plain water during prolonged exercise could lead to overhydration and the dangerous condition of hyponatremia (67, 71, 72). Ingesting plain water could also lead to underhydration caused by poor taste and reduced plasma osmolality.

If intestinal absorption is not compromised during exercise at 70%–80% V̇O₂max (27, 32), this would imply that mucosal blood flow is maintained during exercise. This poses a perplexing problem: splanchnic blood flow is markedly reduced during heavy exercise yet intestinal absorption is maintained.

SPLANCHNIC CIRCULATION

Early studies of splanchnic blood flow during exercise were contradictory. Based on the interpretation by Stickney and Van Liere (103) of the historical review by A.J.

Bridzius (5), local ischemia of the GI tract was postulated to be a concomitant of exercise. This idea was supported by the early human experiments of Wade et al. (109) on fasted and recumbent subjects performing mild exercise but not by studies in dogs (41). Moreover, Lowenthal et al. (57) reported that in the postabsorptive state, exercise reduced estimated hepatic blood flow (EHBF) in men, whereas in the postprandial state, when the gut was active and competing with active skeletal muscle for blood flow, exercise increased EHBF.

Loring B. Rowell has been a pioneer in trying to understand the effects of exercise on the splanchnic circulation (see Fig. 5.8 in Chapter 5). Together with his colleagues (90) at the University of Washington, Seattle, Rowell published a landmark study of postabsorptive subjects showing that an inverse linear relationship exists between splanchnic blood flow and exercise intensity, expressed as a percentage of maximal oxygen uptake. In one subject who had eaten breakfast 2–3 hours earlier, the results were the same (personal communication from Rowell in 1999). The study by Rowell et al. (90) was important because it not only quantified the exercise performed but also evaluated the effects of "maximal" exercise. Although not measured, it could be implied from these observations that intestinal absorption would also decrease as a function of exercise intensity. However, the decrease in splanchnic blood flow could represent a decline in hepatosplenic flow, while mesenteric flow (representing intestinal perfusion) remains unchanged. This is a controversial issue. Qamar and Read (83) reported reduced splanchnic blood flow during treadmill exercise in the fasted and postprandial state in normal human subjects using noninvasive Doppler ultrasound recordings. This technique correlates well with more direct measures of intestinal blood flow (51) and has been used to demonstrate that reductions in blood flow can reduce intestinal mucosal perfusion (107).

In contrast, Eriksen and Waaler (24) reported that after 4 minutes of cycling in the reclining position during mild-to-intense (~75% $\dot{V}O_2$max) exercise in the fasted or postprandial state, superior mesenteric artery blood flow was maintained. Perko and colleagues (80) in 1998 determined mesenteric, celiac, and splanchnic blood flows in humans during semisupine cycle exercise at 70%–85% $\dot{V}O_2$max while fasted and after ingesting a liquid meal of 1000 kcal. They reported a 43% decrease in splanchnic blood flow during exercise. This decrease in blood flow was dominated by a 50% reduction in celiac blood flow, which was more than twice the reduction in superior mesenteric blood flow. Ingesting the 1000 kcal liquid meal elevated superior mesenteric blood flow ~1.0 l/min. During exercise in the postprandial state, exercise reduced superior mesenteric blood flow about 0.5 l/min, but compared with blood flow at rest before meal consumption, blood flow remained elevated by about 1.0 l/min. Thus, digestion-induced vasodilation was essentially maintained during exercise. The emerging concept is that the splanchnic circulation plays a major role in overall cardiovascular regulation during exercise, but not at the expense of digestion/absorption during exercise in the postprandial state.

As indicated previously, Fordtran and Saltin (27) found that treadmill exercise between 64% and 78% $\dot{V}O_2$max did not affect active or passive intestinal transport. As indicated above, intestinal blood flow may not decline during exercise despite the

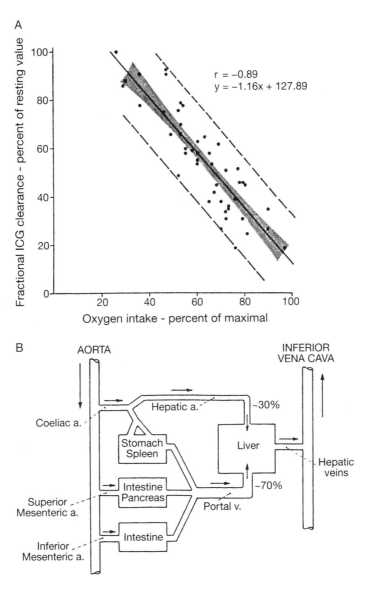

Fig. 12.5. *A*: Correlation between fractional clearance rate of Indocyanine green clearance (ICG) as a percentage of the resting value (100%) and oxygen intake as percentage of maximal oxygen intake. The shaded area about the least-squares regression line represents the 95% confidence interval for the true mean value of y (ICG clearance) for a given value of x (oxygen intake). The dashed lines represent the 95% confidence interval for the true y value of an individual having the given x value. *B*: Schematic representation of the splanchnic vascular bed. Note (1) the parallel arrangement of various arteries supplying splanchnic organs, (2) the series arrangement of all these segments with the liver, and (3) the dual blood supply to the liver via the portal vein and the hepatic artery. (*A*) From Rowell et al. (90). (*B*) From Rowell (89).

fall in splanchnic blood flow. Alternatively, as splanchnic blood flow markedly declines, there is a parallel increase in A–$\dot{V}O_2$ difference and splanchnic oxygen uptake does not change (90). If splanchnic oxygen uptake reflects intestinal oxygen uptake, the energy required to sustain digestion and absorption would not be altered during exercise at 70%–80% $\dot{V}O_2$max despite the decrease in blood flow.

"Gut-Barrier" Function

In 1960, Stickney and Van Liere (103) concluded that mild-to-moderate exercise and relatively short bouts of severe exercise failed to alter normal gastrointestinal function. In the mid-1990s, Brouns and Beckers (7) concluded that the GI tract is not an "athletic organ," citing evidence of GI bleeding and GI symptoms in a significant number of endurance athletes.

In contrast to the latter idea, several reports provide evidence that the GI track can adapt to the effects of chronic exercise. For example, Carrio et al. (13) found that marathon runners had significantly enhanced "basal" gastric emptying of a test meal (99m-Tc human serum albumin egg omelet) compared with control subjects. Heavy exercise at 4.0–4.5 min · km^{-1} for 90 minutes had no effect on gastric emptying. These data support an adaptation to exercise training although the mechanism remains unknown.

In addition to greater gastric emptying, Martin's group (37) found that orocecal transit time (OCTT, mouth-to-colon transit of lactulose) was more rapid in athletes with high caloric intakes than in sedentary subjects with low energy intakes. This increase in OCTT occurs without a decrease in intestinal absorption, suggesting that the hyperphagia of chronic exercise is associated with adaptations within the gastrointestinal tract (37).

In addition to these changes in gastric emptying and intestinal transit, there is also evidence of improved "gut-barrier function" with exercise training. Sakurada and Hales (92) found that core temperature following an IV injection of saline was significantly greater in sedentary than in physically fit sheep when exposed to an environment of 42°C/35°C (dry/wet bulb temperature). However, when both groups of animals were injected with indomethacin, which blocks the pathways involved in endotoxin-induced fever, the rise in core temperature was reduced in the sedentary group to the level observed in the physically fit group. These data provide evidence in animals that endotoxin released from the gut plays a role in determining the difference in heat tolerance between sedentary and physically fit populations.

Can endurance athletes enhance the absorptive capacity of the intestine by altering their diet? In humans and animals, high dietary intake of sucrose and fructose can increase glycolytic enzyme activities in just a few hours, and brush-border disaccharidase activities in only 2–5 days (88). Rapid glucose uptake is presumably due to an increase in the number and/or intrinsic activity of the glucose transporters. Thus, at the present time, there is limited evidence that both the stomach and intestine can adapt to the effects of chronic exercise.

Prolonged Exercise, Heat, and Dehydration

An important observation that emerged in the last two decades was that GI symptoms in endurance athletes is common (104). Forgoros (25) in 1980 was the first to report diarrhea in runners and cautioned against the combined effects of dehydration, hypokalemia, and hyperthermia. This was followed by reports of bloody stools in runners by Cantwell (11) and Keeffe et al. (49). The etiology of this bleeding is uncertain, but is attributed primarily to tissue ischemia, mechanical trauma, and use of nonsteroidal anti-inflammatory drugs (NSAIDs). As participation in endurance events increased, so did gastrointestinal symptoms. Nausea, vomiting, diarrhea, intestinal bleeding, and bloating occurred more frequently with greater duration of competition and degree of dehydration (8). On rare occasions, these complications have been severe (colon resection) and even fatal (106). The mechanism most cited as responsible for GI symptoms related to exercise has been gut ischemia (30, 82).

An equally important observation is that these symptoms and complications, for the most part, dissipate within <72 hours (1, 97). Although the gut has a remarkable capacity for regeneration, there are widespread interindividual differences in susceptibility to these impairments (97).

The role of a gut-barrier function in thermal tolerance, fluid homeostasis, and cardiovascular regulation is emerging. The hypothesis is that exercise, heat, and/or dehydration can produce ischemic and/or hypoxic damage to the intestinal mucosa, leading to increased intestinal permeability and, in turn, to endotoxemia, bacteremia, and the cytokine cascade (35, 36). These events occur presumably because reactive oxygen species are formed and produce large quantities of nitric oxide by activating inducible nitric oxide synthase. Broche-Utne et al. (6) reported plasma endotoxin concentrations in exhausted long-distance runners that would have been fatal in resting patients.

The concept that endotoxin in the blood originates from the intestine and contributes to the increased risk of injury associated with hyperthermia was first presented by Gathiram and coworkers in South Africa (28, 29). This group showed that antibiotics administered prior to heat exposure improved survival from hyperthermia and that plasma anti-LPS concentration fell at a core temperature of only 39°C, long before a rise in plasma endotoxin was observed (29). Broche-Utne et al. (6) also reported that triathletes who trained the hardest for 3 weeks prior to competition had the highest plasma concentration of anti-LPS. Hales et al. (34) suggest that intense training may produce temporary gut ischemia, mucosal damage, increased gut permeability, and leakage of endotoxin, possibly leading to self-immunization. Further support for the above hypothesis was provided by Bouchama et al. (3), who reported endotoxemia and release of tumor necrosis factor and interleukin-l in acute heat-stroke victims.

SUMMARY

For the first half of this century, the effects of exercise on GI function were primarily focused on gastric function. Exercise was not quantitated and, for the most part, it was concluded that mild, moderate, and even severe exercise of limited duration had little detrimental effect. In the last four decades, we have learned considerably more about the effects of exercise on gastric and small intestinal function and are beginning to understand how exercise affects the colon and gut-barrier function. Much of this progress is a result of research designed to formulate the "ideal" fluid replacement beverage during exercise and to physiologists collaborating with gastroenterologists to forge new methods and approaches to GI problems. We are only beginning to realize the importance and integral role of the GI system in cardiovascular function, fluid balance, temperature regulation, and immune responses to exercise.

REFERENCES

1. Argenzio R.A., C.K. Henrickson, and J.A. Liacoz. Restitution of barrier and transport function of porcine colon after acute mucosal injury. *Am. J. Physiol.* 255 (Gastrointest. Liver Physiol. 18):G62–G71, 1988.
2. Barclay G.R. and L.A. Turnberg. Effect of moderate exercise on salt and water transport in the human jejunum. *Gut* 29:816–820, 1988.
3. Bouchama A., R.S. Parhar, A. El-Yazigi, K. Sheth, and S. Al-Sedairy. Endotoxemia and release of tumor necrosis factor and interleukin 1a in acute heat stroke. *J. Appl. Physiol.* 70:2640–2644, 1991.
4. Brener W., R.R. Hendrix, and P.R. McHugh. Regulation of gastric emptying of glucose. *Gastroenterology* 85:76–82, 1983.
5. Bridzius A.J. Einfluss der muskelarbeit auf die magensekretion nach versuchen am hunde. *Z. Ges. Exp. Med.* 51:573–587, 1926.
6. Brock-Utne J.G., S.L. Gaffin, M.T. Wells, P. Gathiram, E. Sohar, M.F. James, D.F. Morrell, and R.J. Norman. Endotoxaemia in exhausted runners following a long-distance race. *S. Afr. Med. J.* 73:533–536, 1988.
7. Brouns F. and E. Beckers. Is the gut an athletic organ? Digestion, absorption and exercise. *Sports Med.* 15:242–257, 1993.
8. Brouns F., W.H.M. Saris, and N.J. Rehrer. Abdominal complaints and gastrointestinal function during long-lasting exercise. *Int. J. Sports Med.* 8:175–189, 1987.
9. Cammack J., N.W. Read, P.A. Cann, B. Greenwood, and A.M. Holgate. Effect of prolonged exercise on the passage of a solid meal through the stomach and small intestine. *Gut* 23:957–961, 1982.
10. Campbell J.M.H., G.O. Mitchell, and A.T.W. Powell. The influence of exercise on digestion. *Guy's Hosp. Rep.* 78:279–293, 1928.
11. Cantwell J.D. Gastrointestinal disorders in runners. *JAMA* 246:1404–1405, 1981.
12. Carlson A.J. *The Control of Hunger in Health and Disease*, Chicago: University of Chicago Press, 1916, pp. 203–211.
13. Carrio I., M. Estorch, R. Serra-Grima, M. Ginjaume, R. Notivol, R. Calabuig, and F. Vilardell. Gastric emptying in marathon runners. *Gut* 30:152–155, 1989.
14. Cooper H., R. Levitan, J.S. Fordtran, and R.J. Ingelfinger. A method for studying absorption of water and solute from the human small intestine. *Gastroenterology* 50:1–7, 1966.
15. Cooper K.H. *Aerobics*, New York: M. Evans and Co., 1968.
16. Cordain L., R.W. Latin, and J.J. Behnke. The effects of an aerobic running program on bowel transit time. *J. Sports Med.* 26:101–104, 1986.

17. Costill D.L. Gastric emptying of fluids during exercise. In: *Perspectives in Exercise Science and Sports Medicine. Fluid Homeostasis During Exercise*, edited by C.V. Gisolfi and D.R. Lamb. Indianapolis: Brown & Benchmark, 1990, pp. 97–127.

18. Costill D.L. and B. Saltin. Factors limiting gastric emptying during rest and exercise. *J. Appl. Physiol.* 37:679–683, 1974.

19. Crane R.K. Hypothesis for mechanism of intestinal active transport of sugars. *Fed. Proc.* 21:891–895, 1962.

20. Crane R.K. Selected topics in the history of biochemistry: personal recollections. In: *Comprehesnive Biochemistry*, edited by G. Semenza. New York: Elsevier Science Publishers, 1983, Vol. 35, pp. 43–69.

21. Csaky T.Z. and M. Thale. Effect of ionic environment on intestinal sugar transport. *J. Physiol. (Lond.)* 151:59–65, 1960.

22. Cunha Ferreira R.M.C., E.J. Elliott, E.A. Brennan, J.A. Walker-Smith, and M.J.G. Farthing. Oral rehydration therapy: a solution to the problem. *Ped. Res.* 22(Abstract): 100, 1987.

23. Curran P.F., Na, Cl and water transport by rat ileum in vitro. *J. Gen. Physiol.* 43:1137–1148, 1960.

24. Eriksen M. and B.A. Waaler. Priority of blood flow to splanchnic organs in humans during pre- and post-meal exercise. *Acta Physiol. Scand.* 150:363–372, 1994.

25. Fogoros R.N. Gastrointestinal disturbances in runners. "Runners' trots." *JAMA* 243: 1743–1744, 1980.

26. Fortran J.S., R. Levitan, V. Bikerman, and B.A. Burrows. The kinetics of water absorption in the human intestine. *Trans. Assoc. Am. Phys.* 74:195–206, 1961.

27. Fortran J.S. and B. Saltin. Gastric emptying and intestinal absorption during prolonged severe exercise. *J. Appl. Physiol.* 23:331–335, 1967.

28. Gathiram P., M.T. Wells, J.G. Brock-Utne, B.C. Wessels, and S.L. Gaffin. Oral administered nonabsorbable antibiotics prevent endotoxemia in primates following intestinal ischemia. *J. Surg. Res.* 45:187–193, 1988.

29. Gathiram P., M.T. Wells, D. Raidoo, J.G. Brock-Utne, and S.L. Gaffin. Portal and systemic plasma lipopolysaccharide concentrations in heat-stressed primates. *Circ. Shock* 25: 223–230, 1988.

30. Gaudin C., E. Zerath, and C.Y. Guezennec. Gastric lesions secondary to long-distance running. *Dig. Dis. Sci.* 35:1239–1243, 1990.

31. George J.D. New clinical method for measuring the rate of gastric emptying: the double sampling test meal. *Gut* 9:237–242, 1968.

32. Gisolfi C.V., K.J. Spranger, R.W. Summers, H.P. Schedl, and T.L. Bleiler. Effects of cycle exercise on intestinal absorption in humans. *J. Appl. Physiol.* 71:2518–2527, 1991.

33. Gisolfi C.V., R.D. Summers, H.P. Schedl, and T.L. Bleiler. Effect of sodium concentration in a carbohydrate-electrolyte solution on intestinal absorption. *Med. Sci. Sports Exerc.* 27:1414–1420, 1995.

34. Hales J.R.S., R.W. Hubbard, and S.L. Gaffin. Limitations of heat tolerance. In: *Handbook of Physiology. Sect. 4 Environmental Physiology*, Vol. I, edited by M.J. Fregly and C.M. Blatteis. New York: Oxford University Press, 1996, pp. 285–355.

35. Hall D.M., K.R. Baumgardner, T.D. Oberley, and C.V. Gisolfi. Splanchnic tissues undergo hypoxic stress during whole body hyperthermia. *Am. J. Physiol.* 276(Gastrointest. Liver Physiol. 39):G1195–G1203, 1999.

36. Hall D.M., G.R. Buettner, R.D. Matthes, and C.V. Gisolfi. Hyperthermia stimulates nitric oxide formation: electron paramagnetic resonance detection of NO-heme in blood. *J. Appl. Physiol.* 77:548–553, 1994.

37. Harris A., A.K. Lindeman, and B.J. Martin. Rapid orocecal transit in chronically active persons with high energy intake. *J. Appl. Physiol.* 70:1550–1553, 1991.

38. Hellebrandt F.A. and L.L. Dimmitt. Studies in the influence of exercise on the digestive work of the stomach. III. Its effect on the relation between secretory and motor function. *Am. J. Physiol.* 107:364–369, 1934.

39. Hellebrandt F.A. and L.L. Hoopes. Studies in the influence of exercise on the digestive work of the stomach. I. Its effect on the secretory cycle. *Am. J. Physiol.* 107:348–354, 1934.

40. Hellebrandt F.A. and R.H. Tepper. Studies in the influence of exercise on the digestive work of the stomach. *Am. J. Physiol.* 107:355–363, 1933.

41. Herrick J.F., J.H. Grindlay, E.J. Baldes, and F.C. Mann. Effect of exercise on the blood flow in the superior mesenteric, renal and common iliac arteries. *Am. J. Physiol.* 128:338–344, 1940.

42. Hunt J.B., E.J. Elliott, P.D. Fairclough, M.L. Clark, and M.J.G. Farthing. Water and solute absorption from hypotonic glucose-electrolyte solutions in human jejunum. *Gut* 33:479–483, 1992.

43. Hunt J.B., E.J. Elliott, and M.J.G. Farthing. Efficacy of a standard United Kingdom oral rehydration solution (ORS) and a hypotonic ORS assessed by human intestinal perfusion. *Aliment. Pharmacol. Therap.* 3:565–571, 1989.

44. Hunt J.N. The site of receptors slowing gastric emptying in response to starch in test meals. *J. Physiol. (Lond.)* 154:270–276, 1960.

45. Hunt J.N. and M. T. Knox. Regulation of gastric emptying. In: *Handbook of Physiology,* edited by C.F. Code. New York: Oxford University Press, 1969, Vol. IV, pp. 1917–1935.

46. Hunt J.N. and M.T. Knox. The slowing of gastric emptying by nine acids. *J. Physiol. (Lond.)* 201:161–179, 1969.

47. Hunt J.N. and J.O. Pathak. The osmotic effect of some simple molecules and ions on gastric emptying. *J. Physiol. (Lond.)* 154:254–269, 1960.

48. Hunt J.N. and D.F. Stubbs. The volume and energy content of meals as a determinant of gastric emptying. *J. Physiol. (Lond.)* 245:209–225, 1975.

49. Keeffe E.B., D.K. Lowe, J.R. Goss, and R. Wayne. Gastrointestinal symptons of marathon runners. *West. J. Med.* 141:481–484, 1984.

50. Keeling W.F., A. Harris, and B. Martin. Orocecal transit during mild exercise in women. *J. Appl. Physiol.* 68:1350–1353, 1990.

51. Kvietys P.R., A.P. Shepherd, and D.N. Granger. Laser-Doppler, H_2 clearance, and microsphere estimates of mucosal blood flow. *Am. J. Physiol.* 249 (Gastrointest. Liver Physiol. 12):G221–G227, 1985.

52. Lamb D.R. and G.R. Brodowicz. Optimal use of fluids of varying formulations to minimize exercise-induced disturbances in homeostasis. *Sports Med.* 3:247–274, 1986.

53. Lambert G.P., R.T. Chang, D. Joensen, X. Shi, R.W. Summers, H.P. Schedl, and C.V. Gisolfi. Simultaneous determination of gastric emptying and intestinal absorption during cycle exercise in humans. *Int. J. Sports Med.* 17:48–55, 1996.

54. Lambert G.P., R.T. Chang, T. Xia, R.W. Summers, and C.V. Gisolfi. Absorption from different intestinal segments during exercise. *J. Appl. Physiol.* 83:204–212, 1997.

55. Lampe J.W., J.L. Slavin, and F.S. Apple. Iron status of active women and the effect of running a marathon on bowel function and gastrointestinal blood loss. *Int. J. Sports Med.* 12:173–179, 1991.

56. Leiper J.B. and R.J. Maughan. Absorption of water and solute from glucose-electrolyte solutions in the human jejunum: effect of citrate or betaine. *Proc. Nat. Soc.* 45:78A, 1986 (abstract).

57. Lowenthal M., K. Harpuder, and S.D. Blatt. Peripheral and visceral vascular effects of exercise and postprandial state in supine position. *J. Appl. Physiol.* 4:689–694, 1952.

58. Marzio L., P. Formica, F. Fabiani, D. LaPenna, L. Vecchiett, and F. Cuccurullo. Influence of physical activity on gastric emptying of liquids in mormal human subjects. *Am. J. Gastro.* 86:1433–1436, 1991.

59. Maughan R. Carbohydrate-electrolyte solutions during prolonged exercise. In: *Perspectives in Exercise Science and Sports Medicine. Vol. 4. Ergogenics: Enhancement of Performance in Exercise and Sport*, edited by D.R. Lamb and M.H. Williams. Indianapolis: Brown & Benchmark, 1991, pp. 35–85.

60. Mellinkoff S.M. and T.E. Machella. Effect of exercise upon liver following partial hepatectomy in albino rats. *Proc. Soc. Exp. Biol. Med.* 74:484–486, 1950.

61. Meshkinpour H., C. Kemp, and R. Fairshter. Effect of aerobic exercise on mouth-to-cecum transit time. *Gastroenterology* 96:938–941, 1989.

62. Mitchell J.B., D.L. Costill, J.A. Houmard, W. J. Fink, R. A. Roberts, and J. A. Davis. Gastric emptying: influence of prolonged exercise and carbohydrate concentration. *Med. Sci. Sports. Exerc.* 21:269–274, 1989.

63. Modigliani R., J.C. Rambaud, and J.J. Bernier. The method of intraluminal perfusion of the human small intestine. *Digestion* 9:176–192, 1973.

64. Murray R. The effects of consuming carbodydrate-electrolyte beverages on gastric emptying and fluid absorption during and following exercise. *Sports Med.* 4:322–351, 1987.

65. Murray R., D.E. Eddy, W.P. Bartoli, and G.L. Paul. Gastric emptying of water and isocaloric carbohydrate solutions consumed at rest. *Med. Sci. Sports Exerc.* 26:725–732, 1994.

66. Myer J.S. *Life and Letters of Dr. William Beaumont*, St. Louis: C.V. Mosby, 1912.

67. Nelson P.B., A.G. Robinson, W. Kapoor, and J. Rinaldo. Hyponatremia in a marathoner. *Phys. Sportsmed.* 16:78–87, 1988.

68. Neufer P.D., D.L. Costill, W.J. Fink, J.P. Kirwan, R.A. Fielding, and M.G. Flynn. Effects of exercise and carbohydrate composition on gastric emptying. *Med. Sci. Sports Exerc.* 18:658–662, 1986.

69. Neufer P.D., A.J. Young, and M.N. Sawka. Gastric emptying during exercise: effects of heat stress and hypohydration. *Eur. J. Appl. Physiol.* 58:433–439, 1989.

70. Neufer P.D., A.J. Young, and M.N. Sawka. Gastric emptying during walking and running: effects of varied exercise intensity. *Eur. J. Appl. Physiol.* 58:440–445, 1989.

71. Noakes T.D. Hyponatraemia of exercise. *Insider. News on Sport Nutrition* 3(4):1–4, 1995.

72. Noakes T.D., N. Goodwin, B.L. Rayner, T. Branken, and R.K.N. Taylor. Water intoxication: a possible complication during endurance exercise. *Med. Sci. Sports Exerc.* 17:370–375, 1985.

73. Oettle G.J. Effect of moderate exercise on bowel habit. *Gut* 32:941–944, 1991.

74. *Official Tract and Field Handbook*. International Amateur Athletic Foundation, Indianapolis: 1961, pp. 22–23, 38.

75. Ollerenshaw K.J., S. Normal, C.G. Wilson, and J.G. Hardy. Exercise and small intestinal transit. *Nuclear Med. Commun.* 8:105–110, 1987.

76. *Olympic USA Track and Field Competition Rules*, Rule 165, pp. 142–143, 1999.

77. Osler W. *William Beaumont. Experiments and Observations on the Gastric Juice and the Physiology of Digestion*, Boston: XIIIth International Physiological Congress, 1929.

78. Owen M.D., K.C. Kregel, P.T. Wall, and C.V. Gisolfi. Effects of ingesting carbohydrate beverages during exercise in the heat. *Med. Sci. Sports Exerc.* 18:568–575, 1986.

79. Patra F.C., D. Mahalanabis, and K.N. Jalan. Bicarbonate enhances sodium absorption from glucose and glycine rehydration solutions. *Acta Paediatr. Scand.* 78:379–383, 1989.

80. Perko M.J., H.B. Nielsen, C. Skak, J.O. Clemmesen, T.V. Schroeder, and N.H. Secher. Mesenteric, coeliac and splanchnic blood flow in humans. *J. Physiol. (Lond.)* 513:907–913, 1998.

81. Peters J., J. Johnston, and J. Edmundson. *Modern Middle- and Long-Distance Running*, edited by Nicholas Kaye Limited. London and Dorking: Adlard and Son Limite, 1957.

82. Putukian M. and C. Potera. Don't miss gastrointestinal disorders in athletes. *Phys. Sportsmed.* 25:80–94, 1997.

83. Qamar M.I. and A.E. Read. Effects of exercise on mesenteric blood flow in man. *Gut* 28: 583–587, 1987.

84. Rao S.S.C., J. Leistikow, M. Chamberlain, and C.V. Gisolfi. Effects of acute graded exercise on human colonic motility. *Am. J. Physiol.* 276 (Gastrointest. Liver Physiol. 39): G1221–G1226, 1999.

85. Rehrer, N.J., E. Beckers, F. Brouns, F. TenHoor, and W.H.M. Saris. Exercise and training effects on gastric emptying of carbohydrate beverages. *Med. Sci. Sports Exerc.* 21:540–549, 1989.

86. Rehrer N.J., E.J. Beckers, F. Brouns, F. TenHoor, and W.H.M. Saris. Effects of dehydration on gastric emptying and gastrointestinal distress while running. *Med. Sci. Sports Exerc.* 22:790–795, 1990.

87. Riklis E. and J. Quastel. Effects of cations on sugar absorption by isolated surviving guinea pig intestine. *Can. J. Biochem. Physiol.* 36:347–362, 1958.

88. Rosenweig N.S., R.H. Herman, and F.B. Stifel. Dietary regulation of small intestinal enzyme activity in man. *Am. J. Clin. Nut.* 24:65–69, 1971.

89. Rowell L.B. The splanchnic circulation. In: *Textbook of Physiology and Biophysics,* edited by T.C. Ruch and H.D. Patton. Philadelphia: W.B. Saunders, 1974, Vol. 2, pp. 215–233.

90. Rowell L.B., J.R. Blackmon, and R.A. Bruce. Indocyanine green clearance and estimated hepatic blood flow during mild to maximal exercise in upright man. *J. Clin. Invest.* 43: 1677–1690, 1964.

91. Ryan A.J., G.P. Lambert, X. Shi, R.T. Chang, R.W. Summers, and C.V. Gisolfi. Effect of hypohydration on gastric emptying and intestinal absorption during exercise. *J. Appl. Physiol.* 84:1581–1588, 1998.

92. Sakurada S. and J.R.S. Hales. A role for gastrointestinal endotoxins in enhancement of heat tolerance by physical fitness. *J. Appl. Physiol.* 84:207–214, 1998.

93. Sandhu D.K., F.L. Christobal, and M.J. Brueton. Optimising oral rehydration solution composition in model systems: studies in normal mammalian small intestine. *Acta Paediatr. Scand.* Suppl. 364:17–22, 1989.

94. Schedl H.P. and J.A. Clifton. Solute and water absorption by the human small intestine. *Nature* 199:1264–1267, 1963.

95. Schedl H.P., R.J. Maughan, and C.V. Gisolfi. Intestinal absorption during rest and exercise: implications for formulating an oral rehydration solution (ORS). *Med. Sci. Sports Exerc.* 26:267–280, 1994.

96. Schultz S.G. and R. Zalusky. Ion transport in isolated rat ileum. II. The interaction between active sodium and active sugar transport. *J. Gen. Physiol.* 47:1043–1059, 1964.

97. Schwartz A.E., A. Vanagunas, and P.L. Kamel. Endoscopy to evaluate gastrointestinal bleeding in marathon runners. *Ann. Internal Med.* 113:632–633, 1990.

98. Shi X., R.W. Summers, H.P. Schedl, R.T. Chang, G.P. Lambert, and C.V. Gisolfi. Effects of solution osmolality on absorption of select fluid replacement solutions in human duodenojejunum. *J. Appl. Physiol.* 77:1178–1184, 1994.

99. Shi X., R.W. Summers, H.P. Schedl, S. W. Flanagan, R.T. Chang, and C.V. Gisolfi. Effects of carbohydrate type and concentration and solution osmolality on water absorption. *Med. Sci. Sports. Exerc.* 27:1607–1615, 1995.

100. Sladen G.E. and A.M. Dawson. Interrelationships between the absorptions of glucose, sodium and water by the normal human jejunum. *J. Clin. Sci.* 36:119–132, 1969.

101. Soffer E.E., R.W. Summers, and C. Gisolfi. The effect of exercise on intestinal motility and transit in trained athletes. *Am. J. Physiol.* 260(Gastrointest. Liver Physiol. 23):G698–G702, 1991.

102. Spickard W.A. Blocking out a tough opponent. *Med. World News* 9:38–74, 1968.

103. Stickney J.C. and E.J. VanLiere. The effects of exercise upon the function of the gastrointestinal tract. In: *Science and Medicine of Exercise and Sports*, edited by W.R. Johnson. New York: Harper & Row, 1960, pp. 236–250.

104. Sullivan S.N. The gastrointestinal symptoms of running. *N. Engl. J. Med.* 304:915, 1981.

105. Thillaninayagam A.V., J.B. Hunt, and M.J.G. Farthing. Enhancing clinical efficacy of oral rehydration therapy: is low osmolality the key? *Gastroenterology* 114:197–210, 1998.

106. Thompson P.D., E.J. Funk, R.A. Carleton, and W.Q. Sturner. Incidence of death during jogging in Rhode Island from 1975 through 1980. *JAMA* 247:2535–2538, 1982.

107. Thoren A., S.-E. Ricksten, S. Lundin, B. Gazelius, and M. Elam. Baroreceptor-mediated reduction of jejunal mucosal perfusion, evaluated with endoluminal laser Doppler flowmetry in conscious humans. *J. Auto. Nerv. Sys.* 68:157–163, 1998.

108. Vist G.E. and R.J. Maughan. The effect of osmolality and carbohydrate content on the rate of gastric emptying. *J. Physiol. (Lond.)* 486:523–531, 1995.

109. Wade O.L., B. Combes, A.W. Childs, H.O. Wheeler, A. Cournand, and S.E. Bradley. The effect of exercise on the splanchnic blood flow and splanchnic blood volume in normal man. *Clin. Sci.* 15:457–463, 1956.

110. Wapnir R.A. and F. Lifshitz. Osmolality and solute concentration—their relationship with an oral hydration solution effectiveness: an experimental assessment. *Pediatr. Res.* 19:894–898, 1985.

INDEX